Recent Developments in Structure-Preserving Algorithms for Oscillatory Differential Equations

Xinyuan Wu · Bin Wang

Recent Developments in Structure-Preserving Algorithms for Oscillatory Differential Equations

Science Press
Beijing

Springer

Xinyuan Wu
Department of Mathematics
Nanjing University
Nanjing, Jiangsu
China

and

School of Mathematical Sciences
Qufu Normal University
Qufu, Shandong
China

Bin Wang
School of Mathematical Sciences
Qufu Normal University
Qufu, Shandong
China

ISBN 978-981-10-9003-5 ISBN 978-981-10-9004-2 (eBook)
https://doi.org/10.1007/978-981-10-9004-2

Jointly published with Science Press, Beijing, China

The print edition is not for sale in China Mainland. Customers from China Mainland please order the print book from: Science Press, Beijing.
ISBN of the China Mainland edition: 978-703-05-5128-3

Library of Congress Control Number: 2018936635

Printed on acid-free paper

This Springer imprint is published by the registered company Springer Nature Singapore Pte Ltd.
part of Springer Nature
The registered company address is: 152 Beach Road, #21-01/04 Gateway East, Singapore 189721, Singapore

2017 Qufu Workshop on Structure-Preserving Algorithms for Differential Equations
June 09-11

2016 Nanjing Workshop on Structure-Preserving Algorithms for Differential Equations
August 8-10,2016,Nanjing University

2015 Nanjing Symposium on Structure-Preserving
Algorithms for Differential Equations Aug. 2-4

Preface

An important area of numerical analysis and scientific computing is geometric numerical integration which is concerned with the discretization of differential equations while respecting their structural invariants and geometry. In the last few decades, numerical simulation for nonlinear oscillators has received a great deal of attention, and many researchers have been concerned with the design and analysis of numerical schemes for solving oscillatory problems. It has been a common practice that a numerical scheme should be designed to preserve as much as possible the (physical/geometric) intrinsic properties of the original continuous systems. Although great advances have by now been made in numerical treatments for oscillatory differential equations, further exploration of novel structure-preserving algorithms is still an active area of research. The objective of this sequel to our previous monograph, which was entitled "Structure-Preserving Algorithms for Oscillatory Differential Equations II", is to study further structure-preserving schemes for oscillatory systems that can be modelled by systems of ordinary and partial differential equations. Problems of this type arise in a variety of fields in science and engineering such as quantum physics, celestial mechanics and molecular dynamics.

Most of the material presented here is drawn from very recent published research work in professional journals by the research group of the authors. The first four chapters of this monograph deal with first-order oscillatory differential equations. The last four chapters consider oscillatory wave equations, and the other chapters address second-order oscillatory differential equations.

Chapter 1 presents in detail functionally fitted energy-preserving methods for solving oscillatory nonlinear Hamiltonian systems. It is known that the exponential integrator is now a very popular numerical method for solving differential equations. Chapter 2 investigates exponential integrators preserving first integrals or Lyapunov functions for conservative or dissipative systems. These methods are applied to solve some highly significant oscillatory problems. In the literature, a principal approach to solving oscillatory problems is based on collocation methods. Therefore, Chap. 3 explores exponential Fourier collocation methods for first-order differential equations. Symplecticity is one of the fundamental properties of

Hamiltonian systems. In Chap. 4, we first develop symplectic exponential Runge–Kutta methods for solving nonlinear Hamiltonian systems. We then consider high-order symplectic and symmetric composition methods for multi-frequency and multidimensional oscillatory Hamiltonian systems in Chap. 5. The construction of high-order ERKN integrators is difficult in practice. Thus, in Chap. 6, we pay our attention to the construction of arbitrary-order ERKN integrators for second-order oscillatory differential equations using group theory. Chapter 7 derives trigonometric collocation methods based on Lagrange basis polynomials for multi-frequency oscillatory second-order differential equations. Butcher's theory of rooted trees is useful in the study of Runge–Kutta type methods. However, the research of this theory for ERKN methods for general multi-frequency and multidimensional oscillatory systems is not satisfied yet due to the existence of a large number of redundant trees. In Chap. 8, we further analyse a compact tri-coloured rooted-tree theory and order conditions for ERKN methods for general multi-frequency oscillatory systems. An important fact is that Klein–Gordon equations can be used to model many nonlinear phenomena. Chapter 9 is concerned with an integral formula adapted to different boundary conditions for arbitrarily high-dimensional nonlinear Klein–Gordon equations. Chapter 10 proposes an energy-preserving and symmetric scheme for nonlinear Hamiltonian wave equations. Chapter 11 describes arbitrarily high-order time-stepping methods for solving nonlinear Klein–Gordon equations based on the operator spectrum theory. The last chapter considers an essential extension of the finite-energy condition for ERKN integrators when applied to nonlinear wave equations.

The presentation in this monograph reflects the current active state of the subject matter which is characterized by mathematical analysis, providing insight into questions of practical calculation and illuminating numerical simulations. All the schemes derived in this monograph have been tested and verified on oscillatory systems from a wide range of applications to show the numerical behaviour of the simulation. Simulations indicate that they are more efficient than the existing high-quality integrators in the scientific literature.

The authors want to thank all their friends and colleagues for their selfless help during the preparation of this monograph. Special thanks go to John Butcher of the University of Auckland, Christian Lubich of Universität Tübingen, Arieh Iserles of the University of Cambridge and Reinout Quispel of La Trobe University for their encouragement.

The authors are also grateful to many friends and colleagues for reading the manuscript and for their valuable suggestions. In particular, the authors take this opportunity to express their sincere appreciation to Robert Peng Kong Chan of the University of Auckland, Qin Sheng of Baylor University, Jichun Li of the University of Nevada Las Vegas, David McLaren of La Trobe University, Adrian Turton Hill of the University of Bath, Xiaowen Chang of McGill University, Jianlin Xia of Purdue University, Marcus David Webb of the University of Cambridge and Xiong You of Nanjing Agricultural University.

Sincere thanks also go to the following people for their help and support in various forms: Zuhe Shen of Nanjing University; Fanwei Meng of Qufu Normal University; Yaolin Jiang and Jing Gao of Xi'an Jiao Tong University; Yongzhong Song, Jinru Chen, Yushun Wang and Qikui Du of Nanjing Normal University; Xinru Wang of Nanjing Medical University; Anders Christian Hansen and Alexei Shadrin of the University of Cambridge; Markus Hegland, Pier Bouwknegt and Lilia Ferrario of the Australian National University; Qiying Wang of the University of Sydney; Shixiao Wang of the University of Auckland; Robert Mclachlan of Massey University; Jialin Hong, Zaijiu Shang, Yifa Tang and Yajuan Sun of the Chinese Academy of Sciences; Yuhao Cong of Shanghai Customs College; Guangda Hu of Shanghai University; Jijun Liu, Zhizhong Sun and Hongwei Wu of Southeast University; Shoufo Li, Aiguo Xiao and Liping Wen of Xiang Tan University; Chuanmiao Chen of Hunan Normal University; Siqing Gan of Central South University; Chengjian Zhang and Chengming Huang of Huazhong University of Science and Technology; Shuanghu Wang of the Institute of Applied Physics and Computational Mathematics; Beijing, Hongjiong Tian and Wansheng Wang of Shanghai Normal University; Yongkui Zou of Jilin University; Jingjun Zhao of Harbin Institute of Technology; Wei Shi of Nanjing Tech Univerity; Qinghong Li of Chuzhou University; Yonglei Fang of Zaozhuang University; Fan Yang, Xianyang Zeng and Hongli Yang of Nanjing Institute of Technology; Kai Liu of Nanjing University of Finance and Economics; and Jiyong Li of Hebei Normal University.

The authors would like to thank Kai Hu, Ji Luo and Lukai Cui for their help with the editing, the editorial and production group of the Science Press, Beijing, and Springer-Verlag, Heidelberg.

The authors also thank their family members for their love and support throughout all these years.

The work on this monograph was supported in part by the Natural Science Foundation of China under Grants 11671200 and 11271186, by NSFC and RS International Exchanges Project under Grant 11411130115, by the Specialized Research Foundation for the Doctoral Program of Higher Education under Grant 20100091110033 and 20130091110041.

Nanjing/Qufu, China — Xinyuan Wu
Qufu, China — Bin Wang

Contents

Chapter 1
Functionally Fitted Continuous Finite Element Methods for Oscillatory Hamiltonian Systems

In recent decades, the numerical simulation for nonlinear oscillators has received much attention and a large number of integrators for oscillatory problems have been developed. In this chapter, based on the continuous finite element approach, we propose and analyse new energy-preserving functionally-fitted, in particular, trigonometrically-fitted methods of an arbitrarily high order for solving oscillatory nonlinear Hamiltonian systems with a fixed frequency. In order to implement these new methods in an accessable and efficient style, they are formulated as a class of continuous-stage Runge–Kutta methods. The numerical results demonstrate the remarkable accuracy and efficiency of the new methods compared with the existing high-order energy-preserving methods in the literature.

1.1 Introduction

It is known that an important area of numerical analysis and scientific computing is geometric numerical integration for differential equations. In this chapter, we consider nonlinear Hamiltonian systems:

$$\dot{y}(t) = f(y(t)) = J^{-1}\nabla H(y(t)), \quad y(t_0) = y_0 \in \mathbb{R}^d, \tag{1.1}$$

where $y \in \mathbb{R}^d$, $d = 2d_1$, $f : \mathbb{R}^d \to \mathbb{R}^d$, $H : \mathbb{R}^d \to \mathbb{R}$ are sufficiently smooth functions and

$$J = \begin{pmatrix} O_{d_1\times d_1} & I_{d_1\times d_1} \\ -I_{d_1\times d_1} & O_{d_1\times d_1} \end{pmatrix}$$

is the canonical symplectic matrix. It is well known that the flow of (1.1) preserves the symplectic form $\mathrm{d}y \wedge J\mathrm{d}y$ and the Hamiltonian or energy $H(y(t))$. In the spirit of geometric numerical integration, it is a natural idea to design schemes that preserve

X. Wu and B. Wang, *Recent Developments in Structure-Preserving Algorithms for Oscillatory Differential Equations*, https://doi.org/10.1007/978-981-10-9004-2_1

both the symplecticity of the flow and the energy. Unfortunately, however, a numerical scheme cannot achieve this goal unless it generates the exact solution (see, e.g. [23], p. 379). Hence, researchers face a choice between preserving symplecticity or preserving energy, and many of them have given more weight to the former in the last decades, and readers are referred to [23] and references therein. Despite insufficient work in the literature on energy-preserving (EP) methods (see, e.g. [1, 4, 6, 10, 11, 17, 19, 21, 29, 38]), EP methods compared with symplectic methods have better nonlinear stability properties, are easier to adapt the time step for, and are more suitable for the integration of chaotic systems (see, e.g. [12, 20, 35, 36]).

On the other hand, in scientific computing and modelling, the design and analysis of methods for periodic or oscillatory systems has been considered by many authors (see, e.g. [2, 18, 22, 34, 42, 46]). Generally, these methods utilize a priori information of special problems and they are more efficient than general-purpose methods. A popular approach to constructing methods suitable for oscillatory problems is using the functionally-fitted (FF) condition, namely, deriving a suitable method by requiring it to integrate members of a given finite-dimensional function space X exactly. If X incorporates trigonometrical or exponential functions, the corresponding methods are also called trigonometrically-fitted (TF) or exponentially-fitted (EF) methods (see, e.g. [15, 27, 32, 37]).

Therefore, combining the ideas of the EF/TF and structure-preserving methods is a promising approach to developing numerical methods which allow long-term computation of solutions to oscillatory Hamiltonian systems (1.1). Just as the research of symplectic and EP methods, EF/TF symplectic methods have been studied extensively by many authors (see, e.g. [7–9, 16, 39, 40, 43]). By contrast, as far as we know, only a few papers paid attention to the EF/TFEP methods (see, e.g. [30, 31, 41]). Usually the existing EF/TFEP methods are derived in the context of continuous-stage Runge–Kutta (RK) methods. The coefficients in these methods are determined by a system of equations resulting from EF/TF, EP and symmetry conditions. As mentioned at the end of [30], it is not easy to find such a system with a unique solution when deriving high-order methods. Furthermore, how to verify the algebraic order of such methods falls into question. A common way is to check the order conditions related to rooted trees. Again, this is inconvenient in the high-order setting since the number of trees increases extremely fast as the order grows. In this chapter, we will construct FFEP methods based on the continuous finite element method, which is inherently energy preserving (see, e.g. [1, 17, 38]). Intuitively, we expect to increase the order of the method through enlarging the finite element space. By adding trigonometrical functions to the space, the corresponding method is naturally trigonometrically fitted. Thus, we are hopeful of constructing FFEP methods, in particular TFEP methods, of arbitrarily high orders.

This chapter is organized as follows. In Sect. 1.2, we construct functionally fitted continuous finite element (FFCFE) methods and present their most important geometric properties. In Sect. 1.3, we interpret them as continuous-stage Runge–Kutta methods and analyse the algebraic order. We then discuss implementation details of these new methods in Sect. 1.4. Numerical results are shown in Sect. 1.5, including the comparison between our new TFEP methods and other prominent structure-preserving methods in the literature. The last section is concerned with the conclusion and discussion.

1.2 Functionally-Fitted Continuous Finite Element Methods for Hamiltonian Systems

Throughout this chapter, we consider the approximation of the solution of the IVP (1.1) on the time interval $I = [t_0, T]$. Let S be a linear space of continuous functions $y(t)$ on the interval I. Let $\{\varphi_i\}_{i=0}^{r-1}$ be a family of sufficiently smooth and linearly independent real-valued functions on I, and let Y be the subspace spanned by $\{\varphi_i\}_{i=0}^{r-1}$:

$$Y = \left\{ w : w(t) = \sum_{i=0}^{r-1} W_i \varphi_i(t),\ W_i \in \mathbb{R}^d \right\}.$$

We assume that the interval $I = [t_0, T]$ is equally partitioned into $t_0 < t_1 < \ldots < t_N = T$, with $t_n = t_0 + nh$ for $n = 0, 1, \ldots, N$. A function w on I is called a *piece-wise Y-type* function if for any $0 \leq n \leq N-1$, there exists a function $g \in Y$, such that

$$w|_{(t_n, t_{n+1})} = g|_{(t_n, t_{n+1})}.$$

It is convenient to introduce the transformation $t = t_0 + \tau h$ for $\tau \in [0, 1]$ in the following analysis. Accordingly, we denote

$$Y_h(t_0) = \{v \text{ on } [0, 1] : v(\tau) = w(t_0 + \tau h),\ w \in Y\}.$$

Hence,

$$Y_h(t_0) = \text{span}\{\tilde{\varphi}_0, \ldots, \tilde{\varphi}_{r-1}\},$$

where $\tilde{\varphi}_i(\tau) = \varphi_i(t_0 + \tau h)$ for $i = 0, 1, \ldots, r-1$. In what follows, lowercase Greek letters such as τ, σ, α always indicate variables on the interval [0,1] unless confusions arise.

Given two integrable functions (scalar-valued or vector-valued) w_1 and w_2 on $[0, 1]$, the inner product $\langle \cdot, \cdot \rangle$ is defined by

$$\langle w_1, w_2 \rangle = \langle w_1(\tau), w_2(\tau) \rangle_\tau = \int_0^1 w_1(\tau) \cdot w_2(\tau) d\tau,$$

where $\cdot$ is the entrywise multiplication operation if w_1, w_2 are both vector-valued functions of the same length.

Given two finite-dimensional function spaces X and Y whose members are $\mathbb{R}^d$-valued, the continuous finite element method for (1.1) is described as follows.

Find a continuous piecewise X-type function $U(t)$ on I with $U(t_0) = y_0$, such that for any piecewise function Y-type function $v(t)$,

$$\int_I v(t) \cdot (U'(t) - f(U(t)))dt = 0, \tag{1.2}$$

where $U(t) \approx y(t)$ on I and $y(t)$ solves (1.1). The term 'continuous finite element'(CFE) comes from the continuity of the finite element solution $U(t)$. Since (1.2) deals with an initial value problem, we only need to consider it on $[t_0, t_0 + h]$.

Find $u \in X_h(t_0)$ with $u(0) = y_0$, such that

$$\langle v, u' \rangle = h \langle v, f \circ u \rangle, \tag{1.3}$$

for any $v \in Y_h(t_0)$, where

$$u(\tau) = U(t_0 + \tau h) \approx y(t_0 + \tau h)$$

for $\tau \in [0, 1]$. Since $U(t)$ is continuous, $y_1 = u(1)$ is the initial value of the local problem on the next interval $[t_1, t_2]$. Thus, we can solve the global variational problem (1.2) on I step by step.

In the special case of

$$X = \text{span}\left\{1, t, \ldots, t^r\right\}, \quad Y = \text{span}\left\{1, t, \ldots, t^{r-1}\right\},$$

Equation (1.2) reduces to the classical continuous finite element method (see, e.g. [1, 25]) denoted by CFEr in this chapter. For the purpose of deriving functionally-fitted methods, we generalise X and Y a little:

$$Y = \text{span}\left\{\varphi_0(t), \ldots, \varphi_{r-1}(t)\right\}, \quad X = \text{span}\left\{1, \int_{t_0}^{t} \varphi_0(s)ds, \ldots, \int_{t_0}^{t} \varphi_{r-1}(s)ds\right\}. \tag{1.4}$$

Then it is sufficient to give X or Y since they can be determined by each other. Furthermore, Y is assumed to be invariant under translation and reflection, namely,

$$\begin{cases} v(t) \in Y \Rightarrow v(t + c) \in Y \text{ for any } c \in \mathbb{R}, \\ v(t) \in Y \Rightarrow v(-t) \in Y. \end{cases} \tag{1.5}$$

Clearly, $Y_h(t_0)$ and $X_h(t_0)$ are irrelevant to t_0 provided (1.5) holds. For convenience, we simplify $Y_h(t_0)$ and $X_h(t_0)$ by Y_h and X_h, respectively. In the remainder of this chapter, we denote the CFE method (1.2) or (1.3) based on the general function spaces (1.4) satisfying the condition (1.5) by FFCFEr.

We note that the FFCFEr method (1.3) is defined by a variational problem, and the well-definedness of this problem has not been confirmed yet. Here we presume the existence and uniqueness of the solution to (1.3). This assumption will be proved in the next section. With this premise, we are able to present three significant properties of the FFCFEr method.

We first conclude that the FFCFEr method is functionally fitted with respect to the space X, from the definition of the variational problem (1.2).

Theorem 1.1 *The FFCFEr method (1.2) solves the IVP (1.1) whose solution is a piecewise X-type function without any error.*

Moreover, the FFCFEr method is an inherently energy-preserving method. The next theorem confirms this point.

Theorem 1.2 *The FFCFEr method (1.3) exactly preserves the Hamiltonian H: $H(y_1) = H(y_0)$.*

Proof Firstly, given a vector V, we denote its ith entry by V_i. For each function $w \in Y_h$, setting $v(\tau) = w(\tau) \cdot e_i \in Y_h$ in (1.3) leads to

$$\int_0^1 w(\tau)_i u'(\tau)_i d\tau = h \int_0^1 w(\tau)_i f(u(\tau))_i d\tau, \quad i = 1, 2, \ldots, d,$$

where e_i is the ith vector of units. Hence,

$$\begin{aligned} \int_0^1 w(\tau)^{\intercal} u'(\tau) d\tau &= \sum_{i=1}^{d} \int_0^1 w(\tau)_i u'(\tau)_i d\tau = \sum_{i=1}^{d} h \int_0^1 w(\tau)_i f(u(\tau))_i d\tau \\ &= h \int_0^1 w(\tau)^{\intercal} f(u(\tau)) d\tau. \end{aligned} \tag{1.6}$$

Since $u(\tau) \in X_h$, $u'(\tau) \in Y_h$ and $J^{-1}u'(\tau) \in Y_h$, taking $w(\tau) = J^{-1}u'(\tau)$ in (1.6), we obtain

$$\begin{aligned} H(y_1) - H(y_0) &= \int_0^1 \frac{d}{d\tau} H(u(\tau)) d\tau = \int_0^1 u'(\tau)^{\intercal} \nabla H(u(\tau)) d\tau \\ &= \int_0^1 (J^{-1}u'(\tau))^{\intercal} f(u(\tau)) d\tau = h^{-1} \int_0^1 u'(\tau)^{\intercal} J u'(\tau) d\tau = 0. \end{aligned}$$

This completes the proof. □

The FFCFEr method can also be viewed as a one-step method $\Phi_h : y_0 \to y_1 = u(1)$. It is well known that reversible methods show a better long-term behaviour than nonsymmetric ones when applied to reversible differential systems such as (1.1) (see, e.g. [23]). This fact motivates the investigation of the symmetry of the FFCFEr method. Since the spaces X and Y satisfy the invariance (1.5), which is a kind of symmetry, the FFCFEr method is expected to be symmetric.

Theorem 1.3 *The FFCFEr method (1.3) is symmetric provided (1.5) holds.*

Proof It follows from (1.5) that we have $X_h = X_{-h}$, $Y_h = Y_{-h}$. Exchanging $y_0 \leftrightarrow y_1$ and replacing h with $-h$ in (1.3) give: $u(0) = y_1$, $y_0 = u(1)$, where

$$\langle v(\tau), u'(\tau)\rangle_\tau = -h\langle v(\tau), f(u(\tau))\rangle_\tau, \quad u(\tau) \in X_{-h} = X_h,$$

for each $v(\tau) \in Y_{-h} = Y_h$. Setting $u_1(\tau) = u(1-\tau) \in X_h$ and $\tau \to 1-\tau$ leads to $u_1(0) = y_0$, $y_1 = u_1(1)$, where

$$\langle v_1(\tau), u_1'(\tau)\rangle_\tau = h\langle v_1(\tau), f(u_1(\tau))\rangle_\tau,$$

for each $v_1(\tau) = v(1-\tau) \in Y_h$. This method is exactly the same as (1.3), which means that the FFCFEr method is symmetric. □

It is well known that polynomials cannot approximate oscillatory functions satisfactorily. If the problem (1.1) has a fixed frequency ω which can be evaluated effectively in advance, then the function space containing the pair $\{\sin(\omega t), \cos(\omega t)\}$ seems to be a more promising candidate for X and Y than a polynomial space. For example, possible Y spaces for deriving the TFCFE method are

$$Y_1 = \begin{cases} \text{span}\{\cos(\omega t), \sin(\omega t)\}, r = 2, \\ \text{span}\left\{1, t, \ldots, t^{r-3}, \cos(\omega t), \sin(\omega t)\right\}, r \geq 3, \end{cases} \tag{1.7}$$

$$Y_2 = \text{span}\{\cos(\omega t), \sin(\omega t), \ldots, \cos(k\omega t), \sin(k\omega t)\}, r = 2k, \tag{1.8}$$

and

$$Y_3 = \text{span}\left\{1, t, \ldots, t^p, t\cos(\omega t), t\sin(\omega t), \ldots, t^k\cos(\omega t), t^k\sin(\omega t)\right\}. \tag{1.9}$$

Correspondingly, by equipping the FFCFE method with the space $Y = Y_1$, Y_2 or Y_3, we obtain three families of TFCFE methods. According to Theorems 1.2 and 1.3, all for them are symmetric energy-preserving methods. To exemplify this framework of the TFCFE method, in numerical experiments, we will test the TFCFE method denoted by TFCFEr and TF2CFEr based on the spaces (1.7) and (1.8). It is noted that TFCFE2 and TF2CFE2 coincide.

1.3 Interpretation as Continuous-Stage Runge–Kutta Methods and the Analysis on the Algebraic Order

An interesting connection between CFE methods and RK-type methods has been shown in several papers (see, e.g. [3, 25, 38]). Since the RK methods are dominant in the numerical integration of ODEs, it is meaningful and useful to transform the FFCFEr method into the corresponding RK-type method which has been widely and conventionally used in applications. After the transformation, the FFCFEr method can be analysed and implemented by standard techniques in ODEs conveniently. To this end, it is helpful to introduce the projection operation P_h. Given a continuous $\mathbb{R}^d$-valued function w on $[0, 1]$, its projection onto Y_h, denoted by $P_h w$, is defined by

$$\langle v, P_h w\rangle = \langle v, w\rangle, \quad \text{for any } v \in Y_h. \tag{1.10}$$

Clearly, $P_h w(\tau)$ can be uniquely expressed as a linear combination of $\{\tilde{\varphi}_i(\tau)\}_{i=0}^{r-1}$:

$$P_h w(\tau) = \sum_{i=0}^{r-1} U_i \tilde{\varphi}_i(\tau), \quad U_i \in \mathbb{R}^d.$$

Taking $v(\tau) = \tilde{\varphi}_i(\tau) e_j$ in (1.10) for $i = 0, 1, \ldots, r-1$ and $j = 1, \ldots, d$, we can observe that the coefficients U_i satisfy the equation

$$M \otimes I_{d\times d} \begin{pmatrix} U_0 \\ \vdots \\ U_{r-1} \end{pmatrix} = \begin{pmatrix} \langle \tilde{\varphi}_0, w\rangle \\ \vdots \\ \langle \tilde{\varphi}_{r-1}, w\rangle \end{pmatrix},$$

where

$$M = (\langle \tilde{\varphi}_i, \tilde{\varphi}_j\rangle)_{0\le i,j\le r-1}.$$

Since $\{\tilde{\varphi}_i\}_{i=0}^{r-1}$ are linearly independent for $h > 0$, the stiffness matrix M is nonsingular. Consequently, the projection can be explicitly expressed by

$$P_h w(\tau) = \langle P_{\tau,\sigma}, w(\sigma)\rangle_\sigma,$$

where

$$P_{\tau,\sigma} = (\tilde{\varphi}_0(\tau), \ldots, \tilde{\varphi}_{r-1}(\tau)) M^{-1} (\tilde{\varphi}_0(\sigma), \ldots, \tilde{\varphi}_{r-1}(\sigma))^\intercal. \tag{1.11}$$

Clearly, $P_{\tau,\sigma}$ can be calculated by a basis other than $\{\tilde{\varphi}_i\}_{i=0}^{r-1}$ since they only differ in a linear transformation. If $\{\phi_0, \ldots, \phi_{r-1}\}$ is an orthonormal basis of X_h under the inner product $\langle\cdot,\cdot\rangle$, then $P_{\tau,\sigma}$ admits a simpler expression:

$$P_{\tau,\sigma} = \sum_{i=0}^{r-1} \phi_i(\tau)\phi_i(\sigma). \tag{1.12}$$

Now, using (1.3) and the definition (1.10) of the operator P_h, we obtain that $u' = hP_h(f \circ u)$ and

$$u'(\tau) = h\langle P_{\tau,\sigma}, f(u(\sigma))\rangle_\sigma. \tag{1.13}$$

Integrating the above equation with respect to τ, we transform the FFCFEr method (1.3) into the continuous-stage RK method:

$$\begin{cases} u(\tau) = y_0 + h\displaystyle\int_0^1 A_{\tau,\sigma} f(u(\sigma))d\sigma, \\ y_1 = u(1), \end{cases} \tag{1.14}$$

where

$$A_{\tau,\sigma} = \int_0^{\tau} P_{\alpha,\sigma} d\alpha = \sum_{i=0}^{r-1} \int_0^{\tau} \phi_i(\alpha) d\alpha \phi_i(\sigma). \tag{1.15}$$

In particular,

$$\phi_i(\tau) = \hat{p}_i(\tau), \tag{1.16}$$

for the CFEr method for $i = 0, 1, \ldots, r-1$, where $\hat{p}_i(\tau)$ is the shifted Legendre polynomial of degree i on $[0, 1]$, scaled in order to be orthonormal. Hence, the CFEr method in the form (1.14) is identical to the energy-preserving collocation method of order $2r$ (see [21]) or the Hamiltonian boundary value method HBVM(∞, r) (see, e.g. [4]). For the FFCFEr method, since $P_{\tau,\sigma}$, $A_{\tau,\sigma}$ are functions of variable h and $u(\tau)$ is implicitly determined by (1.14), it is necessary to analyse their smoothness with respect to h before investigating the analytic property of the numerical solution $u(\tau)$. First of all, it can be observed from (1.11) that $P_{\tau,\sigma} = P_{\tau,\sigma}(h)$ is not defined at $h = 0$ since the matrix M is singular in this case. Fortunately, however, the following lemma shows that the singularity is removable.

Lemma 1.1 *The limit,* $\lim_{h\to 0} P_{\tau,\sigma}$ *exists. Furthermore,* $P_{\tau,\sigma}$ *can be smoothly extended to* $h = 0$ *by setting* $P_{\tau,\sigma}(0) = \lim_{h\to 0} P_{\tau,\sigma}(h)$.

Proof By expanding $\{\varphi_i(t_0 + \tau h)\}_{i=0}^{r-1}$ at t_0, we obtain that

$$(\tilde{\varphi}_0(\tau), \ldots, \tilde{\varphi}_{r-1}(\tau)) = (1, \tau h, \ldots, \frac{\tau^{r-1} h^{r-1}}{(r-1)!}) W + \mathscr{O}(h^r), \tag{1.17}$$

where

$$W = \begin{pmatrix} \varphi_0(t_0) & \varphi_1(t_0) & \cdots & \varphi_{r-1}(t_0) \\ \varphi_0^{(1)}(t_0) & \varphi_1^{(1)}(t_0) & \cdots & \varphi_{r-1}^{(1)}(t_0) \\ \vdots & \vdots & & \vdots \\ \varphi_0^{(r-1)}(t_0) & \varphi_1^{(r-1)}(t_0) & \cdots & \varphi_{r-1}^{(r-1)}(t_0) \end{pmatrix} \tag{1.18}$$

is the Wronskian of $\{\varphi_i(t)\}_{i=0}^{r-1}$ at t_0, and is nonsingular. Post-multiplying the right-hand side of (1.17) by $W^{-1} diag(1, h^{-1}, \ldots, h^{1-r}(r-1)!)$ yields another basis of X_h:

$$\{1 + \mathscr{O}(h), \tau + \mathscr{O}(h), \ldots, \tau^{r-1} + \mathscr{O}(h)\}.$$

Applying the Gram–Schmidt process (with respect to the inner product $\langle \cdot, \cdot \rangle$) to the above basis, we obtain an orthonormal basis $\{\phi_i(\tau) = \hat{p}_i(\tau) + \mathscr{O}(h)\}_{i=0}^{r-1}$. Thus, by (1.12) and defining

$$P_{\tau,\sigma}(0) = \lim_{h\to 0} \sum_{i=0}^{r-1} \phi_i(\tau)\phi_i(\sigma) = \sum_{i=0}^{r-1} \hat{p}_i(\tau)\hat{p}_i(\sigma), \tag{1.19}$$

we can extend $P_{\tau,\sigma}$ to $h = 0$. Since each $\phi_i(\tau) = \hat{p}_i(\tau) + \mathcal{O}(h)$ is smooth with respect to h, $P_{\tau,\sigma}$ is also a smooth function of h. □

From (1.16) and (1.19), it can be observed that the FFCFEr method (1.14) reduces to the CFEr method when $h \to 0$, or equivalently, the energy-preserving collocation method of order $2r$ and HBVM(∞, r) method mentioned above. Since $A_{\tau,\sigma} = \int_0^\tau P_{\alpha,\sigma} d\alpha$ is also a smooth function of h, we can assume that

$$M_k = \max_{\tau,\sigma,h\in[0,1]} \left| \frac{\partial^k A_{\tau,\sigma}}{\partial h^k} \right|, \quad k = 0, 1, \ldots. \tag{1.20}$$

Furthermore, since the right function f in (1.1) maps from $\mathbb{R}^d$ to $\mathbb{R}^d$, the nth-order derivative of f at y denoted by $f^{(n)}(y)$ is a multilinear map from $\underbrace{\mathbb{R}^d \times \ldots \times \mathbb{R}^d}_{n-fold}$ to $\mathbb{R}^d$ defined by

$$f^{(n)}(y)(z_1, \ldots, z_n) = \sum_{1\le\alpha_1,\ldots,\alpha_n\le d} \frac{\partial^n f}{\partial y_{\alpha_1} \cdots \partial y_{\alpha_n}}(y) z_1^{\alpha_1} \ldots z_n^{\alpha_n},$$

where $y = (y_1, \ldots, y_d)^\intercal$ and $z_i = (z_i^1, \ldots, z_i^d)^\intercal$ for $i = 1, \ldots, n$. With this background, we now can give the existence, uniqueness, and especially the smoothness with respect to h for the continuous finite element approximation $u(\tau)$ associated with the FFCFEr method. The proof of the following theorem is based on a fixed-point iteration which is analogous to Picard iteration.

Theorem 1.4 *Given a positive constant R, let*

$$B(y_0, R) = \{y \in \mathbb{R}^d : ||y - y_0|| \le R\}$$

and

$$D_n = \max_{y\in B(y_0,R)} ||f^{(n)}(y)||, \quad n = 0, 1, \ldots, \tag{1.21}$$

where $||\cdot|| = ||\cdot||_\infty$ is the maximum norm for vectors in $\mathbb{R}^d$ or the corresponding induced norm for the multilinear maps $f^{(n)}(y)$, $n \ge 1$. Then the FFCFEr method (1.3) or (1.14) has a unique solution $u(\tau)$ which is smoothly dependent of h provided

$$0 \le h \le \varepsilon < \min\left\{\frac{1}{M_0 D_1}, \frac{R}{M_0 D_0}, 1\right\}. \tag{1.22}$$

Proof Set $u_0(\tau) \equiv y_0$. We construct a function series $\{u_n(\tau)\}_{n=0}^{\infty}$ defined by the relation

$$u_{n+1}(\tau) = y_0 + h \int_0^1 A_{\tau,\sigma} f(u_n(\sigma)) d\sigma, \quad n = 0, 1, \ldots. \tag{1.23}$$

Obviously, $\lim_{n\to\infty} u_n(\tau)$ is a solution to (1.14) provided $\{u_n(\tau)\}_{n=0}^{\infty}$ is uniformly convergent. Thus, we only need to prove the uniform convergence of the infinite series

$$\sum_{n=0}^{\infty}(u_{n+1}(\tau) - u_n(\tau)).$$

It follows from (1.20), (1.22), (1.23) and induction that

$$||u_n(\tau) - y_0|| \leq R, \quad n = 0, 1, \ldots \tag{1.24}$$

Then by using (1.21), (1.22), (1.23), (1.24) and the inequalities

$$||\int_0^1 w(\tau)d\tau|| \leq \int_0^1 ||w(\tau)||d\tau, \quad \text{for } \mathbb{R}^d\text{-valued function } w(\tau),$$
$$||f(y) - f(z)|| \leq D_1||y - z||, \quad \text{for } y, z \in B(y_0, R),$$

we obtain the following inequalities

$$||u_{n+1}(\tau) - u_n(\tau)|| \leq h\int_0^1 M_0 D_1||u_n(\sigma) - u_{n-1}(\sigma)||d\sigma$$
$$\leq \beta||u_n - u_{n-1}||_c, \quad \beta = \varepsilon M_0 D_1,$$

where $||\cdot||_c$ is the maximum norm for continuous functions:

$$||w||_c = \max_{\tau\in[0,1]} ||w(\tau)||, \quad w \text{ is a continuous } \mathbb{R}^d\text{-valued function on } [0, 1].$$

Therefore, we have

$$||u_{n+1} - u_n||_c \leq \beta||u_n - u_{n-1}||_c$$

and

$$||u_{n+1} - u_n||_c \leq \beta^n||u_1 - y_0||_c \leq \beta^n R, \quad n = 0, 1, \ldots. \tag{1.25}$$

Since $\beta < 1$, according to Weierstrass M-test, $\sum_{n=0}^{\infty}(u_{n+1}(\tau) - u_n(\tau))$ is uniformly convergent, and thus, the limit of $\{u_n(\tau)\}_{n=0}^{\infty}$ is a solution to (1.14). If $v(\tau)$ is another solution, then the difference between $u(\tau)$ and $v(\tau)$ satisfies

$$||u(\tau) - v(\tau)|| \leq h\int_0^1 ||A_{\tau,\sigma}(f(u(\sigma)) - f(v(\sigma)))||d\sigma \leq \beta||u - v||_c,$$

and

$$||u - v||_c \leq \beta||u - v||_c.$$

This means $||u - v||_c = 0$, i.e., $u(\tau) \equiv v(\tau)$. Hence, the existence and uniqueness have been proved.

As for the smooth dependence of u on h, since every $u_n(\tau)$ is a smooth function of h, we only need to prove the sequence

$$\left\{ \frac{\partial^k u_n}{\partial h^k}(\tau) \right\}_{n=0}^{\infty}$$

is uniformly convergent for $k \geq 1$. Firstly, differentiating both sides of (1.23) with respect to h yields

$$\frac{\partial u_{n+1}}{\partial h}(\tau) = \int_0^1 (A_{\tau,\sigma} + h\frac{\partial A_{\tau,\sigma}}{\partial h}) f(u_n(\sigma))d\sigma + h\int_0^1 A_{\tau,\sigma} f^{(1)}(u_n(\sigma))\frac{\partial u_n}{\partial h}(\sigma)d\sigma. \quad (1.26)$$

We then have

$$||\frac{\partial u_{n+1}}{\partial h}||_c \leq \alpha + \beta||\frac{\partial u_n}{\partial h}||_c, \quad \alpha = (M_0 + \varepsilon M_1)D_0. \quad (1.27)$$

By induction, it is easy to show that $\left\{\frac{\partial u_n}{\partial h}(\tau)\right\}_{n=0}^{\infty}$ is uniformly bounded:

$$||\frac{\partial u_n}{\partial h}||_c \leq \alpha(1 + \beta + \ldots + \beta^{n-1}) \leq \frac{\alpha}{1-\beta} = C^*, \quad n = 0, 1, \ldots. \quad (1.28)$$

Combining (1.25), (1.26) and (1.28), we obtain

$$\begin{aligned}
&||\frac{\partial u_{n+1}}{\partial h} - \frac{\partial u_n}{\partial h}||_c \\
&\leq \int_0^1 (M_0 + hM_1)||f(u_n(\sigma)) - f(u_{n-1}(\sigma))||d\sigma \\
&+ h\int_0^1 M_0 \Big(||(f^{(1)}(u_n(\sigma)) - f^{(1)}(u_{n-1}(\sigma)))\frac{\partial u_n}{\partial h}(\sigma)|| \\
&\qquad + ||f^{(1)}(u_{n-1}(\sigma))(\frac{\partial u_n}{\partial h}(\sigma) - \frac{\partial u_{n-1}}{\partial h}(\sigma))|| \Big) d\sigma \\
&\leq \gamma\beta^{n-1} + \beta||\frac{\partial u_n}{\partial h} - \frac{\partial u_{n-1}}{\partial h}||_c,
\end{aligned}$$

where

$$\gamma = (M_0 D_1 + \varepsilon M_1 D_1 + \varepsilon M_0 L_2 C^*)R,$$

and L_2 is a constant satisfying

$$||f^{(1)}(y) - f^{(1)}(z)|| \leq L_2||y - z||, \quad \text{for } y, z \in B(y_0, R).$$

Thus, again by induction, we have

$$||\frac{\partial u_{n+1}}{\partial h}-\frac{\partial u_n}{\partial h}||_c \leq n\gamma\beta^{n-1}+\beta^n C^*,\quad n=1,2,\ldots$$

and $\{\frac{\partial u_n}{\partial h}(\tau)\}_{n=0}^{\infty}$ is uniformly convergent. By a similar argument, one can show that other function sequence $\left\{\frac{\partial^k u_n}{\partial h^k}(\tau)\right\}_{n=0}^{\infty}$ for $k\geq 2$ are uniformly convergent as well. Therefore, $u(\tau)$ is smoothly dependent on h. The proof is complete. □

Since our analysis of the algebraic order of the FFCFEr method is mainly based on Taylor's theorem, it is meaningful to investigate the expansion of $P_{\tau,\sigma}(h)$.

Proposition 1.1 *Assume that the Taylor expansion of $P_{\tau,\sigma}(h)$ with respect to h at zero is*

$$P_{\tau,\sigma}=\sum_{n=0}^{r-1}P_{\tau,\sigma}^{[n]}h^n+\mathscr{O}(h^r). \tag{1.29}$$

Then the coefficients $P_{\tau,\sigma}^{[n]}$ satisfy

$$\langle P_{\tau,\sigma}^{[n]},g_m(\sigma)\rangle_\sigma=\begin{cases}g_m(\tau), & n=0,\quad m=r-1,\\ 0, & n=1,\ldots,r-1,\quad m=r-1-n,\end{cases}$$

for any $g_m\in P_m([0,1])$, where $P_m([0,1])$ consists of polynomials of degrees $\leq m$ on $[0,1]$.

Proof It can be observed from (1.11) that

$$\langle P_{\tau,\sigma},\varphi_i(t_0+\sigma h)\rangle_\sigma=\varphi_i(t_0+\tau h),\quad i=0,1,\ldots,r-1. \tag{1.30}$$

Meanwhile, expanding $\varphi_i(t_0+\tau h)$ at t_0 yields

$$\varphi_i(t_0+\tau h)=\sum_{n=0}^{r-1}\frac{\varphi_i^{(n)}(t_0)}{n!}\tau^n h^n+\mathscr{O}(h^r). \tag{1.31}$$

Then by inserting (1.29) and (1.31) into the Eq. (1.30), we obtain that

$$\langle\sum_{n=0}^{r-1}P_{\tau,\sigma}^{[n]}h^n,\sum_{m=0}^{r-1}\frac{\varphi_i^{(m)}(t_0)}{m!}\sigma^m h^m\rangle_\sigma=\sum_{k=0}^{r-1}\frac{\varphi_i^{(k)}(t_0)}{k!}\tau^k h^k+\mathscr{O}(h^r).$$

Considering the terms in h^k leads to

$$\sum_{k=0}^{r-1}\left(\sum_{m+n=k}\frac{\varphi_i^{(m)}(t_0)}{m!}\langle P_{\tau,\sigma}^{[n]},\sigma^m\rangle_\sigma-\frac{\varphi_i^{(k)}(t_0)}{k!}\tau^k\right)h^k=\mathcal{O}(h^r),$$

$$\sum_{m=0}^{k-1}\frac{\varphi_i^{(m)}(t_0)}{m!}P_{m,k-m}+\frac{\varphi_i^{(k)}(t_0)}{k!}(P_{k0}-\tau^k)=0,\quad i,\ k=0,1,\ldots,r-1,$$

and

$$W^\intercal V=0,$$

where $P_{mn}=\langle P_{\tau,\sigma}^{[n]},\sigma^m\rangle_\sigma$, W is the Wronskian (1.18), and $V=(V_{mk})_{0\le m,k\le r-1}$ is an upper triangular matrix with the entries determined by

$$V_{mk}=\begin{cases}\frac{1}{m!}P_{m,k-m}, & m<k,\\ \frac{1}{m!}(P_{m,0}-\tau^m), & m=k.\end{cases}$$

Since W is nonsingular, $V=0$,

$$P_{mn}=\begin{cases}\tau^m, & n=0,\quad m+n\le r-1,\\ 0, & n=1,2,\ldots,r-1,\quad m+n\le r-1.\end{cases}\tag{1.32}$$

Then the statement of the proposition directly follows from (1.32). □

Aside from $P_{\tau,\sigma}$, it is also crucial to analyse the expansion of the solution $u(\tau)$. For convenience, we say that an h-dependent function $w(\tau)$ is regular if it can be expanded as

$$w(\tau)=\sum_{n=0}^{r-1}w^{[n]}(\tau)h^n+\mathcal{O}(h^r),$$

where

$$w^{[n]}(\tau)=\frac{1}{n!}\frac{\partial^n w(\tau)}{\partial h^n}|_{h=0}$$

is a vector-valued function with polynomial entries of degrees $\le n$.

Lemma 1.2 *Given a regular function w and an h-independent sufficiently smooth function g, the composition (if exists) is regular. Moreover, the difference between w and its projection satisfies*

$$P_h w(\tau)-w(\tau)=\mathcal{O}(h^r).$$

Proof Assume that the expansion of $g(w(\tau))$ with respect to h at zero is

$$g(w(\tau)) = \sum_{n=0}^{r-1} p^{[n]}(\tau)h^n + \mathcal{O}(h^r).$$

Then, differentiating $g(w(\tau))$ with respect to h at zero iteratively and using

$$p^{[n]}(\tau) = \frac{1}{n!}\frac{\partial^n g(w(\tau))}{\partial h^n}|_{h=0}, \quad \text{the degree of } \frac{\partial^n w(\tau)}{\partial h^n}|_{h=0} \leq n, \quad n = 0, 1, \ldots, r-1,$$

we can observe that $p^{[n]}(\tau)$ is a vector with polynomial entries of degrees $\leq n$ for $n = 0, 1, \ldots, r-1$ and the first statement is confirmed.

As for the second statement, using Proposition 1.1, we have

$$\begin{aligned}
&P_h w(\tau) - w(\tau)\\
&= \langle \sum_{n=0}^{r-1} P_{\tau,\sigma}^{[n]} h^n, \sum_{k=0}^{r-1} w^{[k]}(\sigma)h^k\rangle_\sigma - \sum_{m=0}^{r-1} w^{[m]}(\tau)h^m + \mathcal{O}(h^r)\\
&= \sum_{m=0}^{r-1}\Big(\sum_{n+k=m} \langle P_{\tau,\sigma}^{[n]}, w^{[k]}(\sigma)\rangle_\sigma - w^{[m]}(\tau)\Big)h^m + \mathcal{O}(h^r)\\
&= \sum_{m=0}^{r-1}(\langle P_{\tau,\sigma}^{[0]}, w^{[m]}(\sigma)\rangle_\sigma - w^{[m]}(\tau))h^m + \mathcal{O}(h^r) = \mathcal{O}(h^r).
\end{aligned}$$

□

Before further discussions, it may be useful to recall some standard results in the theory of ODEs. To emphasize the dependence of the solution to $y'(t) = f(y(t))$ on the initial value, we assume that $y(\cdot, \tilde{t}, \tilde{y})$ solves the IVP:

$$\frac{d}{dt}y(t, \tilde{t}, \tilde{y}) = f(y(t, \tilde{t}, \tilde{y})), \quad y(\tilde{t}, \tilde{t}, \tilde{y}) = \tilde{y}.$$

Clearly, this problem is equivalent to the following integral equation:

$$y(t, \tilde{t}, \tilde{y}) = \tilde{y} + \int_{\tilde{t}}^{t} f(y(\xi, \tilde{t}, \tilde{y}))d\xi.$$

Differentiating it with respect to $\tilde{t}$ and $\tilde{y}$ and using the uniqueness of the solution leads to

$$\frac{\partial y}{\partial \tilde{t}}(t, \tilde{t}, \tilde{y}) = -\frac{\partial y}{\partial \tilde{y}}(t, \tilde{t}, \tilde{y})f(\tilde{y}). \tag{1.33}$$

With the previous analysis results, we are in a position to give the order of FFCFEr.

Theorem 1.5 *The stage order and order of the FFCFEr method (1.3) or (1.14) are r and $2r$, respectively. That is,*

$$u(\tau) - y(t_0 + \tau h) = \mathcal{O}(h^{r+1}),$$

for $0 < \tau < 1$, *and*

$$u(1) - y(t_0 + h) = \mathcal{O}(h^{2r+1}).$$

Proof Firstly, by Theorem 1.4 and Lemma 1.1, we can expand $u(\tau)$ and $A_{\tau,\sigma}$ with respect to h at zero:

$$u(\tau) = \sum_{m=0}^{r-1} u^{[m]}(\tau)h^m + \mathcal{O}(h^r), \quad A_{\tau,\sigma} = \sum_{m=0}^{r-1} A_{\tau,\sigma}^{[m]}h^m + \mathcal{O}(h^r).$$

Then let

$$\delta = u(\sigma) - y_0 = \sum_{m=1}^{r-1} u^{[m]}(\sigma)h^m + \mathcal{O}(h^r) = \mathcal{O}(h).$$

Expanding $f(u(\sigma))$ at y_0 and inserting the above equalities into the first equation of (1.14), we obtain

$$\sum_{m=0}^{r-1} u^{[m]}(\tau)h^m = y_0 + h\int_0^1 \sum_{k=0}^{r-1} A_{\tau,\sigma}^{[k]}h^k \sum_{n=0}^{r-1} F^{(n)}(y_0)(\underbrace{\delta,\ldots,\delta}_{n-fold})d\sigma + \mathcal{O}(h^r), \tag{1.34}$$

where $F^{(n)}(y_0) = f^{(n)}(y_0)/n!$. We claim that $u(\tau)$ is regular, i.e.

$$u^{[m]}(\tau) \in P_m^d = \underbrace{P_m([0,1]) \times \ldots \times P_m([0,1])}_{d-fold}$$

for $m = 0, 1, \ldots, r-1$. This fact can be confirmed by induction. Clearly, $u^{[0]}(\tau) = y_0 \in P_0^d$. If $u^{[n]}(\tau) \in P_n^d$ for $n = 0, 1, \ldots, m$, then by comparing the coefficients of h^{m+1} on both sides of (1.34) and using (1.15) and Proposition 1.1, we obtain that

$$\begin{aligned} u^{[m+1]}(\tau) &= \sum_{k+n=m} \int_0^1 A_{\tau,\sigma}^{[k]} g_n(\sigma)d\sigma = \sum_{k+n=m} \int_0^\tau \int_0^1 P_{\alpha,\sigma}^{[k]} g_n(\sigma)d\sigma d\alpha \\ &= \int_0^\tau \int_0^1 P_{\alpha,\sigma}^{[0]} g_m(\sigma)d\sigma d\alpha = \int_0^\tau g_m(\alpha)d\alpha \in P_{m+1}^d, \quad g_n(\sigma) \in P_n^d. \end{aligned}$$

This completes the induction. By Lemma 1.2, $f(u(\tau))$ is also regular and

$$f(u(\tau)) - P_h(f \circ u)(\tau) = \mathcal{O}(h^r). \tag{1.35}$$

Then it follows from (1.13), (1.33) and (1.35) that

$$\begin{aligned}u(\tau) - y(t_0 + \tau h) &= y(t_0 + \tau h, t_0 + \tau h, u(\tau)) - y(t_0 + \tau h, t_0, y_0) \\ &= \int_0^\tau \frac{d}{d\alpha} y(t_0 + \tau h, t_0 + \alpha h, u(\alpha))d\alpha \\ &= \int_0^\tau (h\frac{\partial y}{\partial \tilde{t}}(t_0 + \tau h, t_0 + \alpha h, u(\alpha)) + \frac{\partial y}{\partial \tilde{y}}(t_0 + \tau h, t_0 + \alpha h, u(\alpha))u'(\alpha))d\alpha \\ &= -h \int_0^\tau \Phi^\tau(\alpha)(f(u(\alpha)) - P_h(f \circ u)(\alpha))d\alpha \\ &= \mathcal{O}(h^{r+1}),\end{aligned} \tag{1.36}$$

where

$$\Phi^\tau(\alpha) = \frac{\partial y}{\partial \tilde{y}}(t_0 + \tau h, t_0 + \alpha h, u(\alpha)).$$

As for the algebraic order, setting $\tau = 1$ in (1.36) leads to

$$\begin{aligned}&u(1) - y(t_0 + h) \\ &= -h \int_0^1 \Phi^1(\alpha)(f(u(\alpha)) - P_h(f \circ u)(\alpha))d\alpha.\end{aligned} \tag{1.37}$$

Since $\Phi^1(\alpha)$ is a matrix-valued function, we partition it as $\Phi^1(\alpha) = (\Phi_1^1(\alpha), \ldots, \Phi_d^1(\alpha))^\intercal$. Using Lemma 1.2 again leads to

$$\Phi_i^1(\alpha) = P_h\Phi_i^1(\alpha) + \mathcal{O}(h^r), \quad i = 1, 2, \ldots, d. \tag{1.38}$$

Meanwhile, setting $w(\alpha) = P_h\Phi_i(\alpha)^\intercal$ in (1.6) and using (1.13) yields

$$\begin{aligned}&\int_0^1 P_h\Phi_i^1(\alpha)f(u(\alpha))d\alpha = h^{-1}\int_0^1 P_h\Phi_i^1(\alpha)u'(\alpha)d\alpha = \int_0^1 P_h\Phi_i^1(\alpha)P_h(f \circ u)(\alpha)d\alpha, \\ &\qquad i = 1, 2, \ldots, d.\end{aligned} \tag{1.39}$$

Therefore, using (1.37)–(1.39) we have

$$\begin{aligned}&u(1) - y(t_0 + h) \\ &= -h\int_0^1 \left(\begin{pmatrix} P_h\Phi_1^1(\alpha) \\ \vdots \\ P_h\Phi_d^1(\alpha) \end{pmatrix} + \mathcal{O}(h^r) \right) (f(u(\alpha)) - P_h(f \circ u)(\alpha))d\alpha \\ &= -h\int_0^1 \begin{pmatrix} P_h\Phi_1^1(\alpha)(f(u(\alpha)) - P_h(f \circ u)(\alpha)) \\ \vdots \\ P_h\Phi_d^1(\alpha)(f(u(\alpha)) - P_h(f \circ u)(\alpha)) \end{pmatrix} d\alpha - h\int_0^1 \mathcal{O}(h^r) \times \mathcal{O}(h^r)d\alpha = \mathcal{O}(h^{2r+1}).\end{aligned}$$

□

According to Theorem 1.5, the TF CFE methods based on the spaces (1.7)–(1.9) are of order $2r$, $4k$ and $2(k + p + 1)$, respectively.

1.4 Implementation Issues

It should be noted that (1.14) is not a practical form for applications. In this section, we will detail the implementation of the FFCFEr method. Firstly, it is necessary to introduce the generalized Lagrange interpolation functions $l_i(\tau) \in X_h$ with respect to $(r+1)$ distinct points $\{d_i\}_{i=1}^{r+1} \subseteq [0,1]$:

$$(l_1(\tau), \ldots, l_{r+1}(\tau)) = (\widetilde{\Phi}_1(\tau), \widetilde{\Phi}_2(\tau), \ldots, \widetilde{\Phi}_{r+1}(\tau))\Lambda^{-1}, \tag{1.40}$$

where $\{\Phi_i(t)\}_{i=1}^{r+1}$ is a basis of X, $\widetilde{\Phi}_i(\tau) = \Phi_i(t_0 + \tau h)$ and

$$\Lambda = \begin{pmatrix} \widetilde{\Phi}_1(d_1) & \widetilde{\Phi}_2(d_1) & \ldots & \widetilde{\Phi}_{r+1}(d_1) \\ \widetilde{\Phi}_1(d_2) & \widetilde{\Phi}_2(d_2) & \ldots & \widetilde{\Phi}_{r+1}(d_2) \\ \vdots & \vdots & & \vdots \\ \widetilde{\Phi}_1(d_{r+1}) & \widetilde{\Phi}_2(d_{r+1}) & \ldots & \widetilde{\Phi}_{r+1}(d_{r+1}) \end{pmatrix}.$$

By means of the expansions

$$\Phi_i(t_0 + d_j h) = \sum_{n=0}^{r} \frac{\Phi_i^{(n)}(t_0)}{n!} d_j^n h^n + \mathscr{O}(h^{r+1}), \quad i, j = 1, 2, \ldots, r+1,$$

we have

$$\Lambda = \begin{pmatrix} 1 & d_1 h & \ldots & \frac{d_1^r h^r}{r!} \\ 1 & d_2 h & \ldots & \frac{d_2^r h^r}{r!} \\ \vdots & \vdots & & \vdots \\ 1 & d_{r+1} h & \ldots & \frac{d_{r+1}^r h^r}{r!} \end{pmatrix} \widetilde{W} + \mathscr{O}(h^{r+1}),$$

where $\widetilde{W}$ is the Wronskian of $\{\Phi_i(t)\}_{i=1}^{r+1}$ at t_0. Since $\widetilde{W}$ is nonsingular, Λ is also nonsingular for h which is sufficiently small but not zero and the Eq. (1.40) makes sense in this case. Then $\{l_i(\tau)\}_{i=1}^{r+1}$ is a basis of X_h satisfying $l_i(d_j) = \delta_{ij}$ and $u(\tau)$ can be expressed as

$$u(\tau) = \sum_{i=1}^{r+1} u(d_i) l_i(\tau).$$

Choosing $d_i = (i-1)/r$ and denoting $y_\sigma = u(\sigma)$, (1.14) now reads

$$\begin{cases} y_\sigma = \displaystyle\sum_{i=1}^{r+1} y_{\frac{i-1}{r}} l_i(\sigma), \\ y_{\frac{i-1}{r}} = y_0 + h \displaystyle\int_0^1 A_{\frac{i-1}{r},\sigma} f(y_\sigma) d\sigma, \quad i = 2, \ldots, r+1. \end{cases} \tag{1.41}$$

When f is a polynomial and $\{\Phi_i(t)\}_{i=1}^{r+1}$ are polynomials, trigonometrical or exponential functions, the integral in (1.41) can be calculated exactly. After solving this algebraic system about variables $y_{1/r}, y_{2/r}, \ldots, y_1$ by iterations, we obtain the numerical solution $y_1 \approx y(t_0 + h)$. Therefore, although the FFCFEr method can be analysed in the form of continuous-stage RK method (1.14), it is indeed an r-stage method in practice. If the integral cannot be directly calculated, we approximate it by a high-order quadrature rule $(b_k, c_k)_{k=1}^{s}$. The corresponding fully discrete scheme for (1.41) is

$$\begin{cases} y_\sigma = \displaystyle\sum_{i=1}^{r+1} y_{\frac{i-1}{r}} l_i(\sigma), \\ y_{\frac{i-1}{r}} = y_0 + h \displaystyle\sum_{k=1}^{s} b_k A_{\frac{i-1}{r}, c_k} f(y_{c_k}), \quad i = 2, \ldots, r+1. \end{cases} \tag{1.42}$$

By an argument which is similar to that stated at the beginning of Sect. 1.3, (1.42) is equivalent to a discrete version of the FFCFEr method (1.3):

$$\begin{cases} u(0) = y_0, \\ \langle v, u' \rangle_\tau = h[v, f \circ u], \quad u(\tau) \in X_h, \text{ for all } v(\tau) \in Y_h, \\ y_1 = u(1), \end{cases}$$

where $[\cdot, \cdot]$ is the discrete inner product:

$$[w_1, w_2] = [w_1(\tau), w_2(\tau)]_\tau = \sum_{k=1}^{s} b_k w_1(c_k) \cdot w_2(c_k).$$

By the proof procedure of Theorem 1.3, one can show that the fully discrete scheme is still symmetric provided the quadrature rule is symmetric, i.e. $c_{s+1-k} = 1 - c_k$ and $b_{s+1-k} = b_k$ for $k = 1, 2, \ldots, s$.

Now it is clear that the practical form (1.41) or (1.42) is determined by the Lagrange interpolation functions $l_i(\tau)$ and the coefficient $A_{\tau,\sigma}$. For the CFEr method,

$$Y_h = \text{span}\left\{1, \tau, \ldots, \tau^{r-1}\right\}, \quad X_h = \text{span}\left\{1, \tau, \ldots, \tau^{r}\right\},$$

and all $l_i(\tau)$ for $i = 1, 2, \ldots, r+1$ are Lagrange interpolation polynomials of degrees r. The $A_{\tau,\sigma}$ for $r = 2, 3, 4$ are given by

$$A_{\tau,\sigma} = \begin{cases} (4 + 6\sigma(-1+\tau) - 3\tau)\tau, \quad r = 2, \\ \tau(9 - 18\tau + 10\tau^2 + 30\sigma^2(1 - 3\tau + 2\tau^2) - 12\sigma(3 - 8\tau + 5\tau^2)), \quad r = 3, \\ \tau(16 - 60\tau + 80\tau^2 - 35\tau^3 + 140\sigma^3(-1 + 6\tau - 10\tau^2 + 5\tau^3) \\ \quad + 60\sigma(-2 + 10\tau - 15\tau^2 + 7\tau^3) - 30\sigma^2(-8 + 45\tau - 72\tau^2 + 35\tau^3)), \quad r = 4. \end{cases}$$

For the TFCFEr method,

$$Y = \text{span}\left\{1, t, \ldots, t^{r-3}, \cos(\omega t), \sin(\omega t)\right\},$$

then

$$Y_h = \text{span}\left\{1, \tau, \ldots, \tau^{r-3}, \cos(\nu\tau), \sin(\nu\tau)\right\},$$
$$X_h = \text{span}\left\{1, \tau, \ldots, \tau^{r-2}, \cos(\nu\tau), \sin(\nu\tau)\right\},$$

where $\nu = h\omega$. The corresponding $A_{\tau,\sigma}$ and $l_i(\tau)$ are more complicated than those of CFEr, but one can calculate them by the formulae (1.15) and (1.40) without any difficulty before solving the IVP numerically. Consequently, the computational cost of the TFCFEr method at each step is comparable to that of the CFEr method. Besides, when ν is small, in order to avoid unacceptable cancellation, it is recommended to calculate variable coefficients in TF methods by their Taylor expansions with respect to ν at zero.

1.5 Numerical Experiments

In this section, we carry out four numerical experiments to test the effectiveness and efficiency of the new methods TFCFEr based on the space (1.7) for $r = 2, 3, 4$ and TF2CFE4 based on the space (1.8) in the long-term computation of structure preservation. These new methods are compared with standard r-stage $2r$th-order EP CFEr methods for $r = 2, 3, 4$. Other methods such as the 2-stage 4th-order EF symplectic Gauss–Legendre collocation method (denoted by EFGL2) derived in [7] and the 2-stage 4th-order EF EP method (denoted by EFCRK2) derived in [30] are also considered. Since all of these structure-preserving methods are implicit, fixed-point iterations are needed to solve the nonlinear algebraic systems at each step. The tolerance error for the iteration solution is set to 10^{-15} in the numerical simulation.

Numerical quantities with which we are mainly concerned are the Hamiltonian error

$$EH = (EH^0, EH^1, \ldots),$$

with

$$EH^n = |H(y_n) - H(y_0)|,$$

and the solution error

$$ME = (ME^0, ME^1, \ldots),$$

with

$$ME^n = ||y_n - y(t_n)||_\infty.$$

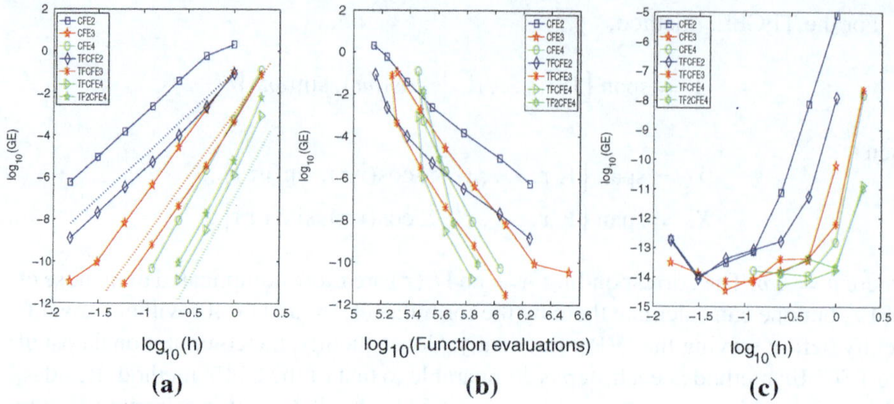

Fig. 1.1 **a** The logarithm of the maximum global error against the logarithm of the stepsize. The dashed lines have slopes four, six and eight. **b** The logarithm of the maximum global error against the logarithm of function evaluations. **c** The logarithm of the maximum global error of Hamiltonian against the logarithm of the stepsize.

Correspondingly, the maximum global errors of Hamiltonian (GEH) and the solution (GE) are defined by:

$$GEH = \max_{n\geq 0} EH^n, \quad GE = \max_{n\geq 0} ME^n,$$

respectively. Here the numerical solution at the time node t_n is denoted by y_n.

Example 1.1 Consider the Perturbed Kepler problem defined by the Hamiltonian:

$$H(p,q) = \frac{1}{2}(p_1^2 + p_2^2) - \frac{1}{(q_1^2 + q_2^2)^{\frac{1}{2}}} - \frac{2\varepsilon + \varepsilon^2}{3(q_1^2 + q_2^2)^{\frac{3}{2}}},$$

with the initial condition $q_1(0) = 1, q_2 = 0, p_1(0) = 0, p_2 = 1 + \varepsilon$, where ε is a small parameter. The exact solution of this IVP is

$$q_1(t) = \cos((1+\varepsilon)t), \quad q_2(t) = \sin((1+\varepsilon)t), \quad p_i(t) = q_i'(t), \ i = 1, 2.$$

Taking $\omega = 1, \varepsilon = 0.001$ and $h = 1/2^i$ for $i = -1, 0, \ldots, 6$, we integrate this problem over the interval $[0, 200\pi]$ by the TF2CFE4, TFCFEr and CFEr methods for $r = 2, 3, 4$. The nonlinear integral in the r-stage method is evaluated by the $(r+1)$-point Gauss–Legendre quadrature rule. Numerical results are presented in Fig. 1.1.

From Fig. 1.1a it can be observed that TFCFEr and TF2CFE4 methods show the expected order. Under the same stepsize, the TF method is more accurate than the non-TF method of the same order. Since the double precision provides only 16 significant digits, the numerical results are polluted significantly by rounding errors

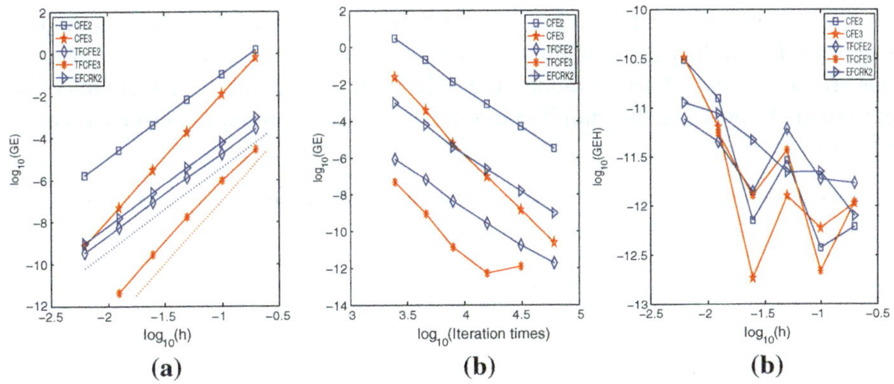

Fig. 1.2 **a** The logarithm of the maximum global error against the logarithm of the stepsize. The dashed lines have slopes four and six. **b** The logarithm of the maximum global error against the logarithm of iteration times. **c** The logarithm of the maximum global error of Hamiltonian against the logarithm of the stepsize. Copyright ©2016 Society for Industrial and Applied Mathematics. Reprinted with permission. All rights reserved.

when the maximum global error attains the magnitude 10^{-11}. Figure 1.1b shows that the efficiency of the TF method is higher than that of the non-TF method of the same order. Besides, high-order methods are more efficient than low-order ones when the stepsize is relatively small.

In Fig. 1.1c, one can see that all of these EP methods preserve the Hamiltonian very well. The errors in the Hamiltonian are mainly contributed by the quadrature error when the stepsize h is large and the rounding error when h is small.

Example 1.2 Consider the Duffing equation defined by the Hamiltonian:

$$H(p, q) = \frac{1}{2}p^2 + \frac{1}{2}(\omega^2 + k^2)q^2 - \frac{k^2}{2}q^4$$

with the initial value $q(0) = 0,\ p(0) = \omega$. The exact solution of this IVP is

$$q(t) = sn(\omega t; k/\omega), \quad p(t) = cn(\omega t; k/\omega)dn(\omega t; k/\omega).$$

where sn, cn and dn are Jacobi elliptic functions. Taking $k = 0.07, \omega = 5$ and $h = 1/5 \times 1/2^i$ for $i = 0, 1, \ldots, 5$, we integrate this problem over the interval $[0, 100]$ by TFCFE2, TFCFE3, CFE2, CFE3 and EFCRK2 methods. Since the nonlinear term f is polynomial, we can calculate the integrals involved in these methods exactly by Mathematica at the beginning of the computation. Numerical results are shown in Fig. 1.2.

In Fig. 1.2a, one can see that the TF method is more accurate than the non-TF method of the same order under the same stepsize. For both as 2-stage $4th$-order methods, TFCFE2 method is more accurate than EFCRK2 method for this problem.

Again, it can be observed from Fig. 1.2b that the efficiency of the CFEr method is lower than that of the EF/TF method of the same order. Although the nonlinear integrals are exactly calculated in theory, Fig. 1.2c shows that all of these methods only approximately preserve the Hamiltonian. It seems that the rounding error increases as $h \to 0$.

Example 1.3 Consider the Fermi–Pasta–Ulam problem studied by Hairer et al. in [22, 23], which is defined by the Hamiltonian

$$H(p,q) = \frac{1}{2}p^{\intercal}p + \frac{1}{2}q^{\intercal}Mq + U(q),$$

where

$$M = \begin{pmatrix} O_{m\times m} & O_{m\times m} \\ O_{m\times m} & \omega^2 I_{m\times m} \end{pmatrix},$$

$$U(q) = \frac{1}{4}\left((q_1 - q_{m+1})^4 + \sum_{i=1}^{m-1}(q_{i+1} - q_{m+i+1} - q_i - q_{m+i})^4 + (q_m + q_{2m})^4\right).$$

In this problem, we choose $m = 2$, $q_1(0) = 1$, $p_1(0) = 1$, $q_3(0) = 1/\omega$, $p_3(0) = 1$, and zero for the remaining initial values. Setting $\omega = 50$, $h = 1/50$ and $\omega = 100$, $h = 1/100$, we integrate this problem over the interval $[0, 100]$ by CFE2, CFE3, TFCFE2, TFCFE3 and EFCRK2 methods. The nonlinear integrals are calculated exactly by Mathematica at the beginning of the computation. We choose the numerical solution obtained by a high-order method with a sufficiently small stepsize as the 'reference solution' in the FPU problem. Numerical results are plotted in Fig. 1.3.

In Fig. 1.3a, c, one can see that the TF methods are more accurate than non-TF ones. Unlike the previous problem, the EFCRK2 method wins slightly over TFCFE2 method in this case. The Fig. 1.3b, d also show that all of these methods display promising EP property, which is especially important in the FPU problem.

Example 1.4 Consider the IVP defined by the nonlinear Schrödinger equation

$$\begin{cases} iu_t + u_{xx} + 2|u|^2 u = 0, \\ u(x,0) = \varphi(x), \end{cases} \tag{1.43}$$

where u is a complex function of x, t, and i is the imaginary unit. Taking the periodic boundary condition $u(x_0, t) = u(x_0 + L, t)$ and discretizing the spatial derivative ∂_{xx} by the pseudospectral method (see e.g. [13]), this problem is converted into a complex system of ODEs:

$$\begin{cases} i\dfrac{d}{dt}U + D^2U + 2|U|^2 \cdot U = 0, \\ U(0) = (\varphi(x_0), \varphi(x_1), \dots, \varphi(x_{d-1}))^{\intercal}, \end{cases}$$

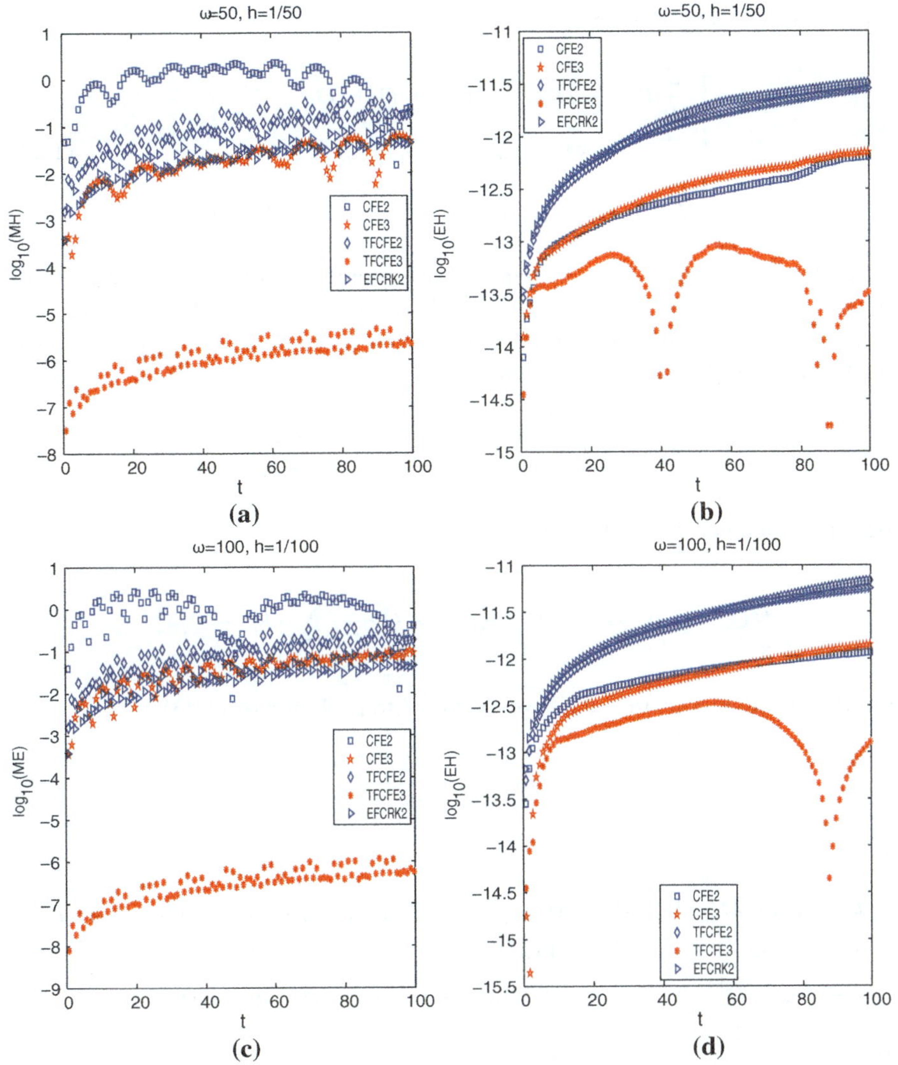

Fig. 1.3 **a, c** The logarithm of the solution error against time t. **b, d** The logarithm of the Hamiltonian error against time t.

or an equivalent Hamiltonian system:

$$\begin{cases} \dfrac{d}{dt}P = -D^2Q - 2(P^2+Q^2)\cdot Q, \\ \dfrac{d}{dt}Q = D^2P + 2(P^2+Q^2)\cdot P, \\ P(0) = real(U(0)), \quad Q(0) = imag(U(0)), \end{cases} \tag{1.44}$$

where the superscript '2' is the entrywise square multiplication operation for vectors, $x_n = x_0 + n\Delta x/d$ for $n = 0, 1, \ldots, d-1$, $U = (U_0(t), U_1(t), \ldots, U_{d-1}(t))^\intercal$, $P(t) = real(U(t))$, $Q(t) = imag(U(t))$ and $D = (D_{jk})_{0\leq j,k\leq d-1}$ is the pseudospectral differential matrix defined by:

$$D_{jk} = \begin{cases} \frac{\pi}{L}(-1)^{j+k}cot(\pi\frac{x_j - x_k}{L}), & j \neq k, \\ 0, & j = k. \end{cases}$$

The Hamiltonian or the total energy of (1.44) is

$$H(P, Q) = \frac{1}{2}P^\intercal D^2P + \frac{1}{2}Q^\intercal D^2Q + \frac{1}{2}\sum_{i=0}^{d-1}(P_i^2 + Q_i^2)^2.$$

In [33], the author constructed a periodic bi-soliton solution of (1.43):

$$u(x, t) = \frac{\Phi}{\Psi}, \tag{1.45}$$

where

$$\begin{aligned} \Phi &= (\exp(iM^2t)M\cosh^{-1}(M(x - A)) - \exp(iN^2t)N\cosh^{-1}(N(x + A))), \\ \Psi &= (\cosh(J) - \sinh(J)(\tanh(M(x - A))\tanh(N(x + A)) \\ &\quad + \cos((M^2 - N^2)t)\cosh^{-1}(M(x - A))\cosh^{-1}(N(x + A)))) \end{aligned}$$

with

$$J = \tanh^{-1}(\frac{2MN}{M^2 + N^2}).$$

This solution can be viewed approximately as the superposition of two single solitons located at $x = A$ and $x = -A$ respectively. Since it decays exponentially when $x \to \infty$, we can take the periodic boundary condition for sufficiently small x_0 and large L with little loss of accuracy. Aside from the total energy, it is well known that the semi-discrete NLS (1.44) has another invariant, the total charge

$$C(P, Q) = \sum_{i=0}^{d-1}(P_i^2 + Q_i^2).$$

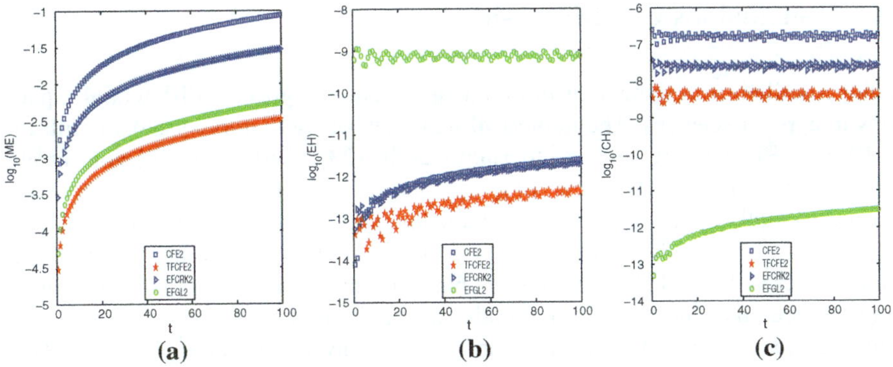

Fig. 1.4 **a** The logarithm of the solution error against time t (left). **b** The logarithm of the Hamiltonian error against time t (middle). **c** The logarithm of the charge error against time t (right).

Thus, we also calculate the error in the charge (EC):

$$EC = (EC^0, EC^1, \ldots)$$

with

$$EC^n = |C^n - C^0|, \quad C^n = C(P^n, Q^n)$$

where $P^n \approx P(t_n)$, $Q^n \approx Q(t_n)$ is the numerical solution at the time node t_n. Taking $x_0 = -50$, $L = 100$, $A = 10$, $M = 1$, $N = 2^{\frac{1}{2}}$, $d = 450$, $h = 0.2$, $\omega = 2$, we integrate the semi-discrete problem (1.44) by the TFCFE2, CFE2 and EFGL2 methods over the time interval [0, 100]. The nonlinear integrals are calculated exactly by Mathematica at the beginning of the computation. Numerical results are presented in Fig. 1.4.

It is noted that the exact solution (1.45) has two approximate frequency M^2 and N^2. By choosing the larger frequency $N^2 = 2$ as the fitting frequency ω, the EF/TF methods still reach higher accuracy than the general-purpose method CFE2, see Fig. 1.4a. Among three EF/TF methods, TFCFE2 is the most accurate. Figure 1.4b shows that three EP methods CFE2, TFCFE2 and EFCRK2 preserve the Hamiltonian (apart from the rounding error). Since EFGL2 is a symplectic method, it preserves the discrete charge, which is a quadratic invariant, see Fig. 1.4c. Although TFCFE2 method cannot preserve the discrete charge, its error in the charge is smaller than the charge errors of CFE2 and EFCRK2.

1.6 Conclusions and Discussions

Highly oscillatory systems constitute an important category of differential equations in applied sciences. The numerical treatment of oscillatory systems is full of challenges. Readers are referred to Hairer et al. [23], Iserles [26], Petzold et al. [14], Cohen et al. [34], Wu et al. [44, 45], and references contained therein. This chapter is mainly concerned with the establishment of high-order functionally-fitted energy-preserving methods for solving oscillatory nonlinear Hamiltonian systems. We have derived new FFCFEr methods based on the analysis of continuous finite element methods. The FFCFEr methods can be thought of as a continuous-stage Runge–Kutta methods, and hence it can be used conveniently in applications. The geometric properties and algebraic orders of the method have been analysed in detail. By equipping FFCFEr with the spaces (1.7) and (1.8), we have developed the TFEP methods denoted by TFCFEr and TF2CFEr which are suitable for solving oscillatory Hamiltonian systems with a fixed frequency ω. Evaluating the nonlinear integrals in the EP methods exactly or approximately, we have compared TFCFEr for $r = 2, 3, 4$ and TF2CFE4 with other structure-preserving methods such as EP methods CFEr for $r = 2, 3, 4$, the EP method EFCRK2 and the symplectic method EFGL2. The numerical results show that the newly derived TFEP methods exhibit definitely a high accuracy, an excellent invariant-preserving property and a prominent long-term behaviour.

In numerical experiments, we are mainly concerned with the TFCFEr methods when applied to oscillatory Hamiltonian systems. However, the FFCFEr methods, by nature, are symmetric and of order $2r$ for the general autonomous system $y'(t) = f(y(t))$. By choosing appropriate function spaces, the FFCFEr methods can be applied to solve a much wider class of dynamic systems in applications. For example, the application of the FF Runge–Kutta method to the stiff system has been shown in [32]. Consequently, we conclude that FFCFEr methods are likely to be a class of highly flexible methods with many potential applications.

In conclusion, in this chapter, from the perspective of the continuous finite element method, we have presented and analysed energy-preserving functionally fitted methods, in particular, trigonometrically fitted methods of an arbitrarily high order for solving oscillatory nonlinear Hamiltonian systems with a fixed frequency. In the next chapters, we will consider multi-frequency highly oscillatory systems.

This chapter is based on the work of Li and Wu [28].

References

1. Betsch, P., Steinmann, P.: Inherently energy conserving time finite element methods for classical mechanics. J. Comput. Phys. **160**, 88–116 (2000)
2. Bettis, D.G.: Numerical integration of products of Fourier and ordinary polynomials. Numer. Math. **14**, 424–434 (1970)

3. Bottasso, C.L.: A new look at finite elements in time : a variational interpretation of Runge-Kutta methods. Appl. Numer. Math. **25**, 355–368 (1997)
4. Brugnano, L., Iavernaro, F., Trigiante, D.: Hamiltonan boundary value methods (Energy preserving discrete line integral methods). J. Numer. Anal. Ind. Appl. Math. **5**, 13–17 (2010)
5. Brugnano, L., Iavernaro, F., Trigiante, D.: A simple framework for the derivation and analysis of effective one-step methods for ODEs. Appl. Math. Comput. **218**, 8475–8485 (2012)
6. Brugnano, L., Iavernaro, F., Trigiante, D.: Energy- and quadratic invariants-preserving integrators based upon Gauss-collocation formulae. SIAM J. Numer. Anal. **50**, 2897–2916 (2012)
7. Calvo, M., Franco, J.M., Montijano, J.I., Rández, L.: Structure preservation of exponentially fitted Runge-Kutta methods. J. Comput. Appl. Math. **218**, 421–434 (2008)
8. Calvo, M., Franco, J.M., Montijano, J.I., Rández, L.: Symmetric and symplectic exponentially fitted Runge-Kutta methods of high order. Comput. Phys. Commun. **181**, 2044–2056 (2010)
9. Calvo, M., Franco, J.M., Montijano, J.I., Rández, L.: On high order symmetric and symplectic trigonometrically fitted Runge-Kutta methods with an even number of stages. BIT Numer. Math. **50**, 3–21 (2010)
10. Celledoni, E., Mclachlan, R.I., Mclaren, D.I., Owren, B., Quispel, G.R.W., Wright, W.M.: Energy-preserving Runge-Kutta methods. ESIAM. Math. Model. Numer. Anal. **43**, 645–649 (2009)
11. Celledoni, E., Mclachlan, R.I., Owren, B., Quispel, G.R.W.: Energy-preserving integrators and the structure of B-series. Found. Comput. Math. **10**, 673–693 (2010)
12. Celledoni, E., Grimm, V., Mclachlan, R.I., Mclaren, D.I., O'Neale, D., Owren, B., Quispel, G.R.W.: Preserving energy resp. dissipation in numerical PDEs using the 'Average Vector Field' method. J. Comput. Phys. **231**, 6770–6789 (2012)
13. Chen, J.B., Qin, M.Z.: Multisymplectic fourier pseudospectral method for the nonlinear Schrödinger equation. Electron. Trans. Numer. Anal. **12**, 193–204 (2001)
14. Cohen, D., Jahnke, T., Lorenz, K., Lubich, C.: Numerical integrators for highly oscillatory Hamiltonian systems: a review. In: Mielke, A. (ed.) Analysis, Modeling and Simulation of Multiscale Problems, pp. 553–576. Springer, Berlin (2006)
15. Coleman, J.P.: P-stability and exponential-fitting methods for $y'' = f(x, y)$. IMA J. Numer. Anal. **16**, 179–199 (1996)
16. Franco, J.M.: Exponentially fitted symplectic integrators of RKN type for solving oscillatory problems. Comput. Phys. Commun. **177**, 479–492 (2007)
17. French, D.A., Schaeffer, J.W.: Continuous finite element methods which preserve energy properties for nonlinear problems. Appl. Math. Comput. **39**, 271–295 (1990)
18. Gautschi, W.: Numerical integration of ordinary differential equations based on trigonometric polynomials. Numer. Math. **3**, 381–397 (1961)
19. Gonzalez, O.: Time integration and discrete hamiltonian systems. J. Nonlinear Sci. **6**, 449–467 (1996)
20. Hairer, E.: Variable time step integration with symplectic methods. Appl. Numer. Math. **25**, 219–227 (1997)
21. Hairer, E.: Energy-preserving variant of collocation methods. J. Numer. Anal. Ind. Appl. Math. **5**, 73–84 (2010)
22. Hairer, E., Lubich, C.: Long-time energy conservation of numerical methods for oscillatory differential equations. SIAM J. Numer. Anal. **38**, 414–441 (2000)
23. Hairer, E., Lubich, C., Wanner, G.: Geometric Numerical Integration, 2nd edn. Springer, Berlin (2006)
24. Huang, N.S., Sidge, R.B., Cong, N.H.: On functionally fitted Runge-Kutta methods. BIT Numer. Math. **46**, 861–874 (2006)
25. Hulme, B.L.: One-step piecewise polynomial Galerkin methods for initial value problems. Math. Comput. **26**, 415–426 (1972)
26. Iserles, A.: On the method of Neumann series for highly oscillatory equations. BIT Numer. Math. **44**, 473–488 (2004)
27. Ixaru, L.G., Vanden Bergehe, G. (eds.): Exponential Fitting. Kluwer Academic Publishers, Dordrecht (2004)

28. Li, Y.W., Wu, X.Y.: Functionally-fitted energy-preserving methods for solving oscillatory nonlinear Hamiltonian systems. SIAM J. Numer. Anal. **54**, 2036–2059 (2016)
29. Mclachlan, R.I., Quispel, G.R.W., Robidoux, N.: Geometric integration using dicrete gradients. Philos. Trans. R. Soc. A **357**, 1021–1046 (1999)
30. Miyatake, Y.: An energy-preserving exponentially-fitted continuous stage Runge-Kutta method for Hamiltonian systems. BIT Numer. Math. **54**, 777–799 (2014)
31. Miyatake, Y.: A derivation of energy-preserving exponentially-fitted integrators for poisson systems. Comput. Phys. Commun. **187**, 156–161 (2015)
32. Ozawa, K.: A functionally fitting Runge-Kutta method with variable coefficients. Jpn. J. Ind. Appl. Math. **18**, 107–130 (2001)
33. Peregrine, D.H.: Water waves, nonlinear Schrödinger equations and their solutions. J. Austral. Math. Soc. Ser B **25**, 16–43 (1983)
34. Petzold, L.R., Jay, L.O., Jeng, Y.: Numerical solution of highly oscillatory ordinary differential equations. Acta Numer. **6**, 437–483 (1997)
35. Simos, J.C.: Assessment of energy-momentum and symplectic schemes for stiff dynamical systems. In: American Sociery for Mechanical Engineers, ASME Winter Annual meeting, New Orleans, Louisiana (1993)
36. Simos, T.E.: Does variable step size ruin a symplectic integrator? Phys. D. Nonlinear Phenom. **60**, 311–313 (1992)
37. Simos, T.E.: An exponentially-fitted Rung-Kutta method for the numerical integration of initial-value problems with periodic or oscillating solutions. Comput. Phys. Commun. **115**, 1–8 (1998)
38. Tang, W., Sun, Y.: Time finite element methods : A unified framework for the numerical discretizations of ODEs. Appl. Math. Comput. **219**, 2158–2179 (2012)
39. Vande Vyver, H.: A fourth order symplectic exponentially fitted integrator. Comput. Phys. Commun. **176**, 255–262 (2006)
40. Vanden Berghe, G., Daele, M., Vande Vyver, H.: Exponentially-fitted Runge-Kutta methods of collocation type : fixed or variable knots? J. Comput. Appl. Math. **159**, 217–239 (2003)
41. Wang, B., Wu, X.Y.: A new high precision energy-preserving integrator for system of second-order differential equations. Phys. Lett. A **376**, 1185–1190 (2012)
42. Wang, B., Iserles, A., Wu, X.Y.: Arbitrary order trigonometric fourier collocation methods for multi-frequency oscillatory systems. Found. Comput. Math. **16**, 151–181 (2016)
43. Wu, X.Y., Wang, B., Xia, J.: Explicit symplectic multidimensional exponential fitting modified Runge-Kutta-Nyström methods. BIT Numer. Math. **52**, 773–791 (2012)
44. Wu, X.Y., You, X., Wang, B.: Structure-Preserving Algorithms for Oscillatory Differential Equations. Springer, Berlin (2013)
45. Wu, X.Y., Liu, K., Shi, W.: Structure-Preserving Algorithms for Oscillatory Differential Equations II. Springer, Berlin (2015)
46. Yang, H., Wu, X.Y., You, X., Fang, Y.: Extended RKN-type methods for numerical integration of perturbed oscillators. Comput. Phys. Commun. **180**, 1777–1794 (2009)

Chapter 2
Exponential Average-Vector-Field Integrator for Conservative or Dissipative Systems

This chapter focuses on discrete gradient integrators intending to preserve the first integral or the Lyapunov function of the original continuous system. Incorporating the discrete gradients with exponential integrators, we discuss a novel exponential integrator for the conservative or dissipative system $\dot{y} = Q(My + \nabla U(y))$, where Q is a $d \times d$ real matrix, M is a $d \times d$ symmetric real matrix and $U : \mathbb{R}^d \to \mathbb{R}$ is a differentiable function. For conservative systems, the exponential integrator preserves the energy, while for dissipative systems, the exponential integrator preserves the decaying property of the Lyapunov function. Two properties of the new scheme are presented. Numerical experiments demonstrate the remarkable superiority of the new scheme in comparison with other structure-preserving schemes in the recent literature.

2.1 Introduction

In this chapter we are interested in the numerical solution of the IVP

$$\dot{y}(t) = Q(My(t) + \nabla U(y(t))), \quad y(t_0) = y^0, \tag{2.1}$$

where the $\dot{y}$ denotes the derivative with respect to time, Q is a $d \times d$ real matrix, M is a $d \times d$ symmetric real matrix and $U : \mathbb{R}^d \to \mathbb{R}$ is a differentiable function. Since M is symmetric, $My(t) + \nabla U(y(t))$ is the gradient of the function

$$H(y(t)) = \frac{1}{2} y(t)^{\intercal} My(t) + U(y(t)).$$

X. Wu and B. Wang, *Recent Developments in Structure-Preserving Algorithms for Oscillatory Differential Equations*, https://doi.org/10.1007/978-981-10-9004-2_2

In physical applications, the quantity H is often referred to as "energy". Two special categories are important in applications:

(i) If Q is skew-symmetric, then (2.1) is a *conservative system* with the first integral H, i.e. $H(y(t))$ is constant.

(ii) If Q is negative semi-definite (denoted by $Q \leq 0$), then (2.1) is a *dissipative system* with the Lyapunov function H, i.e. $H(y(t))$ is monotonically decreasing along the solution $y(t)$.

An even more particular case in the first category is that Q in (2.1) is the identity matrix. The system becomes

$$\dot{y}(t) = My(t) + \nabla U(y(t)), \quad y(t_0) = y^0. \tag{2.2}$$

An algorithm for (2.2) is an exponential integrator if it involves the computation of matrix exponentials (or related matrix functions) and exactly integrates the following system

$$\dot{y}(t) - My(t) = 0.$$

In general, exponential integrators permit larger stepsizes and achieve higher accuracy than non-exponential ones when (2.2) is a very stiff differential equation such as a highly oscillatory ODE or a semi-discrete time-dependent PDE. Therefore, numerous exponential algorithms have been proposed for first-order (see, e.g. [1, 10, 20, 22–26, 31]) and second-order (see e.g. [11, 12, 14, 18, 34]) ODEs.

On the other hand, (2.2) often possesses many important geometrical/physical structures. For example, the canonical Hamiltonian system

$$\dot{y}(t) = J^{-1}\nabla H(y(t)), \quad y(t_0) = y^0, \tag{2.3}$$

is a special case of (2.2), with

$$J = \begin{pmatrix} O_{d\times d} & I_{d\times d} \\ -I_{d\times d} & O_{d\times d} \end{pmatrix}.$$

The flow of (2.3) preserves the symplectic 2-form $\mathrm{d}y \wedge J\mathrm{d}y$ and the function $H(y)$. In the sense of geometric integration, it is a natural idea to design numerical schemes that preserve the two structures. As far as we know, most research papers dealing with exponential integrators up to now focus on the development of high-order explicit schemes but fail to be structure preserving except for symmetric/symplectic/energy-preserving methods for first-order ODEs in [5, 7] and oscillatory second-order ODEs (see, e.g. [18, 32, 33]).

It should be noted that the choice for M in (2.1) or in (2.2) is not unique. In order to take advantage of exponential integrators, the matrix M in (2.1) should be chosen such that $||QM|| \gg ||QHess(U)||$, where $Hess(U)$ is the Hessian matrix of U. For example, highly oscillatory Hamiltonian systems can be characterized by a dominant linear part My, where M implicitly contains the large frequency component. Up to now, many energy-preserving or energy-decaying methods have been proposed in the case of $M = 0$ (see, e.g. [3, 4, 15, 17, 19, 29]). However, these general-purpose methods are not suitable for dealing with (2.1) when $||QM||$ is very large. On the one hand, numerical solutions generated by these methods are far from accurate. On the other hand, they are generally implicit, and iterative solutions are required at each step. But the fixed-point iterations for them are not convergent unless the stepsize is taken very small. As mentioned at the beginning, these two obstacles can hopefully be overcome by introducing exponential integrators. In [32], the authors proposed an energy-preserving AAVF integrator (a trigonometric method) for solving the second-order Hamiltonian system

$$\begin{cases} \ddot{q}(t) + \tilde{M}q(t) = \nabla\tilde{U}(q(t)), & \tilde{M} \text{ is a symmetric matrix,} \\ q(t_0) = q_0, \quad \dot{q}(t_0) = \dot{q}_0, \end{cases}$$

which falls into the class of (2.1) by introducing

$$y = (\dot{q}^\intercal, q^\intercal)^\intercal, \; U(y) = \tilde{U}(q), \; Q = J^{-1},$$

$$M = \begin{pmatrix} I_{d\times d} & 0_{d\times d} \\ 0_{d\times d} & \tilde{M} \end{pmatrix},$$

and

$$U(y) = -\tilde{U}(q).$$

In this chapter, we present and analyse a new exponential integrator for (2.1) which can preserve the first integral or the Lyapunov function.

This chapter is organized as follows. Section 2.2 presents the discrete gradient integrators. In Sect. 2.3, we construct a general structure-preserving scheme for (2.1)–an exponential discrete gradient integrator. Two important properties of the scheme are proven. Symmetry and convergence of the EAVF integrator are investigated in Sect. 2.4. We then present a list of problems which can be solved by this scheme in Sect. 2.5. Numerical results, including the comparison between our new scheme and other structure-preserving schemes in the literature, are shown in Sect. 2.6. Section 2.7 is devoted to concluding remarks.

2.2 Discrete Gradient Integrators

Let $r(z)$ be a holomorphic function in the neighborhood of zero ($r(0) := \lim_{z\to 0} r(z)$ if 0 is a removable singularity)

$$r(z) = \sum_{i=0}^{\infty} \frac{r^{(i)}(0)}{i!} z^i. \tag{2.4}$$

The series (2.4) is assumed to be absolutely convergent. For a matrix A, the matrix-valued function $r(A)$ is defined by

$$r(A) = \sum_{i=0}^{\infty} \frac{r^{(i)}(0)}{i!} A^i.$$

I and O always denote identity and zero matrices of appropriate dimensions respectively. $A^{\frac{1}{2}}$ is a *square root* (not necessarily principal) of a symmetric matrix A. If $r^{(i)}(0) = 0$ for all odd i, then $r(A^{\frac{1}{2}})$ is well defined for every symmetric A (independent of the choice of $A^{\frac{1}{2}}$). For functions of matrices, the reader is referred to [21].

The discrete gradient method is an effective approach to constructing energy-preserving integrators. A *discrete gradient (DG)* of a differentiable function g is a bi-variate mapping $\nabla^{\mathrm{D}} g : \mathbb{R}^d \times \mathbb{R}^d \to \mathbb{R}^d$ satisfying

$$\begin{cases} \nabla^{\mathrm{D}} g(y, \hat{y})^{\intercal}(y - \hat{y}) = g(y) - g(\hat{y}), \\ \nabla^{\mathrm{D}} g(y, y) = \nabla g(y). \end{cases} \tag{2.5}$$

Accordingly, a *DG integrator* for the system (2.3) is defined by

$$y^1 = y^0 + hJ^{-1}\nabla^{\mathrm{D}} H(y^1, y^0). \tag{2.6}$$

Multiplying $\nabla^{\mathrm{D}} g(y^1, y^0)^{\intercal}$ on both sides of (2.6) and using the first identity of (2.5), we obtain $H(y^1) = H(y^0)$, i.e., the scheme (2.6) is energy preservation. For more details on the DG method, readers are referred to [15, 30]. A typical discrete gradient is the *average-vector-field* (AVF) which is defined by

$$\nabla^{\mathrm{D}} g(y, \hat{y}) = \int_0^1 \nabla g((1 - \tau)\hat{y} + \tau y) d\tau. \tag{2.7}$$

Then the AVF integrator for the system (2.3) is given by

$$y^1 = y^0 + hJ^{-1} \int_0^1 \nabla H((1 - \tau)y^0 + \tau y^1) d\tau. \tag{2.8}$$

2.3 Exponential Discrete Gradient Integrators

We next derive the exponential discrete gradient method for the problem (2.1). The starting point is the following variation-of-constants formula for the problem (2.1):

$$y(t_0+h) = \exp(hQM)y(t_0)+h\int_0^1 \exp((1-\xi)hQM)Q\nabla U(y(t_0+\xi h))d\xi. \tag{2.9}$$

Approximating $\nabla U(y(t_0 + \xi h))$ in (2.9) by $\nabla^{\mathrm{D}} U(y^1, y^0)$, we obtain the following *exponential discrete gradient (EDG) integrator*:

$$y^1 = \exp(V)y^0 + h\varphi(V)Q\nabla^{\mathrm{D}}U(y^1, y^0), \tag{2.10}$$

where $V = hQM$,

$$\varphi(V) = (\exp(V) - I)V^{-1},$$

and y^1 is an approximation of $y(t_0 + h)$.

Due to the energy-preserving property of the DG method, we are hopeful of preserving the first integral by (2.10) when Q is skew symmetric. For simplicity, we sometimes write $\nabla^{\mathrm{D}}U(y^1, y^0)$ in brief as $\nabla^{\mathrm{D}}U$. To begin with, we give the following preliminary lemma.

Lemma 2.1 *For any real symmetric matrix M and scalar $h > 0$, the matrix*

$$B = \exp(hQM)^{\intercal} M \exp(hQM) - M$$

satisfies:

$$B = \begin{cases} = 0, & \text{if } Q \text{ is skew-symmetric}, \\ \leq 0, & \text{if } Q \leq 0. \end{cases}$$

Proof Consider the linear ODE:

$$\dot{y}(t) = QMy(t). \tag{2.11}$$

When Q is skew symmetric, (2.11) is a conservative equation with the first integral $\frac{1}{2}y^{\intercal}My$, and its exact solution starting from the initial value $y(0) = y^0$ is $y(t) = \exp(tQM)y^0$. It then follows immediately from

$$\frac{1}{2}y(h)^{\intercal}My(h) = \frac{1}{2}y^{0\intercal}My^0$$

that

$$\frac{1}{2}y^{0\intercal}\exp(hQM)^{\intercal}M\exp(hQM)y^0 = \frac{1}{2}y^{0\intercal}My^0$$

for any vector y^0. Therefore,

$$B = \exp(hQM)^{\intercal}M\exp(hQM) - M$$

is skew-symmetric. Since it is also symmetric, $B = 0$.

Likewise, the case that $Q \leq 0$ can be proved. □

Theorem 2.1 *If Q is skew-symmetric, then the integrator* (2.10) *preserves the first integral H in* (2.1)*:*

$$H(y^1) = H(y^0),$$

where $H(y) = \frac{1}{2}y^\intercal My + U(y)$.

Proof Here we firstly assume that the matrix M is nonsingular. We next calculate $\frac{1}{2}y^{1\intercal}My^1$. Denote $M^{-1}\nabla^D U = \widetilde{\nabla}U$. Replacing y^1 by $\exp(V)y^0 + h\varphi(V)Q\nabla^D U(y^1, y^0)$ leads to

$$\begin{aligned}
&\frac{1}{2}y^{1\intercal}My^1\\
&= \frac{1}{2}(y^{0\intercal}\exp(V)^\intercal + h\nabla^D U^\intercal Q^\intercal\varphi(V)^\intercal)M(\exp(V)y^0 + h\varphi(V)Q\nabla^D U)\\
&= \frac{1}{2}y^{0\intercal}\exp(V)^\intercal M\exp(V)y^0 + hy^{0\intercal}\exp(V)^\intercal M\varphi(V)Q\nabla^D U\\
&\quad + \frac{h^2}{2}\nabla^D U^\intercal Q^\intercal\varphi(V)^\intercal M\varphi(V)Q\nabla^D U\\
&= \frac{1}{2}y^{0\intercal}\exp(V)^\intercal M\exp(V)y^0 + y^{0\intercal}\exp(V)^\intercal M\varphi(V)V\widetilde{\nabla}U\\
&\quad + \frac{1}{2}\widetilde{\nabla}U^\intercal V^\intercal\varphi(V)^\intercal M\varphi(V)V\widetilde{\nabla}U \quad (\text{using } V = hQM)\\
&= \frac{1}{2}y^{0\intercal}\exp(V)^\intercal M\exp(V)y^0 + y^{0\intercal}\exp(V)^\intercal M(\exp(V) - I)\widetilde{\nabla}U\\
&\quad + \frac{1}{2}\widetilde{\nabla}U^\intercal(\exp(V)^\intercal - I)M(\exp(V) - I)\widetilde{\nabla}U \quad (\text{using } \varphi(V)V = \exp(V) - I)\\
&= \frac{1}{2}y^{0\intercal}\exp(V)^\intercal M\exp(V)y^0 + y^{0\intercal}(\exp(V)^\intercal M\exp(V) - \exp(V)^\intercal M)\widetilde{\nabla}U\\
&\quad + \frac{1}{2}\widetilde{\nabla}U^\intercal(\exp(V)^\intercal M\exp(V) - \exp(V)^\intercal M - M\exp(V) + M)\widetilde{\nabla}U.
\end{aligned} \tag{2.12}$$

On the other hand, it follows from the property of the discrete gradient (2.5) that

$$\begin{aligned}
&U(y^1) - U(y^0)\\
&= (y^{1\intercal} - y^{0\intercal})\nabla^D U(y^1, y^0)\\
&= y^{0\intercal}(\exp(V)^\intercal - I)\nabla^D U + h\nabla^D U^\intercal Q^\intercal\varphi(V)^\intercal\nabla^D U\\
&= y^{0\intercal}(\exp(V)^\intercal M - M)\widetilde{\nabla}U + \widetilde{\nabla}U^\intercal V^\intercal\varphi(V)^\intercal M\widetilde{\nabla}U\\
&= y^{0\intercal}(\exp(V)^\intercal M - M)\widetilde{\nabla}U + \widetilde{\nabla}U^\intercal(\exp(V)^\intercal M - M)\widetilde{\nabla}U.
\end{aligned} \tag{2.13}$$

Combining (2.12), (2.13) and collecting terms by types '$y^{0\intercal} * y^0$', '$y^{0\intercal} * \widetilde{\nabla}U$', '$\widetilde{\nabla}U^\intercal * \widetilde{\nabla}U$' lead to

$$
\begin{aligned}
&H(y^1) - H(y^0)\\
&= \frac{1}{2} y^{1\mathsf{T}} M y^1 - \frac{1}{2} y^{0\mathsf{T}} M y^0 + U(y^1) - U(y^0)\\
&= \frac{1}{2} y^{0\mathsf{T}} (\exp(V)^{\mathsf{T}} M \exp(V) - M) y^0 + y^{0\mathsf{T}} (\exp(V)^{\mathsf{T}} M \exp(V) - M) \widetilde{\nabla} U\\
&\quad + \frac{1}{2} \widetilde{\nabla} U^{\mathsf{T}} (\exp(V)^{\mathsf{T}} M \exp(V) - M) \widetilde{\nabla} U + \frac{1}{2} \widetilde{\nabla} U^{\mathsf{T}} (\exp(V)^{\mathsf{T}} M - M \exp(V)) \widetilde{\nabla} U\\
&= \frac{1}{2} (y^0 + \widetilde{\nabla} U)^{\mathsf{T}} B (y^0 + \widetilde{\nabla} U) + \frac{1}{2} \widetilde{\nabla} U^{\mathsf{T}} C \widetilde{\nabla} U = 0,
\end{aligned}
\tag{2.14}
$$

where $B = \exp(V)^{\mathsf{T}} M \exp(V) - M$ and $C = \exp(V)^{\mathsf{T}} M - M \exp(V)$. The last step is from the skew-symmetry of the matrix B (according to Lemma 2.1) and C.

If M is singular, it is easy to find a series of symmetric and nonsingular matrices $\{M_\varepsilon\}$ which converge to M when $\varepsilon \to 0$. Thus, according to the result stated above, it still holds that

$$H_\varepsilon(y_\varepsilon^1) = H_\varepsilon(y^0) \tag{2.15}$$

for all ε, where $H_\varepsilon(y) = \frac{1}{2} y^{\mathsf{T}} M_\varepsilon y + U(y)$ is the first integral of the perturbed problem

$$\dot{y} = Q(M_\varepsilon y + \nabla U(y)), \quad y(t_0) = y^0,$$

and

$$y_\varepsilon^1 = \exp(V_\varepsilon) y^0 + h\varphi(V_\varepsilon) Q \nabla^{\mathrm{D}} U(y_\varepsilon^1, y^0), \quad V_\varepsilon = hQM_\varepsilon.$$

Therefore, when $\varepsilon \to 0$, $y_\varepsilon^1 \to y^1$ and (2.15) lead to

$$H(y^1) = H(y^0).$$

This completes the proof. □

Moreover, the scheme (2.10) can also respect the decay of the first integral when $Q \leq 0$ in (2.1). The next theorem shows this point.

Theorem 2.2 *If Q is negative semi-definite (not necessarily symmetric), then the scheme* (2.10) *preserves the decaying property of the Lyapunov function H in* (2.1)*:*

$$H(y^1) \leq H(y^0),$$

where $H(y) = \frac{1}{2} y^{\mathsf{T}} M y + U(y)$.

Proof If M is nonsingular, the equation in (2.14)

$$H(y^1) - H(y^0) = \frac{1}{2} (y^0 + \widetilde{\nabla} U)^{\mathsf{T}} B (y^0 + \widetilde{\nabla} U)$$

still holds, since the derivation does not depend on the skew-symmetry of Q. By Lemma 2.1, B is negative semi-definite. Thus $H(y^1) \leq H(y^0)$. In the case that M

is singular, this theorem can be easily proved by replacing the equalities

$$H_\varepsilon(y_\varepsilon^1) = H_\varepsilon(y^0), \quad H(y^1) = H(y^0)$$

in the proof of Theorem 2.1 with the inequalities

$$H_\varepsilon(y_\varepsilon^1) \leq H_\varepsilon(y^0), \quad H(y^1) \leq H(y^0).$$

We omit the details. □

2.4 Symmetry and Convergence of the EAVF Integrator

In this chapter, we consider a special type of the discrete gradient in (2.10), the average vector field,

$$\nabla^{\mathrm{D}} U(y, \hat{y}) = \int_0^1 \nabla U((1-\tau)\hat{y} + \tau y) d\tau.$$

The corresponding integrator becomes

$$y^1 = \exp(V)y^0 + h\varphi(V)Q \int_0^1 \nabla U((1-\tau)y^0 + \tau y^1) d\tau, \tag{2.16}$$

where $V = hQM$ and $y^1 \approx y(t_0 + h)$. The scheme (2.16) is called an *exponential AVF integrator* and denoted by EAVF.

In the sequel we present and prove two properties of EAVF—symmetry and convergence.

Theorem 2.3 *The EAVF integrator* (2.16) *is symmetric.*

Proof Exchanging $y^0 \leftrightarrow y^1$ and replacing h by $-h$ in (2.16), we obtain

$$y^0 = \exp(-V)y^1 - h\varphi(-V)Q \int_0^1 \nabla U((1-\tau)y^1 + \tau y^0) d\tau. \tag{2.17}$$

We rewrite (2.17) as:

$$y^1 = \exp(V)y^0 + h\exp(V)\varphi(-V)Q \int_0^1 \nabla U((1-\tau)y^0 + \tau y^1) d\tau. \tag{2.18}$$

Since $\exp(V)\varphi(-V) = \varphi(V)$, (2.18) is the same as (2.16) exactly. This means that EAVF is symmetric. □

It should be noted that the scheme (2.16) is implicit in general, and thus iterative solutions are required. We next discuss the convergence of the fixed-point iteration for the EAVF integrator.

Theorem 2.4 *Suppose that $||\varphi(V)||_2 \leq C$, and that $\nabla U(u)$ satisfies a Lipschitz condition; i.e., there exists a constant L such that*

$$||\nabla U(v) - \nabla U(w)||_2 \leq L||v - w||_2,$$

for all arguments v and $w \in \mathbb{R}^d$. If

$$0 < h \leq \hat{h} < \frac{2}{CL||Q||_2}, \tag{2.19}$$

then the mapping

$$\Psi : z \mapsto \exp(V)y^0 + h\varphi(V)Q\int_0^1 \nabla U((1-\tau)y^0 + \tau z)d\tau$$

has a unique fixed point and the iteration for the EAVF integrator (2.16) *is convergent.*

Proof Since

$$\begin{aligned}
&||\Psi(z_1) - \Psi(z_2)||_2 \\
&= ||h\varphi(V)Q\int_0^1 (\nabla U((1-\tau)y^0 + \tau z_1) - \nabla U((1-\tau)y^0 + \tau z_2))d\tau||_2 \\
&\leq h\|\varphi(V)\|_2\|Q\|_2 \int_0^1 \|\nabla U((1-\tau)y^0 + \tau z_1) - \nabla U((1-\tau)y^0 + \tau z_2)\|_2 d\tau \\
&\leq hCL||Q||_2 \int_0^1 \tau||z_1 - z_2||_2 d\tau \\
&= \frac{h}{2}CL||Q||_2||z_1 - z_2||_2 \\
&\leq \rho||z_1 - z_2||_2,
\end{aligned}$$

where $\rho = \frac{\hat{h}}{2}CL||Q||_2 < 1$, by the Contraction Mapping Theorem, the mapping Ψ has a unique fixed point and the iteration solving the Eq. (2.16) is convergent. □

Remark 2.1 We note two special and important cases in practical applications. If QM is skew-symmetric or symmetric negative semi-definite, then the spectrum of V lies in the left half-plane. Since QM is unitarily diagonalizable and $|\varphi(z)| \leq 1$ for any z satisfying $\mathrm{Re}(z) \leq 0$, we have $||\varphi(V)||_2 \leq 1$.

In many cases, the matrix M has an extremely large norm (e.g., M incorporates high frequency components in oscillatory problems or M is the differential matrix in semi-discrete PDEs), and hence Theorem 2.4 ensures the possibility of choosing relatively large stepsize regardless of M.

In practice, the integral in (2.16) usually cannot be easily calculated. Therefore, we can evaluate it using the s-point Gauss-Legendre (GLs) formula $(b_i, c_i)_{i=1}^s$:

$$\int_0^1 \nabla U((1-\tau)y^0 + \tau y^1)d\tau \approx \sum_{i=1}^{s} b_i \nabla U((1-c_i)y^0 + c_i y^1)).$$

The corresponding scheme is denoted by EAVFGLs. Since the s-point GL quadrature formula is symmetric, EAVFGLs is also symmetric. Due to the fact that $\sum_{i=1}^{s} b_i c_i = 1/2$, the corresponding iteration for EAVFGLs is convergent provided (2.19) holds.

2.5 Problems Suitable for EAVF

2.5.1 *Highly Oscillatory Nonseparable Hamiltonian Systems*

Consider the Hamiltonian

$$H(p,q) = \frac{1}{2}p_1^{\intercal} M_1^{-1} p_1 + \frac{1}{2\varepsilon^2} q_1^{\intercal} A_1 q_1 + S(p,q),$$

where p and q are both d-length vectors, partitioned as

$$p = \begin{pmatrix} p_0 \\ p_1 \end{pmatrix}, \quad q = \begin{pmatrix} q_0 \\ q_1 \end{pmatrix},$$

M_1, A_1 are symmetric positive definite matrices, and $0 < \varepsilon \ll 1$. This Hamiltonian governs oscillatory mechanical systems in 2 or 3 spatial dimensions such as the stiff spring pendulum and the dynamics of the multi-atomic molecule (see, e.g. [8, 9]). With an appropriate canonical transformation (see, e.g. [18]), the Hamiltonian becomes

$$H(p,q) = \frac{1}{2}\sum_{j=1}^{l} \left(p_{1,j}^2 + \frac{\lambda_j^2}{\varepsilon^2} q_{1,j}^2 \right) + S(p,q), \tag{2.20}$$

where $p_1 = (p_{1,1}, \ldots, p_{1,l})^{\intercal}$, $q_1 = (q_{1,1}, \ldots, q_{1,l})^{\intercal}$. The corresponding differential equations are given by

$$\begin{cases} \dot{p}_0 = -\nabla_{q_0} S(p,q), \\ \dot{p}_1 = -\omega^2 q_1 - \nabla_{q_1} S(p,q), \\ \dot{q}_0 = p_0 + (\nabla_{p_0} S(p,q) - p_0), \\ \dot{q}_1 = p_1 + \nabla_{p_1} S(p,q), \end{cases} \tag{2.21}$$

where $\omega^2 = \text{diag}(\omega_1^2, \ldots, \omega_l^2)$, $\omega_j = \lambda_j/\varepsilon$ for $j = 1, \ldots, l$. Equation (2.21) is of the form (2.1) with

$$y = \begin{pmatrix} p \\ q \end{pmatrix}, \quad Q = \begin{pmatrix} O & -I_{d\times d} \\ I_{d\times d} & O \end{pmatrix}, \quad M = \begin{pmatrix} I_{d\times d} & O \\ O & \Omega_{d\times d} \end{pmatrix},$$

and

$$U(p, q) = S(p, q) - \frac{1}{2} p_0^{\intercal} p_0, \quad \Omega = \text{diag}(0, \ldots, 0, \omega_1^2, \ldots, \omega_l^2).$$

Since $q_{11}, \ldots, q_{1l}$ and $p_{11}, \ldots, p_{1l}$ are fast variables, it is favorable to integrate the linear part of them exactly by the scheme (2.16). Note that

$$\varphi(V) = \begin{pmatrix} \text{sinc}(h\Omega^{\frac{1}{2}}) & h^{-1} g_2(h\Omega^{\frac{1}{2}}) \\ h g_1(h\Omega^{\frac{1}{2}}) & \text{sinc}(h\Omega^{\frac{1}{2}}) \end{pmatrix},$$

where $\text{sinc}(z) = \sin(z)/z$, $g_1(z) = (1 - \cos(z))/z^2$, $g_2(z) = \cos(z) - 1$. Unfortunately, the block $h^{-1} g_2(h\Omega^{\frac{1}{2}})$ is not uniformly bounded. In the first experiment, the iteration still works well, perhaps due to the small Lipshitz constant of ∇S.

2.5.2 *Second-Order (Damped) Highly Oscillatory System*

Consider

$$\ddot{q} - N\dot{q} + \Omega q = -\nabla U_1(q), \tag{2.22}$$

where q is a d-length vector variable, $U_1 : \mathbb{R}^d \to \mathbb{R}$ is a differential function, N is a symmetric negative semi-definite matrix, Ω is a symmetric positive semi-definite matrix, $||\Omega||$ or $||N|| \gg 1$. (2.22) stands for highly oscillatory problems such as the dissipative molecular dynamics, the (damped) Duffing and semi-discrete nonlinear wave equations. By introducing $p = \dot{q}$, we write (2.22) as a first-order system of ODEs:

$$\begin{pmatrix} \dot{p} \\ \dot{q} \end{pmatrix} = \begin{pmatrix} N & -\Omega \\ I & O \end{pmatrix} \begin{pmatrix} p \\ q \end{pmatrix} + \begin{pmatrix} -\nabla U_1(q) \\ O \end{pmatrix}, \tag{2.23}$$

which falls into the class of (2.1), where

$$y = \begin{pmatrix} p \\ q \end{pmatrix}, Q = \begin{pmatrix} N & -I \\ I & O \end{pmatrix}, M = \begin{pmatrix} I & O \\ O & \Omega \end{pmatrix}, U(y) = U_1(q).$$

Clearly, $Q \leq 0$ and (2.23) is a dissipative system with the Lyapunov function $H = \frac{1}{2}p^\intercal p + \frac{1}{2}q^\intercal \Omega q + U_1(q)$. Applying the EAVF integrator (2.16) to the Eq. (2.23) yields the scheme:

$$\begin{cases} p^1 = \exp_{11} p^0 + \exp_{12} q^0 - h\varphi_{11} \displaystyle\int_0^1 \nabla U_1((1-\tau)q^0 + \tau q^1)d\tau, \\ q^1 = \exp_{21} p^0 + \exp_{22} q^0 - h\varphi_{21} \displaystyle\int_0^1 \nabla U_1((1-\tau)q^0 + \tau q^1)d\tau, \end{cases} \tag{2.24}$$

where $\exp(hQM)$ and $\varphi(hQM)$ are partitioned into

$$\begin{pmatrix} \exp_{11} & \exp_{12} \\ \exp_{21} & \exp_{22} \end{pmatrix} \text{ and } \begin{pmatrix} \varphi_{11} & \varphi_{12} \\ \varphi_{21} & \varphi_{22} \end{pmatrix},$$

respectively.

It should be noted that only the second equation in the scheme (2.24) needs to be solved by iteration. From the proof procedure of Theorem 2.4, one can find that the convergence of the fixed-point iteration for the second equation in (2.24) is irrelevant to $||QM||$ provided φ_{21} is uniformly bounded.

Theorem 2.5 *Suppose that Ω and N commute and $||\nabla U_1(v) - \nabla U_1(w)||_2 \leq L||v - w||_2$. Then the iteration*

$$\Phi : z \mapsto \exp_{21} p^0 + \exp_{22} q^0 - h\varphi_{21} \int_0^1 \nabla U_1((1-\tau)q^0 + \tau z)d\tau$$

for the scheme (2.24) *is convergent provided*

$$0 < h \leq \hat{h} < \frac{2}{L^{\frac{1}{2}}}.$$

Proof It is crucial here to find a uniform bound of $\|\varphi_{21}\|$. Since Ω and N commute, they can be simultaneously diagonalized:

$$\Omega = F^\intercal \Lambda F, \quad N = F^\intercal \Sigma F,$$

where F is an orthogonal matrix, $\Lambda = \text{diag}(\lambda_1, \ldots, \lambda_d)$, $\Sigma = \text{diag}(\sigma_1, \ldots, \sigma_d)$ and $\lambda_i \geq 0, \sigma_i \leq 0$ for $i = 1, 2, \ldots, d$. It now follows from

$$QM = \begin{pmatrix} F^\intercal & O \\ O & F^\intercal \end{pmatrix} \begin{pmatrix} O & I \\ -\Lambda & \Sigma \end{pmatrix} \begin{pmatrix} F & O \\ O & F \end{pmatrix}$$

that

$$\exp(hQM) = \begin{pmatrix} F^\intercal & O \\ O & F^\intercal \end{pmatrix} \exp\left\{\begin{pmatrix} O & hI \\ -h\Lambda & h\Sigma \end{pmatrix}\right\} \begin{pmatrix} F & O \\ O & F \end{pmatrix}.$$

To show that $\exp_{21}$ and φ_{21} depends on h, we denote them by $\exp_{21}^h$ and φ_{21}^h, respectively. After some calculations, we have

$$\exp_{21}^h = F^\intercal(\Sigma^2 - 4\Lambda)^{-\frac{1}{2}} \cdot 2\sinh(h(\Sigma^2 - 4\Lambda)^{\frac{1}{2}}/2)\exp\left(\frac{h\Sigma}{2}\right)F.$$

We then have

$$\begin{aligned} \|\exp_{21}^h\|_2 &= \|2(\Sigma^2 - 4\Lambda)^{-\frac{1}{2}}\sinh(h(\Sigma^2 - 4\Lambda)^{\frac{1}{2}}/2)\exp\left(\frac{h\Sigma}{2}\right)\|_2 \\ &= h\max_i \left|\frac{\sinh((h^2\sigma_i^2/4 - \lambda_i)^{\frac{1}{2}})}{(h^2\sigma_i^2/4 - \lambda_i)^{\frac{1}{2}}}\exp\left(\frac{h\sigma_i}{2}\right)\right|. \end{aligned} \tag{2.25}$$

In order to estimate $\|\exp_{21}^h\|_2$, the bound of the function

$$g(\lambda, \sigma) = \frac{\sinh((\sigma^2 - 4\lambda)^{\frac{1}{2}})}{(\sigma^2 - 4\lambda)^{\frac{1}{2}}}\exp(\sigma),$$

should be considered for $\sigma \le 0, \lambda \ge 0$. If $\sigma^2 - 4\lambda < 0$, we set $(\sigma^2 - 4\lambda)^{\frac{1}{2}} = \mathrm{i}a$, where i is the imaginary unit and a is a real number. Then we have

$$|g| = \left|\frac{\sin(a)}{a}\exp(\sigma)\right| \le \left|\frac{\sin(a)}{a}\right| \le 1.$$

If $\sigma^2 - 4\lambda \ge 0$, then $a = (\sigma^2 - 4\lambda)^{\frac{1}{2}} \le -\sigma$,

$$|g| = \left|\frac{\sinh(a)}{a}\exp(\sigma)\right| \le \left|\frac{\sinh(a)}{a}\exp(-a)\right| = \left|\frac{1 - \exp(-2a)}{2a}\right| \le 1.$$

Thus,

$$|g(\lambda, \sigma)| \le 1 \quad \text{for } \sigma \le 0, \lambda \ge 0. \tag{2.26}$$

It follows from (2.25) and (2.26) that

$$\|\exp_{21}^h\|_2 = h\max_i \left|g\left(\frac{h\sigma_i}{2}, \lambda_i\right)\right| \le h. \tag{2.27}$$

Therefore, using $\varphi(hQM) = \int_0^1 \exp((1-\xi)hQM)\mathrm{d}\xi$ and (2.27), we obtain

$$\|\varphi_{21}\|_2 = \left\|\int_0^1 \exp_{21}^{(1-\xi)h}\mathrm{d}\xi\right\|_2 \le \int_0^1 \|\exp_{21}^{(1-\xi)h}\|_2\mathrm{d}\xi \le \int_0^1 (1-\xi)h\mathrm{d}\xi = \frac{1}{2}h.$$

The rest of the proof is similar to that of Theorem 2.4 which we omit here. □

It can be observed that in the particular case that $N = 0$, the scheme (2.24) reduces to the AAVF integrator in [32].

2.5.3 Semi-discrete Conservative or Dissipative PDEs

Many time-dependent PDEs are in the form:

$$\frac{\partial}{\partial t} y(x,t) = \mathscr{Q} \frac{\delta \mathscr{H}}{\delta y}, \tag{2.28}$$

where $y(\cdot, t) \in X$ for every $t \geq 0$, X is a Hilbert space like $\mathbf{L}^2(\mathscr{D})$, $\mathbf{L}^2(\mathscr{D}) \times \mathbf{L}^2(\mathscr{D}), \ldots$, $\mathscr{D}$ is a domain in $\mathbb{R}^d$, and $\mathscr{Q}$ is a linear operator on X, the functional $\mathscr{H}[y] = \int_{\mathscr{D}} f(y, \partial_\alpha y) \mathrm{d}x$ (f is smooth, $x = (x_1, \ldots, x_d)$, $\mathrm{d}x = \mathrm{d}x_1 \ldots \mathrm{d}x_d$ and $\partial_\alpha y$ denote the partial derivatives of y with respect to spatial variables x_i, $1 \leq i \leq d$). Under a suitable boundary condition (BC), the *variational derivative* $\dfrac{\delta \mathscr{H}}{\delta y}$ is defined by:

$$\langle \frac{\delta \mathscr{H}}{\delta y}, z \rangle = \frac{\mathrm{d}}{\mathrm{d}\varepsilon}\Big|_{\varepsilon=0} \mathscr{H}[y + \varepsilon z]$$

for any smooth $z \in X$ vanishing on the boundary of $\mathscr{D}$, where $\langle \cdot, \cdot \rangle$ is the inner product of X. If $\mathscr{Q}$ is a skew or negative semi-definite operator with respect to $\langle \cdot, \cdot \rangle$, then the Eq. (2.28) is conservative (e.g., the nonlinear wave, nonlinear Schrödinger, Korteweg–de Vries and Maxwell equations) or dissipative (e.g., the Allen–Cahn, Cahn–Hilliard, Ginzburg–Landau and heat equations), i.e., $\mathscr{H}[y]$ is constant or monotonically decreasing (see, e.g. [6, 13]). In general, after the spatial discretisation, (2.28) becomes a conservative or dissipative system of ODEs in the form (2.1).

A typical example of a conservative system is the nonlinear Schrödinger (NSL) equation:

$$\mathrm{i} \frac{\partial}{\partial t} y + \frac{\partial^2}{\partial x^2} y + V'(|y|^2) y = 0 \tag{2.29}$$

subject to the periodic BC $y(0,t) = y(L,t)$. Denoting $y = p + \mathrm{i}q$ ($\mathrm{i}^2 = -1$), where p, q are the real and imaginary parts of y, the Eq. (2.29) can be written in the form of (2.28):

$$\frac{\partial}{\partial t} \begin{pmatrix} p \\ q \end{pmatrix} = \begin{pmatrix} 0 & -1 \\ 1 & 0 \end{pmatrix} \begin{pmatrix} \dfrac{\partial^2}{\partial x^2} p + V'(p^2 + q^2) p \\ \dfrac{\partial^2}{\partial x^2} q + V'(p^2 + q^2) q \end{pmatrix}, \tag{2.30}$$

where $X = \mathbf{L}^2([0, L]) \times \mathbf{L}^2([0, L])$,

$$\mathscr{H}[y] = \frac{1}{2}\int_0^L \left(V(p^2+q^2) - \left(\frac{\partial}{\partial x}p\right)^2 - \left(\frac{\partial}{\partial x}q\right)^2 \right) \mathrm{d}x.$$

We consider the spatial discretisation of (2.30). It is supposed that the spatial domain is equally partitioned into N intervals: $0 = x_0 < x_1 < \ldots < x_N = L$. Discretizing the spatial derivatives of (2.30) by central differences gives

$$\begin{pmatrix} \dot{\tilde{p}} \\ \dot{\tilde{q}} \end{pmatrix} = \begin{pmatrix} O & -I \\ I & O \end{pmatrix} \begin{pmatrix} D\tilde{p} + V'(\tilde{p}^2+\tilde{q}^2)\tilde{p} \\ D\tilde{q} + V'(\tilde{p}^2+\tilde{q}^2)\tilde{q} \end{pmatrix}, \tag{2.31}$$

where

$$D = \begin{pmatrix} -2 & 1 & & & \\ 1 & -2 & 1 & & \\ & \ddots & \ddots & \ddots & \\ & & 1 & -2 & 1 \\ & & & 1 & -2 \end{pmatrix},$$

is an $N \times N$ symmetric differential matrix, $\tilde{p} = (p_0, \ldots, p_{N-1})^\mathsf{T}, \tilde{q} = (q_0, \ldots, q_{N-1})^\mathsf{T}$, $p_i(t) \approx p(x_i, t)$ and $q_i(t) \approx q(x_i, t)$ for $i = 0, \ldots, N-1$.

As an example of dissipative PDEs we consider the Allen–Cahn (AC) equation

$$\frac{\partial y}{\partial t} = \beta\frac{\partial^2 y}{\partial x^2} + y - y^3, \quad \beta \geq 0, \tag{2.32}$$

subject to the the Neumann BC $\frac{\partial}{\partial x}y(0,t) = \frac{\partial}{\partial x}y(L,t)$. $X = \mathbf{L}^2([0,L])$, $\mathscr{Q} = -1$, $\mathscr{H}[y] = \int_0^L (\frac{1}{2}\beta(\frac{\partial}{\partial x}y)^2 - \frac{1}{2}y^2 + \frac{1}{4}y^4)\mathrm{d}x$. The spatial grids are chosen in the same way as the NLS. Discretizing the spatial derivative with the central difference, we obtain

$$\dot{\tilde{y}} = \beta\hat{D}\tilde{y} + \tilde{y} - \tilde{y}^3, \tag{2.33}$$

where

$$\hat{D} = \begin{pmatrix} -1 & 1 & & & \\ 1 & -2 & 1 & & \\ & \ddots & \ddots & \ddots & \\ & & 1 & -2 & 1 \\ & & & 1 & -1 \end{pmatrix},$$

is the $(N-1) \times (N-1)$ symmetric differential matrix, $\tilde{y} = (y_1, \ldots, y_{N-1})^\mathsf{T}$, $y_i(t) \approx y(x_i, t)$.

Both the semi-discrete NLS equation (2.31) and AC equation (2.33) are of the form (2.1). For the NLS equation, we have

$$Q = \begin{pmatrix} O & -I \\ I & O \end{pmatrix}, \quad M = \begin{pmatrix} D & O \\ O & D \end{pmatrix}, \quad U = \frac{1}{2}\sum_{i=0}^{N-1} V(p_i^2 + q_i^2),$$

while for the AC equation, we have

$$Q = -I, \quad M = -\beta \hat{D}, \quad U = \sum_{i=1}^{N-1}\left(-\frac{1}{2}y_i^2 + \frac{1}{4}y_i^4\right).$$

Therefore, the scheme (2.16) can be applied to solve them. Since the matrix QM is skew or symmetric negative semi-definite in these two cases, according to Remark 2.1, the convergence of fixed-point iterations for them is independent of the differential matrix.

2.6 Numerical Experiments

In this section, we compare the EAVF method (2.16) with the well-known implicit midpoint method which is denoted by MID:

$$y^1 = y^0 + hQ\nabla\widetilde{U}\left(\frac{y^0 + y^1}{2}\right), \tag{2.34}$$

and the traditional AVF method for (2.1) given by

$$y^1 = y^0 + hQ\int_0^1 \nabla\widetilde{U}((1-\tau)y^0 + \tau y^1)d\tau, \tag{2.35}$$

where $\widetilde{U}(y) = U(y) + \frac{1}{2}y^\intercal My$. The authors in [30] showed that (2.35) preserves the first integral or the Lyapunov function $\widetilde{U}$. Our comparison also includes another energy-preserving method of order four for (2.1):

$$\begin{cases} y^{\frac{1}{2}} = y^0 + hQ\displaystyle\int_0^1 \left(\frac{5}{4} - \frac{3}{2}\tau\right)\nabla\widetilde{U}(y_\tau)d\tau, \\ y^1 = y^0 + hQ\displaystyle\int_0^1 \nabla\widetilde{U}(y_\tau)d\tau, \end{cases} \tag{2.36}$$

where

$$y_\tau = (2\tau - 1)(\tau - 1)y^0 - 4\tau(\tau - 1)y^{\frac{1}{2}} + (2\tau - 1)\tau y^1.$$

This method is denoted by CRK since it can be written as a continuous Runge–Kutta method. For details, readers are referred to [17].

Throughout the experiment, the 'reference solution' is computed by high-order methods with a sufficiently small stepsize. We always start to calculate from $t_0 = 0$. $y^n \approx y(t_n)$ is obtained by the time-stepping way $y^0 \to y^1 \to \cdots \to y^n \to \cdots$ for $n = 1, 2, \ldots$ and $t_n = nh$. The error tolerance for iteration solutions of the four methods is set as 10^{-14}. The maximum global error (GE) over the total time interval is defined by:

$$GE = \max_{n\geq 0} ||y^n - y(t_n)||_\infty.$$

The maximum global error of H (EH) on the interval is:

$$EH = \max_{n\geq 0} |H^n - H(y(t_n))|.$$

In our numerical experiments, the computational cost of each method is measured by the number of function evaluations (FE).

Example 2.1 The motion of a triatomic molecule can be modelled by a Hamiltonian system with the Hamiltonian of the form (2.20) (see, e.g. [8]):

$$H(p, q) = S(p, q) + \frac{1}{2}(p_{1,1}^2 + p_{1,2}^2 + p_{1,3}^2) + \frac{\omega^2}{2}(q_{1,1}^2 + q_{1,2}^2 + q_{1,3}^2), \quad (2.37)$$

where

$$S(p, q) = \frac{1}{2}p_0^2 + \frac{1}{4}(q_0 - q_{1,3})^2 - \frac{1}{4}\frac{2q_{1,2} + q_{1,2}^2}{(1 + q_{1,2})^2}(p_0 - p_{1,3})^2 - \frac{1}{4}\frac{2q_{1,1} + q_{1,1}^2}{(1 + q_{1,1})^2}(p_0 + p_{1,3})^2.$$

The initial values are given by:

$$\begin{cases} p_0(0) = p_{1,1}(0) = p_{1,2}(0) = p_{1,3}(0) = 1, \\ q_0(0) = 0.4, q_{1,1}(0) = q_{1,2}(0) = \dfrac{1}{\omega}, q_{1,3} = \dfrac{1}{2^{\frac{1}{2}}\omega}. \end{cases}$$

Setting $h = 1/2^i$ for $i = 6, \ldots, 10$, $\omega = 50$, and $h = 1/100 \times 1/2^i$ for $i = 0, \ldots, 4$, $\omega = 100$, we integrate the problem (2.21) with the Hamiltonian (2.37) over the interval $[0, 50]$. Since the nonlinear term $\nabla S(p, q)$ is complicated to be integrated, we evaluate the integrals in EAVF, AVF and CRK by the 3-point Gauss–Legendre (GL) quadrature formula $(b_i, c_i)_{i=1}^3$:

$$b_1 = \frac{5}{18}, b_2 = \frac{4}{9}, b_3 = \frac{5}{18}; \quad c_1 = \frac{1}{2} - \frac{15^{\frac{1}{2}}}{10}, c_2 = \frac{1}{2}, c_3 = \frac{1}{2} + \frac{15^{\frac{1}{2}}}{10}.$$

The corresponding schemes are denoted by EAVFGL3, AVFGL3 and CRKGL3 respectively. Numerical results are presented in Fig. 2.1.

Figure 2.1a, c show that MID and AVFGL3 lost basic accuracy. It can be observed from Fig. 2.1b, d that AVFGL3, EAVFGL3, CRKGL3 are much more efficient in

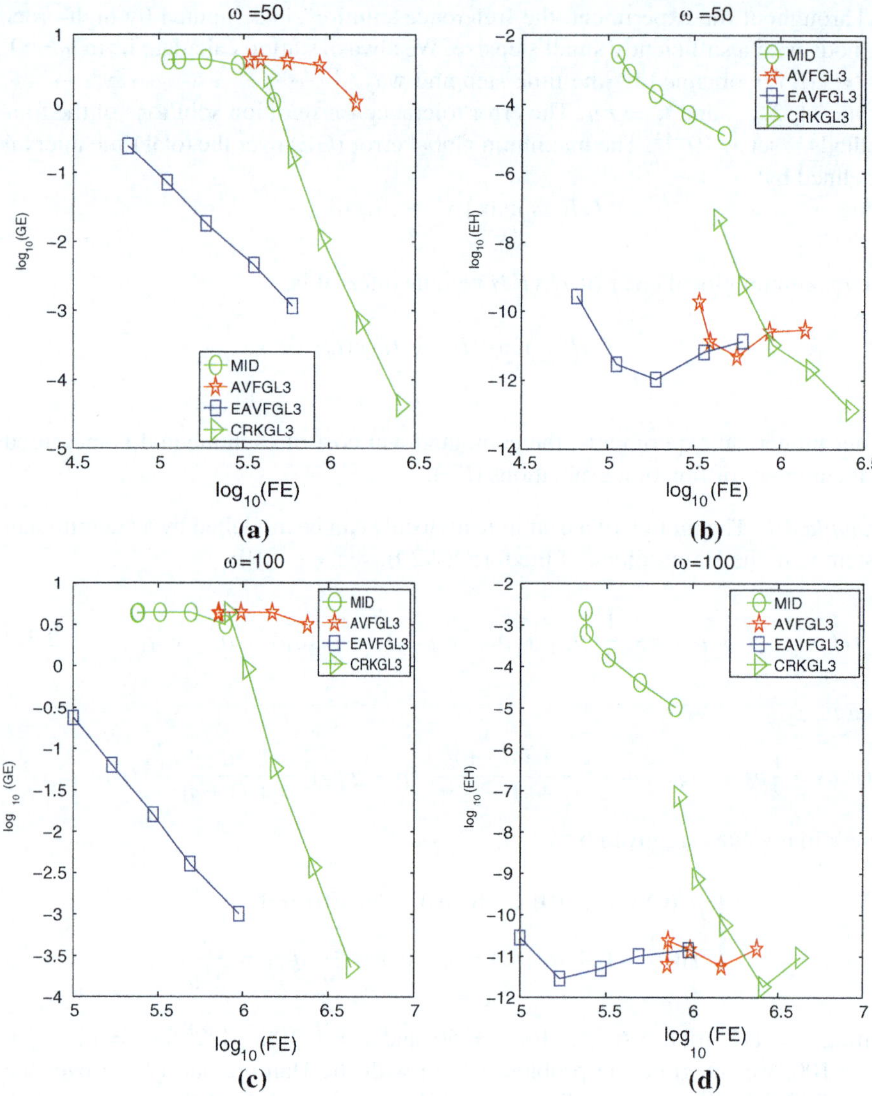

Fig. 2.1 Efficiency curves.

preserving energy than MID. In the aspects of both energy preservation and algebraic accuracy, EAVF is the most efficient among the four methods.

Example 2.2 The equation

$$\begin{aligned}\dot{x}_1 &= -\zeta x_1 - \lambda x_2 + x_1 x_2,\\ \dot{x}_2 &= \lambda x_1 - \zeta x_2 + \frac{1}{2}(x_1^2 - x_2^2),\end{aligned} \tag{2.38}$$

is an averaged system in wind-induced oscillation, where $\zeta \geq 0$ is a damping factor and λ is a detuning parameter (see, e.g. [16]). For convenience, setting $\zeta = rcos(\theta), \lambda = rsin(\theta), r \geq 0, 0 \leq \theta \leq \pi/2$, (see [29]) we write (2.38) as

$$\begin{pmatrix}\dot{x}_1\\ \dot{x}_2\end{pmatrix} = \begin{pmatrix}-\cos(\theta) & -\sin(\theta)\\ \sin(\theta) & -\cos(\theta)\end{pmatrix}\begin{pmatrix} rx_1 - \frac{1}{2}\sin(\theta)(x_2^2 - x_1^2) - \cos(\theta)x_1x_2\\ rx_2 - \sin(\theta)x_1x_2 + \frac{1}{2}\cos(\theta)(x_2^2 - x_1^2)\end{pmatrix}, \tag{2.39}$$

which is of the form (2.1), where

$$\begin{aligned} Q &= \begin{pmatrix}-\cos(\theta) & -\sin(\theta)\\ \sin(\theta) & -\cos(\theta)\end{pmatrix}, \quad M = \begin{pmatrix} r & 0\\ 0 & r\end{pmatrix},\\ U &= -\frac{1}{2}\sin(\theta)\left(x_1x_2^2 - \frac{1}{3}x_1^3\right) + \frac{1}{2}\cos(\theta)\left(\frac{1}{3}x_2^3 - x_1^2x_2\right).\end{aligned} \tag{2.40}$$

Its Lyapunov function (dissipative case, when $\theta < \pi/2$) or the first integral (conservative case, when $\theta = \pi/2$) is:

$$H = \frac{1}{2}r(x_1^2 + x_2^2) - \frac{1}{2}sin(\theta)\left(x_1x_2^2 - \frac{1}{3}x_1^3\right) + \frac{1}{2}cos(\theta)\left(\frac{1}{3}x_2^3 - x_1^2x_2\right).$$

The matrix exponential of the EAVF scheme (2.16) for (2.39) is calculated by:

$$\exp(V) = \begin{pmatrix}\exp(-hcr)cos(hsr) & -\exp(-hcr)sin(hsr)\\ \exp(-hcr)sin(hsr) & \exp(-hcr)cos(hsr)\end{pmatrix},$$

where $c = \cos(\theta), s = \sin(\theta)$, and $\varphi(V)$ can be obtained by $(\exp(V) - I)V^{-1}$. Given the initial values:

$$x_1(0) = 0, x_2(0) = 1,$$

we first integrate the conservative system (2.39) with the parameters $\theta = \pi/2, r = 20$ and stepsizes $h = 1/20 \times 1/2^i$ for $i = -1, \ldots, 4$ over the interval [0, 200]. Setting $\theta = \pi/2 - 10^{-4}, r = 20$, we then integrate the dissipative (2.39) with the stepsizes $h = 1/20 \times 1/2^i$ for $i = -1, \ldots, 4$ over the interval [0, 100]. Numerical errors are presented in Figs. 2.2 and 2.3. It is noted that the integrands appearing in AVF, EAVF are polynomials of degree two and the integrands in CRK are polynomials of degree five. We evaluate the integrals in AVF, EAVF by the 2-point GL quadrature:

$$b_1 = \frac{1}{2}, b_2 = \frac{1}{2}, \quad c_1 = \frac{1}{2} - \frac{3^{\frac{1}{2}}}{6}, c_2 = \frac{1}{2} + \frac{3^{\frac{1}{2}}}{6},$$

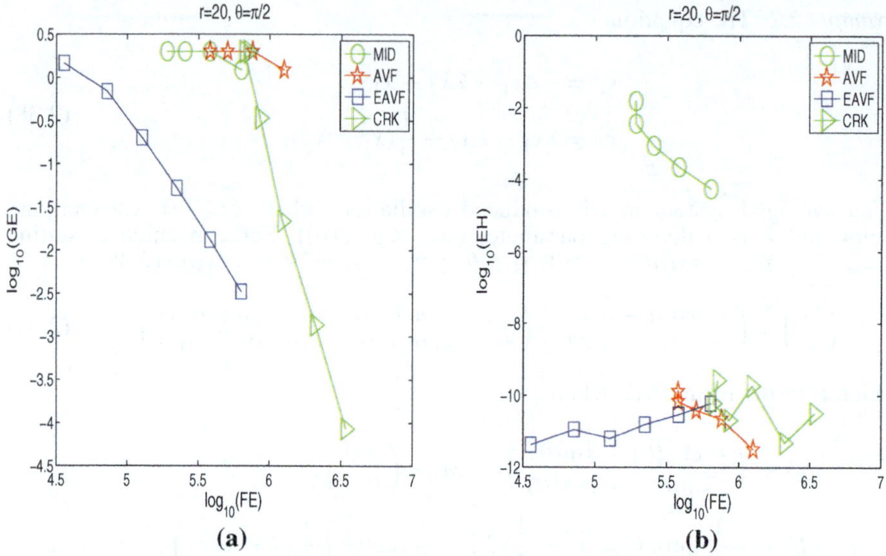

Fig. 2.2 Efficiency curves.

and the integrals appearing in CRK by the 3-point GL quadrature. Then there is no quadrature error.

The efficiency curves of AVF and MID consist of only five points in Figs. 2.2a, b, and 2.3a (two points overlap in Figs. 2.2a and 2.3a), since the fixed-point iterations of MID and AVF are not convergent when $h = 1/10$. Note that QM is skew-symmetric or negative semi-definite, the convergence of iterations for the EAVF method is independent of r by Theorem 2.4 and Remark 2.1. Thus larger stepsizes are allowed for EAVF. The experiment shows that the iterations for EAVF uniformly converge for $h = 1/20 \times 1/2^i$ for $i = -1, \ldots, 4$. Moreover, it can be observed from Fig. 2.3b that MID cannot strictly preserve the decay of the Lyapunov function.

Example 2.3 The PDE:

$$\frac{\partial^2 u}{\partial t^2} = \beta \frac{\partial^3 u}{\partial t \partial x^2} + \frac{\partial^2 u}{\partial x^2}\left(1 + \varepsilon \left(\frac{\partial u}{\partial x}\right)^p\right) - \gamma \frac{\partial u}{\partial t} - m^2 u, \tag{2.41}$$

where $\varepsilon > 0, \beta, \gamma \geq 0$, is a continuous generalization of α-FPU (Fermi–Pasta–Ulam) system (see, e.g. [28]). Taking $\partial_t u = v$ and the homogeneous Dirichlet BC $u(0, t) = u(L, t) = 0$, the Eq. (2.41) is of the type (2.28), where $X = \mathbf{L}^2([0, L]) \times \mathbf{L}^2([0, L])$ and

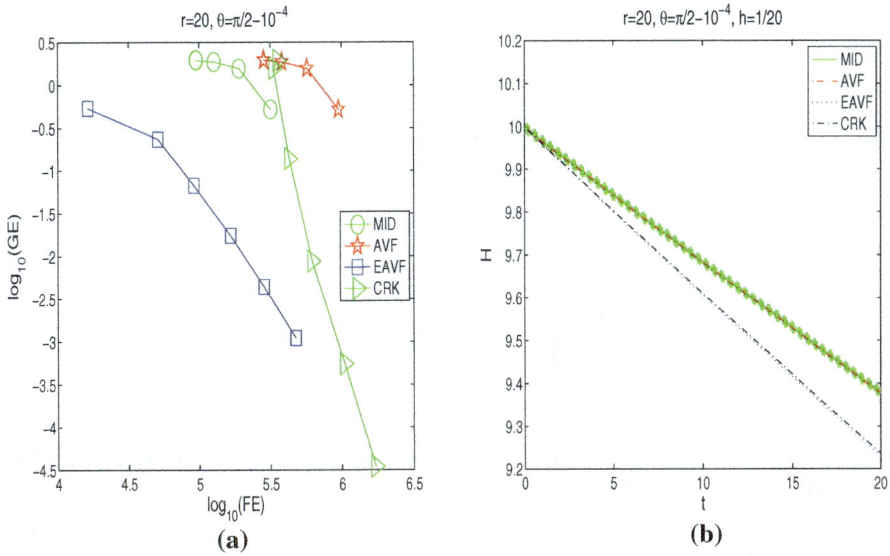

Fig. 2.3 **a** Efficiency curves. **b** The Lyapunov function against time t.

$$y = \begin{pmatrix} u \\ v \end{pmatrix}, \quad \mathscr{Q} = \begin{pmatrix} 0 & 1 \\ -1 & \beta\partial_x^2 - \gamma \end{pmatrix},$$
$$\mathscr{H}[y] = \int_0^L \left(\frac{1}{2}u_x^2 + \frac{m^2}{2}u^2 + \frac{v^2}{2} + \frac{\varepsilon u_x^{p+2}}{(p+2)(p+1)} \right) dx.$$

It is easy to verify that $\mathscr{Q}$ is a negative semi-definite operator, and thus (2.41) is dissipative. The spatial discretization yields a dissipative system of ODEs:

$$\ddot{u}_j(t) - c^2(u_{j-1} - 2u_j + u_{j+1}) + m^2 u_j - \beta'(\dot{u}_{j-1} - 2\dot{u}_j + \dot{u}_{j+1}) + \gamma\dot{u}_j(t)$$
$$= \varepsilon'(V'(u_{j+1} - u_j) - V'(u_j - u_{j-1})),$$

where $c = 1/\Delta x$, $\beta' = c^2\beta$, $\varepsilon' = c^{p+2}\varepsilon$, $V(u) = u^{p+2}/[(p+2)(p+1)]$, $u_j(t) \approx u(x_j, t)$, $x_j = j/\Delta x$ for $j = 1, \ldots, N-1$ and $u_0(t) = u_N(t) = 0$. Note that the nonlinear term $u_{xx}u_x^p$ is approximated by:

$$\frac{\partial^2 u}{\partial x^2}\left(\frac{\partial u}{\partial x}\right)^p |_{x=x_j} = \frac{1}{p+1}\partial_x \left(\frac{\partial u}{\partial x}\right)^{p+1} |_{x=x_j}$$
$$\approx \frac{1}{p+1}\left(\left(\frac{u_{j+1} - u_j}{\Delta x}\right)^{p+1} - \left(\frac{u_j - u_{j-1}}{\Delta x}\right)^{p+1} \right) / \Delta x.$$

We now write it in the compact form (2.22):

$$\ddot{q} - N\dot{q} + \Omega q = -\nabla U_1(q),$$

where $q = (u_1, \ldots, u_{N-1})^\intercal$, $N = \beta' D - \gamma I$, $\Omega = -c^2 D + m^2 I$, $U_1(q) = \varepsilon' \sum_{j=0}^{N-1} V(u_{j+1} - u_j)$ and

$$D = \begin{pmatrix} -2 & 1 & & & \\ 1 & -2 & 1 & & \\ & \ddots & \ddots & \ddots & \\ & & 1 & -2 & 1 \\ & & & 1 & -2 \end{pmatrix}.$$

In this experiment, we set $p = 1, m = 0, c = 1, \varepsilon = \frac{3}{4}$, and $\gamma = 0.005$. Consider the initial conditions in [28]:

$$\phi_j(t) = B \ln \left\{ \left(\frac{1 + \exp[2(\kappa(j - 97) + t \sinh(\kappa))]}{1 + \exp[2(\kappa(j - 96) + t \sinh(\kappa))]} \right) \left(\frac{1 + \exp[2(\kappa(j - 32) + t \sinh(\kappa))]}{1 + \exp[2(\kappa(j - 33) + t \sinh(\kappa))]} \right) \right\}$$

with $B = 5, \kappa = 0.1$, that is,

$$\begin{cases} u_j(0) = \phi_j(0), \\ v_j(0) = \dot{\phi}_j(0). \end{cases}$$

for $j = 1, \ldots, N - 1$. Let $N = 128$, $\beta = 0, 2$. We compute the numerical solution by MID, AVF and EAVF with the stepsizes $h = 1/2^i$ for $i = 1, \ldots, 5$ over the time interval $[0, 100]$. Similarly to EAVF (2.24), the nonlinear systems resulting from MID (2.34) and AVF (2.35) can be reduced to:

$$q^1 = q^0 + hp^0 + \frac{h}{2} N(q^1 - q^0) - \frac{h^2}{4} \Omega (q^1 + q^0) - \frac{h^2}{2} \nabla U_1 \left(\frac{q^0 + q^1}{2} \right),$$

and

$$q^1 = q^0 + hp^0 + \frac{h}{2} N(q^1 - q^0) - \frac{h^2}{4} \Omega (q^1 + q^0) - \frac{h^2}{2} \int_0^1 \nabla U_1((1 - \tau)q^0 + \tau q^1) d\tau$$

respectively. Both the velocity p^1 of MID and AVF can be recovered by

$$\frac{q^1 - q^0}{h} = \frac{p^1 + p^0}{2}.$$

The integrals in AVF and EAVF are exactly evaluated by the 2-point GL quadrature. Since $\exp(hA)$, $\varphi(hA)$ in (2.24) have no explicit expressions, they are calculated by

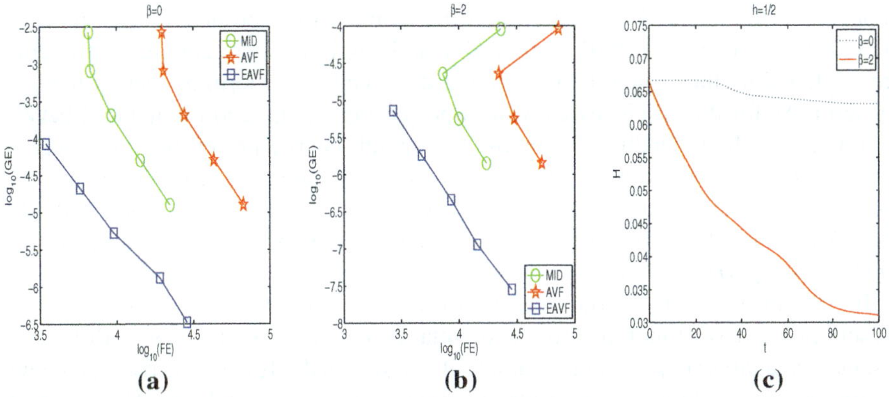

Fig. 2.4 **a**, **b** Efficiency curves. **c** The decay of Lyapunov function obtained by EAVF.

the Matlab package in [2]. The basic idea is to evaluate $\exp(hA), \varphi(hA)$ by their Padé approximations. Numerical results are plotted in Fig. 2.4. Alternatively, there are other popular algorithms such as the contour integral method and the Krylov subspace method for matrix exponentials and φ-functions. Readers are referred to [23] for a summary of algorithms and well-established mathematical software.

According to Theorem 2.5, the convergence of iterations in the EAVF scheme is independent of Ω and N. Iterations of MID and AVF are not convergent when $\beta = 2, h = 1/2$. Thus the efficiency curves of MID and AVF in Fig. 2.4b consist of only 4 points. From Fig. 2.4c, it can be observed that the EAVF method is dissipative even using the relatively large stepsize $h = 1/2$.

2.7 Conclusions and Discussions

Exponential integrators date back to the original work by Hersch [20]. The term "exponential integrators" was coined in the seminal paper by Hochbruck, Lubich and Selhofer [22]. It turns out that exponential integrators have constituted an important class of effective methods for the numerical solution of differential equations in applied sciences and engineering. In this chapter, combining the ideas of the exponential integrator with the average vector field, a new exponential scheme EAVF was proposed and analysed. The EAVF method can preserve the first integral or the Lyapunov function for the conservative or dissipative system (2.1). The symmetry of EAVF is responsible for the good long-term numerical behavior. The implicitness of EAVF means that the solution must be solved iteratively. We have analysed the convergence of the fixed-point iteration and showed that the convergence is free from the influence of a large class of coefficient matrices M. In the dynamics of the triatomic molecule, wind-induced oscillation and the damped FPU problem, we compared the new EAVF method with the MID, AVF and CRK methods. The three problems are modelled by the system (2.1) having a dominant linear term and small nonlinear term. As for the efficiency as well as preserving energy and dissipation, EAVF is superior to the other three methods. In general, energy-preserving and energy-decaying methods are implicit, and iterative solutions are required. With relatively large stepsizes, the iterations of EAVF converge while those of AVF and MID do not. We conclude that EAVF is a promising method for solving the system (2.1) with $||QM|| \gg ||Q\, Hess(U)||$.

In conclusion, exponential integrators are an important class of structure-preserving numerical methods for differential equations. Therefore, we will further discuss and analyse exponential Fourier collocation methods in the next chapter, and symplectic exponential Runge–Kutta methods for solving nonlinear Hamiltonian systems in Chap. 4.

This chapter is based on the work of Li and Wu [27].

References

1. Berland, H., Owren, B., Skaflestad, B.: Solving the nonlinear Schrödinger equations using exponential integrators on the cubic Schrödinger equation. Model. Identif. Control **27**, 201–217 (2006)
2. Berland, H., Skaflestad, B., Wright, W.: EXPINT – A MATLAB package for exponential integrators, ACM Trans. Math. Softw. **33** (2007)
3. Brugnano, L., Iavernaro, F., Trigiante, D.: Hamiltonian boundary value methods (energy preserving discrete line integral methods). J. Numer. Anal. Ind. Appl. Math. **5**, 17–37 (2010)
4. Calvo, M., Laburta, M.P., Montijano, J.I., Rández, L.: Projection methods preserving Lyapunov functions. BIT Numer. Math. **50**, 223–241 (2010)
5. Celledoni, E., Cohen, D., Owren, B.: Symmetric exponential integrators with an application to the cubic Schrödinger equation. Found. Comp. Math. **8**, 303–317 (2008)
6. Celledoni, E., Grimm, V., McLachlan, R.I., Maclaren, D.I., O'Neale, D., Owren, B., Quispel, G.R.W.: Preserving energy resp. dissipation in numerical PDEs using the "Average Vector Field" method. J. Comput. Phys. **231**, 6770–6789 (2012)
7. Cieśliński, J.L.: Locally exact modifications of numerical schemes. Comput. Math. Appl. **62**, 1920–1938 (2013)
8. Cohen, D.: Conservation properties of numerical integrators for highly oscillatory Hamiltonian systems. IMA J. Numer. Anal. **26**, 34–59 (2006)
9. Cohen, D., Jahnke, T., Lorenz, K., Lubich, C.: Numerical integrators for highly oscillatory Hamiltonian systems: a review. In: Mielke, A. (ed.) Analysis, Modeling and Simulation of Multiscale Problems, pp. 553–576. Springer, Berlin (2006)
10. Cox, S.M., Matthews, P.C.: Exponential time differencing for stiff systems. J. Comput. Phys. **176**, 430–455 (2002)
11. Deuflhard, P.: A study of extrapolation methods based on multistep schemes without parasitic solutions. Z. Angew. Math. Phys. **30**, 177–189 (1979)
12. Franco, J.: Runge–Kutta–Nyström methods adapted to the numerical integration of perturbed oscillators. Comput. Phys. Commun. **147**, 770–787 (2002)
13. Furihata, D., Matuso, T.: Discrete Variational Derivative Method: A Structure-Preserving Numerical Method for Partial Differential Equations. Chapman and Hall/CRC, Boca Raton (2010)

14. Gautschi, W.: Numerical integration of ordinary differential equations based on trigonometric polynomials. Numer. Math. **3**, 381–397 (1961)
15. Gonzalez, O.: Time integration and discrete Hamiltonian systems. J. Nonlinear Sci. **6**, 449–467 (1996)
16. Guckenheimer, J., Holmes, P.: Nonlinear Oscillations, Dynamical Systems, and Bifurcations of Vector Fields. Springer, New York (1983)
17. Hairer, E.: Energy-preserving variant of collocation methods. J. Numer. Anal. Ind. Appl. Math. **5**, 73–84 (2010)
18. Hairer, E., Lubich, C., Wanner, G.: Geometric Numerical Integration, vol. XIII, 2nd edn. Springer, Berlin (2006)
19. Hernández-Solano, Y., Atencia, M., Joya, G., Sandoval, F.: A discrete gradient method to enhance the numerical behavior of Hopfield networks. Neurocomputing **164**, 45–55 (2015)
20. Hersch, J.: Contribution à la méthode des équations aux différences. Z. Angew. Math. Phys. **9**, 129–180 (1958)
21. Higham, N.J.: Functions of Matrices: Theory and Computation. SIAM, Philadelphia (2008)
22. Hochbruck, M., Lubich, C., Selhofer, H.: Exponential integrators for large systems of differential equations. SIAM J. Sci. Comput. **19**, 1552–1574 (1998)
23. Hochbruck, M., Ostermann, A.: Exponential integrators. Acta Numerica **19**, 209–286 (2010)
24. Kassam, A.K., Trefethen, L.N.: Fourth order time-stepping for stiff PDEs. SIAM. J. Sci. Comput. **26**, 1214–1233 (2005)
25. Klein, C.: Fourth order time-stepping for low dispersion Korteweg–de Vries and nonlinear Schrödinger equations. Electron. Trans. Numer. Anal. **29**, 116–135 (2008)
26. Lawson, J.D.: Generalized Runge–Kutta processes for stable systems with large Lipschitz constants. SIAM. J. Numer. Anal. Model. **4**, 372–380 (1967)
27. Li, Y.W., Wu, X.Y.: Exponential integrators preserving first integrals or Lyapunov functions for conservative or dissipative systems. SIAM J. SCI. Comput. **38**, A1876–A1895 (2016)
28. Macías-Díaz, J.E., Medina-Ramírez, I.E.: An implicit four-step computational method in the study on the effects of damping in a modified α-Fermi-Pasta-Ulam medium. Commun. Nonlinear Sci. Numer. Simul. **14**, 3200–3212 (2009)
29. McLachlan, R.I., Quispel, G.R.W., Robidoux, N.: A unified approach to Hamiltonian systems, Poisson systems, gradient systems, and systems with Lyapunov functions or first integrals. Phys. Rev. Lett. **81**, 2399–2411 (1998)
30. McLachlan, R.I., Quispel, G.R.W., Robidoux, N.: Geometric integration using dicrete gradients. Philos. Trans. R. Soc. A **357**, 1021–1046 (1999)
31. Pavlov, B.V., Rodionova, O.E.: The method of local linearization in the numerical solution of stiff systems of ordinary differential equations. USSR Comput. Math. Math. Phys. **27**, 30–38 (1987)
32. Wang, B., Wu, X.Y.: A new high precision energy-preserving integrator for system of oscillatory second-order differential equations. Phys. Lett. A **376**, 1185–1190 (2012)
33. Wu, X.Y., Wang, B., Xia, J.: Explicit symplectic multidimensional exponential fitting modified Runge–Kutta–Nyström methods. BIT Numer. Math. **52**, 773–795 (2012)
34. Yang, H., Wu, X.Y., You, X., Fang, Y.: Extended RKN-type methods for numerical integration of perturbed oscillators. Comput. Phys. Commun. **180**, 1777–1794 (2009)

Chapter 3
Exponential Fourier Collocation Methods for First-Order Differential Equations

Commencing from the variation-of-constants formula and incorporating a local Fourier expansion of the underlying problem with collocation methods, this chapter presents a novel class of exponential Fourier collocation methods (EFCMs) for solving systems of first-order ordinary differential equations. We discuss in detail the connections of EFCMs with trigonometric Fourier collocation methods (TFCMs), the well-known Hamiltonian Boundary Value Methods (HBVMs), Gauss methods and Radau IIA methods. It turns out that the novel EFCMs are an extension, in a strict mathematical sense, of these existing methods in the literature.

3.1 Introduction

The subject of this chapter is devoted to analysing and designing novel and efficient numerical integrators for solving the following first-order initial value problem

$$u'(t) + \mathscr{A}u(t) = g(t, u(t)), \qquad u(0) = u_0, \qquad t \in [0, t_{\text{end}}], \tag{3.1}$$

where $g : \mathbb{R} \times \mathbb{R}^d \to \mathbb{R}^d$ is an analytic function, $\mathscr{A}$ is assumed to be a linear operator on a Banach space X with a norm $\|\cdot\|$, and $(-\mathscr{A})$ is the infinitesimal generator of a strongly continuous semigroup $e^{-t\mathscr{A}}$ on X (see, e.g. [27]). These conditions on $\mathscr{A}$ imply that there exist two non-negative real constants C and ω satisfying

$$\left\|e^{-t\mathscr{A}}\right\|_{X \leftarrow X} \leq Ce^{\omega t}, \qquad t \geq 0. \tag{3.2}$$

An analysis of this result can be found in [27]. It is noted that if X is chosen as $X = \mathbb{R}^d$ or $X = \mathbb{C}^d$, then the linear operator $\mathscr{A}$ can be expressed by a $d \times d$ matrix A. Accordingly, e^{-tA} in this case is exactly the matrix exponential function. It also can be observed that the condition (3.2) holds with $\omega = 0$ provided the field of values

X. Wu and B. Wang, *Recent Developments in Structure-Preserving Algorithms for Oscillatory Differential Equations*, https://doi.org/10.1007/978-981-10-9004-2_3

of A is contained in the right complex half-plane. In the special and important case where A is skew-Hermitian or Hermitian positive semidefinite, we have $C = 1$ and $\omega = 0$ in the Euclidean norm, independently of the dimension d. If A originates from a spatial discretisation of a partial differential equation, then this assumption on A leads to temporal convergence results that are independent of the spatial mesh.

As is known, the exact solution of (3.1) can be represented by the variation-of-constants formula

$$u(t) = e^{-tA}u_0 + \int_0^t e^{-(t-\tau)A} g(\tau, u(\tau))d\tau. \tag{3.3}$$

Note that the exponential subsumes the full information on linear oscillations for oscillatory problems. This class of problems (3.1) frequently arises in a wide variety of applications including engineering, mechanics, quantum physics, circuit simulations, flexible body dynamics and other applied sciences (see, e.g. [10, 16, 24, 27, 39, 41, 44, 48]). Parabolic partial differential equations with their spatial discretisations and highly oscillatory problems yield two typical examples of the system (3.1) (see, e.g. [30–34, 42]). Linearizing stiff systems $u'(t) = F(t, u(t))$ also yields examples of the form (3.1) (see, e.g. [15, 25, 28]).

Commencing from the variation-of-constants formula (3.3), the numerical scheme for (3.1) is usually constructed by incorporating the exact propagator of (3.1) in an appropriate way. For example, interpolating the nonlinearity at the known value $g(0, u_0)$ yields the exponential Euler approximation for (3.3). Approximating the functions arising from rational approximation leads to implicit or semi-implicit Runge–Kutta methods, Rosenbrock methods or W-schemes. Recently, the construction, analysis, implementation and application of exponential integrators have been studied by many researchers, and readers are referred to [3, 11–13, 16, 37, 46]. Exponential integrators make explicit use of the quantity Au of (3.1), and a systematic survey of exponential integrators is presented in [27].

Exponential Runge–Kutta methods of collocation type were constructed based on Lagrange interpolation polynomials, and their convergence properties were analysed in [26]. In [40], the authors developed and researched a novel type of trigonometric Fourier collocation methods (TFCMs) for second-order oscillatory differential equations $q''(t) + Mq(t) = f(q(t))$ with a principal frequency matrix $M \in \mathbb{R}^{d\times d}$. These new trigonometric Fourier collocation methods take full advantage of the special structure introduced by the linear term Mq, and their construction analysis incorporates the idea of collocation methods, the variation-of-constants formula and the local Fourier expansion of the system. The results of numerical experiments in [40] show that the trigonometric Fourier collocation methods are much more efficient in comparison with some alternative approaches appeared in the literature. On the basis of the work in [26, 40], in this chapter we make an effort to conduct the research of novel exponential Fourier collocation methods (EFCMs) for efficiently solving first-order differential equations (3.1). The construction of the novel EFCMs incorporates the exponential integrators, the collocation methods, and the local Fourier expansion of the system. Moveover, EFCMs can be of an arbitrarily high order, and when $A \to 0$, EFCMs reduce to the well-known Hamiltonian Boundary Value methods (HBVMs) which have been studied by many researchers (see, e.g. [6–8]). It is also shown in this chapter that EFCMs are an extension of Gauss methods, Radau IIA methods and TFCMs.

The plan of this chapter is as follows. We first formulate the scheme of EFCMs in Sect. 3.2. Section 3.3 discusses the connections of the novel EFCMs with HBVMs, Gauss methods, Radau IIA methods and TFCMs. In Sect. 3.4, we analyse the properties of EFCMs. Section 3.5 is concerned with constructing a practical EFCM and reporting four numerical experiments to demonstrate the excellent qualitative behavior of the novel approximation. The last section includes some concluding comments.

3.2 Formulation of EFCMs

This section presents the formulation of exponential Fourier collocation methods (EFCMs) for systems of first-order differential equations (3.1). We begin with the local Fourier expansion.

3.2.1 *Local Fourier Expansion*

We first restrict the first-order differential equations (3.1) to an interval $[0, h]$ with any $h > 0$:

$$u'(t) + Au(t) = g(t, u(t)), \qquad u(0) = u_0, \qquad t \in [0, h]. \tag{3.4}$$

Consider the shifted Legendre polynomials $\{\widehat{P}_j\}_{j=0}^{\infty}$ satisfying

$$\int_0^1 \widehat{P}_i(x)\widehat{P}_j(x)dx = \delta_{ij}, \qquad \deg\left(\widehat{P}_j\right) = j, \qquad i, j \geq 0,$$

where δ_{ij} is the Kronecker symbol.

We then expand the right-hand-side function of (3.4) as follows:

$$g(\xi h, u(\xi h)) = \sum_{j=0}^{\infty} \widehat{P}_j(\xi)\kappa_j(h, u), \;\; \xi \in [0, 1]; \;\; \kappa_j(h, u) := \int_0^1 \widehat{P}_j(\tau)g(\tau h, u(\tau h))d\tau. \tag{3.5}$$

The system (3.4) now can be rewritten as

$$u'(\xi h) + Au(\xi h) = \sum_{j=0}^{\infty} \widehat{P}_j(\xi)\kappa_j(h, u), \qquad u(0) = u_0. \tag{3.6}$$

Its solution is given by the next theorem.

Theorem 3.1 *The solution of* (3.4) *can be expressed by*

$$u(t) = \varphi_0(-tA)u_0 + t\sum_{j=0}^{\infty} I_j(tA)\kappa_j(t,u), \tag{3.7}$$

where $t \in [0, h]$ *and*

$$I_j(tA) := \int_0^1 \widehat{P}_j(z)e^{-(1-z)tA}dz = \sqrt{2j+1}\sum_{k=0}^{j}(-1)^{j+k}\frac{(j+k)!}{k!(j-k)!}\varphi_{k+1}(-tA). \tag{3.8}$$

The φ*-functions in* (3.8) *(see, e.g. [24, 25, 27, 28]) are defined by:*

$$\varphi_0(z) = e^z, \quad \varphi_k(z) = \int_0^1 e^{(1-\sigma)z}\frac{\sigma^{k-1}}{(k-1)!}d\sigma, \quad k = 1, 2, \ldots.$$

Proof It follows from the variation-of-constants formula (3.3) that

$$\begin{aligned} u(t) &= e^{-tA}u_0 + \int_0^t e^{-(t-\tau)A}g(\tau, u(\tau))d\tau \\ &= \varphi_0(-tA)u_0 + t\int_0^1 e^{-(1-z)tA}g(zt, u(zt))dz. \end{aligned} \tag{3.9}$$

Replacing the function $g(zt, u(zt))$ under the integral in (3.9) by the expansion (3.5) yields

$$\begin{aligned} u(t) &= \varphi_0(-tA)u_0 + t\int_0^1 e^{-(1-z)tA}\sum_{j=0}^{\infty}\widehat{P}_j(z)\kappa_j(t,u)dz \\ &= \varphi_0(-tA)u_0 + t\sum_{j=0}^{\infty}\int_0^1 \widehat{P}_j(z)e^{-(1-z)tA}dz\kappa_j(t,u), \end{aligned}$$

which gives the formula (3.7) by letting $I_j(tA) = \int_0^1 \widehat{P}_j(z)e^{-(1-z)tA}dz$.

It then follows from the definition of shifted Legendre polynomials on the interval $[0, 1]$:

$$\widehat{P}_j(x) = (-1)^j\sqrt{2j+1}\sum_{k=0}^{j}\binom{j}{k}\binom{j+k}{k}(-x)^k, \qquad j = 0, 1, \ldots, \qquad x \in [0, 1], \tag{3.10}$$

that

$$
\begin{aligned}
I_j(tA) &= \int_0^1 \widehat{P}_j(z)e^{-(1-z)tA}dz \\
&= \int_0^1 (-1)^j\sqrt{2j+1}\sum_{k=0}^{j}\binom{j}{k}\binom{j+k}{k}(-z)^k e^{-(1-z)tA}dz \\
&= \sqrt{2j+1}\sum_{k=0}^{j}(-1)^{j+k}\binom{j}{k}\binom{j+k}{k}\int_0^1 z^k e^{-(1-z)tA}dz \\
&= \sqrt{2j+1}\sum_{k=0}^{j}(-1)^{j+k}\frac{(j+k)!}{k!(j-k)!}\varphi_{k+1}(-tA).
\end{aligned}
$$

The proof is complete. □

3.2.2 *Discretisation*

The authors in [8] made use of interpolation quadrature formulae to construct a discretisation for initial value problems. Following [8], two tools are coupled in this subsection. We first truncate the local Fourier expansion after a finite number of terms and then compute the coefficients of the expansion by a suitable quadrature formula.

We now consider truncating the Fourier expansion, a technique which originally appeared in [8]. This can be achieved by truncating the series (3.7) after n ($n \geq 2$) terms with the stepsize h and $V := hA$:

$$
\tilde{u}(h) = \varphi_0(-V)u_0 + h\sum_{j=0}^{n-1} I_j(V)\kappa_j(h,\tilde{u}), \tag{3.11}
$$

which satisfies the following initial value problem:

$$
\tilde{u}'(\xi h) + A\tilde{u}(\xi h) = \sum_{j=0}^{n-1}\widehat{P}_j(\xi)\kappa_j(h,\tilde{u}), \quad \tilde{u}(0) = u_0.
$$

The key challenge in designing practical methods is how to deal with $\kappa_j(h,\tilde{u})$ effectively. To this end, we introduce a quadrature formula using k abscissae $0 \leq c_1 \leq \ldots \leq c_k \leq 1$ and being exact for polynomials of degree up to $m-1$. It is required that $m \geq k$ in this chapter, and we note that many existing quadrature formulae satisfy this requirement, such as the well-known Gauss–Legendre quadrature and Radau quadrature. We thus obtain an approximation of the form

$$\kappa_j(h,\tilde{u}) \approx \sum_{l=1}^{k} b_l \widehat{P}_j(c_l) g(c_l h, \tilde{u}(c_l h)), \quad j = 0, 1, \ldots, n-1, \tag{3.12}$$

where b_l for $l = 1, 2, \ldots, k$ are the quadrature weights. It is noted that since the number of the integrals $\kappa_j(h,\tilde{u})$ is n, it is assumed that $k \geq n$. Thus, we have $m \geq n$.

Therefore, the approximation gives

$$\Delta_j(h,\tilde{u}) := \kappa_j(h,\tilde{u}) - \sum_{l=1}^{k} b_l \widehat{P}_j(c_l) g(c_l h, \tilde{u}(c_l h))$$
$$= \int_0^1 \widehat{P}_j(\tau) g(\tau h, \tilde{u}(\tau h)) d\tau - \sum_{l=1}^{k} b_l \widehat{P}_j(c_l) g(c_l h, \tilde{u}(c_l h)).$$

Since the quadrature is exact for polynomials of degree $m-1$, its remainder depends on the mth derivative of the integrand $\widehat{P}_j(\tau) g(\tau h, u(\tau h))$ with respect to τ. Consequently, we obtain

$$\Delta_j(h,\tilde{u}) = C \int_0^1 \frac{d^m\Big(\widehat{P}_j(\tau) g(\tau h, \tilde{u}(\tau h))\Big)}{d\tau^m}\Big|_{\tau=\zeta} \omega(\tau) d\tau,$$

where C is a constant, ζ ($\zeta \in [0, 1]$) depends on τ, and $\omega(\tau) = \prod_{i=1}^{k}(\tau - c_i)$. Taking account of $\widehat{P}_j^{(k)}(\tau) = 0$ for $k > j$, we obtain

$$\begin{aligned}\Delta_j(h,\tilde{u}) &= C \int_0^1 \widehat{P}_j(\zeta) \hat{g}^{(m)}(\zeta h)\omega(\tau) d\tau h^m + Cm \int_0^1 \widehat{P}_j'(\zeta)\hat{g}^{(m-1)}(\zeta h)\omega(\tau) d\tau h^{m-1} \\ &\quad + \cdots + C\binom{m}{j} \int_0^1 \widehat{P}_j^{(j)}(\zeta)\hat{g}^{(m-j)}(\zeta h)\omega(\tau) d\tau h^{m-j} = \mathcal{O}(h^{m-j}), \\ &\qquad j = 0, 1, \ldots, n-1,\end{aligned}$$

with the notation $\hat{g}^{(k)}(\zeta h) = g^{(k)}(\zeta h, \tilde{u}(\zeta h))$. This guarantees that each $\Delta_j(h,\tilde{u})$ has good accuracy for any $j = 0, 1, \ldots, n-1$. Choosing k large enough, along with a suitable choice of c_l, b_l for $l = 1, 2, \ldots, k$, allows us to approximate the given integral $\kappa_j(h,\tilde{u})$ to any degree of accuracy.

With (3.11) and (3.12), it is natural to consider the following numerical scheme

$$v(h) = \varphi_0(-V)u_0 + h \sum_{j=0}^{n-1} I_j(V) \sum_{l=1}^{k} b_l \widehat{P}_j(c_l) g(c_l h, v(c_l h)),$$

which exactly solves the initial value problem as follows:

$$v'(\xi h) = -Av(\xi h) + \sum_{j=0}^{n-1} \widehat{P}_j(\xi) \sum_{l=1}^{k} b_l \widehat{P}_j(c_l) g(c_l h, v(c_l h)), \quad v(0) = u_0. \tag{3.13}$$

It follows from (3.13) that $v(c_i h)$ for $i = 1, 2, \ldots, k$ satisfy the following first-order differential equations:

$$v'(c_i h) + Av(c_i h) = \sum_{j=0}^{n-1} \widehat{P}_j(c_i) \sum_{l=1}^{k} b_l \widehat{P}_j(c_l) g(c_l h, v(c_l h)), \quad v(0) = u_0. \tag{3.14}$$

Let $v_i = v(c_i h)$. It is clear that (3.14) can be solved by the variation-of-constants formula (3.3). This gives

$$v_i = \varphi_0(-c_i V)u_0 + c_i h \sum_{j=0}^{n-1} I_{j,c_i}(V) \sum_{l=1}^{k} b_l \widehat{P}_j(c_l) g(c_l h, v_l), \quad i = 1, 2, \ldots, k,$$

where

$$\begin{aligned}
I_{j,c_i}(V) &:= \int_0^1 \widehat{P}_j(c_i z) e^{-(1-z)c_i V} dz \\
&= \int_0^1 (-1)^j \sqrt{2j+1} \sum_{k=0}^{j} \binom{j}{k}\binom{j+k}{k} (-c_i z)^k e^{-(1-z)c_i V} dz \\
&= (-1)^j \sqrt{2j+1} \sum_{k=0}^{j} (-c_i)^k \binom{j}{k}\binom{j+k}{k} \int_0^1 z^k e^{-(1-z)c_i V} dz \\
&= (-1)^j \sqrt{2j+1} \sum_{k=0}^{j} (-c_i)^k \frac{(j+k)!}{k!(j-k)!} \varphi_{k+1}(-c_i V).
\end{aligned} \tag{3.15}$$

3.2.3 The Exponential Fourier Collocation Methods

We are now in a position to present the novel exponential Fourier collocation methods for systems of first-order differential equations (3.1).

Definition 3.1 The k-stage exponential Fourier collocation method with an integer n (denoted by EFCM(k,n)) for integrating systems of first-order differential equations (3.1) is defined by

$$\begin{cases} v_i = \varphi_0(-c_i V)u_0 + c_i h \sum_{l=1}^{k} b_l \Big(\sum_{j=0}^{n-1} I_{j,c_i}(V)\widehat{P}_j(c_l) \Big) g(c_l h, v_l), \ i = 1, 2, \ldots, k, \\ v(h) = \varphi_0(-V)u_0 + h \sum_{l=1}^{k} b_l \Big(\sum_{j=0}^{n-1} I_j(V)\widehat{P}_j(c_l) \Big) g(c_l h, v_l), \end{cases} \tag{3.16}$$

where h is the stepsize, $V := hA$, $\widehat{P}_j$ for $j = 0, 1, \ldots, n-1$ are defined by (3.10), and c_l, b_l for $l = 1, 2, \ldots, k$ are the node points and the quadrature weights of a quadrature formula, respectively. Here, n is an integer which is required to satisfy the condition: $2 \le n \le k$. $I_j(V)$ and $I_{j,c_i}(V)$ are determined by

$$\begin{cases} I_j(V) = \sqrt{2j+1} \sum_{k=0}^{j} (-1)^{j+k} \dfrac{(j+k)!}{k!(j-k)!} \varphi_{k+1}(-V), \\ I_{j,c_i}(V) = (-1)^j \sqrt{2j+1} \sum_{k=0}^{j} (-c_i)^k \dfrac{(j+k)!}{k!(j-k)!} \varphi_{k+1}(-c_i V). \end{cases}$$

Remark 3.1 Clearly, it can be observed that the EFCM(k,n) defined by (3.16) exactly integrates the homogeneous linear system $u' + Au = 0$. Therefore, it is trivially A-stable. The EFCM(k,n) (3.16) approximates the solution of (3.1) in the time interval $[0, h]$. Obviously, the solution $v(h)$ after one time-step can be considered as the initial condition for a new initial value problem and $u(t)$ can be approximated in the next time interval $[h, 2h]$. In general, the EFCM(k,n) can be extended to the approximation of the solution in an arbitrary interval $[0, Nh]$, where N is a positive integer.

Remark 3.2 The novel EFCM(k,n) (3.16) developed here is a kind of exponential integrator which requires the approximation of products of φ-functions with vectors. It is noted that if A has a simple structure, it is possible to compute the φ-functions in a fast and reliable way. Moveover, many different approaches to evaluating this action in an efficient way have been proposed in the literature (see, e.g. [1, 2, 4, 22, 23, 27, 35, 36]). Furthermore, all the matrix functions appearing in the EFCM(k,n) (3.16) only need to be calculated once in the actual implementation for the given stepsize h. In Sect. 3.5, we will compare our novel methods with some traditional collocation methods (which do not require the evaluation of matrix functions) by four experiments. For each problem, we will display the work precision diagram in which the global error is plotted versus the execution time. The numerical results given in Sect. 3.5 demonstrate the efficiency of our novel approximation.

3.3 Connections with Some Existing Methods

Various effective methods have been developed so far for solving first-order differential equations and this section is devoted to analysing the connections between our novel EFCMs and some other existing methods in the literature. It turns out that some existing traditional methods can be gained by letting $A \to 0$ in the corresponding EFCMs or by applying EFCMs to special second-order differential equations.

3.3.1 Connections with HBVMs and Gauss Methods

Hamiltonian Boundary Value methods (HBVMs) are an interesting class of integrators, which exactly preserve the energy of polynomial Hamiltonian systems (see, e.g. [6–8]). We first consider the connection between EFCMs and HBVMs.

It can be observed that from (3.15) that when $A \to 0$, $I_j(V)$ and $I_{j,c_i}(V)$ in (3.16) become

$$\tilde{I}_j := I_j(0) = \int_0^1 \widehat{P}_j(z)dz = \begin{cases} 1, & j = 0, \\ 0, & j \geq 1, \end{cases}$$

and

$$\tilde{I}_{j,c_i} := I_{j,c_i}(0) = \int_0^1 \widehat{P}_j(c_i z)dz.$$

This can be summed up in the following result.

Theorem 3.2 *When $A \to 0$, the EFCM(k,n) defined by* (3.16) *reduces to*

$$\begin{cases} v_i = u_0 + c_i h \sum_{l=1}^{k} b_l \Big(\sum_{j=0}^{n-1} \tilde{I}_{j,c_i} \widehat{P}_j(c_l) \Big) g(c_l h, v_l), \quad i = 1, 2, \ldots, k, \\ v(h) = u_0 + h \sum_{l=1}^{k} b_l g(c_l h, v_l), \end{cases} \tag{3.17}$$

which can be rewritten as a k-stage Runge–Kutta method with the following Butcher tableau

$$\begin{array}{c|ccc} c_1 & & & \\ \vdots & \multicolumn{3}{c}{\bar{A} = (\bar{a}_{ij})_{k\times k} = \Big(b_j \sum_{l=0}^{n-1} \widehat{P}_l(c_j) \int_0^{c_i} \widehat{P}_l(\tau)d\tau \Big)_{k\times k}} \\ c_k & & & \\ \hline & b_1 & \cdots & b_k \end{array} \tag{3.18}$$

This method is exactly the Hamiltonian Boundary Value Method HBVM(k,n) by using the discretisation researched in [6–8] for the first-order initial value problem

$$u'(t) = g(t, u(t)),\ u(0) = u_0.$$

Furthermore, it follows from the property of HBVM(k,n) given in [8] that HBVM(k,k) reduces to a k-stage Gauss-Legendre collocation method when a Gaussian distribution of the nodes $(c_1, \ldots, c_k)$ is used. In view of this observation and as a straightforward consequence of Theorem 3.2, we obtain a connection between EFCMs and Gauss methods. This result is described below.

Theorem 3.3 *Under the condition that $c_l,\ b_l$ for $l = 1, 2, \ldots, k$ are chosen respectively as the node points and the quadrature weights of a k-point Gauss–Legendre quadrature over the interval* $[0, 1]$, *then the EFCM(k,k) defined by* (3.16) *reduces to the the corresponding k-stage Gauss method presented in [19] when $A \to 0$.*

3.3.2 Connection Between EFCMs and Radau IIA Methods

Radau collocation methods are well known (see e.g. [20]). The following theorem states the connection between EFCMs and Radau IIA methods.

Theorem 3.4 *Choose $c_l,\ b_l$ for $l = 1, 2, \ldots, k$ respectively as the node points and the weights of the Radau-right quadrature formula. Then the EFCM(k,k) defined by* (3.16) *reduces to a k-stage Radau IIA method presented in [20] when $A \to 0$.*

Proof It follows from Theorem 3.2 that when $A \to 0$, the EFCM(k,k) defined by (3.16) reduces to (3.17) with $n = k$. As is known, the shifted Legendre polynomials $\{\widehat{P}_j\}_{j=0}^{\infty}$ satisfy the following integration formulae (see, e.g. [20])

$$\begin{aligned}
&\int_0^x \widehat{P}_0(t)dt = \xi_1 \widehat{P}_1(x) + \frac{1}{2}\widehat{P}_0(x),\\
&\int_0^x \widehat{P}_m(t)dt = \xi_{m+1} \widehat{P}_{m+1}(x) - \xi_m \widehat{P}_{m-1}(x),\ m = 1, 2, \ldots, k-2,\\
&\int_0^x \widehat{P}_{k-1}(t)dt = \beta_k \widehat{P}_{k-1}(x) - \xi_{k-1} \widehat{P}_{k-2}(x),
\end{aligned}$$

where

$$\xi_m = \frac{1}{2\sqrt{4m^2 - 1}},\quad \beta_k = \frac{1}{4k-2}.$$

These formulae imply

$$\bar{A} = \begin{pmatrix} \int_0^{c_1} \widehat{P}_0(\tau)d\tau & \dots & \int_0^{c_1} \widehat{P}_{k-1}(\tau)d\tau \\ \vdots & & \vdots \\ \int_0^{c_k} \widehat{P}_0(\tau)d\tau & \dots & \int_0^{c_k} \widehat{P}_{k-1}(\tau)d\tau \end{pmatrix} \begin{pmatrix} b_1\widehat{P}_0(c_1) & \dots & b_s\widehat{P}_0(c_s) \\ \vdots & & \vdots \\ b_1\widehat{P}_{k-1}(c_1) & \dots & b_s\widehat{P}_{k-1}(c_s) \end{pmatrix}$$
$$= WX_kQ,$$

where the matrix W is defined by

$$\omega_{ij} = \widehat{P}_{j-1}(c_i), \quad i, j = 1, \dots, k,$$

and the matrices X_k, Q are determined by

$$X_k = \begin{pmatrix} \frac{1}{2} & -\xi_1 & & & \\ \xi_1 & 0 & -\xi_2 & & \\ & \ddots & \ddots & \ddots & \\ & & \xi_{k-2} & 0 & -\xi_{k-1} \\ & & & \xi_{k-1} & \beta_k \end{pmatrix}, \quad Q = \begin{pmatrix} b_1\widehat{P}_0(c_1) & \dots & b_s\widehat{P}_0(c_s) \\ \vdots & & \vdots \\ b_1\widehat{P}_{k-1}(c_1) & \dots & b_s\widehat{P}_{k-1}(c_s) \end{pmatrix}. \tag{3.19}$$

On noticing the fact that the Radau-right quadrature formula is of order $2k-1$, we obtain that polynomials $\widehat{P}_m(x)\widehat{P}_n(x)$ $(m+n \le 2k-2)$ are integrated exactly by this quadrature formula, i.e.,

$$\sum_{i=1}^{k} b_i\widehat{P}_m(c_i)\widehat{P}_n(c_i) = \int_0^1 \widehat{P}_m(x)\widehat{P}_n(x)dx = \delta_{mn},$$

which means $WQ = I$. Therefore,

$$\bar{A} = WX_kW^{-1}.$$

(3.18) now becomes

$$\begin{array}{c|ccc} c_1 & & & \\ \vdots & & \bar{A} = WX_kW^{-1} & \\ c_k & & & \\ \hline & b_1 & \dots & b_k \end{array}$$

which is exactly the same as the scheme of Radau IIA method presented in [5] by using the W-transformation. □

3.3.3 Connection Between EFCMs and TFCMs

A novel type of trigonometric Fourier collocation methods (TFCMs) for second-order oscillatory differential equations

$$q''(t) + Mq(t) = f(q(t)), \qquad q(0) = q_0, \quad q'(0) = q_0' \tag{3.20}$$

has been developed and researched in [40]. These methods can attain arbitrary algebraic order in a very simple way. This is of importance for solving systems of second-order oscillatory ODEs. This subsection is devoted to clarifying the connection between EFCMs and TFCMs.

We apply the TFCMs presented in [40] to (3.20) and denote the numerical solution by $(v_T, u_T)^{\intercal}$. According to the analysis in [40], the numerical solution satisfies the following differential equation

$$\begin{pmatrix} v_T(\xi h) \\ u_T(\xi h) \end{pmatrix}' = \begin{pmatrix} u_T(\xi h) \\ -Mv_T(\xi h) + \sum\limits_{j=0}^{n-1} \widehat{P}_j(\xi) \sum\limits_{l=1}^{k} b_l \widehat{P}_j(c_l) f(v_T(c_l h)) \end{pmatrix} \tag{3.21}$$

with the initial value

$$\big(v_T(0), u_T(0)\big)^{\intercal} = (q_0, q_0')^{\intercal}.$$

By appending the equation $q' = p$, the system (3.20) can be written as

$$\begin{pmatrix} q(t) \\ p(t) \end{pmatrix}' + \begin{pmatrix} 0 & -I \\ M & 0 \end{pmatrix} \begin{pmatrix} q(t) \\ p(t) \end{pmatrix} = \begin{pmatrix} 0 \\ f(q(t)) \end{pmatrix}, \quad \begin{pmatrix} q(0) \\ p(0) \end{pmatrix} = \begin{pmatrix} q_0 \\ q_0' \end{pmatrix}. \tag{3.22}$$

We apply the EFCM(k,n) defined by (3.16) to the system of first-order differential equations (3.22) and denote the corresponding numerical solution by $(v_E, u_E)^{\intercal}$. It follows from the formulation of EFCMs presented in Sect. 3.2 that $(v_E, u_E)^{\intercal}$ is the solution of the system

$$\begin{pmatrix} v_E(\xi h) \\ u_E(\xi h) \end{pmatrix}' + \begin{pmatrix} 0 & -I \\ M & 0 \end{pmatrix} \begin{pmatrix} v_E(\xi h) \\ u_E(\xi h) \end{pmatrix} = \begin{pmatrix} 0 \\ \sum\limits_{j=0}^{n-1} \widehat{P}_j(\xi) \sum\limits_{l=1}^{k} b_l \widehat{P}_j(c_l) f(v_E(c_l h)) \end{pmatrix} \tag{3.23}$$

with the initial value

$$\big(v_E(0), u_E(0)\big)^{\intercal} = (q_0, q_0')^{\intercal}.$$

Clearly, the system (3.23) as well as the initial condition is exactly the same as (3.21). Hence, we have the following theorem.

Theorem 3.5 *The EFCM(k,n) defined by* (3.16) *reduces to a trigonometric Fourier collocation method proposed in [40] when it is applied to solve the special* *first-order differential equations* (3.22); *namely, the second-order oscillatory differential equations* (3.20).

Remark 3.3 It follows from Theorems 3.2, 3.3, 3.4 and 3.5 that EFCMs are an effective extension of HBVMs, Gauss methods, Radau IIA methods and TFCMs. Consequently, EFCMs can be regarded as a generalization of these existing methods in the literature.

3.4 Properties of EFCMs

This section turns to analysing the properties of EFCMs, including their accuracy in preserving the Hamiltonian energy and the quadratic invariants once the underlying problem is a Hamiltonian system, their algebraic order and the convergence condition for fixed-point iteration.

The following result is needed in our analysis, and its proof can be found in [8].

Lemma 3.1 *Let* $f : [0, h] \to \mathbb{R}^d$ *have* j *continuous derivatives in the interval* $[0, h]$. *Then, we obtain* $\int_0^1 \widehat{P}_j(\tau) f(\tau h) d\tau = \mathscr{O}(h^j)$.

As a consequence of this lemma, we have

$$\kappa_j(h, v) = \int_0^1 \widehat{P}_j(\tau) g(\tau h, v(\tau h)) d\tau = \mathscr{O}(h^j).$$

3.4.1 The Hamiltonian Case

Consider the following initial-value Hamiltonian systems

$$u'(t) = J\nabla H(u(t)), \quad u(0) = u_0, \tag{3.24}$$

for the Hamiltonian function $H(u)$ and the skew-symmetric matrix J. Under the condition that

$$J\nabla H(u(t)) = g(t, u(t)) - Au(t), \tag{3.25}$$

the Hamiltonian system (3.24) is identical to first-order initial value problems of the form (3.1). An important example (see, e.g. [14, 40]) is given by

$$H(p, q) = \frac{1}{2} p^\intercal p + \frac{1}{2} q^\intercal M q + U(q),$$

where M is a symmetric and positive semi-definite matrix, and U is a smooth potential with moderately bounded derivatives. This kind of Hamiltonian system frequently arises in applied mathematics, molecular biology, electronics, chemistry, astronomy, classical mechanics and quantum physics, and can be expressed by the following differential equation:

$$\begin{pmatrix} q \\ p \end{pmatrix}' + \begin{pmatrix} 0 & -I \\ M & 0 \end{pmatrix} \begin{pmatrix} q \\ p \end{pmatrix} = \begin{pmatrix} 0 \\ -\nabla U(q) \end{pmatrix},$$

which is exactly a first-order differential system of the form (3.1).

In what follows, we are concerned with the order of preservation for the Hamiltonian energy when EFCMs are applied to solve the Hamiltonian system (3.24)–(3.25).

Theorem 3.6 *Let the quadrature formula in* (3.16) *be exact for polynomials of degree up to* $m-1$*. Then, when EFCM(k,n) is applied to the Hamiltonian system* (3.24)–(3.25), *we have*

$$H(v(h)) = H(u_0) + \mathcal{O}(h^{r+1}) \quad \textit{with} \quad r = \min\{m, 2n\}.$$

Proof It follows from (3.13) and (3.25) that

$$\begin{aligned}
H(v(h)) - H(u_0) &= h\int_0^1 \nabla H(v(\xi h))^\intercal v'(\xi h)d\xi \\
&= h\int_0^1 \nabla H(v(\xi h))^\intercal \Big(\sum_{j=0}^{n-1} \widehat{P}_j(\xi) \sum_{l=1}^{k} b_l \widehat{P}_j(c_l) g(c_l h, v(c_l h)) - Av(\xi h)\Big) d\xi \\
&= h\int_0^1 \Big(g(v(\xi h)) - Av(\xi h)\Big)^\intercal J \Big(\sum_{j=0}^{n-1} \widehat{P}_j(\xi) \sum_{l=1}^{k} b_l \widehat{P}_j(c_l) g(c_l h, v(c_l h)) - Av(\xi h)\Big) d\xi \\
&= h\int_0^1 \Big(g(v(\xi h)) - Av(\xi h)\Big)^\intercal J \Big(g(v(\xi h)) - Av(\xi h) \\
&\quad + \sum_{j=0}^{n-1} \widehat{P}_j(\xi) \sum_{l=1}^{k} b_l \widehat{P}_j(c_l) g(c_l h, v(c_l h)) - g(v(\xi h))\Big) d\xi \\
&= h\int_0^1 \Big(g(v(\xi h)) - Av(\xi h)\Big)^\intercal J \Big(g(v(\xi h)) - Av(\xi h)\Big) d\xi \\
&\quad + h\int_0^1 \Big(g(v(\xi h)) - Av(\xi h)\Big)^\intercal J \Big(\sum_{j=0}^{n-1} \widehat{P}_j(\xi) \sum_{l=1}^{k} b_l \widehat{P}_j(c_l) g(c_l h, v(c_l h)) - g(v(\xi h))\Big) d\xi.
\end{aligned}$$

Since J is skew-symmetric, we have

$$\int_0^1 \Big(g(v(\xi h)) - Av(\xi h)\Big)^\intercal J \Big(g(v(\xi h)) - Av(\xi h)\Big) d\xi = 0.$$

Hence,

$$
\begin{aligned}
&H(v(h)) - H(u_0)\\
&= h\int_0^1 \nabla H(v(\xi h))^T\Big(\sum_{j=0}^{n-1}\widehat{P}_j(\xi)\sum_{l=1}^{k} b_l \widehat{P}_j(c_l) g(c_l h, v(c_l h)) - g(v(\xi h))\Big)d\xi\\
&= h\int_0^1 \nabla H(v(\xi h))^T\Big(\sum_{j=0}^{n-1}\widehat{P}_j(\xi)\sum_{l=1}^{k} b_l \widehat{P}_j(c_l) g(c_l h, v(c_l h)) - \sum_{j=0}^{+\infty}\widehat{P}_j(\xi)\kappa_j(h,v)\Big)d\xi\\
&= -h\int_0^1 \nabla H(v(\xi h))^T\Big(\sum_{j=0}^{n-1}\widehat{P}_j(\xi)\Delta_j(h,v) + \sum_{j=n}^{+\infty}\widehat{P}_j(\xi)\kappa_j(h,v)\Big)d\xi\\
&= -h\sum_{j=0}^{n-1}\int_0^1 \nabla H(v(\xi h))^T\widehat{P}_j(\xi)d\xi\,\Delta_j(h,v) - h\sum_{j=n}^{+\infty}\int_0^1 \nabla H(v(\xi h))^T\widehat{P}_j(\xi)d\xi\,\kappa_j(h,v).
\end{aligned}
$$

From Lemma 3.1, we have

$$
H(v(h)) - H(u_0) = -h\sum_{j=0}^{n-1}\mathcal{O}(h^j \times h^{m-j}) - h\sum_{j=n}^{\infty}\mathcal{O}(h^j \times h^j) = \mathcal{O}(h^{m+1}) + \mathcal{O}(h^{2n+1}),
$$

which shows the result of the theorem. □

3.4.2 The Quadratic Invariants

Quadratic invariants appear often in applications and we thus pay attention to the quadratic invariants of (3.1) in this subsection. Consider the following quadratic function

$$
Q(u) = u^\intercal C u
$$

with a symmetric square matrix C. It is an invariant of (3.1) provided $u^\intercal C(g(t,u) - Au) = 0$ holds.

Theorem 3.7 *Let the quadrature formula in* (3.16) *be exact for polynomials of degree up to* $m-1$. *Then*

$$
Q(v(h)) = Q(u_0) + \mathcal{O}(h^{r+1}) \ \ \textit{with} \ \ r = \min\{m, 2n\}.
$$

Proof It follows from the definition of quadratic function Q that

$$
\begin{aligned}
&Q(v(h)) - Q(u_0)\\
&= \int_0^1 dQ(v(\xi h)) = \int_0^1 \frac{dQ(v(\xi h))}{d\xi}d\xi = 2h\int_0^1 v^\intercal(\xi h) C v'(\xi h) d\xi\\
&= 2h\int_0^1 v^\intercal(\xi h) C\Big(\sum_{j=0}^{n-1}\widehat{P}_j(\xi)\sum_{l=1}^{k} b_l \widehat{P}_j(c_l) g(c_l h, v(c_l h)) - Av(\xi h)\Big)d\xi
\end{aligned}
$$

$$
\begin{aligned}
&= 2h \int_0^1 v^{\mathsf{T}}(\xi h)C\Big(g(\xi h, v(\xi h)) - Av(\xi h) \\
&\quad + \sum_{j=0}^{n-1} \widehat{P}_j(\xi) \sum_{l=1}^{k} b_l \widehat{P}_j(c_l) g(c_l h, v(c_l h)) - g(\xi h, v(\xi h))\Big) d\xi.
\end{aligned}
$$

Since $u^{\mathsf{T}} C(g(t, u) - Au) = 0$, we obtain

$$
\begin{aligned}
&Q(v(h)) - Q(u_0) \\
&= 2h \int_0^1 v^{\mathsf{T}}(\xi h)C\Big(\sum_{j=0}^{n-1} \widehat{P}_j(\xi) \sum_{l=1}^{k} b_l \widehat{P}_j(c_l) g(c_l h, v(c_l h)) - g(\xi h, v(\xi h))\Big) d\xi \\
&= -2h \int_0^1 v^{\mathsf{T}}(\xi h)C\Big(\sum_{j=0}^{n-1} \widehat{P}_j(\xi)\Delta_j(h, v) + \sum_{j=n}^{+\infty} \widehat{P}_j(\xi)\kappa_j(h, v)\Big) d\xi \\
&= -2h \sum_{j=0}^{n-1} \int_0^1 v^{\mathsf{T}}(\xi h)\widehat{P}_j(\xi) d\xi C\Delta_j(h, v) - 2h \sum_{j=n}^{+\infty} \int_0^1 v^{\mathsf{T}}(\xi h)\widehat{P}_j(\xi) d\xi C\kappa_j(h, v) \\
&= -2h \sum_{j=0}^{n-1} \mathcal{O}(h^j \times h^{m-j}) - 2h \sum_{j=n}^{\infty} \mathcal{O}(h^j \times h^j) = \mathcal{O}(h^{m+1}) + \mathcal{O}(h^{2n+1}),
\end{aligned}
$$

which proves the theorem. □

3.4.3 Algebraic Order

Given the importance of different qualitative features, a discussion of the qualitative theory of the underlying ODEs is required. Therefore, in this subsection, we analyse the algebraic order of EFCMs in preserving the accuracy of the solution $u(t)$.

To express the dependence of the solutions of

$$u'(t) = g(t, u(t)) - Au(t)$$

on the initial values, we denote by $u(\cdot, \tilde{t}, \tilde{u})$ the solution satisfying the initial condition $u(\tilde{t}, \tilde{t}, \tilde{u}) = \tilde{u}$ for any given $\tilde{t} \in [0, h]$ and set

$$\Phi(s, \tilde{t}, \tilde{u}) = \frac{\partial u(s, \tilde{t}, \tilde{u})}{\partial \tilde{u}}. \tag{3.26}$$

Recalling the elementary theory of ordinary differential equations, we have the following standard result (see, e.g. [21])

$$\frac{\partial u(s,\tilde{t},\tilde{u})}{\partial \tilde{t}} = -\Phi(s,\tilde{t},\tilde{u})(g(\tilde{t},\tilde{u}) - A\tilde{u}). \tag{3.27}$$

The following theorem states the result on the algebraic order of the novel EFCMs.

Theorem 3.8 *Let the quadrature formula in* (3.16) *be exact for polynomials of degree up to* $m-1$. *We then have*

$$u(h) - v(h) = \mathcal{O}(h^{r+1}) \quad \textit{with} \ \ r = \min\{m, 2n\},$$

for the EFCM(k,n) defined by (3.16).

Proof It follows from Lemma 3.1, (3.26) and (3.27) that

$$\begin{aligned}
u(h) - v(h) &= u(h,0,u_0) - u\big(h,h,v(h)\big) = -\int_0^h \frac{du\big(h,\tau,v(\tau)\big)}{d\tau}d\tau \\
&= -\int_0^h \Big[\frac{\partial u\big(h,\tau,v(\tau)\big)}{\partial \tilde{t}} + \frac{\partial u\big(h,\tau,v(\tau)\big)}{\partial \tilde{u}}v'(\tau)\Big]d\tau \\
&= h\int_0^1 \Phi\big(h,\xi h,v(\xi h)\big)\Big[g\big(\xi h,v(\xi h)\big) - Av(\xi h) - v'(\xi h)\Big]d\xi \\
&= h\int_0^1 \Phi\big(h,\xi h,v(\xi h)\big)\Big[\sum_{j=0}^{+\infty}\widehat{P}_j(\xi)\kappa_j(h,v) - Av(\xi h) \\
&\quad - \sum_{j=0}^{n-1}\widehat{P}_j(\xi)\sum_{l=1}^{k} b_l\widehat{P}_j(c_l)g(c_lh,v(c_lh)) + Av(\xi h)\Big]d\xi \\
&= h\int_0^1 \Phi\big(h,\xi h,v(\xi h)\big)\Big[\sum_{j=0}^{+\infty}\widehat{P}_j(\xi)\kappa_j(h,v) - \sum_{j=0}^{n-1}\widehat{P}_j(\xi)\sum_{l=1}^{k} b_l\widehat{P}_j(c_l)g(c_lh,v(c_lh))\Big]d\xi \\
&= h\int_0^1 \Phi\big(h,\xi h,v(\xi h)\big)\Big[\sum_{j=n}^{+\infty}\widehat{P}_j(\xi)\kappa_j(h,v) + \sum_{j=0}^{n-1}\widehat{P}_j(\xi)\Delta_j(h,v)\Big]d\xi \\
&= h\sum_{j=n}^{+\infty}\int_0^1 \Phi\big(h,\xi h,v(\xi h)\big)\widehat{P}_j(\xi)d\xi\kappa_j(h,v) + h\sum_{j=0}^{n-1}\int_0^1 \Phi\big(h,\xi h,v(\xi h)\big)\widehat{P}_j(\xi)d\xi\Delta_j(h,v) \\
&= h\Big(\sum_{j=n}^{\infty}\mathcal{O}(h^j \times h^j) + \sum_{j=0}^{n-1}\mathcal{O}(h^j \times h^{m-j})\Big) = \mathcal{O}(h^{2n+1}) + \mathcal{O}(h^{m+1}) \\
&= \mathcal{O}(h^{\min\{m,2n\}+1}).
\end{aligned}$$

The proof is complete. □

Remark 3.4 This result means that choosing a suitable quadrature formula as well as a suitable value of n in (3.16) can yield an EFCM of arbitrarily high order. This manipulation is very simple and convenient, and it opens up a new possibility of constructing higher–order EFCMs in a simple and routine manner.

Remark 3.5 It is well known that rth-order numerical methods can preserve the Hamiltonian energy or a quadratic invariant with at least rth degree of accuracy, but unfortunately it follows from the analysis of Sects. 3.4.1 and 3.4.2 that our methods preserve the Hamiltonian energy and quadratic invariants with only rth degree of accuracy.

3.4.4 Convergence Condition of the Fixed-Point Iteration

It is worth noting that usually the EFCM(k,n) defined by (3.16) constitutes a system of implicit equations for the determination of v_i, and an iterative computation is required. In this chapter, we only consider using fixed-point iteration in practical computation. Other iteration methods such as waveform relaxation methods, Krylov subspace methods and preconditioning can be analysed in a similar way. We present the following result for the convergence of fixed-point iteration for the EFCM(k,n) defined by (3.16).

Theorem 3.9 *Assume that g satisfies a Lipschitz condition in the variable u, i.e. there exists a constant L with the property:*

$$\|g(t,u_1)-g(t,u_2)\| \le L\,\|u_1-u_2\|\,.$$

If

$$0 < h < \frac{1}{L\dfrac{Cr^2(e^{\omega}-1)}{\omega}\max\limits_{i,j=1,\cdots,k} c_i|b_j|}, \tag{3.28}$$

then, the fixed-point iteration for the EFCM(k,n) (3.16) *is convergent. Here, C and ω are positive constants independent of A. For a quadrature formula, generally speaking, not all of the node points c_i for $i = 1, 2, \ldots, k$ are equal to zero, and this ensures that* $\max\limits_{i,j=1,\ldots,k} c_i|b_j| \neq 0$.

Proof Following Definition 3.1, the first formula of (3.16) can be rewritten as

$$Q = e^{-cV}u_0 + hA(V)g(ch, Q), \tag{3.29}$$

where $c = (c_1, c_2, \ldots, c_k)^{\intercal}$, $Q = (v_1, v_2, \ldots, v_k)^{\intercal}$, $A(V) = (a_{ij}(V))_{k\times k}$ and $a_{ij}(V)$ are defined as

$$a_{ij}(V) := c_i b_j \sum_{l=0}^{n-1} I_{l,c_i}(V)\widehat{P}_l(c_j).$$

It follows from (3.10) that $|\widehat{P}_j| \le \sqrt{2j+1}$. We then obtain

$$\begin{aligned}\left\|a_{ij}(V)\right\| &\leq c_i|b_j|\sum_{l=0}^{n-1}\sqrt{2l+1}\int_0^1 |\widehat{P}_l(c_iz)|\left\|e^{-(1-z)c_iV}\right\|dz\\ &\leq c_i|b_j|\sum_{l=0}^{n-1}(2l+1)\int_0^1 \left\|e^{-(1-z)c_iV}\right\|dz.\end{aligned}$$

Furthermore, we deduce that

$$\left\|a_{ij}(V)\right\| \leq c_i|b_j|\sum_{l=0}^{n-1}(2l+1)C\int_0^1 e^{\omega(1-z)}dz = Cc_i|b_j|r^2(e^{\omega}-1)/\omega,$$

which yields

$$\|A(V)\| \leq \frac{Cr^2(e^{\omega}-1)}{\omega}\max_{i,j=1,\cdots,k} c_i|b_j|.$$

Let

$$\varphi(x) = e^{-cV}u_0 + hA(V)g(ch,x).$$

We then have

$$\begin{aligned}\|\varphi(x)-\varphi(y)\| &= \|hA(V)g(ch,x)-hA(V)g(ch,y)\| \leq hL\,\|A(V)\|\,\|x-y\|\\ &\leq hL\frac{Cr^2(e^{\omega}-1)}{\omega}\max_{i,j=1,\ldots,k} c_i|b_j|\,\|x-y\|.\end{aligned}$$

This shows that $\varphi(x)$ is a contraction under the assumption (3.28). The well-known Contraction Mapping Theorem then ensures the convergence of fixed-point iteration. □

In what follows, we discuss the convergence of fixed-point iteration for the HBVM(k,n) (3.17) for solving (3.1). When the HBVM(k,n) (3.17) is applied to solve

$$u'(t) = g(t,u(t)) - Au(t), \quad u(0) = u_0,$$

the scheme of HBVM(k,n) becomes

$$\begin{cases} v_i = u_0 + c_ih\displaystyle\sum_{l=1}^{k} b_l\Big(\sum_{j=0}^{n-1}\tilde{I}_{j,c_i}\widehat{P}_j(c_l)\Big)(g(c_lh,v_l)-Av_l), \quad i=1,2,\ldots,k,\\ v(h) = u_0 + h\displaystyle\sum_{l=1}^{k} b_l(g(c_lh,v_l)-Av_l). \end{cases} \tag{3.30}$$

The first formula of (3.30) is also implicit and it requires iterative computation as well. Under the assumption that g satisfies a Lipschitz condition in the variable u, in order to analyse the convergence of fixed-point iteration for the formula (3.30), we denote the iterative function by

$$\psi(x) = u_0 + h\tilde{A}(g(ch, x) - Ax),$$

where $\tilde{A} = (\tilde{a}_{ij})_{k\times k}$ and $\tilde{a}_{ij} = c_i b_j \sum_{l=0}^{n-1} \tilde{I}_{l,c_i} \widehat{P}_l(c_j)$.

We then have

$$\begin{aligned}
\|\psi(x) - \psi(y)\| &= \left\| h\tilde{A}(g(ch, x) - Ax) - h\tilde{A}(g(ch, y) - Ay) \right\| \\
&\le hL \left\|\tilde{A}\right\| \|x - y\| + h \left\|\tilde{A}\right\| \|A\| \|x - y\| \\
&\le h(L + \|A\|) \max_{i,j=1,\dots,k} |\tilde{a}_{ij}| \, \|x - y\| ,
\end{aligned}$$

which means that if

$$0 < h < \frac{1}{(L + \|A\|) \max\limits_{i,j=1,\dots,k} |\tilde{a}_{ij}|},$$

then, fixed-point iteration for HBVM(k,n) is convergent.

Remark 3.6 It is very clear that the convergence of HBVM(k,n) when applied to $u'(t) = g(u(t)) - Au(t)$ depends on $\|A\|$, and the larger $\|A\|$ becomes, the smaller the stepsize required. Whereas, it is of prime importance to note that from (3.28), the convergence of EFCM(k,n) is completely independent of $\|A\|$. This fact implies that EFCMs have better convergence behaviour than HBVMs, especially when $\|A\|$ is large, such as when the problem (3.1) is a stiff system. This point will be numerically demonstrated by the experiments carried out in next section. We also note that an efficient implementation of HBVMs has been considered in [7], and this technique is suitable for stiff first–order and second–order problems.

3.5 A Practical EFCM and Numerical Experiments

As an illustrative example of EFCMs, we choose the 2-point Gauss–Legendre quadrature as the quadrature formula in (3.16), that is exact for all polynomials of degree ≤ 3. This means that $k = 2$ in the k-point Gauss–Legendre quadrature, and this case

$$\begin{aligned}
c_1 &= \frac{3-\sqrt{3}}{6}, \quad c_2 = \frac{3+\sqrt{3}}{6}, \\
b_1 &= \frac{1}{2}, \qquad b_2 = \frac{1}{2}.
\end{aligned} \tag{3.31}$$

We next choose $n = 2$ in (3.16) and denote the corresponding exponential Fourier collocation method as EFCM(2,2). After some calculations, this method can be expressed as

$$
\begin{aligned}
v_1 &= \varphi_0(-c_1V)u_0 + \frac{h}{6}\Big(\sqrt{3}\varphi_1(-c_1V) + (3-2\sqrt{3})\varphi_2(-c_1V)\Big)g(c_1h, v_1)\\
&\quad + \frac{3-2\sqrt{3}}{6}h\Big(\varphi_1(-c_1V) - \varphi_2(-c_1V)\Big)g(c_2h, v_2),\\
v_2 &= \varphi_0(-c_2V)u_0 + \frac{3+2\sqrt{3}}{6}h\Big(\varphi_1(-c_2V) - \varphi_2(-c_2V)\Big)g(c_1h, v_1)\\
&\quad + \frac{h}{6}\Big(-\sqrt{3}\varphi_1(-c_2V) + (3+2\sqrt{3})\varphi_2(-c_2V)\Big)g(c_2h, v_2),\\
v(h) &= \varphi_0(-V)u_0 + \frac{h}{2}\Big((1+\sqrt{3})\varphi_1(-V) - 2\sqrt{3}\varphi_2(-V)\Big)g(c_1h, v_1)\\
&\quad + \frac{h}{2}\Big((1-\sqrt{3})\varphi_1(-V) + 2\sqrt{3}\varphi_2(-V)\Big)g(c_2h, v_2).
\end{aligned}
\tag{3.32}
$$

When $A \to 0$, the method EFCM(2,2) reduces to HBVM(2,2) given in [8], which coincides with the two-stage Gauss method given in [19]. Various examples of EFCMs can be obtained by choosing different quadrature formula and different values of n, and we do not expand on this point in this chapter for brevity.

In order to show the efficiency and robustness of the fourth order method EFCM(2,2), the integrators we select for comparison are also of order four, and we denote them as follows:

- EFCM(2,2): the EFCM(2,2) method of order four derived in this section;
- HBVM(2,2): the Hamiltonian Boundary Value Method of order four in [8] which coincides with the two-stage Gauss method in [19];
- EPCM5s4: the fourth-order energy-preserving collocation method (the case $s = 2$) in [17] with the integrals approximated by the Lobatto quadrature of order eight, which is precisely the "extended Labatto IIIA method of order four" in [29];
- EERK5s4: the explicit five-stage exponential Runge–Kutta method of order four derived in [25].

It is noted that the first three methods are implicit and we use one fixed-point iteration in practical computations for showing the work precision diagram (the global error versus the execution time) as well as energy conservation for a Hamiltonian system. For each problem, we also present the requisite total numbers of iterations for implicit methods when choosing different error tolerances in the fixed-point iteration.

In all the numerical experiments, the matrix exponential is calculated by the algorithm given in [1].

Problem 1 We first consider the Hénon-Heiles Model which is created for describing stellar motion (see, e.g. [9, 19]). The Hamiltonian function of the system is given by

$$
H(p, q) = \frac{1}{2}(p_1^2 + p_2^2) + \frac{1}{2}(q_1^2 + q_2^2) + q_1^2 q_2 - \frac{1}{3}q_2^3.
$$

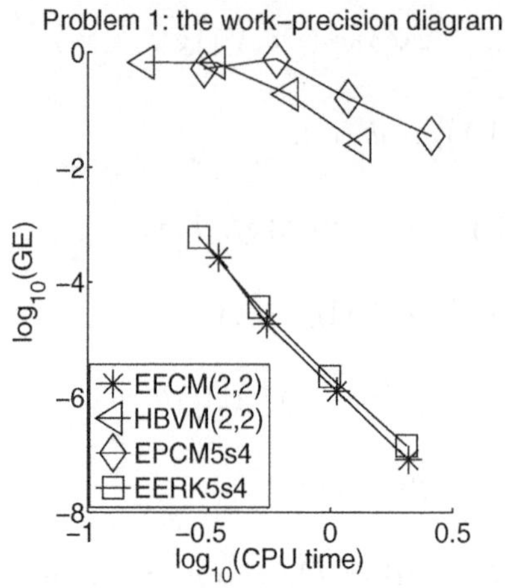

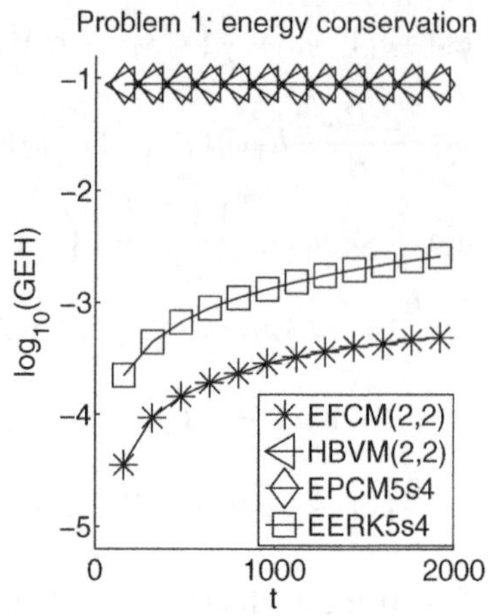

Fig. 3.1 Results for Problem 1. **i**: The log-log plot of the maximum global error (GE) over the integration interval against the execution time. **ii**: The logarithm of the global error of Hamiltonian $GEH = |H_n - H_0|$ against t

This is identical to the following first-order differential equations:

$$\begin{pmatrix} q_1 \\ q_2 \\ p_1 \\ p_2 \end{pmatrix}' + \begin{pmatrix} 0 & 0 & -1 & 0 \\ 0 & 0 & 0 & -1 \\ 1 & 0 & 0 & 0 \\ 0 & 1 & 0 & 0 \end{pmatrix} \begin{pmatrix} q_1 \\ q_2 \\ p_1 \\ p_2 \end{pmatrix} = \begin{pmatrix} 0 \\ 0 \\ -2q_1q_2 \\ -q_1^2 + q_2^2 \end{pmatrix}.$$

The initial values are chosen as

$$\big(q_1(0), q_2(0), p_1(0), p_2(0)\big)^\intercal = \Big(\sqrt{\frac{11}{96}}, 0, 0, \frac{1}{4}\Big)^\intercal.$$

It is noted that we use the result of the standard ODE45 in MATLAB as the true solution for this problem and the next problem. We first solve the problem on the interval [0, 1000] with different stepsizes $h = 1/2^i$ for $i = 2, 3, 4, 5$. The work-precision diagram is presented in Fig. 3.1i. Then, we integrate this problem with the stepsize $h = 1.5$ on the interval [0, 3000]. See Fig. 3.1ii for the energy conservation for different methods. We also solve the problem on [0, 10] with $h = 0.01$ by the three implicit methods and display the total numbers of iterations in Table 3.1 for different error tolerances (tol) chosen in the fixed-point iteration.

Problem 2 The Fermi–Pasta–Ulam problem is an important model for simulations in statistical mechanics which is considered in [14, 18, 19, 43, 47]. It is a Hamiltonian system with the Hamiltonian

Table 3.1 Results for Problem 1. The total numbers of iterations for different error tolerances (tol)

Methods	$tol = 1.0e-006$	$tol = 1.0e-008$	$tol = 1.0e-010$	$tol = 1.0e-012$
EFCM(2,2)	2000	2000	2000	3000
HBVM(2,2)	2000	3000	3769	4000
EPCM5s4	2000	3000	4000	4999

$$H(x,y) = \frac{1}{2}\sum_{i=1}^{2m} y_i^2 + \frac{\omega^2}{2}\sum_{i=1}^{m} x_{m+i}^2 + \frac{1}{4}\Big[(x_1 - x_{m+1})^4 + \sum_{i=1}^{m-1}(x_{i+1} - x_{m+i-1} - x_i - x_{m+i})^4 + (x_m + x_{2m})^4\Big].$$

This leads to

$$\begin{pmatrix} x \\ y \end{pmatrix}' + \begin{pmatrix} \mathbf{0}_{2m\times 2m} & -I_{2m} \\ M & \mathbf{0}_{2m\times 2m} \end{pmatrix}\begin{pmatrix} x \\ y \end{pmatrix} = \begin{pmatrix} 0 \\ -\nabla U(x) \end{pmatrix}, \qquad t \in [0, t_{\text{end}}], \tag{3.33}$$

where

$$M = \begin{pmatrix} \mathbf{0}_{m\times m} & \mathbf{0}_{m\times m} \\ \mathbf{0}_{m\times m} & \omega^2 I_{m\times m} \end{pmatrix},$$

$$U(x) = \frac{1}{4}\Big[(x_1 - x_{m+1})^4 + \sum_{i=1}^{m-1}(x_{i+1} - x_{m+i-1} - x_i - x_{m+i})^4 + (x_m + x_{2m})^4\Big].$$

We choose

$$m = 3,\ \omega = 50,\ x_1(0) = 1,\ y_1(0) = 1,\ x_4(0) = \frac{1}{\omega},\ y_4(0) = 1,$$

and choose zero for the remaining initial values. The system is integrated on the interval [0, 10] with the stepsizes $h = 1/2^k$, $k = 3, 4, 5, 6$. We plot the work-precision diagram in Fig. 3.2i. Then, we solve this problem on the interval [0, 1000] with the stepsize $h = 1/10$ and present the energy conservation in Fig. 3.2ii. Here, it is noted that we do not plot some points in Fig. 3.2 when the errors of the corresponding numerical results are too large. Similar situations occur in the next two problems. Furthermore, we solve the problem on [0, 10] with $h = 0.01$ to show the convergence rate of iterations for the three implicit methods. Table 3.2 lists the total numbers of iterations for different error tolerances.

Problem 3 Consider the semilinear parabolic problem (this problem has been considered in [25])

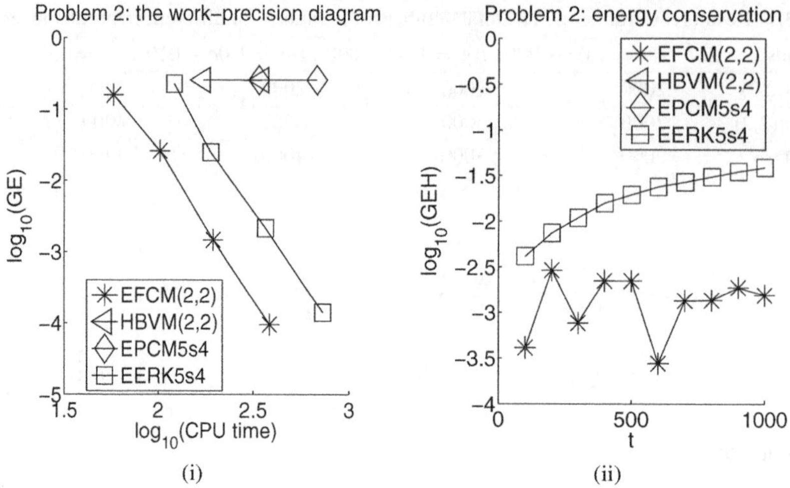

Fig. 3.2 Results for Problem 2. **i**: The log-log plot of the maximum global error (GE) over the integration interval against the execution time. **ii**: The logarithm of the global error of Hamiltonian $GEH = |H_n - H_0|$ against t

Table 3.2 Results for Problem 2. The total numbers of iterations for different error tolerances (tol)

Methods	$tol = 1.0e-006$	$tol = 1.0e-008$	$tol = 1.0e-010$	$tol = 1.0e-012$
EFCM(2,2)	2000	2080	2998	3027
HBVM(2,2)	6801	9291	10980	13912
EPCM5s4	9937	11925	14844	16945

$$\frac{\partial u}{\partial t}(x,t) = \frac{\partial^2 u}{\partial x^2}(x,t) + \frac{1}{1+u(x,t)^2} + \Phi(x,t)$$

for $x \in [0, 1]$ and $t \in [0, 1]$, subject to homogeneous Dirichlet boundary conditions. The source function $\Phi(x,t)$ is chosen in such a way that the exact solution of the problem is $u(x,t) = x(1-x)e^t$.

We discretise this problem in space by using second-order symmetric differences with 1000 grid points. The problem is solved on the interval $[0, 1]$ with different stepsizes $h = 1/2^i$ for $i = 2, 3, 4, 5$. The work-precision diagram is presented in Fig. 3.3. Then, the problem is solved on $[0, 1]$ with $h = \frac{1}{10}$ to show the convergence rate of iterations. See Table 3.3 for the total numbers of iterations for different error tolerances.

Problem 4 Consider a stiff partial differential equation: the Allen–Cahn equation. The Allen–Cahn equation (see, e.g. [15, 32]) is a reaction-diffusion equation of mathematical physics, given by

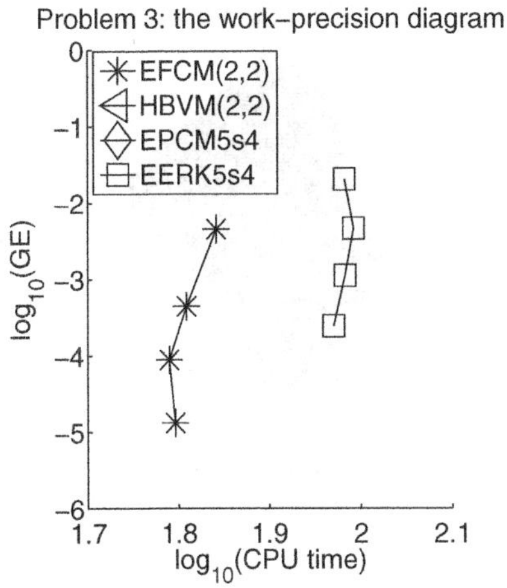

Fig. 3.3 Results for Problem 3. The *log-log* plot of the maximum global error (GE) over the integration interval against the execution time

Table 3.3 Results for Problem 3. The total numbers of iterations for different error tolerances (tol)

Methods	$tol = 1.0e-006$	$tol = 1.0e-008$	$tol = 1.0e-010$	$tol = 1.0e-012$
EFCM(2,2)	40	50	60	73
HBVM(2,2)	86	86	86	86
EPCM5s4	87	87	87	87

$$u_t - \varepsilon u_{xx} = u - u^3, \quad x \in [-1, 1],$$

with $\varepsilon = 0.01$ and initial conditions

$$u(x, 0) = 0.53x + 0.47\sin(-1.5\pi x), \quad u(-1, t) = -1, \quad u(1, t) = 1.$$

We use a 30-point Chebyshev spectral method which yields a system of ordinary differential equations

$$U_t - AU = U - U^3.$$

We apply the MATLAB function *cheb* from [38] for the grid generation and obtain the differentiation matrix A. It is noted that the differentiation matrix A in this example is full.

We first solve this problem on the interval $[0, 70]$ with different stepsizes $h = \frac{1}{100}, \frac{1}{200}, \frac{1}{500}$. The time-evolution of the Allen-Cahn equation for different methods is presented in Figs. 3.4 and 3.5. It is noted that the numerical results of HBVM(2,2) and EPCM5s4 are too large for some stepsizes and thus there is no graph for this case, which means that these methods cannot provide a satisfactory simulation for

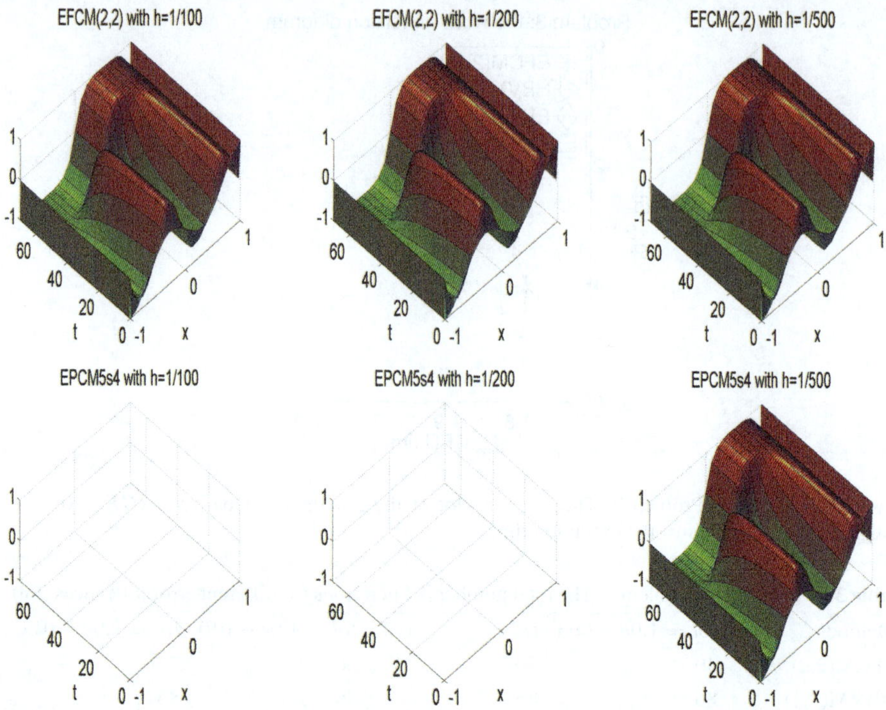

Fig. 3.4 Time evolution for Allen-Cahn equation. The x axis runs from $x = -1$ to $x = 1$ and the t-axis runs from $t = 0$ to $t = 70$

Table 3.4 Results for Problem 4. The total numbers of iterations for different error tolerances (tol)

Methods	$tol = 1.0e-006$	$tol = 1.0e-008$	$tol = 1.0e-010$	$tol = 1.0e-012$
EFCM(2,2)	400	435	608	800
HBVM(2,2)	526	793	1095	1644
EPCM5s4	886	1449	2826	4346

this problem. It can be observed from Fig. 3.5 that EERK5s4 does not produce a satisfactory approximation uniformly, and when the stepsize is decreased, it gives a good approximation. However, our method EFCM(2,2) produces a consistently good approximation no matter which stepsize is chosen. Then, the problem is solved in [0, 100] with $h = 0.16/2^i$ for $i = 0, \ldots, 3$. For this problem, we use the result of the standard ODE15s in MATLAB as the true solution. The work-precision diagram is presented in Fig. 3.6. Finally, we solve the problem on [0, 1] with $h = \frac{1}{200}$ to show the convergence rate of iterations. See Table 3.4 for the total numbers of iterations for different error tolerances.

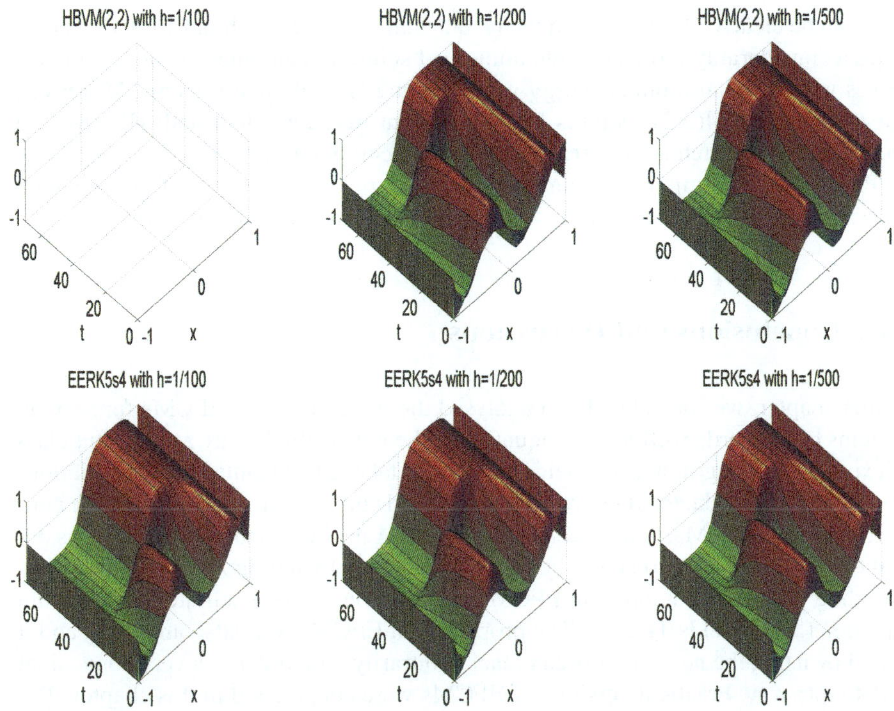

Fig. 3.5 Time evolution for Allen-Cahn equation. The x axis runs from $x = -1$ to $x = 1$ and the t-axis runs from $t = 0$ to $t = 70$

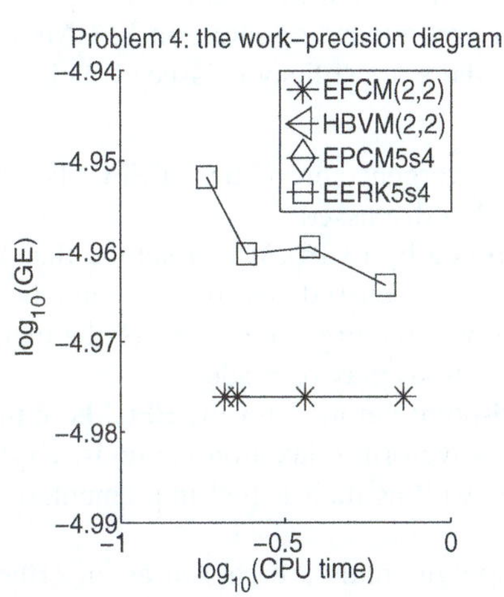

Fig. 3.6 Results for Problem 4. The log-log plot of the global error (GE) over the integration interval against the execution time

It can be clearly observed from the results that the novel method EFCM(2,2) provides a considerably more accurate numerical solution than other methods and preserves well the Hamiltonian energy when solving Hamiltonian systems. Moreover, our method EFCM(2,2) requires less fixed-point iterations than both HBVM(2,2) and EPCM5s4, which is important in long-term computations.

Problem 4 is an important example of the numerical solution of stiff PDEs, which shows that implicit exponential-type integrators are worth studying further.

3.6 Conclusions and Discussions

In this chapter, we formulated and analysed the novel methods EFCMs for solving systems of first-order differential equations. The novel EFCMs are an efficient class of exponential integrators, and their construction takes full advantage of the variation-of-constants formula, local Fourier expansion and collocation. We discussed the connections with HBVMs, Gauss methods, Radau IIA methods and TFCMs. It turns out that the first three traditional methods can be obtained by letting $A \to 0$ in the corresponding EFCMs, and applying EFCMs to the second-order oscillatory differential equation (3.20) yields TFCMs. The properties of EFCMs were also analysed, and it was shown that the new EFCMs can reach arbitrarily high order in a very convenient and simple way. Practical versions of EFCMs were constructed in this chapter. The numerical experiments were carried out and the results affirmatively demonstrate that the novel EFCMs have excellent numerical behaviour in comparison with some existing effective methods in the scientific literature.

Undoubtedly, this is a preliminary research on EFCMs for first-order ordinary differential equations and there are still some issues which could be further considered:

- The error bounds and convergence properties of EFCMs for linear and semilinear problems should be further discussed.
- For the EFCM(k,n) defined by (3.16), it is assumed that $k \geq n$ in this chapter. EFCMs with $k < n$ can be discussed, and this case maybe not affect the computational cost associated with the implementation of the methods for some special systems. Some simplifications may be made.
- We only consider fixed-point iteration for the EFCMs in this chapter. Other iteration methods such as waveform relaxation methods, Krylov subspace methods and preconditioning, as well as their actual implementation for EFCMs, can be analysed.
- The shifted Legendre polynomials are chosen as an orthonormal basis to give the Fourier expansion of the function $g(t, u(t))$. It can be observed that a different choice of orthonormal basis would modify the arguments presented in this chapter. The numerical methods, as well as the corresponding analysis are then modified accordingly. Different choices of the orthonormal basis may be considered.

- Another issue for future exploration is the application of this methodology to other differential equations such as Schrödinger equations and other stiff PDEs.

Symplecticity is an important property, once the underlying system is a Hamiltonian system. Symplectic exponential Runge–Kutta methods for solving nonlinear Hamiltonian systems will be considered in next chapter.

The material of this chapter is based on the work by Wang et al. [45].

References

1. Al-Mohy, A.H., Higham, N.J.: A new scaling and squaring algorithm for the matrix exponential. SIAM J. Matrix Anal. Appl. **31**, 970–989 (2009)
2. Al-Mohy, A.H., Higham, N.J.: Computing the action of the matrix exponential, with an application to exponential integrators. SIAM J. Sci. Comput. **33**, 488–511 (2011)
3. Berland, H., Owren, B., Skaflestad, B.: B-series and order conditions for exponential integrators. SIAM J. Numer. Anal. **43**, 1715–1727 (2005)
4. Berland, H., Skaflestad, B., Wright, W.M.: EXPINT–A MATLAB package for exponential integrators. ACM Trans. Math. Softw. (TOMS) **33**, 4 (2007)
5. Brugnano, L., Iavernaro, F., Magherini, C.: Efficient implementation of Radau collocation methods. Appl. Numer. Math. **87**, 100–113 (2015)
6. Brugnano, L., Iavernaro, F., Trigiante, D.: Hamiltonian boundary value methods (energy preserving discrete line integral methods). J. Numer. Anal. Ind. Appl. Math. **5**, 17–37 (2010)
7. Brugnano, L., Iavernaro, F., Trigiante, D.: A note on the efficient implementation of Hamiltonian BVMs. J. Comput. Appl. Math. **236**, 375–383 (2011)
8. Brugnano, L., Iavernaro, F., Trigiante, D.: A simple framework for the derivation and analysis of effective one-step methods for ODEs. Appl. Math. Comput. **218**, 8475–8485 (2012)
9. Brugnano, L., Iavernaro, F., Trigiante, D.: Energy and quadratic invariants preserving integrators based upon Gauss collocation formulae. SIAM J. Numer. Anal. **50**, 2897–2916 (2012)
10. Brugnano, L., Mazzia, F., Trigiante, D.: Fifty years of stiffness, Recent Advances in Computational and Applied Mathematics, pp. 1–21. Springer, The Netherlands (2011)
11. Caliari, M., Ostermann, A.: Implementation of exponential Rosenbrock-type integrators. Appl. Numer. Math. **59**, 568–581 (2009)
12. Calvo, M.P., Palencia, C.: A class of explicit multistep exponential integrators for semilinear problems. Numer. Math. **102**, 367–381 (2006)
13. Celledoni, E., Cohen, D., Owren, B.: Symmetric exponential integrators with an application to the cubic Schrödinger equation. Found. Comput. Math. **8**, 303–317 (2008)
14. Cohen, D., Jahnke, T., Lorenz, K., Lubich, C.: Numerical integrators for highly oscillatory Hamiltonian systems: a review. In: Mielke, A. (ed.) Analysis, Modeling and Simulation of Multiscale Problems, pp. 553–576. Springer, Berlin (2006)
15. Cox, S.M., Matthews, P.C.: Exponential time differencing for stiff systems. J. Comput. Phys. **176**, 430–455 (2002)
16. Grimm, V., Hochbruck, M.: Error analysis of exponential integrators for oscillatory second-order differential equations. J. Phys. A Math. Gen. **39**, 5495–5507 (2006)
17. Hairer, E.: Energy-preserving variant of collocation methods, JNAIAM J. Numer. Anal. Ind. Appl. Math. **5**, 73–84 (2010)
18. Hairer, E., Lubich, C.: Long-time energy conservation of numerical methods for oscillatory differential equations. SIAM J. Numer. Anal. **38**, 414–441 (2000)
19. Hairer, E., Lubich, C., Wanner, G.: Geometric Numerical Integration: Structure-Preserving Algorithms for Ordinary Differential Equations, 2nd edn. Springer, Berlin (2006)
20. Hairer, E., Nørsett, S.P., Wanner, G.: Solving Ordinary Differential Equations II: Stiff and Differential-Algebraic Problems, 2nd edn. Springer, Berlin (1996)

21. Hale, J.K.: In: Roberte, E. (ed.) Ordinary Differential Equations. Krieger Publishing Company, Huntington (1980)
22. Higham, N.J., Al-Mohy, A.H.: Computing matrix functions. Acta Numer. **19**, 159–208 (2010)
23. Hochbruck, M., Lubich, C.: On Krylov subspace approximations to the matrix exponential operator. SIAM J. Numer. Anal. **34**, 1911–1925 (1997)
24. Hochbruck, M., Lubich, C., Selhofer, H.: Exponential integrators for large systems of differential equations. SIAM J. Sci. Comput. **19**, 1552–1574 (1998)
25. Hochbruck, M., Ostermann, A.: Explicit exponential Runge–Kutta methods for semilineal parabolic problems. SIAM J. Numer. Anal. **43**, 1069–1090 (2005)
26. Hochbruck, M., Ostermann, A.: Exponential Runge–Kutta methods for parabolic problems. Appl. Numer. Math. **53**, 323–339 (2005)
27. Hochbruck, M., Ostermann, A.: Exponential integrators. Acta Numer. **19**, 209–286 (2010)
28. Hochbruck, M., Ostermann, A., Schweitzer, J.: Exponential rosenbrock-type methods. SIAM J. Numer. Anal. **47**, 786–803 (2009)
29. Iavernaro, F., Trigiante, D.: High-order symmetric schemes for the energy conservation of polynomial Hamiltonian problems. JNAIAM J. Numer. Anal. Ind. Appl. Math. **4**, 101–787 (2009)
30. Iserles, A.: On the global error of discretization methods for highly-oscillatory ordinary differential equations. BIT **42**, 561–599 (2002)
31. Iserles, A.: Think globally, act locally: solving highly-oscillatory ordinary differential equations. Appl. Num. Anal. **43**, 145–160 (2002)
32. Kassam, A.K., Trefethen, L.N.: Fourth-order time-stepping for stiff PDEs. SIAM J. Sci. Comput. **26**, 1214–1233 (2005)
33. Khanamiryan, M.: Quadrature methods for highly oscillatory linear and nonlinear systems of ordinary differential equations: part I. BIT Num. Math. **48**, 743–762 (2008)
34. Krogstad, S.: Generalized integrating factor methods for stiff PDEs. J. Comput. Phys. **203**, 72–88 (2005)
35. Lubich, C.: From quantum to classical molecular dynamics: reduced models and numerical analysis, European Mathematical Society (2008)
36. Moler, C., Van Loan, C.: Nineteen dubious ways to compute the exponential of a matrix, twenty-five years later. SIAM Rev. **45**, 3–49 (2003)
37. Ostermann, A., Thalhammer, M., Wright, W.M.: A class of explicit exponential general linear methods. BIT Numer. Math. **46**, 409–431 (2006)
38. Trefethen, L.N.: Spectral Methods in MATLAB. SIAM, Philadelphia (2000)
39. Wang, B., Iserles, A.: Dirichlet series for dynamical systems of first-order ordinary differential equations. Discret. Contin. Dyn. Syst. B **19**, 281–298 (2014)
40. Wang, B., Iserles, A., Wu, X.Y.: Arbitrary-order trigonometric fourier collocation methods for multi-frequency oscillatory systems. Found. Comput. Math. **16**, 151–181 (2016)
41. Wang, B., Li, G.: Bounds on asymptotic-numerical solvers for ordinary differential equations with extrinsic oscillation. Appl. Math. Modell. **39**, 2528–2538 (2015)
42. Wang, B., Liu, K., Wu, X.Y.: A Filon-type asymptotic approach to solving highly oscillatory second-order initial value problems. J. Comput. Phys. **243**, 210–223 (2013)
43. Wang, B., Wu, X.Y.: A new high precision energy-preserving integrator for system of oscillatory second-order differential equations. Phys. Lett. A **376**, 1185–1190 (2012)
44. Wang, B., Wu, X.Y., Meng, F.: Trigonometric collocation methods based on Lagrange basis polynomials for multi-frequency oscillatory second-order differential equations. J. Comput. Appl. Math. **313**, 185–201 (2017)
45. Wang, B., Wu, X.Y., Meng, F.: Exponential Fourier collocation methods for first-order differential equations. J. Comput. Math. **35**, 711–736 (2017)
46. Wu, X.Y., Wang, B., Shi, W.: Efficient energy-preserving integrators for oscillatory Hamiltonian systems. J. Comput. Phys. **235**, 587–605 (2013)
47. Wu, X.Y., Wang, B., Xia, J.: Explicit symplectic multidimensional exponential fitting modified Runge–Kutta–Nyström methods. BIT **52**, 773–795 (2012)
48. Wu, X.Y., You, X., Wang, B.: Structure-Preserving Algorithms for Oscillatory Differential Equations. Springer, Berlin (2013)

Chapter 4
Symplectic Exponential Runge–Kutta Methods for Solving Nonlinear Hamiltonian Systems

Symplecticity is an important property for exponential Runge–Kutta (ERK) methods when the underlying problem $y'(t) = My(t) + f(y(t))$ is a Hamiltonian system. The main theme of this chapter is to present symplectic exponential Runge–Kutta methods. Using the fundamental analysis of geometric integrators, we first derive and analyse the symplectic conditions for ERK methods. These conditions reduce to the conventional ones when $M \to \mathbf{0}$. Furthermore, revised stiff order conditions are proposed and investigated in detail. This chapter is also accompanied by numerical results that demonstrate the potential of the symplectic ERK methods.

4.1 Introduction

The purpose of this chapter is to explore the efficient computation of initial value problems expressed in the autonomous form

$$\begin{cases} y'(t) = My(t) + f(y(t)), \quad t \in [t_0, T], \\ y(t_0) = y_0, \end{cases} \tag{4.1}$$

where the matrix $(-M)$ is symmetric positive definite or skew-Hermitian with eigenvalues of large modulus. Problems of the form (4.1) arise in a wide range of practical problems, such as fluid mechanics, quantum mechanics, electrodynamics, optics, and water waves. Among them, one typical problem originating from the mixed initial-boundary value problems of evolution PDEs, can be written in an abstract form as follows:

$$\begin{cases} \dfrac{\partial u(x,t)}{\partial t} = \mathscr{L}u + \mathscr{N}(u), \quad x \in D,\ t \in [t_0, T], \\ B(x)u(x,t) = 0, \quad x \in \partial D,\ t > t_0, \\ u(x,0) = g(x), \quad x \in D, \end{cases} \tag{4.2}$$

X. Wu and B. Wang, *Recent Developments in Structure-Preserving Algorithms for Oscillatory Differential Equations*, https://doi.org/10.1007/978-981-10-9004-2_4

where D is a spatial domain with boundary ∂D in $\mathbb{R}^d$, $\mathscr{L}$ and $\mathscr{N}$ represent respectively linear and nonlinear operators, and $B(x)$ denotes the boundary operator. Under appropriate discretisation by finite difference approximations, spectral methods or finite elements methods, the problem (4.2) can be converted into (4.1). Stiff problems also yield examples of this type.

It is always challenging to effectively solve the problem (4.1) numerically, since the stiffness occurs due to the linear term My. In light of this point, exponential Runge–Kutta (ERK) methods were proposed for solving this type of problems instead of classical Runge–Kutta (RK) methods. ERK methods have been studied by many authors (see, e.g. [1, 2, 5, 6, 10–13, 15, 17, 18, 20]), and detailed analysis such as the convergence and the construction of these methods can be found therein. It is noted that the extended Runge–Kutta–Nyström (ERKN) methods (see, e.g. [29–32]) can also be classified into the category of exponential integrators, since they are especially designed for efficiently solving second order oscillatory or highly oscillatory problems.

It is well known that both stiff problems and Hamiltonian systems are of prime importance in applications. Much effort has been made in developing a wide variety of approaches to solving each of them. However, it is very clear that problem (4.1) can become identical to a Hamiltonian system if

$$f(y(t)) = J^{-1}\nabla U(y(t))$$

and

$$M = J^{-1}Q,$$

for the skew-symmetric matrix

$$J = \begin{pmatrix} 0 & I \\ -I & 0 \end{pmatrix},$$

where $U(y)$ is a smooth potential function, Q is a symmetric matrix, and I is the identity matrix. This observation motivates the main theme in this chapter, because (4.1) may be a Hamiltonian system. As is known, in the case of Hamiltonian systems, symplectic ERK methods are strongly recommended to preserve the symplecticity of the original problem, since symplectic methods provide long time energy preservation and stability, based on backward error analysis for symplectic methods when applied to Hamiltonian systems [7, 8]. On account of this point, we make a further study on symplectic conditions for ERK methods. Moreover, using the obtained symplectic conditions, we also derive and analyse a class of ERK methods with the important structure-preserving property.

We also note that an important issue for the study of ERK methods is the so-called stiff order. Unfortunately, however, as claimed by Berland et al. in [1], the stiff order conditions are rather restrictive in practice, e.g., the fifth-order ERK method recently constructed by Luan and Ostermann [20] has eight stages provided the full stiff order conditions are considered. Therefore, in this chapter, we deal with the stiff order conditions in a weak form, under which the revised stiff order conditions can be naturally derived from the classical (nonstiff) ones. This process is reasonable based on the fact that no order reduction has been observed, as shown in [15], where ERK methods only need classical (nonstiff) order.

The plan of this chapter is as follows. In Sect. 4.2, we investigate and present sufficient conditions for symplectic ERK methods. In Sect. 4.3, the revised stiff order conditions are investigated and a class of special and important ERK methods are considered, which share the same structure-preserving property as their corresponding RK methods (those corresponding to the underlying ERK methods when $M \to \mathbf{0}$). Section 4.4 is concerned with numerical results to illustrate the efficiency of the symplectic ERK methods. The last section is concerned with conclusions and discussions.

4.2 Symplectic Conditions for ERK Methods

In the study of structure-preserving algorithms, it is an important principle that the construction of numerical schemes for the initial value problem (4.1) should incorporate the structure of the original continuous system in an appropriate way. Taking this point into account, instead of (4.1), we directly consider the following variation-of-constants formula (or the Volterra integral equation) corresponding to (4.1):

$$y(t) = e^{(t-t_0)M} y_0 + \int_{t_0}^{t} e^{(t-\xi)M} f(y(\xi))d\xi. \tag{4.3}$$

It follows from (4.3) that, for any $t, \mu, h \in \mathbb{R}$ with $t, t+\mu h \in [t_0, T]$, the solution to (4.1) satisfies the following integral equation:

$$y(t+\mu h) = e^{\mu h M} y_0 + h \int_{0}^{\mu} e^{(\mu - z)hM} f(y(t+hz))dz, \tag{4.4}$$

which clearly shows the structure of the internal stages and update of an RK-type integrator for solving (4.1). In fact, the case of $0 < \mu < 1$ in (4.4) gives the structure of the internal stages, and $\mu = 1$ in (4.4) presents the structure of the updates of ERK methods. The integral in (4.4) will be approximated by a suitable quadrature formula once the numerical simulation is required for the underlying problem. From this point of view, therefore, ERK methods are generated quite naturally and fundamentally.

It is now easy to formulate ERK methods from the integral equation (4.4). An s-stage ERK method, especially for the stiff problem (4.1), can be written as (see, e.g. [12])

$$\begin{cases} Y_i = e^{c_i hM} y_0 + h \sum_{j=1}^{s} \bar{a}_{ij}(hM) f(Y_j), \quad i = 1, \ldots, s, \\ y_1 = e^{hM} y_0 + h \sum_{i=1}^{s} \bar{b}_i(hM) f(Y_i), \end{cases} \tag{4.5}$$

where $\bar{a}_{ij}(hM)$ and $\bar{b}_i(hM)$ are matrix-valued functions of hM. It is worth mentioning that an ERK method (4.5) reduces to a classical RK method if $M \to \mathbf{0}$. In this sense, the latter is called the *RK method corresponding* to the ERK method (4.5) in this chapter.

It is very clear that (4.1) becomes a Hamiltonian system if $f(y(t)) = J^{-1}\nabla U(y(t))$ and $M = J^{-1}Q$, where $U(y(t))$ is a smooth potential function and Q is a symmetric matrix. With this premise, in the remainder of this chapter we will consider the following Hamiltonian system

$$\begin{cases} y'(t) = J^{-1}Qy(t) + J^{-1}\nabla U(y(t)), \quad t \in [t_0, T], \\ y(t_0) = y_0. \end{cases} \tag{4.6}$$

Hence, the existence of symplectic ERK methods is of great importance for (4.6), but has not received much attention yet in the literature. Consequently, in what follows, we will present and prove the symplectic conditions for ERK methods rigorously. The construction of symplectic ERK methods for solving (4.6) will be analysed in detail in the next section, which definitely confirms the existence of symplectic ERK methods.

Theorem 4.1 *If the coefficients of an s-stage ERK method satisfy the following conditions:*

$$\begin{cases} \bar{b}_i^{\intercal} JSS_i^{-1} = S_i^{-\intercal} S^{\intercal} J\bar{b}_i = \gamma J, \qquad \gamma \in \mathbb{R}, i = 1, \ldots, s, \\ \bar{b}_i^{\intercal} J\bar{b}_j = \bar{b}_i^{\intercal} JSS_i^{-1}\bar{a}_{ij} + \bar{a}_{ji}^{\intercal} S_j^{-\intercal} S^{\intercal} J\bar{b}_j, \qquad i, j = 1, \ldots, s, \end{cases} \tag{4.7}$$

where $S = e^{hM}$ and $S_i = e^{c_i hM}$ for $i = 1, \ldots, s$, then the ERK method is symplectic. Here, γ is an arbitrary real number (independent of i).

Proof We first denote

$$D_i = \frac{\partial f(Y_i)}{\partial y}, \quad X_i = \frac{\partial Y_i}{\partial y_0},$$

for $i = 1, \ldots, s$. If $f = J^{-1}\nabla U(y)$, $M = J^{-1}Q$ in (4.1), then (4.1) is a Hamiltonian system. Thus, M is the infinitesimal symplectic matrix. This leads to the symplecticity of S and S_i as they are exponential forms of λM for some $\lambda \in \mathbb{R}$. Differentiating the scheme (4.5) yields

$$X_i = \frac{\partial Y_i}{\partial y_0} = S_i + h\sum_{j=1}^{s} \bar{a}_{ij} D_j X_j, \tag{4.8}$$

for $i = 1, \ldots, s$, and

$$\frac{\partial y_1}{\partial y_0} = S + h\sum_{i=1}^{s} \bar{b}_i D_i X_i. \tag{4.9}$$

We then have

$$\left(\frac{\partial y_1}{\partial y_0}\right)^\mathsf{T} J\left(\frac{\partial y_1}{\partial y_0}\right) = S^\mathsf{T} J S + h\sum_{i=1}^{s}(\bar{b}_i D_i X_i)^\mathsf{T} J S$$

$$+h\sum_{i=1}^{s} S^\mathsf{T} J\bar{b}_i D_i X_i + h^2\Big(\sum_{i=1}^{s}\bar{b}_i D_i X_i\Big)^\mathsf{T} J\Big(\sum_{i=1}^{s}\bar{b}_i D_i X_i\Big)$$

$$= J + h\sum_{i=1}^{s}(\bar{b}_i D_i X_i)^\mathsf{T} J S + h\sum_{i=1}^{s} S^\mathsf{T} J\bar{b}_i D_i X_i + h^2\sum_{i=1}^{s}\sum_{j=1}^{s}(\bar{b}_i D_i X_i)^\mathsf{T} J(\bar{b}_j D_j X_j). \quad (4.10)$$

Using Eq. (4.8), we obtain

$$(\bar{b}_i D_i X_i)^\mathsf{T} J S S_i^{-1} X_i = (\bar{b}_i D_i X_i)^\mathsf{T} J S + h\sum_{j=1}^{s}(\bar{b}_i D_i X_i)^\mathsf{T} J S S_i^{-1}\bar{a}_{ij} D_j X_j, \quad (4.11)$$

$$(X_i)^\mathsf{T} S_i^{-\mathsf{T}} S^\mathsf{T} J\bar{b}_i D_i X_i = S^\mathsf{T} J\bar{b}_i D_i X_i + h\sum_{j=1}^{s}(\bar{a}_{ij} D_j X_j)^\mathsf{T} S_i^{-\mathsf{T}} S^\mathsf{T} J\bar{b}_i D_i X_i, \quad (4.12)$$

which respectively give

$$(\bar{b}_i D_i X_i)^\mathsf{T} J S = (\bar{b}_i D_i X_i)^\mathsf{T} J S S_i^{-1} X_i - h\sum_{j=1}^{s}(\bar{b}_i D_i X_i)^\mathsf{T} J S S_i^{-1}\bar{a}_{ij} D_j X_j, \quad (4.13)$$

$$S^\mathsf{T} J\bar{b}_i D_i X_i = (X_i)^\mathsf{T} S_i^{-\mathsf{T}} S^\mathsf{T} J\bar{b}_i D_i X_i - h\sum_{j=1}^{s}(\bar{a}_{ij} D_j X_j)^\mathsf{T} S_i^{-\mathsf{T}} S^\mathsf{T} J\bar{b}_i D_i X_i. \quad (4.14)$$

Substituting the new expressions of $(\bar{b}_i D_i X_i)^\mathsf{T} J S$ and $S^\mathsf{T} J\bar{b}_i D_i X_i$ in Eqs. (4.13) and (4.14) into (4.10) yields

$$\left(\frac{\partial y_1}{\partial y_0}\right)^\mathsf{T} J\left(\frac{\partial y_1}{\partial y_0}\right) = J + h\sum_{i=1}^{s}\Big(X_i^\mathsf{T} D_i^\mathsf{T}\bar{b}_i^\mathsf{T} J S S_i^{-1} X_i + X_i^\mathsf{T} S_i^{-\mathsf{T}} S^\mathsf{T} J\bar{b}_i D_i X_i\Big)$$

$$-h^2\sum_{i=1}^{s}\sum_{j=1}^{s}\Big(X_i^\mathsf{T} D_i^\mathsf{T}\bar{b}_i^\mathsf{T} J S S_i^{-1}\bar{a}_{ij} D_j X_j\Big) - h^2\sum_{i=1}^{s}\sum_{j=1}^{s}\Big(X_j^\mathsf{T} D_j^\mathsf{T}\bar{a}_{ij}^\mathsf{T} S_i^{-\mathsf{T}} S^\mathsf{T} J\bar{b}_i D_i X_i\Big)$$

$$+h^2\sum_{i=1}^{s}\sum_{j=1}^{s} X_i^\mathsf{T} D_i^\mathsf{T}\bar{b}_i^\mathsf{T} J\bar{b}_j D_j X_j = J + h\sum_{i=1}^{s} X_i^\mathsf{T}\Big(D_i^\mathsf{T}\bar{b}_i^\mathsf{T} J S S_i^{-1} + S_i^{-\mathsf{T}} S^\mathsf{T} J\bar{b}_i D_i\Big)X_i$$

$$+h^2\sum_{i=1}^{s}\sum_{j=1}^{s} X_i^\mathsf{T} D_i^\mathsf{T}\Big(\bar{b}_i^\mathsf{T} J\bar{b}_j - \bar{b}_i^\mathsf{T} J S S_i^{-1}\bar{a}_{ij} - \bar{a}_{ji}^\mathsf{T} S_j^{-\mathsf{T}} S^\mathsf{T} J\bar{b}_j\Big) D_j X_j\,. \quad (4.15)$$

Since $f = J^{-1}\nabla U(y)$ and $D_i = \dfrac{\partial f(Y_i)}{\partial y}$, a direct calculation gives

$$J D_i + D_i^\mathsf{T} J = 0, \quad i = 1, \ldots, s,$$

on noticing that the Hessian $\dfrac{\partial^2 U}{\partial y^2}$ of U at Y_i is symmetric for $i = 1, \ldots, s$. It then follows from the conditions (4.7) that

$$\Big(\frac{\partial y_1}{\partial y_0}\Big)^{\intercal} J \Big(\frac{\partial y_1}{\partial y_0}\Big) = J.$$

Therefore, the method with coefficients satisfying (4.7) is symplectic. □

Remark 4.1 Here, Theorem 4.1 actually provides a class of sufficient conditions for symplectic ERK methods. Moreover, it can be easily verified that the proposed conditions will reduce to the classical symplectic conditions for RK methods when $M \to \mathbf{0}$. The details are analysed as follows. When $M \to \mathbf{0}$, the matrices $S = e^{hM}$ and $S_i = e^{c_i hM}$ for $i = 1, \ldots, s$ become identity matrices, and $\bar{a}_{ij}, \bar{b}_i$ for $i, j = 1, \ldots, s$ are scalars (more precisely, they are products of the scalars and the identity matrix). In this sense, the first equation of (4.7) holds automatically. The second one is then identical to

$$\bar{b}_i \bar{b}_j J = \bar{b}_i \bar{a}_{ij} J + \bar{a}_{ji} \bar{b}_j J.$$

Hence

$$\bar{b}_i \bar{b}_j = \bar{b}_i \bar{a}_{ij} + \bar{b}_j \bar{a}_{ji},$$

which is exactly the classical symplectic conditions of RK methods [7, 8, 22].

4.3 Symplectic ERK Methods

The direct construction of symplectic ERK methods based on the order conditions accompanying the symplectic conditions is always of high complexity. In spite of this, we make an effort to find a class of ERK methods with the important structure-preserving property. To achieve this goal, the "generalized Runge–Kutta methods", proposed in [17], are helpful and we are hopeful of obtaining some symplectic ERK methods. We first introduce the following theorem, which can be found in [1, 17].

Theorem 4.2 *If* $\mathbf{c} = (c_1, \ldots, c_s)^{\intercal}$, $\mathbf{b} = (b_1, \ldots, b_s)^{\intercal}$ *and* $A = (a_{i,j})_{s\times s}$ *are coefficients of an s-stage RK method of order p, then the ERK method with the same nodes* $\mathbf{c}$, *whose coefficients are defined by*

$$\bar{a}_{ij} = a_{ij} e^{(c_i - c_j)hM}, \qquad \bar{b}_i = b_i e^{(1-c_i)hM}, \quad i, j = 1, \ldots, s, \tag{4.16}$$

is also of order p when applied to the stiff problem (4.1).

The mapping (4.16) actually gives an effective approach for constructing ERK methods based on classical RK methods. It is rather attractive since RK methods have already been well developed in the literature. It is noted that the order is obtained in the sense of the classical (nonstiff) order. However, we will show below that the classical (nonstiff) order conditions are sufficient for the convergence order provided a class of modified stiff order conditions is admitted.

As claimed by Berland et al. [1], the stiff order conditions are rather restrictive. Here, we reconsider the stiff order conditions in a revised version, which does not affect the convergence order of the ERK methods. In fact, the stiff order conditions are derived from the estimation of the global error bound (see [10] for details). Using the explicit form of (4.24) in [10], the expression of the global errors e_n for ERK methods (4.5) can be written as

$$\begin{aligned} e_{n+1} =& e^{hM} e_n + h\mathscr{N}(e_n)e_n - h^2\psi_2(hM)f'(t_n) \\ & - h^3\psi_3(hM)f''(t_n) - h^3\sum_{i=1}^{s}\bar{b}_i(hM)J_n\psi_{2,i}(hM)f'(t_n) \\ & - h^4\psi_4(hM)f'''(t_n) - h^4\sum_{i=1}^{s}\bar{b}_i(hM)J_n\psi_{3,i}(hM)f''(t_n) \\ & - h^4\sum_{i=1}^{s}\bar{b}_i(hM)J_n\sum_{j=1}^{s}\bar{a}_{ij}J_n\psi_{2,j}(hM)f'(t_n) \\ & - h^4\sum_{i=1}^{s}\bar{b}_i(hM)c_iK_n\psi_{2,j}(hM)f'(t_n) + h^5\mathscr{R}_n, \end{aligned} \tag{4.17}$$

where J_n and K_n denote arbitrary square matrices, $\psi_i(hM)$ and $\psi_{i,j}(hM)$ are matrix-valued functions of hM respectively defined by

$$\psi_i(hM) = \varphi_i(hM) - \sum_{k=1}^{s}\bar{b}_k(hM)\frac{c_k^{i-1}}{(i-1)!}, \tag{4.18}$$

$$\psi_{j,i}(hM) = \varphi_j(c_ihM)c_i^j - \sum_{k=1}^{s}\bar{a}_{ik}(hM)\frac{c_k^{j-1}}{(j-1)!}, \tag{4.19}$$

and $\varphi_k(z)$ is defined by

$$\varphi_k(z) = \int_0^1 e^{(1-\theta)z}\frac{\theta^{k-1}}{(k-1)!}d\theta, \quad k \geq 1, \tag{4.20}$$

which has the recurrence relation

$$\varphi_{k+1}(z) = \frac{\varphi_k(z) - \varphi_k(0)}{z}, \quad \varphi_0(z) = e^z. \tag{4.21}$$

By setting some terms in (4.17) as zero, the stiff order conditions can be derived accordingly, just as the authors did in [10]. However, this results in restrictive algebraic conditions which are very difficult to satisfy in practice. Fortunately, a careful observation from (4.17) can help us deal with the stiff order in a modified version, in which the stiff order conditions are approximated satisfactorily by the required order of the underlying integrator instead of the stiff order conditions. In light of this approach, the revised stiff order conditions up to order four are obtained and listed in Table 4.1. It is quite reasonable in applications to admit the new revised stiff order conditions which can be thought of an extension of the conventional ones. With the revised stiff order conditions, (4.17) can be simplified as

$$e_{n+1} = e^{hM} e_n + h\mathscr{N}(e_n)e_n + h^5\widetilde{\mathscr{R}}_n, \tag{4.22}$$

which has no obvious reduction effect on the convergence order. This approach also can be found in determining the order conditions for ERKN methods [30, 32].

The most important advantage of admitting the revised stiff order conditions is that these conditions can be naturally deduced from the classical order conditions. Here, we give an example to show how to achieve the fifth condition in Table. 4.1, based on the fourth (classical) order conditions. For convenience, we formally express $\bar{a}_{ij}(hM)$ and $\bar{b}_i(hM)$ as

$$\bar{a}_{ij}(hM) = \sum_{k=0}^{\infty} \bar{a}_{ij}^{(k)} \cdot (hM)^k, \quad \bar{b}_i(hM) = \sum_{k=0}^{\infty} \bar{b}_i^{(k)} \cdot (hM)^k, \quad i, j = 1, \ldots, s, \tag{4.23}$$

where the coefficients $\bar{a}_{ij}^{(k)}$ and $\bar{b}_i^{(k)}$ are real numbers. Moreover, it can be derived from the recurrence relation (4.21) that

$$\varphi_k(V) = \sum_{j=0}^{\infty} \frac{V^j}{(j+k)!}, \tag{4.24}$$

for any matrix V and $k \geq 0$. Hence, taking (4.19), (4.23) and (4.24) into account, we have

$$\sum_{i=0}^{s} \bar{b}_i(hM) J_n \psi_{2,i}(hM) = \sum_{\mu=0}^{\infty} \sum_{\nu=0}^{\mu} \sum_{i=1}^{s} \bar{b}_i^{(\nu)} \left(\frac{c_i^{2+\mu-\nu}}{(2+\mu-\nu)!} - \sum_{k=1}^{s} \bar{a}_{ik}^{(\mu-\nu)} c_k \right) \cdot h^{\mu} (M^{\nu} J_n M^{\mu-\nu}). \tag{4.25}$$

The fifth condition $\sum_{i=0}^{s} \bar{b}_i(hM) J_n \psi_{2,i}(hM) = \mathscr{O}(h^2)$ in Table. 4.1 then becomes identical to

$$\sum_{i=1}^{s} \bar{b}_i^{(0)} \left(\frac{c_i^2}{2!} - \sum_{k=1}^{s} \bar{a}_{ik}^{(0)} c_k \right) = 0, \tag{4.26}$$

Table 4.1 The revised stiff order conditions up to order four

No.	*Order*	*Order conditions*
1	1	$\psi_1(hM) = \mathcal{O}(h^4)$
2	2	$\psi_2(hM) = \mathcal{O}(h^3)$
3	2	$\psi_{1,i}(hM) = \mathcal{O}(h^3)$
4	3	$\psi_3(hM) = \mathcal{O}(h^2)$
5	3	$\sum_{i=0}^{s} \bar{b}_i(hM) J_n \psi_{2,i}(hM) = \mathcal{O}(h^2)$
6	4	$\psi_4(hM) = \mathcal{O}(h)$
7	4	$\sum_{i=1}^{s} \bar{b}_i(hM) J_n \psi_{3,i}(hM) = \mathcal{O}(h)$
8	4	$\sum_{i=1}^{s} \bar{b}_i(hM) J_n \sum_{j=1}^{s} \bar{a}_{ij} J_n \psi_{2,j}(hM) = \mathcal{O}(h)$
9	4	$\sum_{i=1}^{s} \bar{b}_i(hM) c_i K_n \psi_{2,j}(hM) = \mathcal{O}(h)$

$$\sum_{\nu=0}^{1}\sum_{i=1}^{s}\bar{b}_i^{(\nu)}\left(\frac{c_i^{3-\nu}}{(3-\nu)!}-\sum_{k=1}^{s}\bar{a}_{ik}^{(1-\nu)}c_k\right)=\sum_{i=1}^{s}\bar{b}_i^{(0)}\left(\frac{c_i^3}{3!}-\sum_{k=1}^{s}\bar{a}_{ik}^{(1)}c_k\right)$$
$$+\bar{b}_i^{(1)}\left(\frac{c_i^2}{2!}-\sum_{k=1}^{s}\bar{a}_{ik}^{(0)}c_k\right)=0. \tag{4.27}$$

In can be easily verified that the two Eqs. (4.26) and (4.27) are satisfied, based on the following conditions of order four [1, 10]:

$$\sum_{i=1}^{s}\bar{b}_i^{(0)}c_i^2=\frac{1}{3},\quad \sum_{i=1}^{s}\sum_{k=1}^{s}\bar{b}_i^{(0)}\bar{a}_{ik}^{(0)}c_k=\frac{1}{6},\quad \sum_{i=1}^{s}\bar{b}_i^{(0)}c_i^3=\frac{1}{4},$$
$$\sum_{i=1}^{s}\sum_{k=1}^{s}\bar{b}_i^{(0)}\bar{a}_{ik}^{(1)}c_k=\frac{1}{24},\quad \sum_{i=1}^{s}\bar{b}_i^{(1)}c_i^2=\frac{1}{12},\quad \sum_{i=1}^{s}\sum_{k=1}^{s}\bar{b}_i^{(1)}\bar{a}_{ik}^{(0)}c_k=\frac{1}{24}.$$

The other conditions in Table. 4.1 can be verified in a similar way and the details are omitted here.

The discussions about the stiff order conditions is not pursued further here, since we are mainly devoted to investigating the symplectic conditions for ERK methods, and developing symplectic ERK integrators in this chapter. In the sequel, we will denote the coefficients of classical RK methods by $\mathbf{c} = (c_1, \ldots, c_s)^\intercal$, $\mathbf{b} = (b_1, \ldots, b_s)^\intercal$ and $A = (a_{ij})_{s\times s}$ for convenience. The following theorem states the main result of this chapter.

Theorem 4.3 *If an s-stage RK method is symplectic, then the ERK method yielded by (4.16) is also symplectic.*

Proof Inserting (4.16) into each term in (4.7) yields

$$\begin{cases} \bar{b}_i^{\mathsf{T}} J S S_i^{-1} = \big(b_i e^{(1-c_i)hM}\big)^{\mathsf{T}} J \big(e^{(1-c_i)hM}\big), \\ S_i^{-\mathsf{T}} S^{\mathsf{T}} J \bar{b}_i = \big(e^{(1-c_i)hM}\big)^{\mathsf{T}} J \big(b_i e^{(1-c_i)hM}\big), \\ \bar{b}_i^{\mathsf{T}} J \bar{b}_j = \big(b_i e^{(1-c_i)hM}\big)^{\mathsf{T}} J \big(b_j e^{(1-c_j)hM}\big), \\ \bar{b}_i^{\mathsf{T}} J S S_i^{-1} \bar{a}_{ij} + \bar{a}_{ji}^{\mathsf{T}} S_j^{-\mathsf{T}} S^{\mathsf{T}} J \bar{b}_j = \big(b_i e^{(1-c_i)hM}\big)^{\mathsf{T}} J \big(e^{(1-c_i)hM}\big)\big(a_{ij} e^{(c_i-c_j)hM}\big) \\ \qquad + \big(a_{ji} e^{(c_j-c_i)hM}\big)^{\mathsf{T}} \big(e^{(1-c_j)hM}\big)^{\mathsf{T}} J \big(b_j e^{(1-c_j)hM}\big). \end{cases} \tag{4.28}$$

Noting that the following identity holds

$$P^{\mathsf{T}} J P = J, \tag{4.29}$$

provided P is symplectic, and that $e^{\beta hM}$ is symplectic for any real β and infinitesimal symplectic matrix M, it follows from (4.28) that

$$\begin{cases} \bar{b}_i^{\mathsf{T}} J S S_i^{-1} = S_i^{-\mathsf{T}} S^{\mathsf{T}} J \bar{b}_i = b_i \big(e^{(1-c_i)hM}\big)^{\mathsf{T}} J \big(e^{(1-c_i)hM}\big) = b_i J, \\ \bar{b}_i^{\mathsf{T}} J \bar{b}_j = \big(b_i e^{(1-c_i)hM}\big)^{\mathsf{T}} J \big(b_j e^{(1-c_j)hM}\big) = b_i b_j \big(e^{-c_i hM}\big)^{\mathsf{T}} J \big(e^{-c_j hM}\big) = b_i b_j J \big(e^{(c_i-c_j)hM}\big), \\ \bar{b}_i^{\mathsf{T}} J S S_i^{-1} \bar{a}_{ij} + \bar{a}_{ji}^{\mathsf{T}} S_j^{-\mathsf{T}} S^{\mathsf{T}} J \bar{b}_j = (b_i a_{ij} + b_j a_{ji}) J \big(e^{(c_i-c_j)hM}\big), \end{cases} \tag{4.30}$$

which immediately leads to the satisfaction of the symplectic conditions (4.7) based on those conditions for RK methods, i.e.,

$$b_i b_j = b_i a_{ij} + b_j a_{ji}.$$

This completes the proof. □

Another interesting result about the "generalized Runge–Kutta methods" of [17] is that if the corresponding RK method is symmetric, i.e., the coefficients of an s-stage ERK method satisfy the following conditions

$$\begin{cases} 1 - c_{s+1-i} = c_i, \quad i = 1, \ldots, s, \\ \bar{b}_i(hM) = e^{hM} \bar{b}_{s+1-i}(-hM), \quad i = 1, \ldots, s, \\ e^{(1-c_{s+1-i})hM} \bar{b}_{s+1-j}(-hM) = \bar{a}_{ij}(hM) + \bar{a}_{s+1-i,s+1-j}(-hM), \quad i, j = 1, \ldots, s, \end{cases} \tag{4.31}$$

then the ERK method yielded by (4.16) is symmetric as well. We refer the reader to [3] for more details on this result.

Theorem 4.4 *If the coefficients of an s-stage ERK method satisfy both the symplectic conditions (4.7) and symmetric conditions (4.31), then the ERK method is symplectic and symmetric.*

Proof Under the assumptions of the theorem, the conclusion is quite clear. We therefore omit the details of the proof here. □

4.4 Numerical Experiments

In this section, we implement some numerical experiments to show the high accuracy and good energy preservation of symplectic ERK methods stated in the previous section. In our experiments, the corresponding RK methods are selected as follows:

- RK2: the implicit midpoint method of order two;
- RK4: the Legendre-Gauss collocation method of order four [8].

It should be noted here that both RK2 and RK4 are symplectic, and then ERK2 and ERK4 obtained by the formula (4.16) share the same order as them by Theorem 4.2 and the same symplecticity as their corresponding RK methods by Theorem 4.3. Since all these underlying methods are implicit, iterations are required in the implementation of these methods. The study of existence and uniqueness of numerical solutions of ERK methods is entirely similar to that of implicit RK methods (see, e.g. [4, 16]), and we shall therefore assume unique existence of solution in the remainder of this chapter. As recommended by Hairer et al. (see VIII.6 in [8]), fixed-point iteration is used for the solution of the implicit RK methods, whereas the Newton iteration should be considered for the two implicit ERK methods (see, e.g. [24, 27]). The iteration will be stopped once the norm of the difference of two successive approximations is smaller than 10^{-16}. In all the experiments, the maximum norm is used for both the global errors (GE) and the difference of two successive approximations during the iterations. Throughout the numerical experiments, we point out that the matrix-valued functions $\varphi_k(V)$ $(k \geq 0)$ are exactly evaluated. For larger problems, the Krylov subspace method is well known, and recommended in this case due to its fast convergence. The details about Krylov subspace methods can be found in [9, 13].

As emphasised by the authors in [9, 12, 13], we are hopeful of showing higher accuracy for the ERK methods than their corresponding RK methods in numerical experiments, since they can exactly solve the homogeneous equation $y'(t) = My$, and M always has eigenvalues of large modulus. Meanwhile, good energy preservation is also expected due to the symplecticity of the underlying ERK methods. Another point is that the convergence of iterations for the implicit ERK methods is much better than that for the corresponding RK methods. The main reason is that the occurrence of My in the RK methods when applied to system (4.1) will obviously decrease the convergence due to the large norm of M. Consequently, the faster convergence for the ERK methods results in less consumed CPU time. On the basis of the analysis stated above, we will focus on the previously mentioned advantages of the symplectic ERK methods over their corresponding traditional symplectic RK methods during our numerical experiments.

Problem 4.1 Consider the Duffing equation (see, e.g. [19])

$$\begin{cases} \ddot{q} + \omega^2 q = k^2(2q^3 - q), \\ q(0) = 0, \quad \dot{q}(0) = \omega, \end{cases}$$

with $0 \leq k < \omega$.

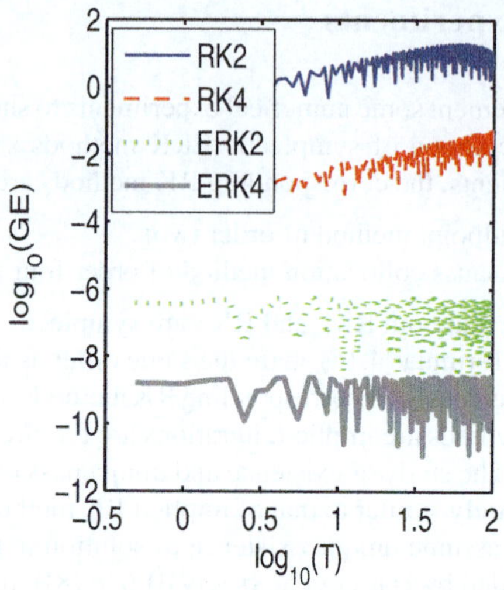

Fig. 4.1 Results for Problem 4.1: the global errors with $h = 1/40$

Let $p = q'$, $z = (p, q)^{\intercal}$. Then the Duffing equation can be rewritten as

$$z'(t) = Mz + f(z),$$

where

$$M = \begin{pmatrix} 0 & -\omega^2 \\ 1 & 0 \end{pmatrix},$$

and

$$f = \left(k^2(2q^3 - q),\ 0\right)^{\intercal}.$$

This is a Hamiltonian system with the Hamiltonian

$$H(p, q) = \frac{1}{2}p^2 + \frac{1}{2}\omega^2 q^2 + \frac{k^2}{2}(q^2 - q^4).$$

The analytic solution is given by

$$q(t) = sn(\omega t, k/\omega),$$

where *sn* is the Jacobian elliptic function.

This problem is solved on the interval [0, 100] with $\omega = 10$, $k = 0.03$ and the stepsize $h = 1/40$. The global errors for these methods are shown in Fig. 4.1. It can be observed from Fig. 4.1 that these two ERK methods significantly display better numerical behaviour in terms of accuracy than their corresponding RK methods.

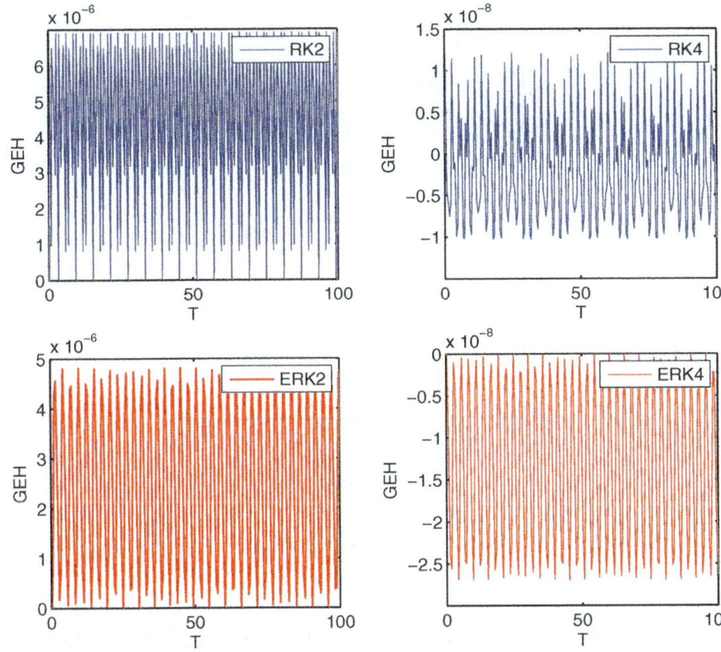

Fig. 4.2 Results for Problem 4.1: the energy preservation

Energy preservation behaviour is shown in Fig. 4.2, from which it can be observed that the obtained ERK methods show comparable energy preservation in comparison with their corresponding RK methods. The CPU times (in seconds) are 0.52, 0.91, 3.26 and 3.34, respectively, for ERK2, ERK4, RK2 and RK4. This shows the faster convergence and higher efficiency of ERK methods than the corresponding RK methods. This also indicates the superiority of the two symplectic ERK methods.

Problem 4.2 Consider the Fermi–Pasta–Ulam problem (see, e.g. [8]) which is an important nonlinear model for research on nonlinear dynamical systems in physics:

$$x''(t) + Ax(t) = -\nabla_x U(x(t)), \tag{4.32}$$

where

$$A = \begin{pmatrix} \mathbf{0}_{m\times m} & \mathbf{0}_{m\times m} \\ \mathbf{0}_{m\times m} & \omega^2 I_{m\times m} \end{pmatrix},$$

$$U(x) = \frac{1}{4}\left((x_1 - x_{m+1})^4 + \sum_{i=1}^{m-1}(x_{i+1} - x_{m+i-1} - x_i - x_{m+i})^4 + (x_m + x_{2m})^4\right).$$

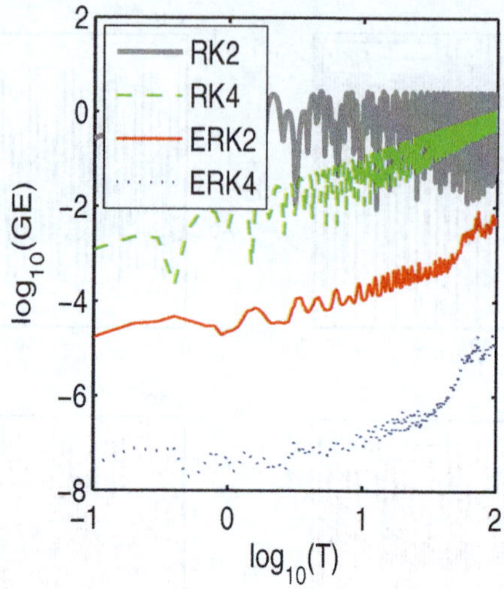

Fig. 4.3 Results for Problem 4.2: the global errors with $h = 1/200$

With $y = x'$, this problem can be expressed by the following Hamiltonian system:

$$z'(t) = Mz(t) + f(z(t)),$$

where $z = (y^\intercal, x^\intercal)^\intercal$,

$$M = \begin{pmatrix} \mathbf{0} & -A \\ E & \mathbf{0} \end{pmatrix},$$

and

$$f = \Big(-(\nabla_x U(x))^\intercal, \, \mathbf{0}^\intercal \Big)^\intercal,$$

with the Hamiltonian

$$H(z) = \frac{1}{2} \sum_{i=1}^{2m} y_i^2 + \frac{\omega^2}{2} \sum_{i=1}^{m} x_{m+i}^2 + U(x). \tag{4.33}$$

Here, E is the identity matrix.

In this experiment, we choose

$$m = 3, \ x_1(0) = 1, \ y_1(0) = 1, \ x_4(0) = \frac{1}{\omega}, \ y_1(0) = 1, \ \omega = 100,$$

and zero for the remaining initial data. This problem is integrated on the interval $[0, 100]$ with the stepsize $h = 1/200$. As shown in Fig. 4.3, the ERK methods give much better accuracy than their corresponding RK methods in global errors. Good

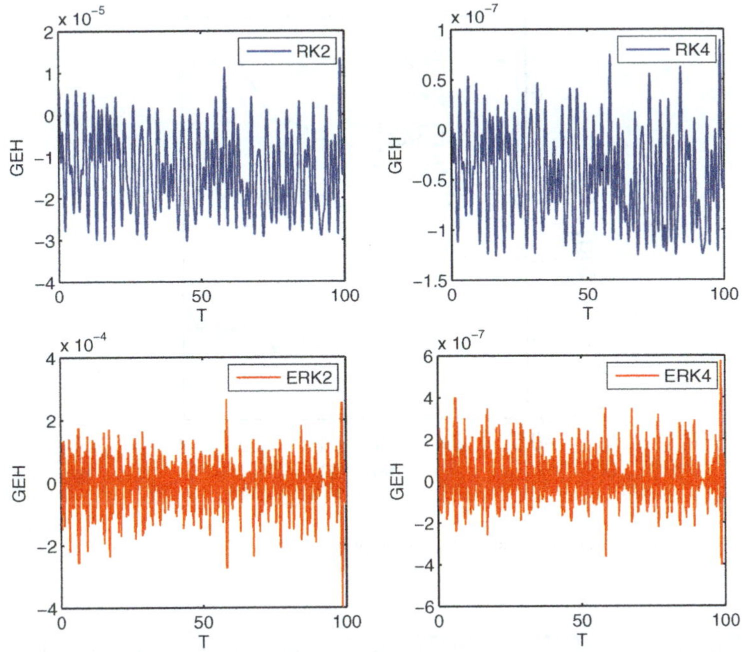

Fig. 4.4 Results for Problem 4.2: the energy preservation

energy preservation behaviour is also displayed by ERK2 and ERK4 in Fig. 4.4. The higher efficiency of the symplectic ERK methods than RK methods is supported by their smaller CPU times (seconds), which are 2.09, 8.43, 29.20 and 30.20, respectively, for ERK2, ERK4, RK2, and RK4.

Problem 4.3 Consider the sine-Gordon equation with the periodic boundary conditions (see, e.g. [23])

$$\begin{cases} \dfrac{\partial^2 u}{\partial t^2} = \dfrac{\partial^2 u}{\partial x^2} - \sin u, \ -5 \leq x \leq 5, \ t \geq 0, \\ u(-5,t) = u(5,t). \end{cases} \tag{4.34}$$

Here, we use the Fourier pseudo-spectral discretisation (see e.g. [26]) for the spatial derivative. Then it can be converted into the following ordinary differential equations:

$$\frac{d}{dt}\begin{pmatrix} U' \\ U \end{pmatrix} = \begin{pmatrix} \mathbf{0} & M \\ E & \mathbf{0} \end{pmatrix}\begin{pmatrix} U' \\ U \end{pmatrix} + \begin{pmatrix} -\sin(U) \\ 0 \end{pmatrix}, \tag{4.35}$$

where $U(t) = (u_1(t), \dots, u_N(t))^\intercal$ with $u_i(t) \approx u(x_i, t)$, $x_i = -5 + i\Delta x$ for $i = 1, \dots, N$, $\Delta x = 10/N$, E is the identity matrix and the second-order spectral differentiation matrix M can be found in [26]. It can be verified that $-M$ is symmetric positive semi-definite. The Hamiltonian corresponding to (4.35) is given by

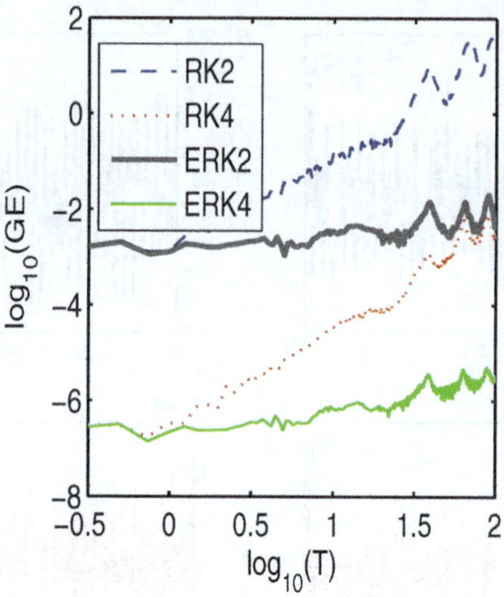

Fig. 4.5 Results for Problem 4.3: the global errors with $h = 1/40$

$$H(U', U) = \frac{1}{2}U'^{\intercal}U' + \frac{1}{2}U^{\intercal}(-M)U - \big(\cos u_1 + \ldots + \cos u_N\big).$$

For this problem, we set the initial conditions as

$$U(0) = (\pi)_{i=1}^{N},\ U'(0) = \sqrt{N}\left(0.01 + \sin\left(\frac{2\pi i}{N}\right)\right)_{i=1}^{N},$$

with $N = 64$. Again, Fig. 4.5 shows the much better accuracy of the two ERK methods than their corresponding RK methods. The detailed behaviour of energy conservation for each method is shown in Fig. 4.6, which clearly displays comparable performance in qualitative behaviour between the ERK integrators and their corresponding RK methods. The CPU times (in seconds) are 0.86, 3.77, 5.59 and 6.50, respectively, for ERK2, ERK4, RK2 and RK4.

Problem 4.4 Consider the nonlinear Klein–Gordon equation with the periodic boundary condition (see, e.g. [14, 28])

$$\begin{cases} u_{tt} + u_{xx} + u + u^3 = 0, \quad 0 < x < L, \quad t \in (0, T), \\ u(0, t) = u(L, t). \end{cases}$$

The initial conditions are given by

$$u(x, 0) = A\left[1 + \cos\left(\frac{2\pi}{L}x\right)\right], \quad u_t(x, 0) = 0,$$

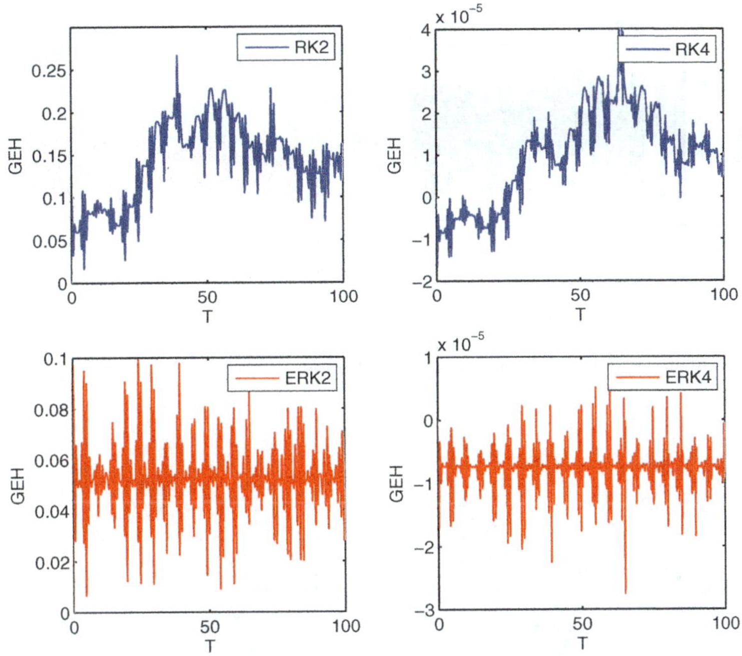

Fig. 4.6 Results for Problem 4.3: the energy preservation

where $L = 1.28$ and A is the amplitude. Similarly to Problem 4.3, if the Fourier pseudo-spectral discretisation is applied to this problem, the semi-discrete ODEs can be obtained:

$$\frac{d}{dt}\begin{pmatrix} U' \\ U \end{pmatrix} = \begin{pmatrix} \mathbf{0} & M \\ E & \mathbf{0} \end{pmatrix}\begin{pmatrix} U' \\ U \end{pmatrix} + \begin{pmatrix} -U - U^3 \\ 0 \end{pmatrix}, \tag{4.36}$$

whose Hamiltonian is given by

$$H(U', U) = \frac{1}{2}U'^{\intercal}U' + \frac{1}{2}U^{\intercal}(-M)U + \frac{1}{2}U^2 + \frac{1}{4}U^4.$$

For this problem, we set $A = 20$. As claimed in [14, 28], this equation is challenging for numerical methods, since the solution shows abrupt changes in both time and space directions with a large amplitude. Similarly to [28], we also carry out numerical simulations with the space stepsize $\Delta x = 0.02$ and the time stepsize $h = 0.01$. The good energy preservation for the two symplectic ERK methods is shown in Fig. 4.7, where the relative errors $RGEH = \frac{GEH}{H_0}$ are plotted for the large value of $H_0 = 1.14 \times 10^7$ and amplitude $A = 20$. Moreover, we display the numerical wave forms from the two ERK methods in Figs. 4.8 and 4.9, respectively. It is shown that both the two ERK methods perform very well, since they preserve

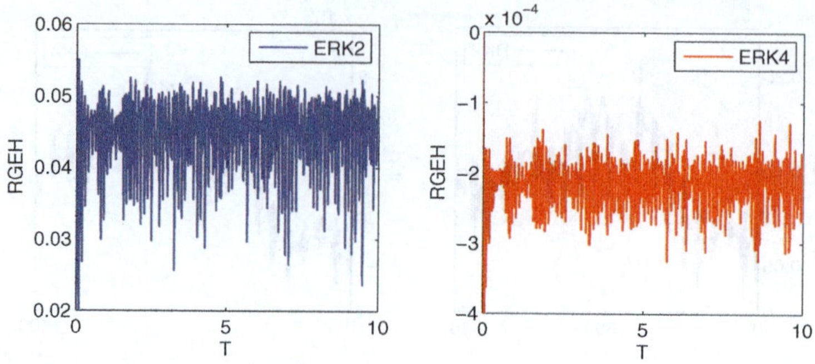

Fig. 4.7 Results for Problem 4.4: the energy preservation

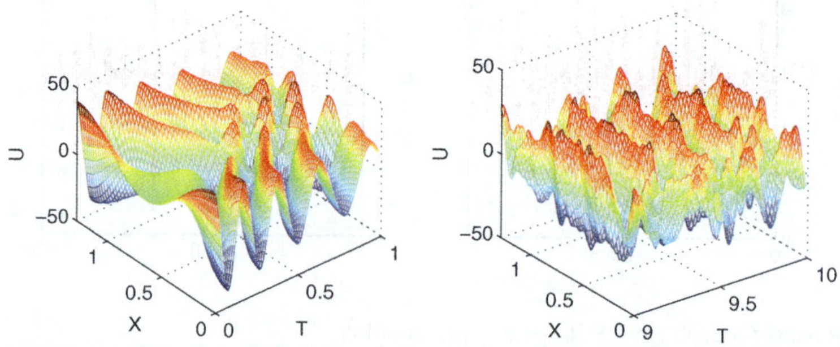

Fig. 4.8 Results for Problem 4.4: the numerical wave forms of ERK2 on different time intervals: the left $t \in [0, 1]$ and the right $t \in [9, 10]$

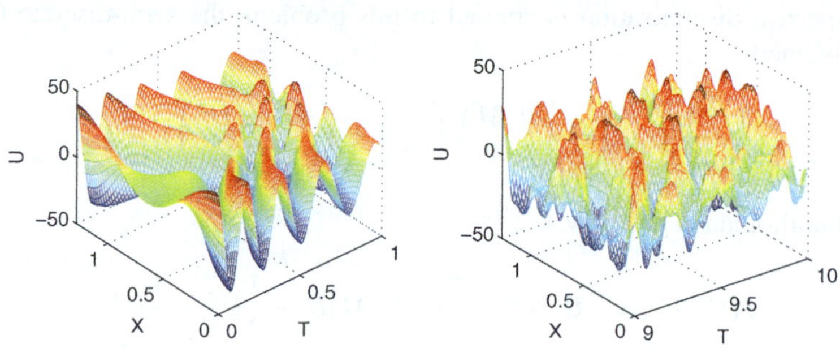

Fig. 4.9 Results for Problem 4.4: the numerical wave forms of ERK4 on different time intervals: the left $t \in [0, 1]$ and the right $t \in [9, 10]$

spatial symmetry as well as the continuity of the solution. Unfortunately, however, the two corresponding RK methods cannot give effective numerical results, since the iterations in the case of $\Delta x = 0.02$ and $h = 0.01$ are not convergent for both RK2 and RK4. The CPU times (in seconds) are 0.23 and 2.37, respectively, for ERK2 and ERK4.

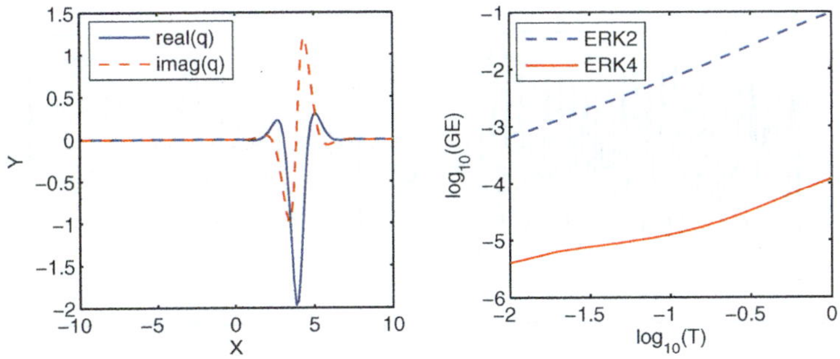

Fig. 4.10 Results for Problem 4.5: exact solutions and the global errors with $h = 0.01$ at $T = 1$

Finally, we turn to the important nonlinear Schrödinger (NLS) equations.

Problem 4.5 Consider the nonlinear Schrödinger (NLS) equation (see, e.g. [25])

$$iq_t = q_{xx} + 2|q|^2 q, \quad t \in (0, T),$$

on the interval $x \in [-10, 10]$ with periodic boundary conditions. The exact solution is given by

$$q(x, t) = 2\eta e^{-i[2\zeta x - 4(\zeta^2 - \eta^2)t + (\Phi_0 + \frac{\pi}{2})]} \operatorname{sech}(2\eta x - 8\zeta \eta t - x_0),$$

where x_0, Φ_0, ζ and η are some constants.

For this problem, we respectively denote u and v as the real and imaginary parts of q. If the Fourier pseudo-spectral method is used for the spatial discretization, this problem can be converted into the Hamiltonian system of the form

$$\frac{d}{dt}\begin{pmatrix} U \\ V \end{pmatrix} = \begin{pmatrix} \mathbf{0} & M \\ -M & \mathbf{0} \end{pmatrix}\begin{pmatrix} U \\ V \end{pmatrix} + \begin{pmatrix} 2(U^2 + V^2)V \\ -2(U^2 + V^2)U \end{pmatrix}, \tag{4.37}$$

whose Hamiltonian reads

$$H(U, V) = \frac{1}{2}U^{\intercal}(-M)U + \frac{1}{2}V^{\intercal}(-M)V - \frac{1}{2}\left(U^{\intercal}U + V^{\intercal}V\right)^2, \tag{4.38}$$

where M is the second-order spectral differentiation matrix approximating the spatial derivative, $U(t) = (u_1(t), \dots, u_N(t))^{\intercal}$, and $V(t) = (v_1(t), \dots, v_N(t))^{\intercal}$. Note that the multiplication of two vectors occurring in (4.37) is in the componentwise sense.

This problem is numerically solved with the given parameters $x_0 = 0$, $\Phi_0 = 0$, $\zeta = 1$, $\eta = 1$, $N = 128$ and $T = 1$. The real and imaginary parts of the exact solution and the global errors for ERK2 and ERK4 with $h = 0.01$ at the endpoint are plotted in Fig. 4.10. It is worth mentioning that only numerical results for ERK methods are plotted in Fig. 4.10, as their corresponding RK methods do not work due

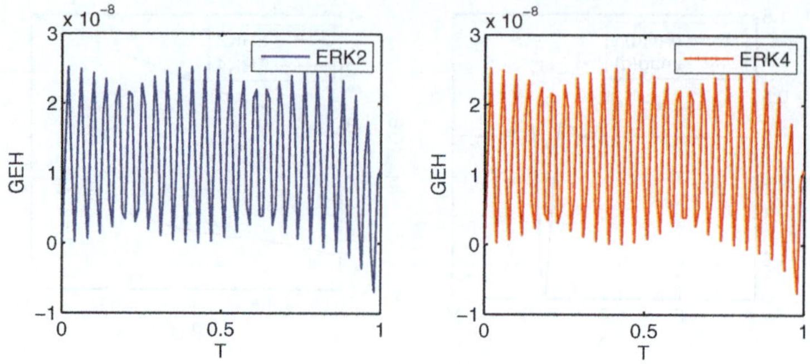

Fig. 4.11 Results for Problem 4.5: the energy preservation

to the appearance of non-convergence during the interactive process. This shows the better performance and broader applicability of symplectic ERK methods over their corresponding symplectic RK methods. Moreover, it can be observed from Fig. 4.11 that both ERK methods show good energy preservation behaviour as well. The CPU times (in seconds) are 0.20 and 0.57, respectively, for ERK2 and ERK4.

4.5 Conclusions and Discussions

Exponential Runge–Kutta methods are very attractive and practical in applications since they always show better performance than classical RK methods in dealing with stiff problems. However, when the underlying problem (4.1) is a Hamiltonian system, the research work has not received much attention up to now. This motivates the main theme of this work: *effective integrators for this kind of Hamiltonian systems using ERK integrators*. With respect to the construction of effective high-order ERK methods, we investigated the structure-preserving property of the novel ERK integrators such as the symplecticity in this chapter. To this end, sufficient conditions for symplecticity were derived by a fundamental analysis of geometric integrators. Furthermore, we presented a novel class of *structure-preserving ERK methods*; that is the structure-preserving "generalized Runge–Kutta methods" (see Lawson [17]), which can preserve symplecticity in the same way as their corresponding RK methods. In order to dispose of the restriction of the conventional stiff order conditions, revised stiff conditions were proposed and investigated in detail. After the establishment of the associated theory for structure-preserving ERK methods, we derived a class of symplectic ERK methods. We took ERK2 and ERK4 as examples in this chapter. Finally, we conducted some numerical experiments, including the approximation of a nonlinear Schrödinger equation, in comparison with the corresponding symplectic Gauss-Legendre RK methods: RK2 and RK4, and the numerical results (both the accuracy and behaviour of energy preservation) are quite promising, and strongly support our theoretical analysis in this chapter. The numerical experiments demonstrate that our symplectic ERK methods are more efficient in many settings than classical methods for the computation of nonlinear Hamiltonian systems.

It is noted that a new exponential scheme EAVF was proposed in the recent paper [18] and summarised in Chap. 2, which preserves the first integral or the Lyapunov function for the conservative or dissipative system. Therefore, it seems that the other properties of structure preservation such as energy preservation and symmetry should be investigated further for the development of ERK integrators. This is the point we also wish to emphasise here.

In the previous four chapters we paid attention to first-order differential equations. In the next four chapters, we will turn to structure-preserving algorithms for multi-frequency and multi-dimensional highly oscillatory second-order differential equations which frequently occur in a wide variety of science and engineering applications.

The material of this chapter is based on the work by Mei and Wu [21].

References

1. Berland, H., Owren, B., Skaflestad, B.: B-series and order conditions for exponential integrators. SIAM J. Numer. Anal. **43**, 1715–1727 (2005)
2. Cox, S., Matthews, P.: Exponential time differencing for stiff systems. J. Comput. Phys. **176**, 430–455 (2002)
3. Celledoni, E., Cohen, D., Owren, B.: Symmetric exponential integrators with an application to the cubic Schröinger equation. Found. Comput. Math. **8**, 303–317 (2008)
4. Crouzeix, M., Hundsdorfer, W.H., Spijker, M.N.: On the existence of solutions to the algebraic equations in Runge–Kutta methods. BIT **23**, 84–91 (1983)
5. Dimarco, G., Pareschi, L.: Exponential Runge–Kutta methods for stiff kinetic equations. SIAM J. Numer. Anal. **49**, 2057–2077 (2011)
6. Dujardin, G.: Exponential Runge–Kutta methods for the Schröinger equation. Appl. Numer. Math. **59**, 1839–1857 (2009)
7. Feng, K., Qin, M.: Symplectic Geometric Algorithms for Hamiltonian Systesms. Springer, Berlin (2010)
8. Hairer, E., Lubich, C., Wanner, G.: Geometric Numerical Integration: Structure-Preserving Algorithms for Ordinary Differential Equations, 2nd edn. Springer, Berlin (2006)
9. Hochbruck, M., Lubich, C.: On Krylov subspace approximations to the matrix exponential operator. SIAM J. Numer. Anal. **34**, 1911–1925 (1997)
10. Hochbruck, M., Ostermann, A.: Explicit exponential Runge–Kutta methods for semilinear parabolic problems. SIAM J. Numer. Anal. **43**, 1069–1090 (2005)
11. Hochbruck, M., Ostermann, A.: Exponential Runge-Kutta methods for parabolic problems. Appl. Numer. Math. **53**, 323–339 (2005)
12. Hochbruck, M., Ostermann, A.: Exponential integrators. Acta Numer. **19**, 209–286 (2010)
13. Hochbruck, M., Lubich, C., Selhofer, H.: Exponential integrators for large systems of differential equations. SIAM J. Sci. Comput. **19**, 1552–1574 (1998)
14. Jiménez, S., Vázquez, L.: Analysis of four numerical schemes for a nonlinear Klein–Gordon equation. Appl. Math. Comput. **35**, 61–94 (1990)
15. Kassam, A.K., Trefethen, L.N.: Fourth-order time stepping for stiff PDEs. SIAM J. Sci. Comput. **26**, 1214–1233 (2005)
16. Kraaijevanger, J.F.B.M., Schneid, F.: On the unique solvability of the Runge–Kutta equations. Numer. Math. **59**, 129–157 (1991)

17. Lawson, J.D.: Generalized Runge–Kutta processes for stable systems with large Lipschitz constants. SIAM J. Numer. Anal. **4**, 372–380 (1967)
18. Li, Y.W., Wu, X.Y.: Exponential integrators preserving first integrals or Lyapunov functions for conservative or dissipative systems. SIAM J. Sci. Comput. **38**, 1876–1895 (2016)
19. Liu, K., Shi, W., Wu, X.Y.: An extended discrete gradient formula for oscillatory Hamiltonian systems. J. Phys. A: Math. Theor. **46**, 165203(1–19) (2013)
20. Luan, V.T., Ostermann, A.: Explicit exponential Runge-Kutta methods of high order for parabolic problems. J. Comput. Appl. Math. **256**, 168–179 (2014)
21. Mei, L.J., Wu, X.Y.: Symplectic exponential Runge-Kutta methods for solving nonlinear Hamiltonian systems. J. Comput. Phys. **338**, 567–584 (2017)
22. Sanz-Serna, J.M.: Runge–Kutta schemes for Hamiltonian systems. BIT Numer. Anal. **28**, 877–883 (1988)
23. Schiesser, W.E., Griffiths, G.W.: A Compendium of Partial Differential Equation Models: Method of Lines Analysis with Matlab. Cambridge University Press, Cambridge (2009)
24. Spijker, M.N.: Stiffness in numerical initial-value problems. J. Comput. Appl. Math. **72**, 393–406 (1996)
25. Taha, T.: A numerical scheme for the nonlinear Schrödinger equation. Comput. Math. Appl. **22**, 77–84 (1991)
26. Trefethen, L.N.: Spectral Methods in MATLAB. SIAM, Philadelphia (2000)
27. van Dorsselaer, J.L.M., Spijker, M.N.: The error committed by stopping the Newton iteration in the numerical solution of stiff initial value problems. IMA J. Numer. Anal. **14**, 183–209 (1994)
28. Wang, Y., Wang, B.: High-order multi-symplectic schemes for the nonlinear Klein–Gordon equation. Appl. Math. Comput. **166**, 608–632 (2005)
29. Wu, X.Y., Wang, B., Xia, J.: Explicit symplectic multidimensional exponential fitting modified Runge–Kutta–Nystrom methods. BIT Numer. Math. **52**, 773–791 (2012)
30. Wu, X.Y., You, X., Wang, B.: Structure-Preserving Algorithms for Oscillatory Differential Equations. Science Press Beijing and Springer, Berlin (2013)
31. Wu, X.Y., Liu, K., Shi, W.: Structure-Preserving Algorithms for Oscillatory Differential Equations II. Springer, Berlin (2015)
32. Yang, H., Wu, X.Y., You, X., Fang, Y.: Extended RKN-type methods for numerical integration of perturbed oscillators. Comput. Phys. Commun. **180**, 1777–1794 (2009)

Chapter 5
High-Order Symplectic and Symmetric Composition Integrators for Multi-frequency Oscillatory Hamiltonian Systems

This chapter presents symplectic and symmetric composition methods based on Adapted Runge–Kutta–Nyström (ARKN) and extended Runge–Kutta–Nyström (ERKN) integrators for solving multi-frequency and multi-dimensional oscillatory Hamiltonian systems with the Hamiltonian $H(p, q) = \frac{1}{2}p^{\intercal}p + \frac{1}{2}q^{\intercal}Kq + U(q)$, where $p = q'$ and K is a symmetric and positive semi-definite matrix. We first consider the symplecticity conditions for multi-frequency and multi-dimensional ARKN integrators. We then analyse the symplecticity of the adjoint integrators of the multi-frequency and multi-dimensional symplectic ARKN and ERKN integrators, respectively. On the basis of the theoretical analysis, and using the idea of composition methods, we derive four new high-order symplectic and symmetric integrators. The numerical results quantitatively show the advantage and efficiency of the high-order symplectic and symmetric integrators.

5.1 Introduction

Geometric numerical integrators are designed specially for the numerical solution of differential equations which possess some geometric/physical properties (Hamiltonian, divergence-free, symmetry, symplecticity, etc.) that should be respected by numerical methods as much as possible. Readers are referred to [3, 4, 10, 13, 16, 19, 31] for this topic. Oscillation is also an important physical property. In fact, differential equations having oscillatory solutions are frequently encountered in many fields of the applied sciences and engineering, such as celestial mechanics, theoretical physics, quantum chemistry and molecular dynamics. The study of modelling and simulation for oscillatory problems is of particular interest in applications. A lot of theoretical and numerical analysis has been done in this field [10, 31]. A variety of methods and analytical tools arise in this area such as stroboscopic

X. Wu and B. Wang, *Recent Developments in Structure-Preserving Algorithms for Oscillatory Differential Equations*, https://doi.org/10.1007/978-981-10-9004-2_5

averaging methods, heterogeneous multiscale methods, the technique of modified Fourier expansions. For these methods, readers are referred to [5–7, 9, 18] and the references therein.

Among typical topics is the numerical integration of an oscillatory system of the form

$$\begin{cases} q''(t) + Kq(t) = f(q(t)), & t \in [t_0, T], \\ q(t_0) = q_0, \quad q'(t_0) = q_0', \end{cases} \tag{5.1}$$

where K is a $d \times d$ positive semi-definite matrix that implicitly contains the dominant frequencies of the oscillatory problem and $f : \mathbb{R}^d \to \mathbb{R}^d, q \in \mathbb{R}^d, q' \in \mathbb{R}^d$. It should be noted that (5.1) is a *multi-frequency and multi-dimensional nonlinear oscillatory problem*. The design and analysis of numerical methods for nonlinear oscillators is an important problem that has received a great deal of attention in the last few years.

It has now become a common practice in geometric numerical integration that numerical algorithms should take advantage of the special structure of the underlying problem. In [29], the authors took account of the special structure of system (5.1) brought by the linear term Kq and proposed the so-called multi-frequency and multi-dimensional ARKN integrators. An outstanding advantage of multi-frequency and multi-dimensional ARKN integrators for (5.1) is that their updates are incorporated with the special structure of the system (5.1) so that they naturally integrate exactly the multi-frequency oscillatory homogeneous system $y'' + Ky = 0$. Very recently, Wu et al. (see [30]) formulated a standard form of the multi-frequency and multi-dimensional ERKN methods in which both the internal stages and updates are incorporated with the special structure of the system (5.1). Therefore, the multi-frequency oscillatory homogeneous system $y'' + Ky = 0$ can be exactly integrated, not only by the updates but also by the internal stages of an ERKN integrator. The ERKN integrators exhibit the correct qualitative behaviour much better than classical RKN methods (see, e.g. [20, 22, 24, 26–29, 32]).

On the other hand, the idea of composition methods is quite useful to improve the order of a basic method while preserving some desirable properties. It is well known that numerical integrators of arbitrarily high orders can be achieved by composition of an integrator with low order. Let φ_h be a basic method and $\gamma_1, \ldots, \gamma_s$ real numbers. We then call its composition

$$\psi_h = \varphi_{\gamma_s h} \circ \cdots \circ \varphi_{\gamma_1 h} \tag{5.2}$$

the corresponding *composition method*.

A more general case is to consider the composition of both the basic integrators and their adjoint integrators with different stepsizes, i.e., to replace composition (5.2) by the more general formula

$$\psi_h = \varphi_{\alpha_s h} \circ \varphi^*_{\beta_s h} \circ \cdots \circ \varphi^*_{\beta_2 h} \circ \varphi_{\alpha_1 h} \circ \varphi^*_{\beta_1 h}. \tag{5.3}$$

The adjoint method of a method is defined as follows [11].

Definition 5.1 The adjoint method Φ_h^* of a method Φ_h is defined as the inverse map of the original method with reversed time step $-h$, i.e., $\Phi_h^* := \Phi_{-h}^{-1}$. A method with $\Phi_h^* = \Phi_h$ is called symmetric.

With regard to composition methods, readers are referred to [1, 2, 15, 17, 23, 33]. A systematic introduction of the idea for composition methods, including the order conditions, can be found in [10].

This chapter focuses on the compositions of multi-frequency and multi-dimensional symplectic ARKN and ERKN integrators. The remainder of this chapter is organized as follows. In Sects. 5.2 and 5.3, we derive some properties for ARKN and ERKN integrators. Based on these properties, we derive four novel high-order symplectic and symmetric methods by using the composition of multi-frequency and multi-dimensional symplectic ARKN and ERKN integrators, respectively. In Sect. 5.4, numerical experiments are carried out, and the advantage and the efficiency of the new methods is shown by the numerical results. The last section is devoted to conclusions and discussions.

5.2 Composition of Multi-frequency ARKN Methods

To begin with, we consider the unconditionally convergent matrix-valued functions which were first defined in [32]

$$\phi_l(K) := \sum_{k=0}^{\infty} \frac{(-1)^k K^k}{(2k+l)!}, \quad l = 0, 1. \tag{5.4}$$

Some properties of the matrix-valued functions (5.4) are given in the following proposition, which can be proved in a straightforward way.

Proposition 5.1 *For a symmetric and positive semi-definite matrix K, the ϕ-functions $\phi_l(K)$ for $l = 0, 1$, defined by* (5.4) *satisfy:*
(*i*)

$$\phi_0^2(K) + K\phi_1^2(K) = I, \tag{5.5}$$

where I is the identity matrix with the same dimension as M.
(*ii*)

$$\begin{aligned}
&\phi_0(a^2K)\phi_0(b^2K) \pm abK\phi_1(a^2K)\phi_1(a^2K) = \phi_0((a \mp b)^2K),\\
&b\phi_1(b^2K)\phi_0(a^2K) \pm a\phi_1(a^2K)\phi_0(b^2K) = (b \pm a)\phi_1((b \pm a)^2K), \forall a, b \in \mathbb{R}.
\end{aligned} \tag{5.6}$$

In the recent paper (see [32]), the authors presented the following variation-of-constants formula for the exact solution and its derivative for the multi-frequency oscillatory system (5.1).

Theorem 5.1 *If $K \in \mathbb{R}^{d\times d}$ is a positive semi-definite matrix and $f : \mathbb{R}^d \to \mathbb{R}^d$ is continuous in* (5.1)*, then the solution of* (5.1) *and its derivative satisfy*

$$\begin{cases} q(t) = \phi_0((t-t_0)^2K)q_0 + (t-t_0)\phi_1((t-t_0)^2K)q_0' + \displaystyle\int_{t_0}^{t}(t-\xi)\phi_1((t-\xi)^2K)\hat{f}(\xi)d\xi, \\ q'(t) = -(t-t_0)K\phi_1((t-t_0)^2K)q_0 + \phi_0((t-t_0)^2K)q_0' + \displaystyle\int_{t_0}^{t}\phi_0((t-\xi)^2K)\hat{f}(\xi)d\xi, \end{cases} \tag{5.7}$$

for any real number t_0, t, where $\hat{f}(\xi) = f(q(\xi))$.

We note that if K is a symmetric and positive semi-definite matrix, K has the decomposition: $K = P^{\intercal}\Omega^2 P = W^2$ with $W = P^{\intercal}\Omega P$. However, (5.7) does not involve the decomposition of matrix K. This point is important, especially for the computational issues of an integrator based on (5.7), since K is not necessarily diagonal or symmetric in (5.1) and the decomposition $K = W^2$ is not always feasible.

It follows immediately from (5.7) that

$$\begin{cases} q(t_n + \nu h) = \phi_0(\nu^2h^2K)q_n + h\phi_1(\nu^2h^2K)q_n' \\ \qquad + h^2\int_0^{\nu}(\nu-\gamma)\phi_1((\nu-\gamma)^2h^2K)\hat{f}(t_n+h\gamma)d\gamma, \\ q'(t_n+\nu h) = -\nu hK\phi_1(\nu^2h^2K)q_n + \phi_0(\nu^2h^2K)q_n' \\ \qquad + \nu h\int_0^{\nu}\phi_0((\nu-\gamma)^2h^2K)\hat{f}(t_n+h\gamma)d\gamma, \end{cases} \tag{5.8}$$

for $0 < \nu < 1$, and

$$\begin{cases} q(t_n + h) = \phi_0(h^2K)q_n + h\phi_1(h^2K)q_n' \\ \qquad + h^2\int_0^{1}(1-\gamma)\phi_1((1-\gamma)^2h^2K)\hat{f}(t_n+h\gamma)d\gamma, \\ q'(t_n+h) = -hK\phi_1(h^2K)q_n + \phi_0(h^2K)q_n' \\ \qquad + h\int_0^{1}\phi_0((1-\gamma)^2h^2K)\hat{f}(t_n+h\gamma)d\gamma. \end{cases} \tag{5.9}$$

From (5.9), revising only the updates of classical RKN methods obtains the following s-stage multi-frequency and multi-dimensional ARKN integrators proposed in [32]:

$$\begin{cases} Q_i = q_n + c_ihq_n' + h^2\displaystyle\sum_{j=1}^{s}\bar{a}_{ij}\big(f(Q_j) - KQ_j\big), \qquad i = 1,\ldots,s, \\ q_{n+1} = \phi_0(V)q_n + h\phi_1(V)q_n' + h^2\displaystyle\sum_{i=1}^{s}\bar{b}_i(V)f(Q_i), \\ q_{n+1}' = \phi_0(V)q_n' - hK\phi_1(V)q_n + h\displaystyle\sum_{i=1}^{s}b_i(V)f(Q_i), \end{cases} \tag{5.10}$$

where, $\bar{a}_{ij} \in \mathbb{R}$ for $i, j=1,\ldots,s$, the weights $b_i, \bar{b}_i : \mathbb{R}^{d\times d} \to \mathbb{R}^{d\times d}$ for $i = 1,\ldots,s$ are matrix-valued functions of $V = h^2K$.

A numerical method has *order* p, if for a sufficiently smooth Problem (5.1) the conditions:

$$e_{n+1} := q_{n+1} - q(t_n+h) = \mathcal{O}(h^{p+1}) \quad \text{and} \quad e_{n+1}' := q_{n+1}' - q'(t_n+h) = \mathcal{O}(h^{p+1}) \tag{5.11}$$

are satisfied, where $q(t_n+h)$ and $q'(t_n+h)$ are the exact solution of (5.1) and its derivative at t_n+h, respectively, q_{n+1} and q'_{n+1} are the one step numerical results obtained by the method from the exact starting values $q_n=q(t_n)$ and $q'_n=q'(t_n)$ (the local assumptions). The order conditions of the multi-frequency and multi-dimensional ARKN integrators (5.10) have been investigated in [32].

The following theorem gives the adjoint integrator of a multi-frequency and multi-dimensional ARKN method.

Theorem 5.2 *The adjoint integrator of an s-stage multi-frequency and multi-dimensional ARKN method* (5.10) *with stepsize h has the following form*

$$\begin{cases} Q_i=\big(\phi_0(V)+c_{s+1-i}V\phi_1(V)\big)q_n+\big(\phi_1(V)-c_{s+1-i}\phi_0(V)\big)hq'_n \\ \quad +h^2\sum\limits_{j=1}^{s}\bar{a}_{s+1-i,s+1-j}\big(f(Q_j)-KQ_j\big)+h^2\sum\limits_{j=1}^{s}\big(\bar{b}^*_j(V)-c_{s+1-i}b^*_j(V)\big)f(Q_j), \\ q_{n+1}=\phi_0(V)q_n+h\phi_1(V)q'_n+h^2\sum\limits_{i=1}^{s}\bar{b}^*_i(V)f(Q_i), \\ q'_{n+1}=-hK\phi_1(V)q_n+\phi_0(V)q'_n+h\sum\limits_{i=1}^{s}b^*_i(V)f(Q_i), \end{cases} \tag{5.12}$$

where

$$\begin{cases} \bar{b}^*_i(V)=\phi_1(V)b_{s+1-i}(V)-\phi_0(V)\bar{b}_{s+1-i}(V), \\ b^*_i(V)=V\phi_1(V)\bar{b}_{s+1-i}(V)+\phi_0(V)b_{s+1-i}(V),\ i=1,2,\ldots,s. \end{cases} \tag{5.13}$$

Proof Exchanging $q_{n+1}\leftrightarrow q_n$, $q'_{n+1}\leftrightarrow q'_n$ and replacing h by $-h$ in the ARKN method (5.10) yields

$$\begin{cases} Q_i=q_{n+1}-c_ihq'_{n+1}+h^2\sum\limits_{j=1}^{s}\bar{a}_{ij}(f(Q_j)-KQ_j), \quad i=1,2,\ldots,s, \\ q_n=\phi_0(V)q_{n+1}-h\phi_1(V)q'_{n+1}+h^2\sum\limits_{i=1}^{s}\bar{b}_i(V)f(Q_i), \\ q'_n=hK\phi_1(V)q_{n+1}+\phi_0(V)q'_{n+1}-h\sum\limits_{i=1}^{s}b_i(V)f(Q_i). \end{cases} \tag{5.14}$$

It follows from (5.14) that

$$\begin{cases} Q_i=\big(\phi_0(V)+c_iV\phi_1(V)\big)q_n+\big(\phi_1(V)-c_i\phi_0(V)\big)hq'_n-h^2\sum\limits_{j=1}^{s}\bar{a}_{ij}KQ_j \\ \quad +h^2\sum\limits_{j=1}^{s}\big[\bar{a}_{ij}-c_i\big(\phi_0(V)b_j(V)+V\phi_1(V)\bar{b}_j(V)\big)+\big(\phi_1(V)b_j(V)-\phi_0(V)\bar{b}_j(V)\big)\big]f(Q_j), \\ q_{n+1}=\phi_0(V)q_n+h\phi_1(V)q'_n+h^2\sum\limits_{i=1}^{s}\big(\phi_1(V)b_i(V)-\phi_0(V)\bar{b}_i(V)\big)f(Q_i), \\ q'_{n+1}=-hK\phi_1(V)q_n+\phi_0(V)q'_n+h\sum\limits_{i=1}^{s}\big(\phi_0(V)b_i(V)+V\phi_1(V)\bar{b}_i(V)\big)f(Q_i). \end{cases} \tag{5.15}$$

Replacing all indices i and j by $s+1-i$ and $s+1-j$, respectively, we obtain the result of the theorem. □

From Theorem 5.2, it is easy to see that the adjoint integrator of an ARKN integrator is not again an ARKN method.

If $f(q)$ in (5.1) is the negative gradient of a real-valued function $U(q)$ and K is a symmetric and positive semi-definite matrix, let $p = q'$, and then (5.1) is in fact identical to the following *multi-frequency and multi-dimensional Hamiltonian system*

$$\begin{cases} q' = H_p(p,q), \\ p' = -H_q(p,q), \\ q(t_0) = q_0, \ p(t_0) = p_0, \end{cases} \tag{5.16}$$

with the Hamiltonian

$$H(p,q) = \frac{1}{2}p^{\intercal}p + \frac{1}{2}q^{\intercal}Kq + U(q). \tag{5.17}$$

The following theorem gives the symplectic conditions of a multi-frequency and multi-dimensional ARKN method.

Theorem 5.3 *If the coefficients of an s-stage multi-frequency and multi-dimensional ARKN method* (5.10) *satisfy the following conditions*

$$\begin{cases} b_i(V)\phi_0(V) + \bar{b}_i(V)V\phi_1(V) = d_i I, & i = 1,\ldots,s, \\ b_i(V)\phi_1(V) - \bar{b}_i(V)\phi_0(V) = c_i d_i I, & i = 1,\ldots,s, \\ d_i\bar{a}_{ij} = 0, & i,j = 1,\ldots,s, \\ b_i(V)\bar{b}_j(V) = b_j(V)\bar{b}_i(V), & i,j = 1,\ldots,s, \end{cases} \tag{5.18}$$

where $d_i \in \mathbb{R}$ for $i = 1,\ldots,s$, then the integrator is symplectic.

Proof With the notation of a differential 2-form used in [19], the symplecticity of the methods (5.10) for (5.1) is identical to

$$\sum_{J=1}^{d} \mathrm{d}q_{n+1}^J \wedge \mathrm{d}{q'}_{n+1}^J = \sum_{J=1}^{d} \mathrm{d}q_n^J \wedge \mathrm{d}{q'}_n^J.$$

We first consider the special case, where K is a diagonal matrix with non-negative entries, i.e., $K = \mathrm{diag}(k_{11}, k_{22}, \cdots, k_{dd})$, then $V = \mathrm{diag}(v_{11}, v_{22}, \cdots, v_{dd})$ with $v_{jj} = h^2 k_{jj}$ for $j = 1,\ldots,d$. Accordingly, $\phi_0(V)$, $\phi_1(V)$, $b_i(V)$ and $\bar{b}_i(V)$ are all diagonal matrices. Denote $f_i = f(Q_i)$. The ARKN method (5.10) then becomes

$$\begin{cases} Q_i^J = q_n^J + c_i h q_n^{'J} + h^2 \sum\limits_{j=1}^{s} a_{ij}(f_j^J - k_{JJ} Q_j^J), \ i = 1, \ldots, s, \\ q_{n+1}^J = \phi_0(v_{JJ}) q_n^J + \phi_1(v_{JJ}) h q_n^{'J} + h^2 \sum\limits_{i=1}^{s} \bar{b}_i(v_{JJ}) f_i^J, \\ q_{n+1}^{'J} = -h k_{JJ} \phi_1(v_{JJ}) q_n^J + \phi_0(v_{JJ}) q_n^{'J} + h \sum\limits_{i=1}^{s} b_i(v_{JJ}) f_i^J, \end{cases} \tag{5.19}$$

where the superscript indices $J = 1, 2, \ldots, d$ denote the Jth component of a vector.

Differentiating q_{n+1}^J and $q_{n+1}^{'J}$ and taking external products, we have

$$\begin{aligned} \mathrm{d}q_{n+1}^J \wedge \mathrm{d}q_{n+1}^{'J} &= [\phi_0^2(v_{JJ}) + v_{JJ}\phi_1^2(v_{JJ})]\mathrm{d}q_n^J \wedge \mathrm{d}q_n^{'J} \\ &+ h \sum_{i=1}^{s} [b_i(v_{JJ})\phi_0(v_{JJ}) + \bar{b}_i(v_{JJ}) v_{JJ} \phi_1(v_{JJ})]\mathrm{d}q_n^J \wedge \mathrm{d}f_i^J \\ &+ h^2 \sum_{i=1}^{s} [b_i(v_{JJ})\phi_1(v_{JJ}) - \bar{b}_i(v_{JJ})\phi_0(v_{JJ})]\mathrm{d}q_n^{'J} \wedge \mathrm{d}f_i^J \\ &+ h^3 \sum_{i,j=1}^{s} \bar{b}_i(v_{JJ}) b_j(v_{JJ}) \mathrm{d}f_i^J \wedge \mathrm{d}f_j^J. \end{aligned}$$

It follows from the identity $\phi_0^2(v_{JJ}) + v_{JJ}\phi_1^2(v_{JJ}) = 1$ that

$$\begin{aligned} \mathrm{d}q_{n+1}^J \wedge \mathrm{d}q_{n+1}^{'J} &= \mathrm{d}q_n^J \wedge \mathrm{d}q_n^{'J} + h \sum_{i=1}^{s} [b_i(v_{JJ})\phi_0(v_{JJ}) + \bar{b}_i(v_{JJ}) v_{JJ} \phi_1(v_{JJ})]\mathrm{d}q_n^J \wedge \mathrm{d}f_i^J \\ &+ h^2 \sum_{i=1}^{s} [b_i(v_{JJ})\phi_1(v_{JJ}) - \bar{b}_i(v_{JJ})\phi_0(v_{JJ})]\mathrm{d}q_n^{'J} \wedge \mathrm{d}f_i^J \\ &+ h^3 \sum_{i,j=1}^{s} \bar{b}_i(v_{JJ}) b_j(v_{JJ}) \mathrm{d}f_i^J \wedge \mathrm{d}f_j^J. \end{aligned} \tag{5.20}$$

Differentiating the first formula of (5.19) yields

$$\mathrm{d}q_n^J = \mathrm{d}Q_i^J - c_i h \mathrm{d}q'^J_n - h^2 \sum_{j=1}^{s} \bar{a}_{ij}(\mathrm{d}f_j^J - k_{JJ}\mathrm{d}Q_j^J), \qquad i = 1, \ldots, s.$$

Thus,

$$\mathrm{d}q_n^J \wedge \mathrm{d}f_i^J = \mathrm{d}Q_i^J \wedge \mathrm{d}f_i^J - c_i h \mathrm{d}q'^J_n \wedge \mathrm{d}f_i^J - h^2 \sum_{j=1}^{s} \bar{a}_{ij}(\mathrm{d}f_j^J - k_{JJ}\mathrm{d}Q_j^J) \wedge \mathrm{d}f_i^J. \tag{5.21}$$

Substituting (5.21) into (5.20) gives

$$\begin{aligned} &\mathrm{d}q_{n+1}^J \wedge \mathrm{d}q'^J_{n+1} \\ &= \mathrm{d}q_n^J \wedge \mathrm{d}q'^J_n + h \sum_{i=1}^{s} [(b_i(v_{JJ})\phi_0(v_{JJ}) + \bar{b}_i(v_{JJ}) v_{JJ} \phi_1(v_{JJ}))]\mathrm{d}Q_i^J \wedge \mathrm{d}f_i^J \end{aligned}$$

$$
\begin{aligned}
&+h\sum_{i,j=1}^{s}[v_{JJ}b_i(v_{JJ})\bar{a}_{ij}\phi_0(v_{JJ})+v_{JJ}^2\bar{b}_i(v_{JJ})\bar{a}_{ij}\phi_1(v_{JJ})]\mathrm{d}Q_j^J\wedge \mathrm{d}f_i^J\\
&+h^2\sum_{i=1}^{s}[b_i(v_{JJ})\phi_1(v_{JJ})-\bar{b}_i(v_{JJ})c_iv_{JJ}\phi_1(v_{JJ})-b_i(v_{JJ})c_i\phi_0(v_{JJ})\\
&\quad-\bar{b}_i(v_{JJ})\phi_0(v_{JJ})]\mathrm{d}{q'}_n^J\wedge \mathrm{d}f_i^J\\
&+h^3\sum_{i,j=1}^{s}[b_i(v_{JJ})\bar{a}_{ij}\phi_0(v_{JJ})+\bar{b}_i(v_{JJ})\bar{a}_{ij}v_{JJ}\phi_1(v_{JJ})+\bar{b}_i(v_{JJ})b_j(v_{JJ})]\mathrm{d}f_i^J\wedge \mathrm{d}f_j^J
\end{aligned}
\tag{5.22}
$$

for each $J=1,\dots,d$. Summing over all J yields

$$
\begin{aligned}
&\sum_{J=1}^{d}\mathrm{d}q_{n+1}^J\wedge \mathrm{d}{q'}_{n+1}^J\\
&=\sum_{J=1}^{d}\mathrm{d}q_n^J\wedge \mathrm{d}{q'}_n^J+h\sum_{i=1}^{s}\sum_{J=1}^{d}[(b_i(v_{JJ})\phi_0(v_{JJ})+\bar{b}_i(v_{JJ})v_{JJ}\phi_1(v_{JJ}))]\mathrm{d}Q_i^J\wedge \mathrm{d}f_i^J\\
&\quad+h\sum_{i,j=1}^{s}\sum_{J=1}^{d}[v_{JJ}b_i(v_{JJ})\bar{a}_{ij}\phi_0(v_{JJ})+v_{JJ}^2\bar{b}_i(v_{JJ})\bar{a}_{ij}\phi_1(v_{JJ})]\mathrm{d}Q_j^J\wedge \mathrm{d}f_i^J\\
&\quad+h^2\sum_{i=1}^{s}\sum_{J=1}^{d}[b_i(v_{JJ})\phi_1(v_{JJ})-\bar{b}_i(v_{JJ})c_iv_{JJ}\phi_1(v_{JJ})-b_i(v_{JJ})c_i\phi_0(v_{JJ})\\
&\quad-\bar{b}_i(v_{JJ})\phi_0(v_{JJ})]\mathrm{d}{q'}_n^J\wedge \mathrm{d}f_i^J\\
&\quad+h^3\sum_{i,j=1}^{s}\sum_{J=1}^{d}[b_i(v_{JJ})\bar{a}_{ij}\phi_0(v_{JJ})+\bar{b}_i(v_{JJ})\bar{a}_{ij}v_{JJ}\phi_1(v_{JJ})+\bar{b}_i(v_{JJ})b_j(v_{JJ})]\mathrm{d}f_i^J\wedge \mathrm{d}f_j^J.
\end{aligned}
\tag{5.23}
$$

By the first and second equations in (5.18), the h^2 term in the right-hand side vanishes and (5.23) can be rewritten as

$$
\begin{aligned}
&\sum_{J=1}^{d}\mathrm{d}q_{n+1}^J\wedge \mathrm{d}{q'}_{n+1}^J\\
&=\sum_{J=1}^{d}\mathrm{d}q_n^J\wedge \mathrm{d}{q'}_n^J+h\sum_{i=1}^{s}d_i\sum_{J=1}^{d}\mathrm{d}Q_i^J\wedge \mathrm{d}f_i^J+h\sum_{i,j=1}^{s}d_i\bar{a}_{ij}\sum_{J=1}^{d}v_{JJ}\mathrm{d}Q_j^J\wedge \mathrm{d}f_i^J\\
&\quad+h^3\sum_{i,j=1}^{s}\sum_{J=1}^{d}[\bar{a}_{ij}d_i+\bar{b}_i(v_{JJ})b_j(v_{JJ})]\mathrm{d}f_i^J\wedge \mathrm{d}f_j^J.
\end{aligned}
\tag{5.24}
$$

Since

$$
\mathrm{d}f_i^J\wedge \mathrm{d}Q_i^J=\Big(\sum_{I=1}^{d}\frac{\partial f^J}{\partial q^I}(Q_i)\mathrm{d}Q_i^I\Big)\wedge \mathrm{d}Q_i^J=\sum_{I=1}^{d}\frac{\partial f^J}{\partial q^I}(Q_i)\mathrm{d}Q_i^I\wedge \mathrm{d}Q_i^J,
$$

it follows from $f = -\nabla U$ that

$$
\begin{aligned}
\sum_{J=1}^{d} \mathrm{d}f_i^J \wedge \mathrm{d}Q_i^J &= \sum_{I,J=1}^{d} \Big(\frac{\partial f^J}{\partial q^I}(Q_i)\mathrm{d}Q_i^I\Big) \wedge \mathrm{d}Q_i^J \\
&= \sum_{I<J} \Big(\frac{\partial f^J}{\partial q^I}(Q_i) - \frac{\partial f^I}{\partial q^J}(Q_i)\Big)\mathrm{d}Q_i^I \wedge \mathrm{d}Q_i^J \\
&= \sum_{I<J} \Big[-\frac{\partial^2 U}{\partial q^J \partial q^I}(Q_i) + \frac{\partial^2 U}{\partial q^I \partial q^J}(Q_i)\Big]\mathrm{d}Q_i^I \wedge \mathrm{d}Q_i^J \\
&= 0,
\end{aligned}
\tag{5.25}
$$

where the last term vanishes by the symmetry of the Hessian matrix.

Therefore, it follows from (5.25) and the assumptions (5.18) that

$$
\sum_{J=1}^{d} \mathrm{d}q_{n+1}^J \wedge \mathrm{d}q'^J_{n+1} = \sum_{J=1}^{d} \mathrm{d}q_n^J \wedge \mathrm{d}q'^J_n.
$$

For the general case, since K is a symmetric and positive semi-definite matrix, K has the following decomposition:

$$
K = P^{\intercal}\Omega^2 P = W^2 \text{ with } W = P^{\intercal}\Omega P, \tag{5.26}
$$

where P is an orthogonal matrix and Ω is a diagonal matrix with nonnegative diagonal entries which are the square roots of the eigenvalues of K. Accordingly, by using the variable substitution $z(t) = Pq(t)$, the system (5.1) is identical to the system

$$
\begin{cases}
z''(t) + \Omega^2 z(t) = Pf(P^T z(t)), & t \in [t_0, T], \\
z(t_0) = z_0 = Py_0, \quad q'(t_0) = q_0' = Pq_0', &
\end{cases}
\tag{5.27}
$$

where $f(q) = -\nabla_q U(q),\ Pf\big(P^{\intercal}z(t)\big) = -P\nabla_q U\big(P^{\intercal}z(t)\big) = -\nabla_z U\big(P^{\intercal}z(t)\big)$.

Then the symplectic multi-frequency and multi-dimensional ARKN method, for the case where K is diagonal with nonnegative entries, can be applied to the transformed system. Furthermore, the methods are invariant under linear transformation. Hence, the methods applied to the system (5.1) can be expressed in terms of $z(t)$ via the multiplication by P and with the notation $Z_i = PQ_i,\ z_n = Pq_n$.

We then have

$$\begin{aligned}\sum_{J=1}^{d} \mathrm{d}z_{n+1}^{J} \wedge \mathrm{d}z'^{J}_{n+1} &= \sum_{J=1}^{d} \mathrm{d}\sum_{i=1}^{d}(p_{Ji}q_{n+1}^{i}) \wedge \sum_{J=1}^{d} \mathrm{d}\sum_{k=1}^{d}(p_{Jk}q'^{k}_{n+1}) \\ &= \sum_{J=1}^{d}\sum_{i=1}^{d}(p_{Ji}\mathrm{d}q_{n+1}^{i}) \wedge \sum_{J=1}^{d}\sum_{k=1}^{d}(p_{Jk}\mathrm{d}q'^{k}_{n+1}) \\ &= \sum_{J=1}^{d}\sum_{i=1}^{d}\sum_{k=1}^{d} p_{Ji}p_{Jk}(\mathrm{d}q_{n+1}^{i} \wedge \mathrm{d}q'^{k}_{n+1}) \\ &= \sum_{J=1}^{d} \mathrm{d}q_{n+1}^{J} \wedge \mathrm{d}q'^{J}_{n+1}.\end{aligned} \tag{5.28}$$

Likewise,

$$\sum_{J=1}^{d} \mathrm{d}z_{n} \wedge \mathrm{d}z'^{J}_{n} = \sum_{J=1}^{d} \mathrm{d}q_{n}^{J} \wedge \mathrm{d}q'^{J}_{n}. \tag{5.29}$$

Therefore, it follows from (5.28) and (5.29) that

$$\sum_{J=1}^{d} \mathrm{d}q_{n+1}^{J} \wedge \mathrm{d}q'^{J}_{n+1} = \sum_{J=1}^{d} \mathrm{d}q_{n}^{J} \wedge \mathrm{d}q'^{J}_{n}.$$

The orthogonality of the matrix $P = (p_{Ji})_{d\times d}$ is used in the proof of (5.28). The proof is complete. □

Remark 5.1 It is clear from Theorem 5.3 that a symplectic ARKN method for a single-frequency oscillatory Hamiltonian system cannot ensure itself to be again a symplectic method when applied to multi-frequency and multi-dimensional oscillatory Hamiltonian system, since a symplectic multi-frequency and multi-dimensional ARKN method requires additional conditions in comparison with a symplectic single-frequency ARKN method as shown in (5.18). With regard to symplecticity conditions for single-frequency ARKN methods, readers are referred to [20, 21]. For exactly the same reason, a symplectic ERKN method for a single-frequency oscillatory Hamiltonian system cannot be guaranteed to be a symplectic method when applied to multi-frequency and multi-dimensional oscillatory Hamiltonian systems.

Remark 5.2 From the first two equations in symplecticity conditions (5.18), we can solve $b_i(V), \bar{b}_i(V)$ explicitly:

$$\begin{cases} b_i(V) = d_i\big(\phi_0(V) + c_i V\phi_1(V)\big), \\ \bar{b}_i(V) = d_i\big(\phi_1(V) - c_i\phi_0(V)\big), \quad i = 1, 2, \ldots, s. \end{cases} \tag{5.30}$$

Thus, if $d_i \neq 0$ for some i, then $b_i(V), \bar{b}_i(V) \neq 0$, and the third equation of conditions (5.18) indicates $\bar{a}_{ij} = 0$ for $j = 1, 2, \ldots, s$. Therefore, stage i with nonzero $b_i(V)$ and $\bar{b}_i(V)$, is independent of internal stages. On the other hand, if $d_i = 0$ for some i, then $b_i(V) = \bar{b}_i(V) = 0$ and the i-stage can neither contribute to the updates of ARKN methods, nor to any other internal stages with nonzero $b_i(V)$ and $\bar{b}_i(V)$. Hence, in the rest of this section, we always assume that $d_i \neq 0$ for $i = 1, 2, \ldots, s$.

Before going on to the analysis of symplecticiy of the adjoint integrator of a multi-frequency and multi-dimensional ARKN method, we present the following theorem.

Theorem 5.4 *If the coefficients of an s-stage multi-frequency and multi-dimensional ARKN method satisfy conditions* (5.18), *i.e., the method is symplectic, then,* $c_1 = c_2 = \cdots = c_s$.

Proof From (5.30), choosing arbitrary i and j, we have

$$b_i(V)\bar{b}_j(V) = d_i d_j \big(\phi_0(V) + c_i V\phi_1(V)\big)\big(\phi_1(V) - c_j\phi_0(V)\big),$$
$$b_j(V)\bar{b}_i(V) = d_i d_j \big(\phi_0(V) + c_j V\phi_1(V)\big)\big(\phi_1(V) - c_i\phi_0(V)\big).$$

On noticing the fourth equation of conditions (5.18), we obtain

$$\big(\phi_0(V) + c_i V\phi_1(V)\big)\big(\phi_1(V) - c_j\phi_0(V)\big) = \big(\phi_0(V) + c_j V\phi_1(V)\big)\big(\phi_1(V) - c_i\phi_0(V)\big),$$

or

$$(c_i - c_j)\big(\phi_0^2(V) + V\phi_1^2(V)\big) = 0.$$

Using (5.5), we have $c_i = c_j$ for arbitrary i and j. □

By Remark 5.2 and Theorem 5.4, we present the following conclusion.

Theorem 5.5 *The multi-frequency and multi-dimensional symplectic ARKN method has only one stage.*

Proof From Remark 5.2 we have $\bar{a}_{ij} = 0$ for $i, j = 1, 2, \ldots, s$, and from Theorem 5.4 we have $c_1 = c_2 = \cdots = c_s$. □

By Theorem 5.5, it can be verified that the order of a multi-frequency and multi-dimensional symplectic ARKN cannot exceed two (see [20]).

We are now in a position to present the analysis on the symplecticiy of the adjoint integrator for a multi-frequency and multi-dimensional ARKN method. It follows from Theorems 5.3 and 5.5 that a multi-frequency, multi-dimensional, symplectic ARKN method has the following form

$$\begin{cases} Q_1 = q_n + c_1 h q_n', \\ q_{n+1} = \phi_0(V)q_n + h\phi_1(V)q_n' + h^2\bar{b}_1(V)f(Q_1), \\ q_{n+1}' = \phi_0(V)q_n' - hK\phi_1(V)q_n + hb_1(V)f(Q_1), \end{cases} \tag{5.31}$$

where

$$b_1(V) = d_1\big(\phi_0(V) + c_1 V\phi_1(V)\big),$$
$$\bar{b}_1(V) = d_1\big(\phi_1(V) - c_1\phi_0(V)\big),$$

and the corresponding adjoint integrator is given by

$$\begin{cases} Q_1 = \big(\phi_0(V) + c_1 V\phi_1(V)\big)q_n + \big(\phi_1(V) - c_1\phi_0(V)\big)hq_n' \\ q_{n+1} = \phi_0(V)q_n + h\phi_1(V)q_n' + h^2\bar{b}_1^*(V)f(Q_1), \\ q_{n+1}' = -hK\phi_1(V)q_n + \phi_0(V)q_n' + hb_1^*(V)f(Q_1), \end{cases} \tag{5.32}$$

where

$$\begin{cases} \bar{b}_1^*(V) = \phi_1(V)b_1(V) - \phi_0(V)\bar{b}_1(V) = c_1 d_1 I, \\ b_1^*(V) = V\phi_1(V)\bar{b}_1(V) + \phi_0(V)b_1(V) = d_1 I. \end{cases}$$

By (5.32), we have the following result on the symplecticiy of the adjoint integrator for a multi-frequency and multi-dimensional ARKN method.

Theorem 5.6 *The adjoint integrator* (5.32) *of a multi-frequency and multi-dimensional symplectic ARKN method* (5.31) *is symplectic.*

Proof By Theorem 5.5, a symplectic ARKN method has only one stage. Hence, it is easy to verify that its adjoint method is symplectic and we omit details. □

From the analysis described above, it is clear that high-order symplectic ARKN methods do not exist. Furthermore, since the adjoint integrator of an ARKN method is not again an ARKN method, an ARKN method cannot be symmetric. To get through this barrier, we can resort to the composition of ARKN methods to obtain high-order symplectic and symmetric methods, although they are not ARKN methods any more.

Consider the following one-stage ARKN method of order two, which can be thought of as an extended version of the Störmer–Verlet method [25]:

$$\begin{cases} Q_1 = q_n + \dfrac{1}{2}hq_n', \\ q_{n+1} = \phi_0(V)q_n + \phi_1(V)(hq_n') + h^2\big(\phi_1(V) - \dfrac{1}{2}\phi_0(V)\big)f(Q_1), \\ q_{n+1}' = -hK\phi_1(V)q_n + \phi_0(V)q_n' + h\big(\phi_0(V) + \dfrac{1}{2}V\phi_1(V)\big)f(Q_1). \end{cases} \tag{5.33}$$

It can be observed that (5.33) is a symplectic method and its adjoint integrator is also symplectic by Theorem 5.6. If we let $V \to \mathbf{0}$ in (5.33), then (5.33) reduces to the Störmer–Verlet formula for (5.1) or (5.16):

$$\begin{cases} Q_1 = q_n + \dfrac{h}{2}q_n', \\ q_{n+1} = q_n + hq_n' + \dfrac{h^2}{2}g(Q_1), \\ q_{n+1}' = q_n' + hg(Q_1), \end{cases} \tag{5.34}$$

where $g(q) = f(q) - Kq$.

Using (5.33) as the basic method, we consider the fourth order symmetric composition of the form (5.3) with the coefficients (see, e.g. [10])

$$\alpha_1 = \beta_3 = \frac{1}{2(2-2^{1/3})}, \quad \alpha_2 = \beta_2 = -\frac{2^{1/3}}{2(2-2^{1/3})}, \quad \alpha_3 = \beta_1 = \frac{1}{2(2-2^{1/3})}. \tag{5.35}$$

We denote the composition of (5.33) with coefficients (5.35) by CARKNp4s6.

As pointed out in [10], for achieving a composition method of high order, the solutions with the minimal number of stages do not give the best methods. Thus, we consider the following fourth order symmetric composition coefficients given in [2]:

$$\begin{aligned}
&\alpha_1 = \beta_6 = 0.16231455076687, \ \alpha_2 = \beta_5 = 0.37087741497958,\\
&\alpha_3 = \beta_4 = 0.059762097006575, \ \alpha_4 = \beta_3 = -0.40993371990193,\\
&\alpha_5 = \beta_2 = 0.23399525073150, \ \alpha_6 = \beta_1 = 0.082984406417405.
\end{aligned} \tag{5.36}$$

We denote the composition of (5.33) with the coefficients (5.36) by CARKNp4s12. Both CARKNp4s6 and CARKNp4s12 methods are symplectic and symmetric and of order four.

5.3 Composition of ERKN Integrators

Another class of efficient methods for the oscillatory system (5.1) is the so-called multi-frequency and multi-dimensional ERKN integrators (see [30]). ERKN integrators are designed by taking advantage of the special structure brought by Kq in both the updates and the internal stages. In light of the variation-of-constants formula (5.8)–(5.9) for (5.1), improving both the internal stages and updates of a classical RKN method leads to the definition of ERKN integrators.

Definition 5.2 An s-stage multi-frequency and multi-dimensional ERKN integrator with stepsize h for oscillatory system (5.1) is defined by

$$\begin{cases}
Q_i = \phi_0(c_i^2 V)q_n + hc_i\phi_1(c_i^2 V)q_n' + h^2 \sum\limits_{j=1}^{s} a_{ij}(V) f(Q_j), \quad i = 1, \ldots, s,\\
q_{n+1} = \phi_0(V)q_n + h\phi_1(V)q_n' + h^2 \sum\limits_{i=1}^{s} \bar{b}_i(V) f(Q_i),\\
q_{n+1}' = -hK\phi_1(V)q_n + \phi_0(V)q_n' + h \sum\limits_{i=1}^{s} b_i(V) f(Q_i),
\end{cases} \tag{5.37}$$

where $b_i, \bar{b}_i$ for $i = 1, \ldots, s$, and a_{ij} for $i, j = 1, \ldots, s$ are matrix-valued functions of $V = h^2 K$.

The order conditions for multi-frequency and multi-dimensional ERKN integrators can be found in [26]. We are now concerned with the adjoint integrators of ERKN integrators.

Theorem 5.7 *The adjoint integrator of an s-stage multi-frequency and multi-dimensional ERKN integrator* (5.37) *with stepsize h has the following form:*

$$\begin{cases} Q_i = \phi_0(c_i^{*2}V)q_n + c_i^*\phi_1(c_i^{*2}V)hq_n' + h^2\sum\limits_{j=1}^{s} a_{ij}^*(V)f(Q_j), i = 1, 2, \ldots, s, \\ q_{n+1} = \phi_0(V)q_n + h\phi_1(V)q_n' + h^2\sum\limits_{i=1}^{s} \bar{b}_i^*(V)f(Q_i), \\ q_{n+1}' = -hM\phi_1(V)q_n + \phi_0(V)q_n' + h\sum\limits_{i=1}^{s} b_i^*(V)f(Q_i), \end{cases} \tag{5.38}$$

where

$$\begin{cases} c_i^* = 1 - c_{s+1-i}, \\ \bar{b}_i^*(V) = \phi_1(V)b_{s+1-i}(V) - \phi_0(V)\bar{b}_{s+1-i}(V), \\ b_i^*(V) = V\phi_1(V)\bar{b}_{s+1-i}(V) + \phi_0(V)b_{s+1-i}(V), \\ a_{ij}^*(V) = \phi_0(c_{s+1-i}^2 V)\bar{b}_j^*(V) - c_{s+1-i}\phi_1(c_{s+1-i}^2 V)b_j^*(V) + a_{s+1-i,s+1-j}(V), \\ \quad i,\ j = 1, 2, \ldots, s. \end{cases} \tag{5.39}$$

Proof The proof is similar to that of Theorem 5.2 and we omit the details. □

From Theorem 5.7, it can be observed that the adjoint integrator of a multi-frequency and multi-dimensional ERKN integrator is again an ERKN integrator.

With regard to the symplecticity conditions of ERKN integrators, we have the following theorem [28].

Theorem 5.8 *An s-stage multi-frequency and multi-dimensional ERKN integrator* (5.37) *is symplectic if its coefficients satisfy the following conditions:*

$$\begin{cases} \phi_0(V)b_i(V) + V\phi_1(V)\bar{b}_i(V) = d_i\phi_0(c_i^2V),\ d_i \in \mathbb{R},\ i = 1, 2, \ldots, s, \\ \phi_1(V)b_i(V) - \phi_0(V)\bar{b}_i(V) = c_i d_i \phi_1(c_i^2V),\ i = 1, 2, \ldots, s, \\ \bar{b}_i(V)b_j(V) + d_i a_{ij}(V) = \bar{b}_j(V)b_i(V) + d_j a_{ji}(V),\ i,\ j = 1, 2, \ldots, s, \end{cases} \tag{5.40}$$

where $V = h^2K$.

Remark 5.3 From the first two equations in symplectic conditions (5.40), we can solve $b_i(V), \bar{b}_i(V)$ explicitly:

$$\begin{aligned} b_i(V) &= d_i\big(\phi_0(V)\phi_0(c_i^2V) + c_iV\phi_1(V)\phi_1(c_i^2V)\big), \\ \bar{b}_i(V) &= d_i\big(\phi_1(V)\phi_0(c_i^2V) - c_i\phi_0(V)\phi_1(c_i^2V)\big),\ i = 1, 2, \ldots, s. \end{aligned}$$

Thus, by (5.6), we have

$$b_i(V) = d_i\phi_0((1-c_i)^2V),\ \bar{b}_i(V) = d_i(1-c_i)\phi_1((1-c_i)^2V),\ i = 1, 2, \ldots, s. \tag{5.41}$$

In what follows, we give an analysis of the symplecticiy of the adjoint integrator of a multi-frequency and multi-dimensional ERKN integrator.

Theorem 5.9 *If the coefficients of an s-stage multi-frequency and multi-dimensional ERKN integrator satisfy conditions* (5.40), *i.e., the integrator is symplectic, then its adjoint integrator* (5.38) *is also symplectic.*

Proof It is sufficient to prove that the coefficients of (5.38) satisfy the symplecticity conditions (5.40). In fact, by (5.39) and (5.40), the coefficients of the adjoint integrator of a symplectic ERKN integrator satisfy

$$\begin{cases} c_i^* = 1 - c_{s+1-i}, \\ \bar{b}_i^*(V) = c_{s+1-i}d_{s+1-i}\phi_1(c_{s+1-i}^2V), \\ b_i^*(V) = d_{s+1-i}\phi_0(c_{s+1-i}^2V), \\ a_{ij}^*(V) = c_{s+1-j}d_{s+1-j}\phi_0(c_{s+1-i}^2V)\phi_1(c_{s+1-j}^2V) \\ \qquad - c_{s+1-i}d_{s+1-j}\phi_1(c_{s+1-i}^2V)\phi_0(c_{s+1-j}^2V) + a_{s+1-i,s+1-j}(V), \\ \qquad i, j = 1, 2, \ldots, s. \end{cases}$$

We then have

$$\begin{cases} \phi_0(V)b_i^*(V) + V\phi_1(V)\bar{b}_i^*(V) \\ = \phi_0(V)d_{s+1-i}\phi_0(c_{s+1-i}^2V) + V\phi_1(V)c_{s+1-i}d_{s+1-i}\phi_1(c_{s+1-i}^2V) \\ = d_{s+1-i}\phi_0((1-c_{s+1-i})^2V) = d_{s+1-i}\phi_0(c_i^{*2}V),\ i = 1, 2, \ldots, s, \\ \phi_1(V)b_i^*(V) - \phi_0(V)\bar{b}_i^*(V) \\ = \phi_1(V)d_{s+1-i}\phi_0(c_{s+1-i}^2V) - \phi_0(V)c_{s+1-i}d_{s+1-i}\phi_1(c_{s+1-i}^2V) \\ = d_{s+1-i}(1-c_{s+1-i})\phi_1((1-c_{s+1-i})^2V) = d_{s+1-i}c_i^*\phi_1(c_i^{*2}V), \\ i = 1, 2, \ldots, s. \end{cases} \tag{5.42}$$

Let $d_i^* = d_{s+1-i}$. Then, the first two conditions are satisfied in (5.40). Concerning the third condition in (5.40), we have

$$\begin{aligned} &\bar{b}_i^*(V)b_j^*(V) + d_i^*a_{ij}^*(V) - (\bar{b}_j^*(V)b_i^*(V) + d_j^*a_{ji}^*(V)) \\ &= c_{s+1-i}d_{s+1-i}\phi_1(c_{s+1-i}^2V)d_{s+1-j}\phi_0(c_{s+1-j}^2V) \\ &\quad - c_{s+1-j}d_{s+1-j}\phi_1(c_{s+1-j}^2V)d_{s+1-i}\phi_0(c_{s+1-i}^2V) \\ &\quad + d_{s+1-i}\big(c_{s+1-j}d_{s+1-j}\phi_0(c_{s+1-i}^2V)\phi_1(c_{s+1-j}^2V) \\ &\quad - c_{s+1-i}d_{s+1-j}\phi_1(c_{s+1-i}^2V)\phi_0(c_{s+1-j}^2V) + a_{s+1-i,s+1-j}(V)\big) \\ &\quad - d_{s+1-j}\big(c_{s+1-i}d_{s+1-i}\phi_0(c_{s+1-j}^2V)\phi_1(c_{s+1-i}^2V) \end{aligned} \tag{5.43}$$

$$
\begin{aligned}
&- c_{s+1-j}d_{s+1-i}\phi_1(c^2_{s+1-j}V)\phi_0(c^2_{s+1-i}V) + a_{s+1-j,s+1-i}(V)\big)\\
&= d_{s+1-i}\big(c_{s+1-j}d_{s+1-j}\phi_0(c^2_{s+1-i}V)\phi_1(c^2_{s+1-j}V) + a_{s+1-i,s+1-j}(V)\big)\\
&- d_{s+1-j}\big(c_{s+1-i}d_{s+1-i}\phi_0(c^2_{s+1-j}V)\phi_1(c^2_{s+1-i}V) + a_{s+1-j,s+1-i}(V)\big)\\
&= d_{s+1-i}d_{s+1-j}(c_{s+1-j} - c_{s+1-i})\phi_1((c_{s+1-j} - c_{s+1-i})^2V)\\
&+ d_{s+1-i}a_{s+1-i,s+1-j}(V) - d_{s+1-j}a_{s+1-j,s+1-i}(V).
\end{aligned}
$$

By (5.41) and the third equation of (5.40), we obtain

$$
\begin{aligned}
&d_{s+1-i}a_{s+1-i,s+1-j}(V) - d_{s+1-j}a_{s+1-j,s+1-i}(V)\\
&= \bar{b}_{s+1-j}(V)b_{s+1-i}(V) - \bar{b}_{s+1-i}(V)b_{s+1-j}(V)\\
&= d_{s+1-i}d_{s+1-j}(1 - c_{s+1-j})\phi_1((1 - c_{s+1-j})^2V)\phi_0((1 - c_{s+1-i})^2V)\\
&- d_{s+1-i}d_{s+1-j}(1 - c_{s+1-i})\phi_1((1 - c_{s+1-i})^2V)\phi_0((1 - c_{s+1-j})^2V)\\
&= d_{s+1-i}d_{s+1-j}(c_{s+1-i} - c_{s+1-j})\phi_1((c_{s+1-i} - c_{s+1-j})^2V).
\end{aligned}
\tag{5.44}
$$

Substituting (5.44) into (5.43) yields

$$
\bar{b}^*_i(V)b^*_j(V) + d^*_i a^*_{ij}(V) - (\bar{b}^*_j(V)b^*_i(V) + d^*_j a^*_{ji}(V)) = 0.
$$

The proof is complete. □

Section 5.2 remarks that the composition of an ARKN method is not again an ARKN method. However, the composition of an ERKN integrator is still an ERKN integrator, as shown in the next theorem.

Theorem 5.10 *The composition of an s_1-stage ERKN integrator with stepsize αh and an s_2-stage ERKN integrator with stepsize βh is a new ERKN integrator with $s = s_1 + s_2$ stages and stepsize $(\alpha + \beta)h$.*

Proof Let the two ERKN integrators with stepsizes αh and βh be

$$
\begin{cases}
Q_i = \phi_0(c_i^2\alpha^2h^2K)q_n + \alpha h c_i\phi_1(c_i^2\alpha^2h^2K)q'_n + \alpha^2h^2\sum\limits_{j=1}^{s_1} a_{ij}(\alpha^2h^2K)f(Q_j),\\
\qquad\qquad\qquad\qquad\qquad\qquad\qquad\qquad\qquad\qquad i = 1,\ldots,s_1,\\
q_{n+1} = \phi_0(\alpha^2h^2K)q_n + \alpha h\phi_1(\alpha^2h^2K)q'_n + \alpha^2h^2\sum\limits_{i=1}^{s_1}\bar{b}_i(\alpha^2h^2K)f(Q_i),\\
q'_{n+1} = -\alpha hK\phi_1(\alpha^2h^2K)q_n + \phi_0(\alpha^2h^2K)q'_n + \alpha h\sum\limits_{i=1}^{s_1} b_i(\alpha^2h^2K)f(Q_i),
\end{cases}
\tag{5.45}
$$

and

$$
\begin{cases}
Q_i^* = \phi_0(c_i^{*2}\beta^2h^2K)q_n + \beta hc_i^*\phi_1(c_i^{*2}\beta^2h^2K)q_n' + \beta^2h^2\sum\limits_{j=1}^{s_2} a_{ij}^*(\beta^2h^2K)f(Q_j^*), \\
\qquad\qquad\qquad\qquad\qquad\qquad\qquad\qquad\qquad i = 1,\dots,s_2, \\
q_{n+1} = \phi_0(\beta^2h^2K)q_n + \beta h\phi_1(\beta^2h^2K)q_n' + \beta^2h^2\sum\limits_{i=1}^{s_2} \bar{b}_i^*(\beta^2h^2K)f(Q_i^*), \\
q_{n+1}' = -\beta hK\phi_1(\beta^2h^2K)q_n + \phi_0(\beta^2h^2K)q_n' + \beta h\sum\limits_{i=1}^{s_2} b_i^*(\beta^2h^2K)f(Q_i^*).
\end{cases}
\tag{5.46}
$$

Denote them by $\varphi_{h_1}^1$ and $\varphi_{h_2}^2$, respectively. Then $\varphi_{h_2}^2 \circ \varphi_{h_1}^1$ has the following form:

$$
\begin{cases}
Q_i = \phi_0(c_i^2\alpha^2h^2K)q_n + \alpha hc_i\phi_1(c_i^2\alpha^2h^2K)q_n' + \alpha^2h^2\sum\limits_{j=1}^{s_1} a_{ij}(\alpha^2h^2K)f(Q_j), \\
\qquad\qquad\qquad\qquad\qquad\qquad\qquad i = 1,\dots,s_1, \\
\tilde{q}_{n+1} = \phi_0(\alpha^2h^2K)q_n + \alpha h\phi_1(\alpha^2h^2K)q_n' + \alpha^2h^2\sum\limits_{i=1}^{s_1} \bar{b}_i(\alpha^2h^2K)f(Q_i), \\
\tilde{q}_{n+1}' = -\alpha hK\phi_1(\alpha^2h^2K)q_n + \phi_0(\alpha^2h^2K)q_n' + \alpha h\sum\limits_{i=1}^{s_1} b_i(\alpha^2h^2K)f(Q_i), \\
Q_i^* = \phi_0(c_i^{*2}\beta^2h^2K)\tilde{q}_{n+1} + \beta hc_i^*\phi_1(c_i^{*2}\beta^2h^2K)\tilde{q}_{n+1}' + \beta^2h^2\sum\limits_{j=1}^{s_2} a_{ij}^*(\beta^2h^2K)f(Q_j^*), \\
\qquad\qquad\qquad\qquad\qquad\qquad\qquad i = 1,\dots,s_2, \\
q_{n+1} = \phi_0(\beta^2h^2K)\tilde{q}_{n+1} + \beta h\phi_1(\beta^2h^2K)\tilde{q}_{n+1}' + \beta^2h^2\sum\limits_{i=1}^{s_2} \bar{b}_i^*(\beta^2h^2K)f(Q_i^*), \\
q_{n+1}' = -\beta hK\phi_1(\beta^2h^2K)\tilde{q}_{n+1} + \phi_0(\beta^2h^2K)\tilde{q}_{n+1}' + \beta h\sum\limits_{i=1}^{s_2} b_i^*(\beta^2h^2K)f(Q_i^*).
\end{cases}
\tag{5.47}
$$

Let $Q_{s_1+i} = Q_i^*$ for $i = 1,\dots,s_2$ and $s = s_1 + s_2$. Substituting the second and third terms into the last three terms of (5.47), with some tedious computations and manipulations, we obtain

$$
\begin{cases}
Q_i = \phi_0(c_i^2\alpha^2h^2K)q_n + \alpha hc_i\phi_1(c_i^2\alpha^2h^2K)q_n' + \alpha^2h^2\sum\limits_{j=1}^{s_1} a_{ij}(\alpha^2h^2K)f(Q_j), \\
\qquad\qquad\qquad\qquad\qquad\qquad\qquad i = 1,\dots,s_1, \\
Q_i = \phi_0((\alpha + c_i^*\beta)^2h^2K)q_n + (\alpha + c_i^*\beta)h\phi_1((\alpha + c_i^*\beta)^2h^2K)q_n' + h^2\sum\limits_{j=1}^{s} \tilde{a}_{ij}f(Q_j), \\
\qquad\qquad\qquad\qquad i = s_1 + 1,\dots,s, \\
q_{n+1} = \phi_0((\alpha+\beta)^2h^2K)q_n + (\alpha+\beta)h\phi_1((\alpha+\beta)^2h^2K)q_n' + (\alpha+\beta)^2h^2\sum\limits_{i=1}^{s} \tilde{\bar{b}}_i f(Q_i), \\
q_{n+1}' = -(\alpha+\beta)hK\phi_1((\alpha+\beta)^2h^2K)q_n + \phi_0((\alpha+\beta)^2h^2K)q_n' + (\alpha+\beta)h\sum\limits_{i=1}^{s} \tilde{b}_i f(Q_i),
\end{cases}
\tag{5.48}
$$

where $\tilde{a}_{ij}, \tilde{\bar{b}}_i, \tilde{b}_i$ for $i, j = 1,\dots,s$ are the algebraic compositions of the coefficients of the two ERKN integrators, that makes them the matrix-valued functions of $V = h^2K$. □

Table 5.1 The number of equations from order conditions for order $p = 1, \dots, 8$

Order p	1	2	3	4	5	6	7	8
Number of equations	1	3	6	11	21	40	79	157

It should be noted here that although there exist high-order symplectic and symmetric ERKN integrators, the order conditions together with the symmetry and symplecticity conditions are very important for achieving a high-order symplectic and symmetric ERKN integrator. Table 5.1 shows the number of equations that need to be solved from order conditions. It can be observed that, as the order of the method grows, the number of equations from the order conditions increases rapidly, not to mention the equations from the symmetry and symplecticity conditions (the number of equations depends on how many stages the integrator uses; the number of equations may be reduced but the number is still very large). Therefore, in practice, the derivation of a high-order symplectic and symmetric ERKN integrator based on order conditions, symmetry conditions and symplecticity conditions is very difficult. However, it will be useful to generate high-order symplectic and symmetric ERKN integrators by using a procedure of composition.

Consider the following one-stage ERKN integrator of order two, which is another extended version of the Störmer–Verlet method [25]

$$\begin{cases} Q_1 = \phi_0(\frac{V}{4})q_n + \frac{1}{2}h\phi_1(\frac{V}{4})q_n', \\ q_{n+1} = \phi_0(V)q_n + \phi_1(V)(hq_n') + \frac{h^2}{2}\phi_1(\frac{V}{4})f(Q_1), \\ q_{n+1}' = -hK\phi_1(V)q_n + \phi_0(V)q_n' + h\phi_0(\frac{V}{4})f(Q_1), \end{cases} \tag{5.49}$$

It can be verified that (5.49) is symplectic and symmetric. We note that letting $V \to \mathbf{0}$ in (5.49) also gives the Störmer–Verlet formula (5.34).

Using (5.49) as the basic method, since it is symmetric, we consider the sixth order symmetric composition of the form (5.2) with coefficients given in [33]

$$\begin{aligned} &\gamma_1 = \gamma_7 = 0.78451361047755726381949763, \ \gamma_2 = \gamma_6 = 0.23557321335935813368479318, \\ &\gamma_3 = \gamma_5 = -1.17767998417887100694641568, \ \gamma_4 = 1.31518632068391121888424973. \end{aligned} \tag{5.50}$$

We denote the composition of (5.49) with the coefficients (5.50) by CERKNp6s7. The method CERKNp6s7 is symplectic and symmetric of order six. Moreover, it can be seen from Theorem 5.10 that CERKNp6s7 is a seven-stage multi-frequency and multi-dimensional ERKN integrator of order six.

We also consider the following eighth order symmetric composition coefficients [10]:

$$\begin{aligned}
&\gamma_1 = \gamma_{15} = 0.74167036435061295344822780, \ \gamma_2 = \gamma_{14} = -0.40910082580003159399730010, \\
&\gamma_3 = \gamma_{13} = 0.19075471029623837995387626, \ \gamma_4 = \gamma_{12} = -0.57386247111608226665638773, \\
&\gamma_5 = \gamma_{11} = 0.29906418130365592384446354, \ \gamma_6 = \gamma_{10} = 0.33462491824529818378495798, \\
&\gamma_7 = \gamma_9 = 0.31529309239676659663205666, \ \gamma_8 = -0.79688793935291635401978884.
\end{aligned} \tag{5.51}$$

We denote the composition of (5.49) with the coefficients (5.51) by CERKNp8s15 which is a symplectic and symmetric ERKN integrator of order eight.

Remark 5.4 The second-order symmetric Gautschi-type method is also an alternative basic method for oscillatory problem. Take Deuflhard's Trigonometric Method (see [8]) for example, and let $\Omega = K^{1/2}$. The one-step form of Deuflhard's method for (5.1) reads

$$\begin{cases}
q_{n+1} = \cos(h\Omega)q_n + \Omega^{-1}\sin(h\Omega)q'_n + \dfrac{h^2}{2}(h\Omega)^{-1}\sin(h\Omega)f(q_n), \\
q'_{n+1} = -\Omega\sin(h\Omega)q_n + \cos(h\Omega)q'_n + \dfrac{1}{2}h(\cos(h\Omega)f(q_n) + f(q_{n+1})).
\end{cases} \tag{5.52}$$

The method is also symmetric and symplectic, which makes it a good option as the basic method. It is noted from the definition of the ϕ-functions that Deuflhard's method can be reformulated as

$$\begin{cases}
Q_1 = q_n, \\
Q_2 = \phi_0(V)q_n + \phi_1(V)(hq'_n) + \dfrac{h^2}{2}\phi_1(V)f(Q_1), \\
q_{n+1} = \phi_0(V)q_n + \phi_1(V)(hq'_n) + \dfrac{h^2}{2}\phi_1(V)f(Q_1), \\
q'_{n+1} = -hK\phi_1(V)q_n + \phi_0(V)q'_n + \dfrac{h}{2}(\phi_0(V)f(Q_1) + f(Q_2)).
\end{cases} \tag{5.53}$$

In other words, Deuflhard's method can be viewed as a two-stage ERKN integrator of order two with FSAL property (the last evaluation at any step is the same as the first evaluation at the next step). Thus it only needs one function evaluation per step.

5.4 Numerical Experiments

In order to show the robustness and efficiency of the symplectic and symmetric methods proposed in this chapter in comparison with the existing methods in the scientific literature, we use four problems in numerical experiments. The methods used for comparison are:

- CRKNp6s7: the composition of Störmer–Verlet method with coefficients (5.50);
- CRKNp8s15: the composition of Störmer–Verlet method with coefficients (5.51);
- CDeuflhardp6s7: the composition of Deuflhard's method with coefficients (5.50);

- CDeuflhardp8s15: the composition of Deuflhard's method with coefficients (5.51);
- SRKNp4s3: the three-stage symplectic Runge–Kutta–Nyström method of order four given in [11].

For each experiment, we will display the efficiency curves: accuracy versus the computational cost measured by the number of function evaluations required by each method and the energy error of each method. If the error is very large, we do not plot the points in the figure of the numerical results.

Problem 1 Consider the orbital problem with perturbation

$$
\begin{aligned}
q_1'' + q_1 &= -\frac{2\varepsilon + \varepsilon^2}{r^5} q_1, \quad q_1(0) = 1, \; q_1'(0) = 0, \\
q_2'' + q_2 &= -\frac{2\varepsilon + \varepsilon^2}{r^5} q_2, \quad q_2(0) = 0, \; q_2'(0) = 1 + \varepsilon,
\end{aligned}
$$

where $r = \sqrt{q_1^2 + q_2^2}$. This is a Hamiltonian system with the Hamiltonian

$$
H = \frac{1}{2} p^T p + \frac{1}{2} q^T K q + U(q),
$$

where

$$
U(q) = -\frac{2\varepsilon + \varepsilon^2}{3(q_1^2 + q_2^2)^{\frac{3}{2}}}, \qquad K = \begin{pmatrix} 1 & 0 \\ 0 & 1 \end{pmatrix}.
$$

The analytic solution is given by

$$
q_1(t) = \cos(t + \varepsilon t), \quad q_2(t) = \sin(t + \varepsilon t).
$$

The problem is solved on the interval [0, 1000] with $\varepsilon = 10^{-3}$. We take stepsizes $h = 1/2^j$ for the methods CERKNp8s15, CRKNp8s15, CDeuflhardp8s15, $h = 1/2^{j+1}$ for CERKNp6s7, CRKNp6s7, CDeuflhardp6s7, $h = 6/(15 \times 2^j)$ for CARKNp4s6, $h = 12/(15 \times 2^j)$ for CARKNp4s12, and $h = 3/(15 \times 2^j)$ for SRKNp4s3, where $j = 0, \ldots, 3$.

Figure 5.1(i) shows the error of the position q at $t_{end} = 1000$ versus the computational effort. We integrate this problem with stepsize $h = 1.5$ in the interval $[0, t_{end}]$, $t_{end} = 5 \times 10^i$ for $i = 0, \ldots, 3$. Figure 5.1(ii) shows the energy errors of different methods.

Problem 2 Consider the Fermi–Pasta–Ulam problem, which can be expressed by a Hamiltonian system with the Hamiltonian

$$
H(p, q) = \frac{1}{2} p^T p + \frac{1}{2} q^T K q + U(q),
$$

where

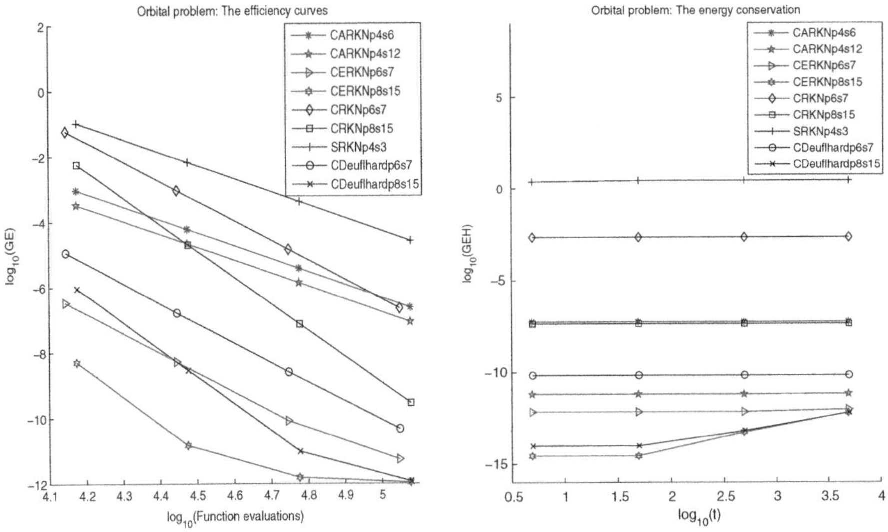

(i) (ii)

Fig. 5.1 Results for Problem 1. (i): The logarithm of the global error (GE) over the integration interval against the logarithm of the number of function evaluations. (ii): The logarithm of the maximum global error of Hamiltonian $GEH = \max |H_n - H_0|$ against $\log_{10}(t_{end})$

$$K = \begin{pmatrix} \mathbf{0}_{m\times m} & \mathbf{0}_{m\times m} \\ \mathbf{0}_{m\times m} & \omega^2 I_{m\times m} \end{pmatrix},$$

$$U(q) = \frac{1}{4}\left((q_1 - q_{m+1})^4 + \sum_{i=1}^{m-1} (q_{i+1} - q_{m+i+1} - q_i - q_{m+i})^4 + (q_m + q_{2m})^4\right).$$

Following [10], we choose $\omega = 100$ and

$$m = 3,\ q_1(0) = 1,\ p_1(0) = 1,\ q_4(0) = \frac{1}{\omega},\ p_4(0) = 1, \tag{5.54}$$

and choose zero for the remaining initial values.

The problem is integrated on the interval $[0, 20]$ with stepsizes $h = 0.08/2^j$ for the methods CERKNp8s15, CRKNp8s15, CDeuflhardp8s15, $h = 0.08/2^{j+1}$ for CERKNp6s7, CRKNp6s7, CDeuflhardp6s7, $h = 0.08 \times 6/(15 \times 2^j)$ for CARKN p4s6, $h = 0.08 \times 12/(15 \times 2^j)$ for CARKNp4s12, and $h = 0.08 \times 3/(15 \times 2^j)$ for SRKNp4s3, where $j = 0, \ldots, 3$.

Figure 5.2 (i) shows the error of the position q at $t_{end} = 20$ versus the computational effort. We integrate this problem with stepsize $h = 0.01$ on the interval $[0, t_{end}]$, $t_{end} = 10 \times 5^i$ for $i = 0, \ldots, 3$. Figure 5.2(ii) shows the energy errors of different methods.

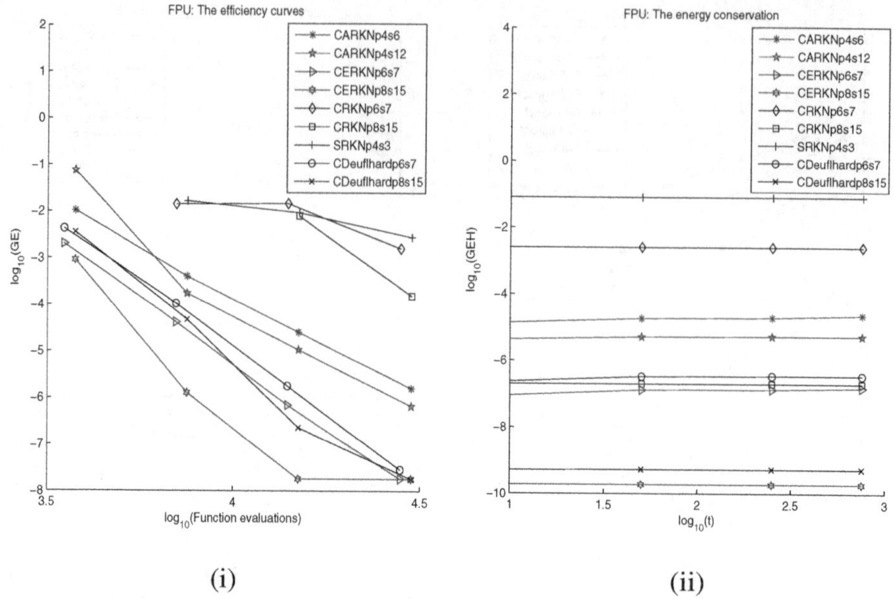

Fig. 5.2 Results for Problem 2. (i): The logarithm of the global error (GE) over the integration interval against the logarithm of the number of function evaluations. (ii): The logarithm of the maximum global error of Hamiltonian $GEH = \max |H_n - H_0|$ against $\log_{10}(t_{end})$

Problem 3 Consider the sine-Gordon equation with periodic boundary conditions

$$\begin{cases} \dfrac{\partial^2 u}{\partial t^2} = \dfrac{\partial^2 u}{\partial x^2} - \sin u, \quad -1 < x < 1, \quad t > 0, \\ u(-1, t) = u(1, t). \end{cases}$$

By semi-discretization on the spatial variable with second-order symmetric differences, and introducing generalized momenta $p = q'$, we obtain a Hamiltonian system with the Hamiltonian

$$H(p, q) = \frac{1}{2} p^\intercal p + \frac{1}{2} q^\intercal K q + U(q),$$

where $q(t) = (u_1(t), \ldots, u_d(t))^\intercal$ and $U(q) = -(\cos(u_1) + \ldots + \cos(u_d))$ with $u_i(t) \approx u(x_i, t)$, $x_i = -1 + i\Delta x$ for $i = 1, \ldots, d$, $\Delta x = 2/d$, and

$$K = \frac{1}{\Delta x^2} \begin{pmatrix} 2 & -1 & & & -1 \\ -1 & 2 & -1 & & \\ & \ddots & \ddots & \ddots & \\ & & -1 & 2 & -1 \\ -1 & & & -1 & 2 \end{pmatrix}. \tag{5.55}$$

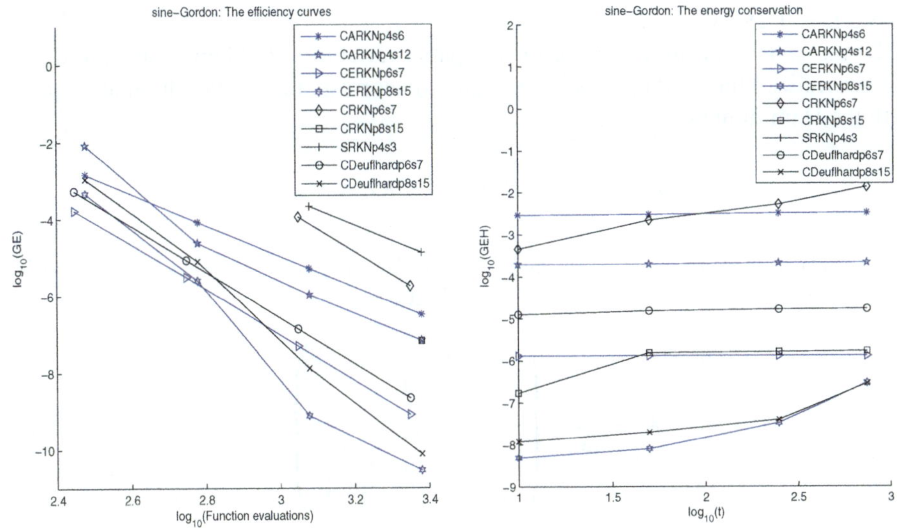

(i) (ii)

Fig. 5.3 Results for Problem 3. (i): The logarithm of the global error (GE) over the integration interval against the logarithm of the number of function evaluations. (ii): The logarithm of the maximum global error of Hamiltonian $GEH = \max |H_n - H_0|$ against $\log_{10}(t_{end})$

We take $d = 32$ and the initial conditions as

$$q(0) = (\pi)_{i=1}^{d}, \quad p(0) = \sqrt{d}\left(0.01 + \sin(\frac{2\pi i}{d})\right)_{i=1}^{d}.$$

The problem is integrated on the interval [0, 10] with stepsizes $h = 1/2^j$ for the methods CERKNp8s15, CRKNp8s15, CDeuflhardp8s15, $h = 1/2^{j+1}$ for CERKNp6s7, CRKNp6s7, CDeuflhardp6s7, $h = 6/(15 \times 2^j)$ for CARKNp4s6, $h = 12/(15 \times 2^j)$ for CARKNp4s12, and $h = 3/(15 \times 2^j)$ for SRKNp4s3, where $j = 1, \ldots, 4$.

Figure 5.3 (i) shows the error of the position q at $t_{end} = 10$ versus the computational effort. We integrate this problem with stepsize $h = 0.08$ in the interval $[0, t_{end}]$, $t_{end} = 10 \times 5^i$ for $i = 0, \ldots, 3$. Figure 5.3(ii) shows the energy errors of different methods.

Problem 4 Consider the nonlinear Klein-Gordon equation (see, e.g. [12])

$$\begin{cases} \dfrac{\partial^2 u}{\partial t^2} - \dfrac{\partial^2 u}{\partial x^2} + u + u^3 = 0, \quad 0 < x < L, \ t > 0, \\ u(x, 0) = A\big(1 + \cos(\frac{2\pi}{L}x)\big), \\ u_t(x, 0) = 0, \\ u(0, t) = u(L, t), \end{cases}$$

where $L = 1.28$, $A = 0.9$.

By the same semi-discretization on the spatial variable as Problem 3, and introducing generalized momenta $p = q'$, we obtain the corresponding Hamiltonian system with the Hamiltonian

$$H(p, q) = \frac{1}{2}p^{\intercal}p + \frac{1}{2}q^{\intercal}Kq + U(q),$$

where $q(t) = (u_1(t), \ldots, u_d(t))^{\intercal}$ and $U(q) = \frac{1}{2}u_1^2 + \frac{1}{4}u_1^4 + \ldots + \frac{1}{2}u_d^2 + \frac{1}{4}u_d^4$ with $u_i(t) \approx u(x_i, t)$, $x_i = i\Delta x$ for $i = 1, \ldots, d$, $\Delta x = L/d$,

$$K = \frac{1}{\Delta x^2}\begin{pmatrix} 2 & -1 & & & -1 \\ -1 & 2 & -1 & & \\ & \ddots & \ddots & \ddots & \\ & & -1 & 2 & -1 \\ -1 & & & -1 & 2 \end{pmatrix}. \tag{5.56}$$

and the initial conditions are

$$q(0) = \Big(0.9\big(1 + \cos(\frac{2\pi i}{d})\big)\Big)_{i=1}^{d}, \quad p(0) = \big(0\big)_{i=1}^{d}.$$

We take $d = 32$, and integrate the problem on the interval $[0, 10]$ with stepsizes $h = 1/2^j$ for the methods CERKNp8s15, CRKNp8s15, CDeuflhardp8s15, $h = 1/2^{j+1}$ for CERKNp6s7, CRKNp6s7, CDeuflhardp6s7, $h = 6/(15 \times 2^j)$ for CARKNp4s6, $h = 12/(15 \times 2^j)$ for CARKNp4s12, and $h = 3/(15 \times 2^j)$ for SRKNp4s3, where $j = 3, \ldots, 6$.

Figure 5.4(i) shows the error of the position q at $t_{end} = 10$ versus the computational effort. We then integrate the problem with stepsize $h = 0.08$ on the interval $[0, t_{end}]$, $t_{end} = 10 \times 5^i$ for $i = 0, \ldots, 3$. Figure 5.4(ii) shows the energy errors of the different methods.

It follows from the numerical results shown in Figs. 5.1, 5.2, 5.3 and 5.4 that, overall, for the problems under consideration, the symmetric and symplectic composition methods based on ARKN and ERKN methods (extended Störmer–Verlet method, Deuflhard's method) give more efficient and accurate qualitative features than the classical symmetric and symplectic composition methods based on the RKN method (Störmer–Verlet method). In addition, the classical method SRKNp4s3 gives the poorest results for both efficiency and energy conservation. This fact shows that taking account of the oscillatory structure is a significant factor to be considered in the numerical integration of oscillatory Hamiltonian systems. On the other hand, from Figs. 5.1 and 5.4, it is clear that for Problems 1 and 4, the symmetric and symplectic composition method CRKNp8s15 of order eight is more efficient than the ARKN-based symmetric and symplectic composition methods CARKNp4s6 and CARKNp4s12 of order four. This means that apart from symplecticity, symmetry,

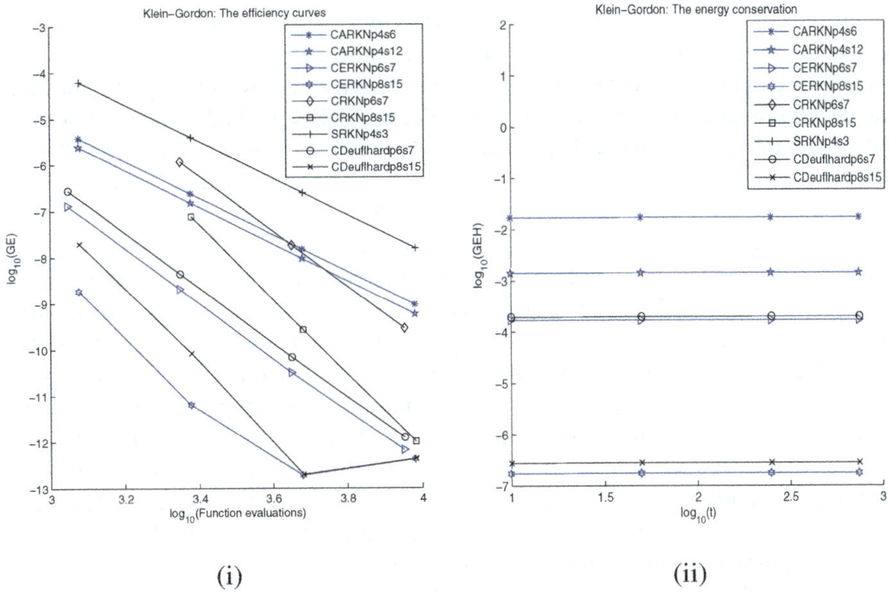

Fig. 5.4 Results for Problem 4. (i): The logarithm of the global error (GE) over the integration interval against the logarithm of the number of function evaluations. (ii): The logarithm of the maximum global error of Hamiltonian $GEH = \max |H_n - H_0|$ against $\log_{10}(t_{end})$

adaption to oscillation, the algebraic order cannot be ignored when designing efficient numerical methods.

For a large stepsize, it can be observed from the results of the numerical experiments that the composition methods based on ARKN methods and ERKN integrators perform very well. However, the traditional composition methods give unsatisfactory qualitative behavior. Therefore, the symmetric and symplectic composition methods based on ARKN methods and ERKN integrators are more suitable for the long-term integration of oscillatory Hamiltonian systems.

5.5 Conclusions and Discussions

For solving a multi-frequency and multi-dimensional oscillatory second-order initial value problem of the form $q'' + Kq = f(q)$, multi-frequency and multi-dimensional ARKN methods and multi-frequency and multi-dimensional ERKN integrators are incorporated with the special structure brought by the linear term Kq. These integrators exactly integrate the multi-frequency oscillatory homogeneous system $q'' + Kq = 0$. The sympletic conditions of multi-frequency and multi-dimensional ERKN integrators were presented for second-order oscillatory Hamiltonian systems (see [28]). This chapter derived the symplectic conditions for multi-frequency and multi-dimensional ARKN methods. Furthermore, the symplectic conditions for the adjoint integrators of multi-frequency and multi-dimensional symplectic ARKN methods and symplectic ERKN integrators were analysed, respectively. We showed that the adjoint method of the one-stage symplectic ARKN method is still symplectic. The adjoint integrator of a multi-frequency and multi-dimensional symplectic ERKN integrator is also symplectic. With these properties, we analysed and derived four new high-order symplectic and symmetric methods by the composition of ARKN and ERKN integrators. The numerical results support the theoretical analysis and show that these new composition methods are more efficient than the composition methods of traditional RKN methods when applied to multi-frequency and multi-dimensional oscillatory Hamiltonian systems. Last but not least, we again point out that a symplectic ARKN method or a symplectic ERKN integrator for a single-frequency oscillatory Hamiltonian system cannot ensure itself to be again a symplectic method when applied to a multi-frequency and multi-dimensional oscillatory Hamiltonian system.

This chapter focused on symplectic and symmetric composition integrators of RKN-type methods for multi-frequency oscillatory Hamiltonian systems. The next chapter will be concerned with the construction of arbitrary order ERKN integrators, including symplectic and symmetric ERKN methods.

The material of this chapter is based on the work by Liu and Wu [14].

References

1. Blanes, S., Casas, F., Murua, A.: Splitting and composition methods in the numerical integration of differential equations. Bol. Soc. Esp. Mat. Apl. **45**, 89–145 (2008)
2. Blanes, S., Moan, P.C.: Practical symplectic partitioned Runge-Kutta and Runge-Kutta(-Nyström) methods. J. Comput. Appl. Math. **142**, 313–330 (2002)
3. Bridges, T.J., Reich, S.: Multi-symplectic integrators: numerical schemes for Hamiltonian PDEs that conserve symplecticity. Phys. Lett. A **284**, 184–193 (2001)
4. Bridges, T.J., Reich, S.: Numerical methods for Hamiltonian PDEs. J. Phys. A: Math. Gen. **39**, 5287–5320 (2006)
5. Calvo, M.P., Chartier, P., Murua, A., Sanz-Serna, J.M.: Numerical stroboscopic averaging for ODEs and DAEs. Appl. Numer. Math. **61**, 1077–1095 (2011)
6. Calvo, M.P., Chartier, P., Murua, A., Sanz-Serna, J.M.: A stroboscopic numerical method for highly oscillatory problems. In: Engquist, B., Runborg, O., Tsai, R. (eds.) Numerical Analysis and Multiscale Computations. Lecture Notes in Computational Science and Engineering, vol. 82, pp. 73–87. Springer, Berlin (2011)
7. Calvo, M.P., Sanz-Serna, J.M.: Heterogeneous multiscale methods for mechanical systems with vibrations. SIAM J. Sci. Comput. **32**, 2029–2046 (2011)
8. Deuflhard, P.: A study of extrapolation methods based on multistep schemes without parasitic solutions. Z. angew. Math. Phys. **30**, 177–189 (1979)
9. Engquist, W.E.B., Li, X., Ren, W., Vanden-Eijnden, E.: Heterogeneous multiscale methods: a review. Appl. Numer. Math. **2**, 367–450 (2007)
10. Hairer, E., Lubich, C., Wanner, G.: Geometric Numerical Integration: Structure-Preserving Algorithms for Ordinary Differential Equations, 2nd edn. Springer, Berlin (2006)
11. Hairer, E., Nørsett, S.P., Wanner, G.: Solving Ordinary Differential Equations I: Nonstiff Problems, 2nd edn. Springer, Berlin (1993)
12. Jimánez, S., Vázquez, L.: Analysis of four numerical schemes for a nonlinear Klein-Gordon equation. Appl. Math. Comput. **35**, 61–93 (1990)

13. Leimkuhler, B., Reich, S.: Simulating Hamiltonian Dynamics. Cambridge Monographs on Applied and Computational Mathematics, vol. 14. Cambridge University Press, Cambridge (2005)
14. Liu, K., Wu, X.Y.: High-order symplectic and symmetric composition methods for multi-frequency and multi-dimensional oscillatory Hamiltonian systems. J. Comput. Math. **33**, 355–377 (2015)
15. McLachlan, R.I.: On the numerical integration of ordinary differential equations by symmetric composition methods. SIAM J. Sci. Comput. **16**, 151–168 (1995)
16. McLachlan, R.I., Quispel, G.R.W.: Splitting methods. Acta Numer. **11**, 341–434 (2002)
17. Reich, S.: Symplectic integration of constrained Hamiltonian systems by composition methods. SIAM J. Numer. Anal. **33**, 475–491 (1996)
18. Sanz-Serna, J.M.: Modulated Fourier expansions and heterogeneous multiscale methods. IMA J. Numer. Anal. **29**, 595–605 (2009)
19. Sanz-Serna, J.M., Calvo, M.P.: Numerical Hamiltonian Problem. Applied Mathematics and Mathematical Computation, 2nd edn. Springer, Chapman & Hall, London (1994)
20. Shi, W., Wu, X.Y.: On symplectic and symmetric ARKN methods. Comput. Phys. Commun. **183**, 1250–1258 (2012)
21. Shi, W., Wu, X.Y.: A note on symplectic and symmetric ARKN methods. Comput. Phys. Commun. **184**, 2408–2411 (2013)
22. Shi, W., Wu, X.Y., Xia, J.: Explicit multi-symplectic extended leap-frog methods for Hamiltonian wave equations. J. Comput. Phys. **231**, 7671–7694 (2012)
23. Suzuki, M.: General theory of higher-order decomposition of exponential operators and symplectic integrators. Phys. Lett. A **165**, 387–395 (1992)
24. Wang, B., Wu, X.Y.: A new high precision energy-preserving integrator for system of oscillatory second-order differential equations. Phys. Lett. A. **376**, 1185–1190 (2012)
25. Wang, B., Wu, X.Y., Zhao, H.: Novel improved multidimensional Störmer-Verlet formulas with applications to four aspects in scientific computation. Math. Comput. Model. **57**, 857–872 (2013)
26. Wu, X.Y., Wang, B.: Multidimensional adapted Runge-Kutta-Nyström methods for oscillatory systems. Comput. Phys. Commun. **181**, 1955–1962 (2010)
27. Wu, X.Y., Wang, B., Shi, W.: Efficient energy-preserving integrators for oscillatory Hamiltonian systems. J. Comput. Phys. **235**, 587–605 (2013)
28. Wu, X.Y., Wang, B., Xia, J.: Explicit symplectic multidimensional exponential fitting modified Runge-Kutta-Nyström method. BIT **52**, 773–795 (2012)
29. Wu, X.Y., You, X., Li, J.: Note on derivation of order conditions for ARKN methods for perturbed oscillators. Comput. Phys. Commun. **180**, 1545–1549 (2009)
30. Wu, X.Y., You, X., Shi, W., Wang, B.: ERKN integrators for systems of oscillatory second-order differential equations. Comput. Phys. Commun. **181**, 1873–1887 (2010)
31. Wu, X.Y., You, X., Wang, B.: Structure-Preserving Algorithms for Oscillatory Differential Equations. Springer, Berlin (2013)
32. Wu, X.Y., You, X., Xia, J.: Order conditions for ARKN methods solving oscillatory systems. Comput. Phys. Commun. **180**, 2250–2257 (2009)
33. Yoshida, H.: Construction of higher order symplectic integrators. Phys. Lett. A **150**, 262–268 (1990)

Chapter 6
The Construction of Arbitrary Order ERKN Integrators via Group Theory

This chapter presents the construction of arbitrary order extended Runge–Kutta–Nyström (ERKN) integrators. In general, ERKN methods are more effective than traditional Runge–Kutta–Nyström (RKN) methods in dealing with oscillatory Hamiltonian systems. However, the theoretical analysis for ERKN methods, such as the order conditions, the symplecticity conditions and the symmetric conditions, becomes much more complicated than that for RKN methods. Therefore, it is a bottleneck to construct high-order ERKN methods efficiently. This chapter first establishes the ERKN group Ω for ERKN methods and the RKN group G for RKN methods, respectively, and then shows that ERKN methods are a natural extension of RKN methods. That is, there exists an epimorphism η of the ERKN group Ω onto the RKN group G. This epimorphism gives a global insight into the structure of the ERKN group by the analysis of its kernel and the corresponding RKN group G. We also establish a particular mapping φ of G into Ω that each image element is an ideal representative element of the congruence class in Ω. Furthermore, an elementary theoretical analysis shows that this mapping φ can preserve many structure-preserving properties, such as the order, the symmetry and the symplecticity. From the epimorphism η together with its section φ, we may gain knowledge about the structure of the ERKN group Ω through the RKN group G.

6.1 Introduction

We are concerned in this chapter with initial value problems (IVP) of second-order oscillatory differential equations

$$\begin{cases} y''(t) + My(t) = f(y(t)), \\ y(t_0) = y_0, \ y'(t_0) = y'_0, \end{cases} \tag{6.1}$$

X. Wu and B. Wang, *Recent Developments in Structure-Preserving Algorithms for Oscillatory Differential Equations*, https://doi.org/10.1007/978-981-10-9004-2_6

with M a (symmetric) positive semi-definite matrix and $\|M\| \gg 1$, which frequently arise in many aspects of scientific and engineering computing, such as celestial mechanics, theoretical physics, chemistry and electronics. Effective numerical methods for solving this type of problems are of great importance (see, e.g. [4, 7–10, 13, 14]). Using the oscillatory structure introduced by the linear term My in (6.1), Yang et al. [34] proposed extended Runge–Kutta–Nyström (ERKN) methods. Much research effort on ERKN methods has been made and ERKN methods show notable efficiency and higher accuracy than the traditional Runge–Kutta–Nyström (RKN) methods in dealing with (6.1) (see, e.g. [28, 29, 31, 32, 35, 36]). It is clear that (6.1) becomes a Hamiltonian system once $f(y) = -\nabla U(y)$, where $U(y)$ is a smooth potential function. The symmetric conditions and symplectic conditions for ERKN methods have also been investigated [15, 16, 25, 27, 30]. However, it is very difficult to obtain a high-order ERKN method with some important structure properties, even though the order conditions, the symmetric conditions and the symplectic conditions have been well established.

On the one hand, we have known an important property of ERKN methods, that is, when $M \to \mathbf{0}$, each ERKN method reduces to a classical RKN method. This property implies that there exists an intrinsic relation between ERKN and RKN methods. On the other hand, the structural properties such as symmetry and symplecticity of RKN methods have been studied by many authors and very useful results have been achieved [1–3, 19–21, 23, 24, 26]. Taking account of these two points, in this chapter we attempt to clarify this intrinsic relation between ERKN and RKN methods by introducing an epimorphism η from the ERKN group Ω to the RKN group G. In particular, we establish a particular mapping η from G to Ω. Consequently, the properties of ERKN methods including the order, the symmetry, and the symplecticity, are inherited from the classical RKN methods via the mapping φ.

The plan of this chapter is as follows. In Sect. 6.2 we briefly review the classical RKN methods and then construct the RKN group G. In Sect. 6.3, the theories associated with the ERKN group Ω are established. Especially, we show that there exists an epimorphism η from Ω to G. In Sect. 6.4, we address the particular mapping φ from G to Ω in detail. It turns out that this mapping preserves the order, the symplecticity, and almost the symmetry. In Sect. 6.5 we carry out some numerical experiments for the high-order structure-preserving ERKN methods derived from the theoretical analysis in Sect. 6.4. The last section is concerned with conclusions and discussions.

6.2 Classical RKN Methods and the RKN Group

This section begins with an overview of the results on classical RKN methods for second-order initial value problems

$$\begin{cases} y''(t) = f(y(t)), \\ y(t_0) = y_0, \ y'(t_0) = y_0', \end{cases} \tag{6.2}$$

where the right-hand-side function f does not depend on the derivative y' and time t. As is well known, to approximate this autonomous system more efficiently than with traditional Runge–Kutta (RK) methods, the so-called Runge–Kutta–Nyström (RKN) methods were proposed [18]. An s-stage classical RKN method with a stepsize h for the problem (6.2) is defined as

$$\begin{cases} Y_i = y_n + c_i h y'_n + h^2 \sum_{j=1}^{s} a_{ij} f(Y_j), \quad i = 1, \ldots, s, \\ y_{n+1} = y_n + h y'_n + h^2 \sum_{i=1}^{s} \bar{b}_i f(Y_i), \\ y'_{n+1} = y'_n + h \sum_{i=1}^{s} b_i f(Y_i), \end{cases} \tag{6.3}$$

where $a_{ij}, \bar{b}_i, b_i$ for $i, j = 1, \ldots, s$ are real constants. Usually, the RKN method (6.3) can be briefly expressed in a Butcher Tableau

$$\begin{array}{c|c} c & A \\ \hline & \bar{b}^{\mathsf{T}} \\ \hline & b^{\mathsf{T}} \end{array} \doteq \begin{array}{c|ccc} c_1 & a_{11} & \cdots & a_{1s} \\ \vdots & \vdots & & \vdots \\ c_s & a_{s1} & \cdots & a_{ss} \\ \hline & \bar{b}_1 & \cdots & \bar{b}_s \\ \hline & b_1 & \cdots & b_s \end{array} . \tag{6.4}$$

In order to establish an RKN group conveniently, we will specify an RKN method Φ with a stepsize h by Φ_h. Then $\Phi_{\gamma h}$ and $\Phi_{\beta h}$ are regarded as two different elements once $\gamma \neq \beta$, even though they share the same coefficients. To construct a group related to RKN methods, a binary composition is needed. Similarly to Hairer and Wanner [11] in 1974, we consider the composition of two RKN methods but matching allowance for their corresponding stepsizes.

We then introduce the following definition.

Definition 6.1 Suppose that Φ_h is an s-stage RKN method defined by (6.3) for the problem (6.2), Φ'_h is called an *essential* 0-*stepsize* form of Φ_h if the formula for Φ'_h reads

$$\begin{cases} Y_i = y_n + c_i h y'_n + h^2 \sum_{j=1}^{s} a_{ij} f(Y_j), \quad i = 1, \ldots, s, \\ y_{n+1} = y_n + h^2 \sum_{i=1}^{s} \bar{b}_i f(Y_i), \\ y'_{n+1} = y'_n + h \sum_{i=1}^{s} b_i f(Y_i). \end{cases} \tag{6.5}$$

Accordingly, Φ_h is called an *h-stepsize* form of Φ'_h.

Remark 6.1 From Definition 6.1, the only difference is in the second equation compared with (6.3). This is essential. That is, the equation

$$y_{n+1} = y_n + hy'_n + h^2 \sum_{i=1}^{s} \bar{b}_i f(Y_i) ,$$

for Φ_h has been changed into

$$y_{n+1} = y_n + h^2 \sum_{i=1}^{s} \bar{b}_i f(Y_i) ,$$

in (6.5) for Φ'_h. This means that the numerical solution $(y_{n+1}, y'_{n+1}) = \Phi_h(y_n, y'_n)$ approximates the exact solution at $t_n + h$, whereas $(y_{n+1}, y'_{n+1}) = \Phi'_h(y_n, y'_n)$ can only approximate to $(y(t_n), y'(t_n))$ at t_n. It is noted that if Ψ_h is a classical RKN method, then $\Psi_{0\cdot h}$ is just the identity I. This implies that an RKN method $\Psi_{0\cdot h}$ with the stepsize 0 is totally different from its *essential* 0-*stepsize* form Ψ'_h under our new definition.

Suppose that Φ^1_h and Φ^2_h are two RKN methods with s_1 stages and s_2 stages, respectively. Their coefficients are respectively denoted by $c = (c_1, \ldots, c_{s_1})^{\intercal}$, $b = (b_1, \ldots, b_{s_1})^{\intercal}$, $\bar{b} = (\bar{b}_1, \ldots, \bar{b}_{s_1})^{\intercal}$, $A = (a_{ij})_{s_1 \times s_1}$ and $c^* = (c^*_1, \ldots, c^*_{s_2})^{\intercal}$, $b^* = (b^*_1, \ldots, b^*_{s_2})^{\intercal}$, $\bar{b}^* = (\bar{b}^*_1, \ldots, \bar{b}^*_{s_2})^{\intercal}$, $A^* = (a^*_{ij})_{s_2 \times s_2}$. We next consider the composition of $\Phi^1_{\gamma h}$ and $\Phi^2_{\beta h}$. Taking (y_0, y'_0) as the starting value at t_0 and (y_1, y'_1) as the updated value after one step, we can express the composition law of $(y_1, y'_1) = (\Phi^2_{\beta h} \circ \Phi^1_{\gamma h})(y_0, y'_0)$ as

$$\begin{cases} Y_i = y_0 + \gamma c_i h y'_0 + \gamma^2 h^2 \sum_{j=1}^{s_1} a_{ij} f(Y_j), \quad i = 1, \ldots, s_1, \\ \widetilde{y}_1 = y_0 + \gamma h y'_0 + \gamma^2 h^2 \sum_{i=1}^{s_1} \bar{b}_i f(Y_i), \\ \widetilde{y}'_1 = y'_0 + \gamma h \sum_{i=1}^{s_1} b_i f(Y_i), \\ \widetilde{Y}_k = \widetilde{y}_1 + \beta c^*_k h \widetilde{y}'_1 + \beta^2 h^2 \sum_{j=1}^{s_2} a^*_{kj} f(\widetilde{Y}_j), \quad k = 1, \ldots, s_2, \\ y_1 = \widetilde{y}_1 + \beta h \widetilde{y}'_1 + \beta^2 h^2 \sum_{i=1}^{s_2} \bar{b}^*_i f(\widetilde{Y}_i), \\ y'_1 = \widetilde{y}'_1 + \sum_{i=1}^{s_2} \beta h b^*_i f(\widetilde{Y}_i). \end{cases} \tag{6.6}$$

Canceling $\widetilde{y}_1$ and $\widetilde{y}_1'$ from (6.6), we obtain the following simplified form

$$\begin{cases} Y_i = y_0 + \gamma c_i h y_0' + h^2 \sum_{j=1}^{s_1} \gamma^2 a_{ij} f(Y_j), \quad i = 1, \ldots, s_1 \\ \widetilde{Y}_k = y_0 + (\gamma + \beta c_k^*) h y_0' + h^2 \Big(\sum_{j=1}^{s_1} (\gamma^2 \bar{b}_j + \gamma\beta c_k^* b_j) f(Y_j) + \sum_{j=1}^{s_2} \beta^2 a_{kj}^* f(\widetilde{Y}_j) \Big), \quad k = 1, \ldots, s_2 \\ y_1 = y_0 + (\gamma + \beta) h y_0' + h^2 \Big(\sum_{i=1}^{s_1} (\gamma^2 \bar{b}_i + \gamma\beta b_i) f(Y_i) + \sum_{i=1}^{s_2} \beta^2 \bar{b}_i^* f(\widetilde{Y}_i) \Big), \\ y_1' = y_0' + h \Big(\sum_{i=1}^{s_1} \gamma b_i f(Y_i) + \sum_{i=1}^{s_2} \beta b_i^* f(\widetilde{Y}_i) \Big). \end{cases} \tag{6.7}$$

Now let us have a further discussion on the formula (6.7). If $\gamma + \beta \neq 0$, we observe that (6.7) is just an RKN method $\Psi_{(\gamma+\beta)h}$ with the stepsize $(\gamma + \beta)h$. Meanwhile, by a careful calculation the Butcher tableau of RKN method Ψ_h reads

$$\begin{array}{c|cc} \gamma c/\delta & \gamma^2 A/\delta^2 & \\ (\gamma e + \beta c^*)/\delta & \tilde{A}/\delta^2 & \beta^2 A^*/\delta^2 \\ \hline & \tilde{b}^{\mathsf{T}}/\delta^2 & \beta^2 \bar{b}^{*\mathsf{T}}/\delta^2 \\ \hline & \gamma b^{\mathsf{T}}/\delta & \beta b^{*\mathsf{T}}/\delta \end{array}, \tag{6.8}$$

where $\delta = \gamma + \beta$, $\tilde{A}_{ij} = \gamma^2 \bar{b}_j + \gamma\beta c_i^* b_j$, $\tilde{b}_j = \gamma^2 \bar{b}_j + \gamma\beta b_j$ for $i = 1, \ldots, s_2$, $j = 1, \ldots, s_1$, and $e = (1, \ldots, 1)^{\mathsf{T}}$ is the $s_2 \times 1$ vector of units. It is clear that the updated value (y_1, y_1') just approximates the exact value at $t_0 + (\gamma + \beta)h$.

However, for the case of $\gamma + \beta = 0$, the formula (6.7) is no longer of classical RKN type. In this case, $\Phi_{\beta h}^1 \circ \Phi_{\gamma h}^1$ is just an *essential* 0-*stepsize* RKN method, whose corresponding h-*stepsize* form can be expressed in the following Butcher tableau

$$\begin{array}{c|cc} \gamma c & \gamma^2 A & \\ \gamma e + \beta c^* & \tilde{A} & \beta^2 A^* \\ \hline & \tilde{b}^{\mathsf{T}} & \beta^2 \bar{b}^{*\mathsf{T}} \\ \hline & \gamma b^{\mathsf{T}} & \beta b^{*\mathsf{T}} \end{array}, \tag{6.9}$$

where $\tilde{A}_{ij}$ and $\tilde{b}_j$ are the same as in formula (6.8). In this case, it should be noted that $\sum_i \gamma b_i + \sum_i \beta b_i^* = 0$ when $\gamma + \beta = 0$ and $\sum_i b_i = \sum_i b_i^* = 1$. Although this case is not significant in practice, it will be indispensable in the construction of an RKN group in the remainder of this chapter.

Define

$G_1 := \{\Phi_{\alpha h} \mid \Phi_h \text{ is a classical RKN method for } \alpha \in \mathbb{R}\}$,

$G_0 := \{\Phi'_{\alpha h} \mid \Phi'_{\alpha h}$ is the essential 0-stepsize form of $\Phi_{\alpha h}$ and $\Phi_{\alpha h} \in G_1$ *with* $\sum_i b_i = 0\}$, and $G = G_1 \bigcup G_0$.

We then have the following result.

Theorem 6.1 *$(G, \circ, I)$ is a group with respect to the composition $\circ$ and the identity I.*

Proof It is clear that the composition $\circ$ is associative, and for each element $\Theta \in G$ we certainly have $\Theta \circ I = I \circ \Theta = \Theta$. Moreover, if Φ and Ψ are two arbitrary elements in G, from the formula (6.7) and the above analysis we know that $\Phi \circ \Psi \in G$. This shows the closure property of G under the product $\circ$. We next show that each element in G is invertible.

For an s-stage RKN method Λ_h defined by (6.3), the existing results [19] have revealed the existence of its adjoint method Λ_h^*. If the coefficients of the adjoint method are denoted by $c^* = (c_1^*, \ldots, c_s^*)^\intercal$, $b^* = (b_1^*, \ldots, b_s^*)^\intercal$, $\bar{b}^* = (\bar{b}_1^*, \ldots, \bar{b}_s^*)^\intercal$, and $A^* = \left(a_{ij}^*\right)_{s\times s}$, then they satisfy

$$\begin{cases} c_i^* = 1 - c_{s+1-i}\ , \\ a_{ij}^* = (1 - c_{s+1-i})b_{s+1-j} - \bar{b}_{s+1-j} + a_{s+1-i,s+1-j}\ , \\ \bar{b}_j^* = b_{s+1-j} - \bar{b}_{s+1-j}\ , \\ b_j^* = b_{s+1-j}\ , \end{cases} \tag{6.10}$$

for $1 \leqslant i, j \leqslant s$. Certainly Λ_h^* belongs to G, and hence $\Lambda_{-h}^* \in G$. Furthermore, from the definition of adjoint methods, we have $\Lambda_h^{-1} = \Lambda_{-h}^*$ straightforwardly. Consequently, we have $\Lambda_h^{-1} \in G$, so does its essential 0-*stepsize* form Λ'_h, namely, $\Lambda_h^{'-1} \in G$. This completes the proof. □

Remark 6.2 Here, the above way of defining an RKN group has some nonessential differences from that of the RK group defined by Hairer and Wanner [11]. These differences actually rely on the following fact. If Φ_h and Ψ_h are two different RKN methods and they are not adjoint to each other, then the composition $\Phi_h \circ \Psi_{-h}^*$ does *not* belong to G_1 any more. Here Ψ_h^* denotes the adjoint method of Ψ_h. That is why we have additionally introduced Definition 6.1 and the set G_0. Likewise, it is also needed to introduce another new definition (Definition 6.2) when constructing the ERKN group in the next section.

6.3 ERKN Group and Related Issues

6.3.1 Construction of ERKN Group

In this section, we are concerned with the group-structure analysis of the efficient integrator for the oscillatory second-order initial problem (6.1). It seems that classical

RKN methods could still be applied to these problems as numerical integrators, since one may move the term My from the left-hand side to the right-hand side of the differential equation and then the problem (6.1) can be also transformed to the type of (6.2). However, when $||M|| \gg 1$, RKN methods may not be very effective methods for solving (6.1) and show bad numerical behavior. This is mainly caused by the highly oscillatory effect introduced by the linear term My. Taking account of this point, the extended Runge–Kutta–Nyström (ERKN) methods were proposed and designed especially for the oscillatory problem (6.1).

Based on the matrix-variation-of-constants formula [33], an s-stage ERKN method [34] for IVP (6.1) is defined by

$$\begin{cases} Y_i = \phi_0(c_i^2 V)y_n + c_i h\phi_1(c_i^2 V)y_n' + h^2 \sum\limits_{j=1}^{s} a_{ij}(V)f(Y_j), \quad i = 1, \ldots, s, \\ y_{n+1} = \phi_0(V)y_n + h\phi_1(V)y_n' + h^2 \sum\limits_{i=1}^{s} \bar{b}_i(V)f(Y_i), \\ y_{n+1}' = -hM\phi_1(V)y_n + \phi_0(V)y_n' + h\sum\limits_{i=1}^{s} b_i(V)f(Y_i). \end{cases} \tag{6.11}$$

Here, $c_1, \ldots, c_s$ are real constants, $b_i(V), \bar{b}_i(V)$ and $a_{ij}(V)$ for $i, j = 1, \ldots, s$ are matrix-valued functions of $V \equiv h^2 M$ which are usually expressed in formal series in terms of V

$$b_i(V) = \sum_{k=0}^{\infty} \frac{b_i^{(2k)}}{(2k)!}V^k, \quad \bar{b}_i(V) = \sum_{k=0}^{\infty} \frac{\bar{b}_i^{(2k)}}{(2k)!}V^k, \quad a_{ij}(V) = \sum_{k=0}^{\infty} \frac{a_{ij}^{(2k)}}{(2k)!}V^k, \tag{6.12}$$

and

$$\phi_j(V) := \sum_{k=0}^{\infty} \frac{(-1)^k V^k}{(2k+j)!}, \quad j = 0, 1, \ldots. \tag{6.13}$$

The properties related to $\phi_j(V)$ for $j = 0, 1, \ldots$ can be found in [31] and the details are omitted here. We can also express the ERKN method (6.11) in a Butcher tableau

$$\begin{array}{c|c} c & A(V) \\ \hline & \bar{b}(V)^{\mathsf{T}} \\ \hline & b(V)^{\mathsf{T}} \end{array} = \begin{array}{c|ccc} c_1 & a_{11}(V) & \cdots & a_{1s}(V) \\ \vdots & \vdots & & \vdots \\ c_s & a_{s1}(V) & \cdots & a_{ss}(V) \\ \hline & \bar{b}_1(V) & \cdots & \bar{b}_s(V) \\ \hline & b_1(V) & \cdots & b_s(V) \end{array}. \tag{6.14}$$

Proceeding in the same spirit as for RKN methods, we also introduce a new definition for ERKN methods as follows.

Definition 6.2 Suppose that Ψ_h is an s-stage ERKN method defined by (6.11) for the problem (6.1), Ψ'_h is called *essential* 0-*stepsize* form of Ψ_h if the formula for Ψ'_h reads

$$\begin{cases} Y_i = \phi_0(c_i^2 V)y_n + c_i h\phi_1(c_i^2 V)y'_n + h^2 \sum_{j=1}^{s} a_{ij}(V)f(Y_j), \quad i = 1, \ldots, s, \\ y_{n+1} = y_n + h^2 \sum_{i=1}^{s} \bar{b}_i(V)f(Y_i), \\ y'_{n+1} = y'_n + h \sum_{i=1}^{s} b_i(V)f(Y_i). \end{cases} \tag{6.15}$$

Then Ψ_h is called an h-*stepsize* form of Ψ'_h.

Suppose that Υ_h^1 and Υ_h^2 are two ERKN methods with s_1 stages and s_2 stages, respectively. The coefficients of Υ_h^1 are denoted as $(c, b, \bar{b}, A)$, and those of Υ_h^2 are additionally denoted with a star $(c^*, b^*, \bar{b}^*, A^*)$. We now consider the composition of $\Upsilon_{\gamma h}^1$ and $\Upsilon_{\beta h}^2$. After a careful calculation, we derive the scheme $\Upsilon_{\beta h}^2 \circ \Upsilon_{\gamma h}^1$ as follows

$$\begin{cases} Y_i = \phi_0(\gamma^2 c_i^2 V)y_0 + \gamma c_i h\phi_1(\gamma^2 c_i^2 V)y'_0 + h^2 \sum_{j=1}^{s_1} \gamma^2 A_{ij}(\gamma^2 V)f(Y_j), \quad i = 1, \ldots, s_1, \\ \widetilde{Y}_k = \phi_0((\gamma + \beta c_k^*)^2 V)y_0 + (\gamma + \beta c_k^*)h\phi_1((\gamma + \beta c_k^*)^2 V)y'_0 \\ \quad + h^2\Big(\sum_{j=1}^{s_1} (\gamma^2 \bar{b}_j(\gamma^2 V)\phi_0(\beta^2 c_k^{*2} V) + \gamma\beta c_k^* b_j(\gamma^2 V)\phi_1(\beta^2 c_k^{*2} V))f(Y_j) \\ \quad + \sum_{j=1}^{s_2} \beta^2 A_{kj}^*(\beta^2 V)f(\widetilde{Y}_j)\Big), \quad k = 1, \ldots, s_2, \\ y_1 = \phi_0((\gamma + \beta)^2 V)y_0 + (\gamma + \beta)h\phi_1((\gamma + \beta)^2 V)y'_0 \\ \quad + h^2\Big(\sum_{j=1}^{s_1} (\gamma^2 \bar{b}_j(\gamma^2 V)\phi_0(\beta^2 V) + \gamma\beta b_j(\gamma^2 V)\phi_1(\beta^2 V))f(Y_j) + \sum_{i=1}^{s_2} \beta^2 \bar{b}_i^*(\beta^2 V)f(\widetilde{Y}_i)\Big), \\ y'_1 = -(\gamma + \beta)hM\phi_1((\gamma + \beta)^2 V)y_0 + \phi_0((\gamma + \beta)^2 V)y'_0 \\ \quad + h\Big(\sum_{j=1}^{s_1} (-\gamma^2 \beta V \bar{b}_j(\gamma^2 V)\phi_1(\beta^2 V) + \gamma b_j(\gamma^2 V)\phi_0(\beta^2 V))f(Y_j) + \sum_{i=1}^{s} \beta b_i(\beta^2 V)f(\widetilde{Y}_i)\Big). \end{cases}$$

For the case $\gamma + \beta \neq 0$, gives that the composition $\Upsilon_{\beta h}^2 \circ \Upsilon_{\gamma h}^1$ indicates a new $\Upsilon_{\delta h}$, namely, an $(s_1 + s_2)$-stage ERKN method with the stepsize $\delta h = (\gamma + \beta)h$, and Butcher tableau

$$\begin{array}{c|cc} \gamma c/\delta & \gamma^2 A(\gamma^2/\delta^2 V)/\delta^2 & \\ (\gamma e+\beta c^*)/\delta & \bar{A}(V/\delta^2)/\delta^2 & \beta^2 A^*(\beta^2/\delta^2 V)/\delta^2 \\ \hline & & \\ & \bar{B}^{\intercal}(V/\delta^2)/\delta^2 & \beta^2 \bar{b}^{*\intercal}(\beta^2/\delta^2 V)/\delta^2 \\ \hline & B^{\intercal}(V/\delta^2)/\delta & \beta b^{*\intercal}(\beta^2/\delta^2 V)/\delta \end{array}, \tag{6.16}$$

where

$$\begin{cases} \bar{A}_{ij}(V) = \gamma^2 \bar{b}_j(\gamma^2 V)\phi_0(\beta^2 c_i^{*2} V) + \gamma\beta c_i^* b_j(\gamma^2 V)\phi_1(\beta^2 c_i^{*2} V), \\ \bar{B}_j(V) = \gamma^2 \bar{b}_j(\gamma^2 V)\phi_0(\beta^2 V) + \gamma\beta b_j(\gamma^2 V)\phi_1(\beta^2 V), \\ B_j(V) = -\gamma^2 \beta V \bar{b}_j(\gamma^2 V)\phi_1(\beta^2 V) + \gamma b_j(\gamma^2 V)\phi_0(\beta^2 V), \end{cases} \tag{6.17}$$

for $i = 1, \ldots, s_2,\ j = 1, \ldots, s_1$.

If $\gamma + \beta = 0$, the composition $\Upsilon' = \Upsilon^2_{\beta h} \circ \Upsilon^1_{\gamma h}$ is also *essential* 0-*stepsize*, whose corresponding *h-stepsize* form can be expressed in the following Butcher tableau

$$\begin{array}{c|cc} \gamma c & \gamma^2 A(\gamma^2 V) & \\ \gamma e+\beta c^* & \bar{A}(V) & \beta^2 A^*(\beta^2 V) \\ \hline & & \\ & \bar{B}^{\intercal}(V) & \beta^2 \bar{b}^{*\intercal}(\beta^2 V) \\ \hline & B^{\intercal}(V) & \beta b^{*\intercal}(\beta^2 V) \end{array}, \tag{6.18}$$

where $\bar{A}_{ij}(V)$, $\bar{B}_j(V)$, $B_j(V)$ have the same expression as (6.17) with $\sum_i \gamma b_i^{(0)} + \sum_i \beta b_i^{*(0)} = 0$.

Define

$\Omega_1 := \{\Phi_{\alpha h} \mid \Phi_h \ is\ an\ ERKN\ method\ for\ \alpha \in \mathbb{R}\}$,

$\Omega_0 := \{\Phi'_{\alpha h} \mid \Phi'_{\alpha h}$ is the essential 0-stepsize form of $\Phi_{\alpha h},\ \Phi_{\alpha h} \in \Omega_1$ *with* $\sum_i b_i^{(0)} = 0\}$, and $\Omega = \Omega_1 \bigcup \Omega_0$.

Then we have the following theorem.

Theorem 6.2 *$(\Omega, \circ, I)$ is a group with respect to the composition $\circ$ and the identity I.*

The proof is similar to that of Theorem 6.1, except that the coefficients of the adjoint method of (6.11) can be expressed as

$$\begin{cases} c_i^* = 1 - c_{s+1-i}\,, \\ a_{ij}^*(V) = \phi_0(c_{s+1-i}^2 V)\bar{b}_j(V) - c_{s+1-i}\phi_1(c_{s+1-i}^2 V)b_j(V) + a_{s+1-i,s+1-j}(V)\,, \\ \bar{b}_j^*(V) = \phi_1(V)b_{s+1-j}(V) - \phi_0(V)\bar{b}_{s+1-j}(V)\,, \\ b_j^*(V) = V\phi_1(V)\bar{b}_{s+1-j}(V) + \phi_0(V)b_{s+1-j}(V)\,, \end{cases} \tag{6.19}$$

for $1 \leqslant i, j \leqslant s$. Hence, we omit the details here.

6.3.2 The Relation Between the RKN Group G and the ERKN Group Ω

In the previous sections, we have established the RKN group $(G, \circ, I)$ and the ERKN group $(\Omega, \circ, I)$. A direct observation shows that when $M \to \mathbf{0}$, the oscillatory problem (6.1) becomes the traditional second-order initial value problem (6.2), and the ERKN method (6.11) hence reduces to the RKN method (6.3). This point indicates that there exists an inherent relationship between RKN methods and ERKN methods. Thus, ERKN methods are usually regarded as an extension of RKN methods. In the following, we will rigorously demonstrate this extension relationship.

In what follows, we will denote the coefficients of an RKN method in lower-case by $(c, b, \bar{b}, a)$ and those of an ERKN method in upper-case by $(C, B, \bar{B}, A)$. As ERKN methods depend on the matrix M, for each element $\Psi \in \Omega$ we will denote it as $\Psi(M)$ to show this relevance if necessary. Then the word *reduces* can be defined as a map

$$\eta : \quad \Omega \longrightarrow G, \quad \eta(\Psi) = \lim_{M \to \mathbf{0}} \Psi(M), \quad \forall\ \Psi \in \Omega.$$

As a continuation, we arrive at the following useful theorem.

Theorem 6.3 *The map η is an epimorphism of the group Ω onto the group G.*

Proof Suppose that Ψ^1 and Ψ^2 are two elements of Ω respectively with the stepsizes γh and βh, $\Phi^1 = \eta(\Psi^1)$, and $\Phi^2 = \eta(\Psi^2)$. From the composition laws (6.8) and (6.9) of RKN methods and those of ERKN methods (6.16) and (6.18), it can be easily verified that $\eta(\Psi^2 \circ \Psi^1) = \Phi^2 \circ \Phi^1 = \eta(\Psi^2) \circ \eta(\Psi^1)$. In addition, from the fact that $\eta(I) = I$, we conclude that η is a homomorphism of Ω into G.

We next show that η is surjective. For each element $\Phi \in G$, which is denoted by the coefficients $(c, b, \bar{b}, a)$, there exists $\Psi \in \Omega$, whose coefficients can be expressed as

$$C = c, \ B(V) = b \otimes E_n, \ \bar{B}(V) = \bar{b} \otimes E_n, \ A(V) = a \otimes E_n, \tag{6.20}$$

where $\otimes$ is the Kronecker product, E_n is an $n \times n$ identity matrix and n is the dimension of square matrix M. Obviously the coefficients $(C, B, \bar{B}, A)$ define an element in Ω, and thus η is surjective. This completes the proof. □

Corollary 6.1 *Let K be the kernel of η, i.e. $K = \eta^{-1}(I)$, then K is a normal subgroup of Ω. Moreover, the induced map $\bar{\eta}$ is an isomorphism of the quotient group $\overline{\Omega} = \Omega / K$ onto the group G.*

Theorem 6.3 actually gives a global view of ERKN methods by connecting them with classical RKN methods via the epimorphism map η. From Corollary 6.1, the map η defines a congruence relation $\equiv$ by the normal subgroup K, where

$$\Phi \equiv \Psi (mod\ K) \quad if \quad \Phi^{-1} \circ \Psi \in K.$$

Then by finding a representative element Ψ for each congruence class $\overline{\Psi} \in \overline{\Omega}$, we can theoretically give all the elements in $\overline{\Psi}$, since for each $\Theta \in \overline{\Psi}$ there exists $\Gamma \in K$ such that $\Theta = \Psi \circ \Gamma$. This fact indicates that $\overline{\Psi}$ is the coset of Ψ relative to K, i.e. $\overline{\Psi} = \Psi \circ K$. Hence it only remains to describe the normal subgroup K in detail. This can be easily obtained from the following definition of K

$$K = \{\Psi \in \Omega_0 | b_j^{(0)} = 0 \ and\ \bar{b}_j^{(0)} = 0,\ \forall j\}.$$

6.4 A Particular Mapping of *G* into *Ω*

In Sect. 6.3, we have investigated the ERKN group as a whole. However, as mentioned in the previous section, we can just have a theoretical description for each congruence class $\overline{\Psi} \in \overline{\Omega}$, and this is not associated with the important properties of the method, such as the symplecticity, the symmetry and the order. Recalling Corollary 6.1 again, and taking account of the fact that $\overline{\Psi} = \Psi \circ K$, it is of great importance to select a representative element Ψ with favourable properties for the congruence class $\overline{\Psi}$, even though we cannot give a detailed analysis for each element in $\overline{\Psi}$.

Meanwhile, because $\eta(\overline{\Psi}) = \Phi \in G$, Φ inherits all the advantages of the ERKN elements in $\overline{\Psi}$. Hence all the ERKN elements in $\overline{\Psi}$ cannot have better structural properties than the reduced RKN element Φ. Taking account of this point, we may find this appropriate representative element Ψ with the help of the corresponding reduced RKN element Φ. In fact, what we are considering is to find a normal mapping φ from G into Ω, so that $\varphi(\Phi)$ can preserve as many properties as the original RKN method Φ does. A direct result about the potential mapping φ is that it should be the section of η. That is, the composition $\eta \circ \varphi = I$ is the identity on G. In this sense, the underlying mapping defined by (6.20) may be a straightforward candidate for φ. Unfortunately, most properties cannot be preserved in this easy way, and we have to reconsider a proper mapping φ.

From the variation-of-constants formula (see, e.g. [31, 32]) for the problem (6.2) and the problem (6.1), as well as the corresponding RKN method for (6.2) and ERKN integrator (6.11), we consider the following mapping:

$$\varphi: G \longrightarrow \Omega, \qquad \begin{array}{c|c} c & a \\ \hline & \bar{b}^{\mathsf{T}} \\ \hline & b^{\mathsf{T}} \end{array} \longmapsto \begin{array}{c|c} C & A(V) \\ \hline & \bar{B}(V)^{\mathsf{T}} \\ \hline & B(V)^{\mathsf{T}} \end{array}, \tag{6.21}$$

where

$$\begin{cases} C_i = c_i, \\ A_{ij}(V) = a_{ij}\phi_1((c_i - c_j)^2 V), \\ \bar{B}_i(V) = \bar{b}_i\phi_1((1 - c_i)^2 V), \\ B_i(V) = b_i\phi_0((1 - c_i)^2 V), \end{cases}$$

for $1 \leqslant i, j \leqslant s$ and s is the stage of the RKN method. This mapping naturally maps a classical RKN method Φ to an ERRK method $\varphi(\Phi)$. Meanwhile, being a representative element for the congruence class $\overline{\varphi(\Phi)}$, we will show by the following several theorems that $\varphi(\Phi)$ almost preserves all the properties that the original RKN method Φ has.

Theorem 6.4 *If $\Phi \in G$ is symplectic, then $\varphi(\Phi) \in \Omega$ is symplectic.*

Proof From the definition of G, it is needed to verify that the result holds for all RKN methods. Hence we can suppose that Φ is an s-stage symplectic RKN method. The results from Suris [26] and Okunbor and Skeel [19] show that the coefficients of Φ should satisfy the following symplectic conditions:

$$\begin{cases} \bar{b}_i = (1 - c_i)b_i, \quad 1 \le i \le s, \\ \bar{b}_i b_j + b_i a_{ij} = \bar{b}_j b_i + b_j a_{ji}, \quad 1 \le i, j \le s. \end{cases} \tag{6.22}$$

We next show that the ERKN method $\varphi(\Phi)$, whose coefficients C, $A(V)$, $\bar{B}(V)$, $B(V)$ defined by (6.21), is symplectic. Although there is no sufficient and necessary conditions for the symplecticity of ERKN methods, we will prove that $\varphi(\Phi)$ satisfies the following conditions:

$$\begin{cases} \phi_0(V)B_i(V) + V\phi_1(V)\bar{B}_i(V) = d_i\phi_0(c_i^2 V), \quad d_i \in \mathbb{R}, i = 1, 2, ..., s, \\ \phi_1(V)B_i(V) - \phi_0(V)\bar{B}_i(V) = c_i d_i \phi_1(c_i^2 V), \quad i = 1, 2, ..., s, \\ \bar{B}_i B_j + d_i A_{ij} = \bar{B}_j B_i + d_j A_{ji}, \quad i, j = 1, 2, ..., s, \end{cases} \tag{6.23}$$

which are sufficient conditions for symplectic ERKN methods originally proposed by Wu et al. [30].

The equations $\bar{B}_i(V) = \bar{b}_i \phi_1((1-c_i)^2 V)$ and $B_i(V) = b_i \phi_0((1-c_i)^2 V)$ exactly give the first two equations of (6.23), with $d_i = b_i$. Then, inserting the expression of $A(V)$, $\bar{B}(V)$, $B(V)$ into the the third equation of (6.23), we obtain

$$\begin{aligned} &\bar{B}_i B_j + d_i A_{ij} - (\bar{B}_j B_i + d_j A_{ji}) \\ =&(1 - c_i)b_i b_j \phi_1((1 - c_i)^2 V)\phi_0((1 - c_j)^2 V) + b_i a_{ij} \phi_1((c_i - c_j)^2 V) \\ &- \big((1 - c_j)b_i b_j \phi_1((1 - c_j)^2 V)\phi_0((1 - c_i)^2 V) + b_j a_{ji} \phi_1((c_i - c_j)^2 V)\big) \\ =&\big(b_i b_j (c_j - c_i) + b_i a_{ij} - b_j a_{ji}\big)\phi_1((c_i - c_j)^2 V) \\ =&0. \end{aligned}$$

The last equation directly follows from (6.22). This completes the proof. □

Theorem 6.5 *If $\Phi \in G$ is symmetric and the coefficients satisfy the simplifying assumption $\bar{b}_i = b_i(1 - c_i)$, then $\varphi(\Phi) \in \Omega$ is symmetric.*

Proof Similarly to Theorem 6.4, we only need to verify the case that Φ is an s-stage RKN method. Hence, we need to derive the symmetric conditions of ERKN methods [16, 27]

$$\begin{cases} c_i = 1 - c_{s+1-i} \,, \\ A_{ij}(V) = \phi_0(c_{s+1-i}^2 V)\bar{B}_j(V) - c_{s+1-i}\phi_1(c_{s+1-i}^2 V)B_j(V) + A_{s+1-i,s+1-j}(V) \,, \\ \bar{B}_j(V) = \phi_1(V)B_{s+1-j}(V) - \phi_0(V)\bar{B}_{s+1-j}(V) \,, \\ B_j(V) = V\phi_1(V)\bar{B}_{s+1-j}(V) + \phi_0(V)B_{s+1-j}(V) \,, \end{cases} \tag{6.24}$$

from the symmetric conditions of RKN methods [12],

$$\begin{cases} c_i = 1 - c_{s+1-i} \,, \\ a_{ij} = (1 - c_{s+1-i})b_{s+1-j} - \bar{b}_{s+1-j} + a_{s+1-i,s+1-j} \,, \\ \bar{b}_j = b_{s+1-j} - \bar{b}_{s+1-j} \,, \\ b_j = b_{s+1-j} \,. \end{cases} \tag{6.25}$$

The first equation of (6.24) naturally holds. On noting that $b_j = b_{s+1-j}$ and $\bar{b}_j = b_j(1 - c_j)$, we have

$$\begin{aligned} &\phi_1(V)B_{s+1-j} - \phi_0(V)\bar{B}_{s+1-j} \\ =&b_{s+1-j}\phi_1(V)\phi_0((1-c_{s+1-j})^2V) - b_{s+1-j}(1-c_{s+1-j})\phi_0(V)\phi_1((1-c_{s+1-j})^2V) \\ =&b_j\phi_1(V)\phi_0(c_j^2V) - b_jc_j\phi_1(V)\phi_0(c_j^2V) \\ =&b_j(1-c_j)\phi_1((1-c_j)^2V) \\ =&\bar{B}_j(V), \end{aligned}$$

and

$$\begin{aligned} &V\phi_1(V)\bar{B}_{s+1-j} + \phi_0(V)B_{s+1-j} \\ =&b_{s+1-j}(1-c_{s+1-j})V\phi_1(V)\phi_1((1-c_{s+1-j})^2V) + b_{s+1-j}\phi_0(V)\phi_0((1-c_{s+1-j})^2V) \\ =&b_j\big(c_jV\phi_1(V)\phi_1(c_j^2V) + \phi_0(V)\phi_0(c_j^2V)\big) \\ =&b_j\phi_0((1-c_j)^2V) \\ =&B_j(V). \end{aligned}$$

These give the third and fourth equations of (6.24). Furthermore, it follows from (6.25) and the simplifying assumption $a_{ij} = b_j(c_i - c_j) + a_{s+1-i,s+1-j}$. We thus have

$$\begin{aligned} &\phi_0(c_{s+1-i}^2V)\bar{B}_j(V) - c_{s+1-i}\phi_1(c_{s+1-i}^2V)B_j(V) + A_{s+1-i,s+1-j}(V) \\ =&b_j(1-c_j)\phi_0((1-c_i)^2V)\phi_1((1-c_j)^2V) - b_j(1-c_i)\phi_1((1-c_i)^2V)\phi_0((1-c_j)^2V) \\ &+ a_{s+1-i,s+1-j}\phi_1((c_i-c_j)^2V) \\ =&\big(b_j(c_i-c_j) + a_{s+1-i,s+1-j}\big)\phi_1((c_i-c_j)^2V), \\ =&a_{ij}\phi_1((c_i-c_j)^2V) \\ =&A_{ij}(V), \end{aligned}$$

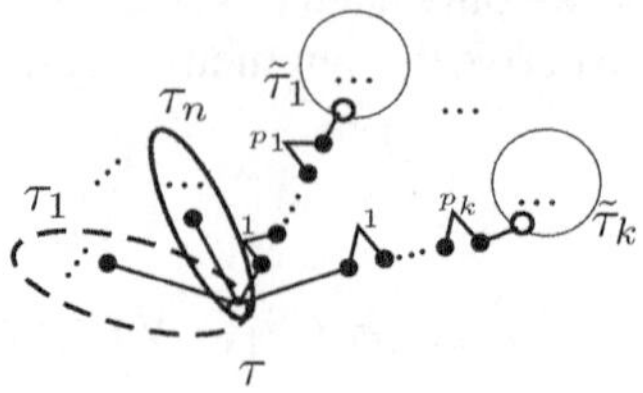

Fig. 6.1 Figure of tree $\tau = \tau_1 \times \tau_2 \times \cdots \times \tau_n \times \big(W_+ b_+ (b_+ B_+)^{p_1}(\tilde{\tau}_1)\big) \times \cdots \times \big(W_+ b_+ (b_+ B_+)^{p_k}(\tilde{\tau}_k)\big)$

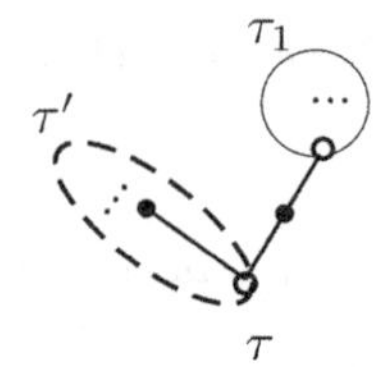

Fig. 6.2 Figure of tree $\tau = (W_+ b_+ (b_+ B_+)^0(\tau_1)) \times \tau'$

and consequently the second equation of (6.24) is satisfied. This completes the proof. □

Remark 6.3 Although the condition $\bar{b}_i = b_i(1 - c_i)$ required by Theorem 6.5, looks like an additional simplifying condition, in fact this assumption is already contained in the symplectic conditions for RKN methods in Theorem 6.4.

The following theorem is related to the order of a numerical method and the corresponding order conditions. Hence it seems plausible to gain some knowledge of special Nyström tree (SNT) and simplified special extended Nyström tree (SSENT), which is respectively designed to deal with order conditions of RKN and ERKN methods. Further details concerning SNT and SSENT can be found in [12, 35]. For the convenience of the proof, we introduce the following two definitions and a basic lemma, which will be used in the proof of the theorem later.

Definition 6.3 The degree of merge node $d(\tau)$ on SSENT are recursively defined as follows.
1. $d(\tau) = 0$, if $\tau \in SNT$;
2. $d(\tau) = k + \sum_{j=1}^{n} d(\tau_j) + \sum_{i=1}^{k} d(\tilde{\tau}_i)$, if $\tau = \tau_1 \times \tau_2 \times \cdots \times \tau_n \times \big(W_+ b_+ (b_+ B_+)^{p_1}(\tilde{\tau}_1)\big) \times \cdots \times \big(W_+ b_+ (b_+ B_+)^{p_k}(\tilde{\tau}_k)\big)$ and $\tau_i, \tilde{\tau}_j \in SSENT$, $p_i \in \mathbb{N}_+$ (see Fig. 6.1).

Definition 6.4 If $\tau = \big(W_+ b_+ (b_+ B_+)^0(\tau_1)\big) \times \tau'$ (see Fig. 6.2), then we define τ_1 to be the first generation of τ. We recursively define that τ_n is the nth ($n \geq 2$) generation of τ, if there exists $\tau_0 \in SSENT$ that τ_n is the first generation of τ_0 and τ_0 is the $(n-1)$th generation of τ.

Lemma 6.1 *If $\tau = \tau_1 \times \tau_2,\ \tau_1, \tau_2 \in SSENT$, then the order $\rho(\tau)$, the sign $s(\tau)$, the density $\gamma(\tau)$, and the weight $\Phi_i(\tau)$ satisfy*

$$\begin{aligned} \rho(\tau) =& \rho(\tau_1) + \rho(\tau_2) - 1, & s(\tau) = s(\tau_1) \cdot s(\tau_2), \\ \gamma(\tau) =& \rho(\tau) \cdot \frac{\gamma(\tau_1)}{\rho(\tau_1)} \cdot \frac{\gamma(\tau_2)}{\rho(\tau_2)}, & \Phi_i(\tau) = \Phi_i(\tau_1) \cdot \Phi_i(\tau_2). \end{aligned} \tag{6.26}$$

This lemma can be directly obtained from the definition of order, density and sign of an SSENT tree. Hence, we omit the remaining details of the proof here.

Theorem 6.6 *If $\Psi \in G$ is of order p ($p \geq 1$), then $\varphi(\Psi) \in \Omega$ is also of order p.*

Proof Suppose that Ψ is an s-stage RKN method. The theorem can be stated as follows.

If the order conditions [12]

$$\begin{cases} \displaystyle\sum_{i=1}^{s} \bar{b}_i \Phi_i(\tau) = \frac{1}{(\rho(\tau)+1)\gamma(\tau)}, & \forall \tau \in SNT_m, \quad m \leq p-1, \\ \displaystyle\sum_{i=1}^{s} b_i \Phi_i(\tau) = \frac{1}{\gamma(\tau)}, & \forall \tau \in SNT_m, \quad m \leq p, \end{cases} \tag{6.27}$$

hold for Ψ, then the order conditions [35]

$$\begin{cases} \displaystyle\sum_{i=1}^{s} \bar{B}_i \Phi_i(\tau) = \frac{\rho(\tau)!}{\gamma(\tau)s(\tau)} \phi_{\rho(\tau)+1} + \mathcal{O}(h^{p-\rho(\tau)}), & \forall \tau \in SSENT_m, \quad m \leq p-1, \\ \displaystyle\sum_{i=1}^{s} B_i \Phi_i(\tau) = \frac{\rho(\tau)!}{\gamma(\tau)s(\tau)} \phi_{\rho(\tau)} + \mathcal{O}(h^{p-\rho(\tau)+1}), & \forall \tau \in SSENT_m, \quad m \leq p, \end{cases} \tag{6.28}$$

also hold for $\varphi(\Psi)$ under the mapping (6.21).

We will prove this theorem by induction. To this end, the degree of merge node $d(\tau)$ is used as an indicator. To show this in detail, we split the proof in two parts with $d(\tau) = 0$ and $d(\tau) > 0$ for all $\tau \in SSENT$. As stated in [35], we should first note that SNT is in fact a subset of SSENT. In the first part of the proof we will show that for each $\tau \in SNT$, i.e. $d(\tau) = 0$, the statement of (6.28) holds.

Noting that $s(\tau) = 1$ holds for all $\tau \in SNT$, we can rewrite (6.28) as an equivalent form

$$\begin{cases} \displaystyle\sum_{i=1}^{s} \bar{B}_i^{(2l)} \Phi_i(\tau) = \frac{\rho(\tau)!}{\gamma(\tau)} \frac{(-1)^l (2l)!}{(\rho(\tau)+1+2l)!}, & \forall \tau \in SNT_m, \quad 2l \leq p-m-2, \\ \displaystyle\sum_{i=1}^{s} B_i^{(2l)} \Phi_i(\tau) = \frac{\rho(\tau)!}{\gamma(\tau)} \frac{(-1)^l (2l)!}{(\rho(\tau)+2l)!}, & \forall \tau \in SNT_m, \quad 2l \leq p-m-1, \end{cases} \tag{6.29}$$

with the definitions of matrix-valued functions (6.12–6.13). Furthermore, taking account of the mapping (6.21) and (6.12–6.13), we obtain the following equations

$$\begin{cases} A_{ij}^{(2k)} = a_{ij}(c_i - c_j)^k \dfrac{(-1)^k}{2k+1}, \\ \bar{B}_j^{(2k)} = \bar{b}_j(1 - c_j)^k \dfrac{(-1)^k}{2k+1}, \\ B_j^{(2k)} = b_j(1 - c_j)^k (-1)^k, \end{cases} \tag{6.30}$$

for the constants $A_{ij}^{(2k)}$, $B_j^{(2k)}$, $\bar{B}_j^{(2k)}$ by comparing the corresponding coefficients of each term V^k. Inserting the new expressions of (6.30) into (6.29) gives

$$\begin{cases} \sum_{i=1}^{s} \bar{b}_i (1 - c_i)^{2l} \Phi_i(\tau) = \dfrac{1}{\gamma(\tau)} \dfrac{\rho(\tau)!(2l+1)!}{(\rho(\tau)+1+2l)!}, & \forall \tau \in SNT_m, \quad m + 2l + 1 \le p - 1, \\ \sum_{i=1}^{s} b_i (1 - c_i)^{2l} \Phi_i(\tau) = \dfrac{1}{\gamma(\tau)} \dfrac{\rho(\tau)!(2l)!}{(\rho(\tau)+2l)!}, & \forall \tau \in SNT_m, \quad m + 2l + 1 \le p. \end{cases} \tag{6.31}$$

This means that we only need show the correctness of (6.31) instead of (6.28).

Noting that $\Phi_i(\tau)$ is the weight of SNT tree τ, then $c_i^k \Phi_i(\tau)$ will be the weight of a new SNT tree $\tau' = \tau_0 \times \tau$, where

$\tau_0 =$ 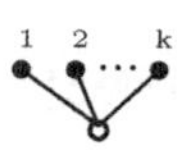

.

Considering Lemma 6.1, and noting that $\gamma(\tau_0) = \rho(\tau_0) = k + 1$, we then have

$$\rho(\tau') = \rho(\tau) + k, \quad \gamma(\tau') = \rho(\tau) \cdot \frac{\gamma(\tau)}{\rho(\tau_0)} \cdot \frac{\gamma(\tau)}{\rho(\tau_0)} = \gamma(\tau) \frac{\rho(\tau) + k}{\rho(\tau)}, \quad \Phi_i(\tau') = c_i^k \Phi_i(\tau). \tag{6.32}$$

For any $k \le 2l$, it can be deduced that $k + \rho(\tau) \le p$, i.e. $\rho(\tau') \le p$. Thus, together with the order conditions (6.27) for the special SNT tree τ' and (6.32), the following equations

$$\begin{cases} \sum_{i=1}^{s} \bar{b}_i \big(c_i^k \Phi_i(\tau)\big) = \dfrac{\rho(\tau)}{\gamma(\tau)(\rho(\tau)+k)(\rho(\tau)+k+1)}, & \forall \tau \in SNT_m, \ m + k + 1 \le p - 1, \\ \sum_{i=1}^{s} b_i \big(c_i^k \Phi_i(\tau)\big) = \dfrac{\rho(\tau)}{\gamma(\tau)(\rho(\tau)+k)}, & \forall \tau \in SNT_m, \ m + k + 1 \le p. \end{cases} \tag{6.33}$$

are satisfied. Multiplying by $(-1)^k C_{2l}^k$ the two sides of (6.33) and summing over k from 0 to $2l$, we obtain

$$
\begin{cases}
\sum_{k=0}^{2l}(-1)^k C_{2l}^k \sum_{i=1}^{s} \bar{b}_i\left(c_i^k \Phi_i(\tau)\right) = \sum_{i=1}^{s} \bar{b}_i(1-c_i)^{2l}\Phi_i(\tau) \\
\qquad = \sum_{k=0}^{2l}(-1)^k C_{2l}^k \dfrac{\rho(\tau)}{\gamma(\tau)(\rho(\tau)+k)(\rho(\tau)+k+1)}, \\
\sum_{k=0}^{2l}(-1)^k C_{2l}^k \sum_{i=1}^{s} b_i\left(c_i^k \Phi_i(\tau)\right) = \sum_{i=1}^{s} b_i(1-c_i)^{2l}\Phi_i(\tau) \\
\qquad = \sum_{k=0}^{2l}(-1)^k C_{2l}^k \dfrac{\rho(\tau)}{\gamma(\tau)(\rho(\tau)+k)}.
\end{cases}
\tag{6.34}
$$

Comparing (6.31) with (6.34), it can be concluded that if the two conditions,

$$
\begin{cases}
\sum_{k=0}^{2l}(-1)^k C_{2l}^k \dfrac{\rho(\tau)}{(\rho(\tau)+k)(\rho(\tau)+k+1)} = \dfrac{\rho(\tau)!(2l+1)!}{(\rho(\tau)+1+2l)!} \\
\sum_{k=0}^{2l}(-1)^k C_{2l}^k \dfrac{\rho(\tau)}{(\rho(\tau)+k)} = \dfrac{\rho(\tau)!(2l)!}{(\rho(\tau)+2l)!},
\end{cases}
\tag{6.35}
$$

hold for any $\rho(\tau) \le p$, the Eq. (6.31) will be satisfied. Here C_{2l}^k denotes the binomial coefficient $\frac{(2l)!}{k!(2l-k)!}$. It is clear that (6.35) is just a special case of the two identical equations

$$
\begin{cases}
\sum_{k=0}^{2l}(-1)^k C_{2l}^k \dfrac{n}{(n+k)(n+k+1)} = \dfrac{n!(2l+1)!}{(n+1+2l)!}, \quad \forall n \in \mathbb{N}_+, \\
\sum_{k=0}^{2l}(-1)^k C_{2l}^k \dfrac{n}{(n+k)} = \dfrac{n!(2l)!}{(n+2l)!}, \quad \forall n \in \mathbb{N}_+.
\end{cases}
\tag{6.36}
$$

Hence, the proof of this part is complete.

For the second part of the proof, we suppose that the order conditions for $\varphi(\Psi)$ hold for any $d(\tau) = K$ $(\rho(\tau) \le p)$. This means that the equations

$$
\begin{cases}
\sum_{i=1}^{s} \bar{B}_i \Phi_i(\tau) = \dfrac{\rho(\tau)!}{\gamma(\tau)s(\tau)}\phi_{\rho(\tau)+1} + \mathcal{O}(h^{p-\rho(\tau)}), \\
\sum_{i=1}^{s} B_i \Phi_i(\tau) = \dfrac{\rho(\tau)!}{\gamma(\tau)s(\tau)}\phi_{\rho(\tau)} + \mathcal{O}(h^{p-\rho(\tau)+1}),
\end{cases}
\tag{6.37}
$$

are satisfied for $\tau \in SSENT$, $d(\tau) = K$ $(\rho(\tau) \le p)$. We turn to showing that (6.37) also holds for any $\tau \in SSENT$ with $d(\tau) = K+1$ $(\rho(\tau) \le p)$.

Suppose that $\tau \in SSENT$, $\rho(\tau) \leq p$ and $d(\tau) = K + 1$. It follows from Definitions 6.3 and 6.4 that there must exist two integers $l \geq 1$, $n \geq 0$ and a corresponding $SSENT$ tree τ_n in τ, where τ_n with the particular form $\tau_n = \big(W_+ b_+ (b_+ B_+)^l(\tau_0)\big)$ is the nth generation of τ. Here, it is convenient to suppose that τ_{k+1} is the first generation of τ_k for $1 \leq k \leq n-1$ and τ_1 is the the first generation of τ, that is

$$\tau = \big(W_+ b_+ (b_+ B_+)^0(\tau_1)\big) \times \tau_1', \quad \tau_k = \big(W_+ b_+ (b_+ B_+)^0(\tau_k + 1)\big) \times \tau_{k+1}',$$

where τ_k' may be some $SSENT$ tree depending on τ. Using Lemma 6.1, we have the following formula

$$\begin{cases} s(\tau_k) = s(\tau_{k+1}) \cdot s(\tau_{k+1}'), \\ \rho(\tau_k) = \rho(\tau_{k+1}) + \rho(\tau_{k+1}') + 1, \\ \gamma(\tau_k) = \rho(\tau_k)\big(\rho(\tau_{k+1}) + 1\big)\gamma(\tau_{k+1})\dfrac{\gamma(\tau_{k+1}')}{\rho(\tau_{k+1}')}. \end{cases} \tag{6.38}$$

Recursively iterating (6.38) implies that

$$\begin{cases} s(\tau) = s(\tau_n) \cdot \displaystyle\prod_{k=1}^{n} s(\tau_{k+1}'), \\ \rho(\tau) = \rho(\tau_{k+1}) + n + \displaystyle\sum_{k=1}^{n} \rho(\tau_{k+1}'), \\ \gamma(\tau) = \rho(\tau)\big(\rho(\tau_n) + 1\big)\gamma(\tau_n) \cdot \displaystyle\prod_{k=1}^{n-1} \rho(\tau_k)\big(\rho(\tau_k) + 1\big) \cdot \prod_{k=1}^{n} \dfrac{\gamma(\tau_{k+1}')}{\rho(\tau_{k+1}')}. \end{cases} \tag{6.39}$$

Modifying the $SSENT$ tree τ by merely replacing τ_n with $\tilde{\tau}_n$, we obtain a new tree $\tilde{\tau}$ (certainly τ_k will become a new one $\tilde{\tau}_k$ and τ_k' remains the same). Let $\delta = \rho(\tau_n) - \rho(\tilde{\tau}_n)$. Then it follows from (6.39) that

$$\begin{cases} s(\tilde{\tau}) = s(\tau) \cdot \dfrac{s(\tilde{\tau}_n)}{s(\tau_n)}, \\ \rho(\tau) - \rho(\tilde{\tau}) = \rho(\tau_1) - \rho(\tilde{\tau}_1) = \cdots = \rho(\tau_n) - \rho(\tilde{\tau}_n) = \delta, \\ \gamma(\tilde{\tau}) = \gamma(\tau)\dfrac{\gamma(\tilde{\tau}_n)}{\gamma(\tau_n)}\Big(1 - \dfrac{\delta}{\rho(\tau)}\Big)\Big(1 - \dfrac{\delta}{\rho(\tau_n) + 1}\Big) \cdot \displaystyle\prod_{k=1}^{n-1}\Big(1 - \dfrac{\delta}{\rho(\tau_k)}\Big)\Big(1 - \dfrac{\delta}{\rho(\tau_k) + 1}\Big). \end{cases} \tag{6.40}$$

For $\tau_n = \big(W_+ b_+ (b_+ B_+)^l(\tau_0)\big)$ (see Fig. 6.3), we can derive

$$\begin{cases} s(\tau_n) = (-1)^l s(\tau_0), \\ \rho(\tau_n) = \rho(\tau_0) + 2l + 2, \\ \gamma(\tau_n) = \gamma(\tau_0)\dfrac{(\rho(\tau_0) + 2l + 2)!}{\rho(\tau_0)!(2l)!}. \end{cases} \tag{6.41}$$

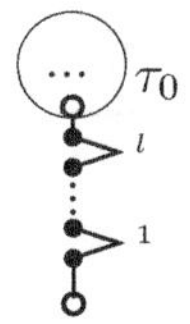

Fig. 6.3 Figure of tree $\tau_n = \big(W_+ b_+ (b_+ B_+)^l (\tau_0)\big)$

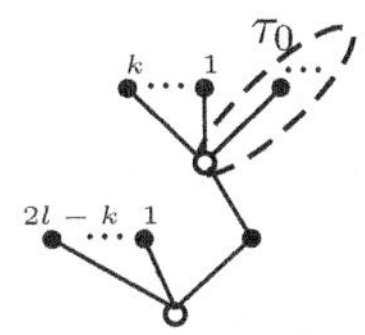

Fig. 6.4 Figure of tree $\tilde{\tau}_n$

We now consider $\tilde{\tau}_n$ (see Fig. 6.4) with the particular form

$$\tilde{\tau}_n = \overbrace{\overset{\bullet}{\circ} \times \overset{\bullet}{\circ} \times \cdots \times \overset{\bullet}{\circ}}^{2l-k\ folds} \times \big(W_+ b_+ (b_+ B_+)^0 (\tau_0 \times \overbrace{\overset{\bullet}{\circ} \times \cdots \times \overset{\bullet}{\circ}}^{k\ folds})\big), \quad 0 \leq k \leq 2l.$$

We then have

$$\begin{cases} s(\tilde{\tau}_n) = s(\tau_0), \\ \rho(\tilde{\tau}_n) = \rho(\tau_0) + 2l + 2, \\ \gamma(\tilde{\tau}_n) = \big(\rho(\tau_0) + k\big)\big(\rho(\tau_0) + k + 1\big)\big(\rho(\tau_0) + 21 + 2\big)\dfrac{\gamma(\tau_0)}{\rho(\tau_0)}. \end{cases} \tag{6.42}$$

Combining (6.40) with (6.41–6.42), we derive the following equations

$$\begin{cases} s(\tilde{\tau}) = s(\tau) \cdot (-1)^l, \\ \rho(\tilde{\tau}) = \rho(\tau), \ i.e.\ \delta = 0, \\ \gamma(\tilde{\tau}) = \gamma(\tau)\dfrac{\big(\rho(\tau_0) + k\big)\big(\rho(\tau_0) + k + 1\big)(\rho(\tau_0) - 1)!(2l)!}{(\rho(\tau_0) + 2l + 1)!}. \end{cases} \tag{6.43}$$

Keep in mind that the weights of τ and $\tilde{\tau}(k)$ (here we concretely denote the new tree $\tilde{\tau}$ as $\tilde{\tau}(k)$ since it really depends on k) can be respectively expressed as

$$\begin{cases} \Phi_i(\tau) = \sum_{\mu=1}^{s}\sum_{\nu=1}^{s} \Delta_\mu A_{\mu\nu}^{(2l)} \Xi_\nu = \sum_{\mu=1}^{s}\sum_{\nu=1}^{s} \Delta_\mu a_{\mu\nu}(c_\mu - c_\nu)^{2l} \dfrac{(-1)^l}{2l+1} \Xi_\nu, \\ \Phi_i(\tilde{\tau}(k)) = \sum_{\mu=1}^{s}\sum_{\nu=1}^{s} c_\mu^{2l-k} \Delta_\mu a_{\mu\nu} c_\nu^k \Xi_\nu, \end{cases} \tag{6.44}$$

where Δ_μ, Ξ_ν are some summation depending on other branches of τ. It follows from (6.44) that

$$\Phi_i(\tau) = \sum_{k=0}^{2l} \frac{(-1)^{l+k} C_{2l}^k}{2l+1} \Phi_i(\tilde{\tau}(k)). \tag{6.45}$$

Moreover, from Definition 6.3, the equation $d(\tilde{\tau}(k)) = d(\tau) - 1 = K$ holds for any $0 \le k \le 2l$. By the assumption in this part we know that the order conditions (6.37) are satisfied for such $\tilde{\tau}(k)$ $(0 \le k \le 2l)$. Combining the Eq. (6.45) with the order conditions for $\tilde{\tau}(k)$, we have

$$\begin{cases} \sum_{i=1}^{s} \bar{B}_i \Phi_i(\tau) = \sum_{k=0}^{2l} \dfrac{(-1)^{l+k} C_{2l}^k}{2l+1} \sum_{i=1}^{s} \bar{B}_i \Phi_i(\tilde{\tau}(k)) \\ \qquad = \sum_{k=0}^{2l} \dfrac{(-1)^{l+k} C_{2l}^k}{2l+1} \dfrac{\rho(\tilde{\tau}(k))!}{\gamma(\tilde{\tau}(k)) s(\tilde{\tau}(k))} \phi_{\rho(\tilde{\tau}(k))+1} + \mathcal{O}(h^{p-\rho(\tilde{\tau}(k))}), \\ \sum_{i=1}^{s} B_i \Phi_i(\tau) = \sum_{k=0}^{2l} \dfrac{(-1)^{l+k} C_{2l}^k}{2l+1} \sum_{i=1}^{s} B_i \Phi_i(\tilde{\tau}(k)) \\ \qquad = \sum_{k=0}^{2l} \dfrac{(-1)^{l+k} C_{2l}^k}{2l+1} \dfrac{\rho(\tilde{\tau}(k))!}{\gamma(\tilde{\tau}(k)) s(\tilde{\tau}(k))} \phi_{\rho(\tilde{\tau}(k))} + \mathcal{O}(h^{p-\rho(\tilde{\tau}(k))+1}). \end{cases} \tag{6.46}$$

Taking account of the formula (6.43), which is related to τ and the new $SSENT$ tree $\tilde{\tau}(k)$, the equations in (6.46) imply that

$$\begin{cases} \sum_{i=1}^{s} \bar{B}_i \Phi_i(\tau) = \sum_{k=0}^{2l} \dfrac{(-1)^k C_{2l}^k (\rho(\tau_0) + 2l + 1)!}{(\rho(\tau_0) + k)(\rho(\tau_0) + k + 1)(\rho(\tau_0) - 1)!(2l+1)!} \cdot \dfrac{\rho(\tau)!}{\gamma(\tau)s(\tau)} \phi_{\rho(\tau)+1} \\ \qquad + \mathcal{O}(h^{p-\rho(\tau)}), \\ \sum_{i=1}^{s} B_i \Phi_i(\tau) = \sum_{k=0}^{2l} \dfrac{(-1)^k C_{2l}^k (\rho(\tau_0) + 2l + 1)!}{(\rho(\tau_0) + k)(\rho(\tau_0) + k + 1)(\rho(\tau_0) - 1)!(2l+1)!} \cdot \dfrac{\rho(\tau)!}{\gamma(\tau)s(\tau)} \phi_{\rho(\tau)} \\ \qquad + \mathcal{O}(h^{p-\rho(\tau)+1}), \end{cases} \tag{6.47}$$

by replacing $s(\tilde{\tau}(k)), \gamma(\tilde{\tau}(k)), \rho(\tilde{\tau}(k))$ with $s(\tau), \gamma(\tau), \rho(\tau)$. Comparing (6.47) with (6.37), we observe that whether the order conditions are satisfied for τ depends on the following equation

$$\sum_{k=0}^{2l} \frac{(-1)^k C_{2l}^k (\rho(\tau_0) + 2l + 1)!}{(\rho(\tau_0) + k)(\rho(\tau_0) + k + 1)(\rho(\tau_0) - 1)!(2l+1)!} = 1. \tag{6.48}$$

It has also been known that (6.48) is just a special case of the identity

$$\sum_{k=0}^{2l} \frac{(-1)^k C_{2l}^k (n+2l+1)!}{(n+k)(n+k+1)(n-1)!(2l+1)!} = 1, \quad n, l \in \mathbb{N}_+.$$

Hence, (6.37) also holds for any $\tau \in SSENT$ that $d(\tau) = K + 1$ ($\rho(\tau) \leq p$).

Since both the base case and the inductive step have been demonstrated by the above two processes, we have completed the proof of this theorem. □

The theorems established in this section essentially reveal the relation between classical RKN methods and ERKN methods. An original and natural way to construct certain high-order ERKN methods is based on the order conditions (6.28), by which only general fifth/sixth order ERKN methods have now been found and it is quite difficult to find an arbitrarily high order ERKN method due to the high complexity. However, the theoretical results stated above can provide us with another simple way to construct high-order ERKN methods. In this way, we only need to find its corresponding reduced RKN method and these have been well studied in the literature. Furthermore, ERKN methods with particular properties, such as symmetry and symplecticity, can also be obtained via the mapping (6.21) and their reduced RKN methods. Finally, we are able to obtain knowledge of ERKN methods by studying RKN methods instead of ERKN methods themselves, especially in the construction of high-order ERKN methods.

6.5 Numerical Experiments

In order to show applications of the results presented in the previous section, we conduct some numerical experiments. First, we select some classical RKN methods as follows:

- RKN3s4: the three-stage symmetric symplectic Runge–Kutta–Nyström method of order four proposed by Forest and Ruth [6];
- RKN7s6: the seven-stage symmetric symplectic Runge–Kutta–Nyström method of order six given by Okunbor and Skeel [21];
- RKN6s6: the six-stage Runge–Kutta–Nyström method of order six given by Papakostas and Tsitourasy [22];
- RKN16s10: the sixteen-stage Runge–Kutta–Nyström method of order ten presented by Dormand, El-Mikkawy and Prince [5].

Then from the mapping (6.21), their corresponding ERKN methods are also obtained with the individual properties maintained. We denote their corresponding ERKN methods as ERKN3s4, ERKN7s6, ERKN6s6, and ERKN16s10, respectively.

During the numerical experiments, we will display the efficiency curves and the conservation of energy for each Hamiltonian system. It should be noted that, the numerical solution obtained by RKN16s10 with a small stepsize is used as the standard reference solution, if the analytical solution cannot be explicitly given.

Problem 6.1 We first consider an orbital problem with perturbation [29]

$$\begin{cases} q_1'' = -q_1 - \dfrac{2\varepsilon+\varepsilon^2}{r^5} q_1, & q_1(0) = 1, \quad q_1'(0) = 0, \\ q_2'' = -q_2 - \dfrac{2\varepsilon+\varepsilon^2}{r^5} q_2, & q_2(0) = 0, \quad q_1'(0) = 1+\varepsilon, \end{cases}$$

where $r = \sqrt{q_1^2 + q_2^2}$, and the analytical solution is given by

$$q_1(t) = \cos(t + \varepsilon t), \quad q_2(t) = \sin(t + \varepsilon t),$$

with the Hamiltonian

$$H = \frac{p_1^2 + p_2^2}{2} + \frac{q_1^2 + q_2^2}{2} - \frac{2\varepsilon + \varepsilon^2}{3(q_1^2 + q_2^2)^{\frac{3}{2}}}.$$

We numerically integrate the problem on the interval [0, 1000] with $\varepsilon = 10^{-3}$. It is clear from the efficiency curves in Fig. 6.5a that ERKN methods are usually superior to their corresponding reduced RKN methods with respect to the global error (GE), and a high-order RKN/ERKN method also shows better performance than a low-order RKN/ERKN method in dealing with an oscillatory problem. Figure 6.5b demonstrates that symplectic methods (RKN3s4, RKN7s6, ERKN3s4, ERKN7s6) show their good energy-conservation property for the Hamiltonian, while the other methods without symplecticity lead to a linear energy dissipation on a long-term scale. The detailed results on the energy conservation for ERKN3s4 and ERKN7s6 are shown in Fig. 6.6. All these results from Figs. 6.5 and 6.6 are consistent with those of classical numerical methods, and show that ERKN methods obtained by the map φ with the reduced RKN methods are remarkably efficient and effective.

Problem 6.2 We consider the Hénon–Heilse system

$$\begin{cases} q_1'' + q_1 = -2q_1 q_2, \\ q_2'' + q_2 = -q_1^2 + q_2^2, \end{cases} \tag{6.49}$$

with the initial conditions $q_1(0) = \sqrt{\frac{5}{48}}$, $p_2(0) = \frac{1}{4}$, $q_2(0) = p_1(0) = 0$. The Hamiltonian of the system is given by

$$H(p, q) = \frac{1}{2}(p_1^2 + p_2^2) + \frac{1}{2}(q_1^2 + q_2^2) + q_1^2 q_2 - \frac{1}{3} q_2^3.$$

We first integrate this problem on the interval [0, 1000] with different stepsizes. The efficiency curves for each method are shown in Fig. 6.7a, which indicate the comparable efficiency for ERKN methods to their corresponding reduced RKN ones, since $||M||$ now nearly has the same magnitude as $||\frac{\partial f}{\partial q}||$. This phenomenon also

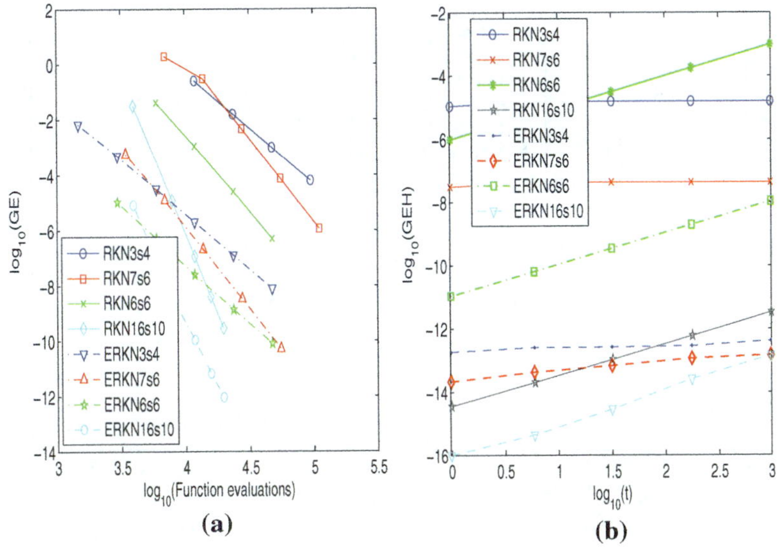

Fig. 6.5 Results for Problem 6.1: **a** The log-log plot of maximum global error GE against number of function evaluations; **b** the logarithm of the maximum global error of Hamiltonian $GEH = max|H_n - H_0|$ against $\log_{10}(t)$ with the stepsize $h = 1$

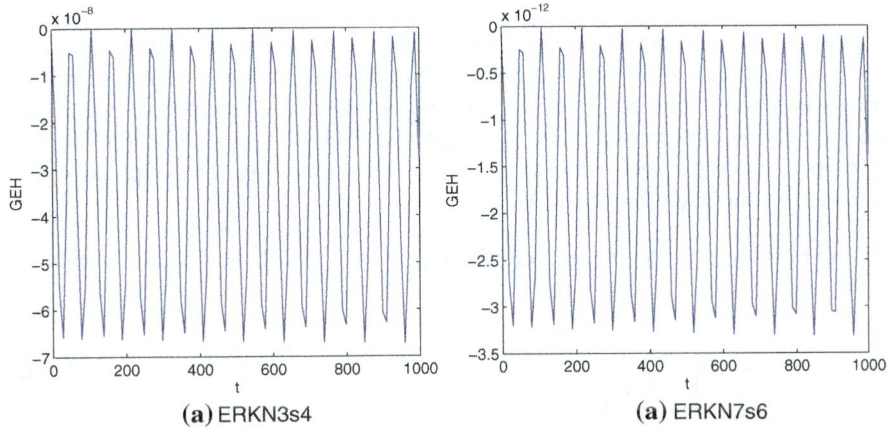

Fig. 6.6 Results for Problem 6.1: the global error for Hamiltonian of symplectic methods ERKN3s4 and ERKN7s6 with the stepsize $h = 2$

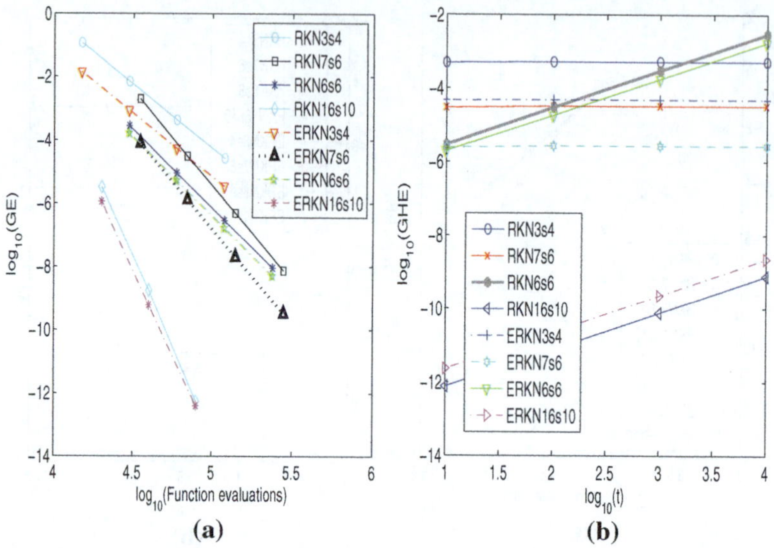

Fig. 6.7 Results for Problem 6.2: **a** The log-log plot of maximum global error GE against the number of function evaluations; **b** the logarithm of the maximum global error of Hamiltonian $GEH = max|H_n - H_0|$ against $\log_{10}(t)$ with the stepsize $h = 0.5$

occurs in Fig. 6.7b, where the energy-conservation curve for each method is plotted. Besides, we can also observe from Fig. 6.7 that the difference between symplectic methods is a little more remarkable than that between non-symplectic ones. The good energy-conservation property of symplectic ERKN methods (ERKN3s4 and ERKN7s6) is clearly shown in Fig. 6.8, which demonstrates that the symplecticity is maintained by the mapping φ very well.

Problem 6.3 We consider the sine-Gordon equation [13] with the periodic boundary conditions

$$\begin{cases} \dfrac{\partial^2 u}{\partial t^2} = \dfrac{\partial^2 u}{\partial x^2} - \sin u, \ -5 \le x \le 5, \ t \ge 0, \\ u(-5, t) = u(5, t). \end{cases} \tag{6.50}$$

A semi-discretization on the spatial variable with the second-order symmetric differences gives the following differential equations in time

$$\frac{d^2 U}{dt^2} + MU = F(U), \tag{6.51}$$

where $U(t) = (u_1(t), \ldots, u_N(t))^\intercal$ with $u_i(t) \approx u(x_i, t)$, $x_i = -5 + i\Delta x$ for $i = 1, \ldots, N$, $\Delta x = 10/N$, and

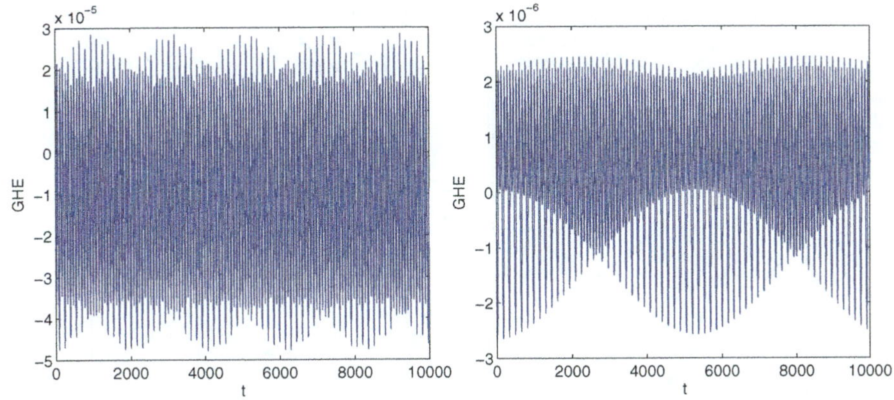

Fig. 6.8 Results for Problem 6.2: the global error for Hamiltonian of symplectic methods ERKN3s4 and ERKN7s6 with the stepsize $h = 0.5$

$$M = \frac{1}{\Delta x^2}\begin{pmatrix} 2 & -1 & & & -1 \\ -1 & 2 & -1 & & \\ & \ddots & \ddots & \ddots & \\ & & -1 & 2 & -1 \\ -1 & & & -1 & 2 \end{pmatrix}$$

$$F(U) = -\sin(U) = -(\sin u_1, \ldots, \sin u_N)^{\intercal}.$$

The corresponding Hamiltonian is given by

$$H(U', U) = \frac{1}{2}U'^{\intercal}U' + \frac{1}{2}U^{\intercal}MU - \big(\cos u_1 + \ldots + \cos u_N\big).$$

For this problem, we take the initial conditions as

$$U(0) = (\pi)_{i=1}^{N},\ U_t(0) = \sqrt{N}\big(0.01 + \sin(\frac{2\pi i}{N})\big)_{i=1}^{N},$$

with $N = 64$. For the efficiency curves in Fig. 6.9a, we integrate the problem for $t_{end} = 10$ with the different stepsizes. Figure 6.9a shows the good efficiency and accuracy of all the ERKN methods. In Fig. 6.9b, all methods give rise to energy dissipation even if the method is symplectic. This phenomenon is mainly caused by the chaotic behavior of the problem, in which a sufficiently small perturbation may lead to a significant error after a long time, and this increase is always exponential. It can be observed from Fig. 6.10 that the numerical reference solution obtained by RKN16s10 obviously shows notable difference between the initial interval [0, 100] and the terminal interval [900, 1000]. Figure 6.11 gives a further demonstration that the global errors of RKN7s6 and ERKN7s6 increase nearly in an exponential fashion

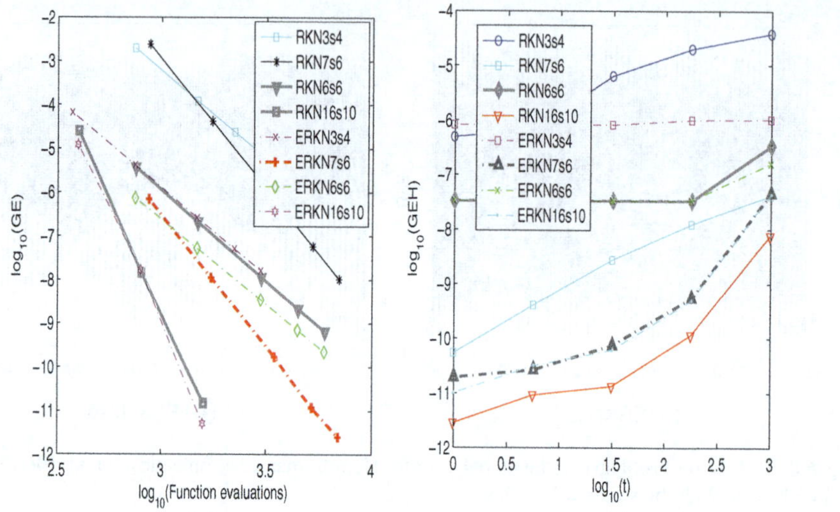

(a) (b)

Fig. 6.9 Results for Problem 6.3: **a** The log-log plot of maximum global error GE against number of function evaluations; **b** the logarithm of the maximum global error of Hamiltonian $GEH = max|H_n - H_0|$ against $\log_{10}(t)$ with the stepsize $h = 0.01$

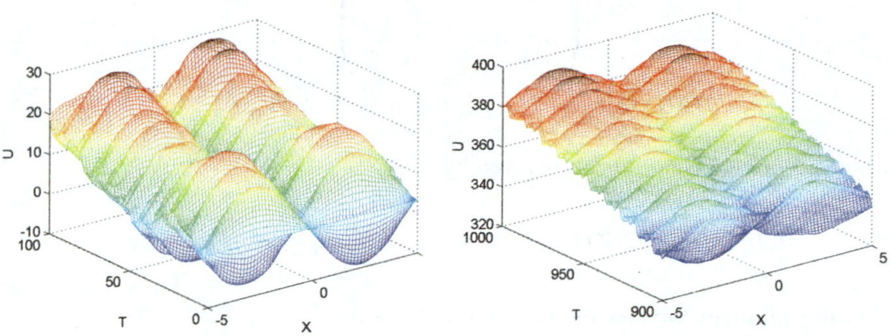

(a) [0, 100] (b) [900, 1000]

Fig. 6.10 Results for Problem 6.3: The numerical reference solution in different interval obtained by RKN16s10 with the stepsize $h = 0.001$

with time t. This may lead to non-conservation for symplectic methods in practical numerical computations.

Problem 6.4 We consider the Fermi-Pasta-Ulam problem (see, e.g. [10]), which can be expressed by a Hamiltonian system with the Hamiltonian

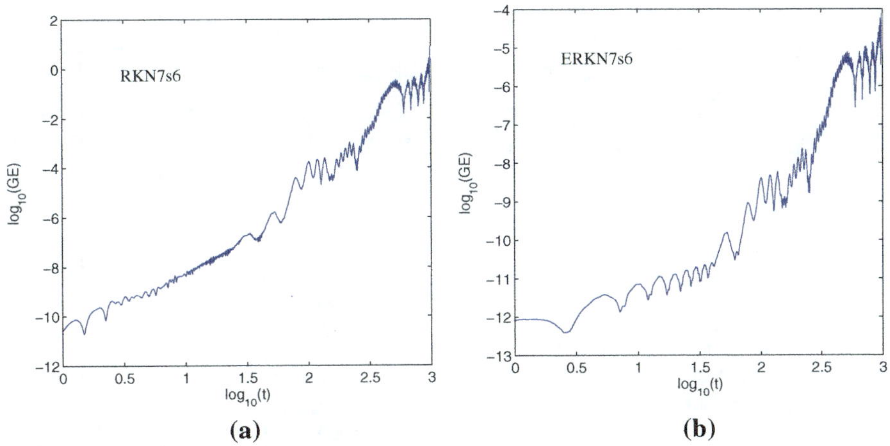

Fig. 6.11 Results for Problem 6.3: the global error for RKN7s6 and ERKN7s6 with the stepsize $h = 0.01$, respectively

$$
\begin{aligned}
H(y,x) = {} & \frac{1}{2}\sum_{i=1}^{2m} y_i^2 + \frac{\omega^2}{2}\sum_{i=1}^{m} x_{m+i}^2 + \frac{1}{4}\Big((x_1 - x_{m+1})^4 \\
& + \sum_{i=1}^{m-1}(x_{i+1} - x_{m+i-1} - x_i - x_{m+i})^4 + (x_m + x_{2m})^4\Big),
\end{aligned}
\tag{6.52}
$$

where x_i represents a scaled displacement of the ith stiff spring, x_{m+i} is a scaled expansion (or compression) of the ith stiff spring, and y_i, y_{m+i} are their velocities (or momenta).

The corresponding Hamiltonian system is given by

$$
\begin{cases}
x' = H_y(y,x), \\
y' = -H_x(y,x),
\end{cases}
\tag{6.53}
$$

which can be also written in the equivalent form of the oscillatory second-order differential equations

$$
x''(t) + Mx(t) = -\nabla_x U(x), \tag{6.54}
$$

where

$$
y = x', \; M = \begin{pmatrix} \mathbf{0}_{m\times m} & \mathbf{0}_{m\times m} \\ \mathbf{0}_{m\times m} & \omega^2 I_{m\times m} \end{pmatrix},
$$

$$
U(x) = \frac{1}{4}\Big((x_1 - x_{m+1})^4 + \sum_{i=1}^{m-1}(x_{i+1} - x_{m+i-1} - x_i - x_{m+i})^4 + (x_m + x_{2m})^4\Big).
$$

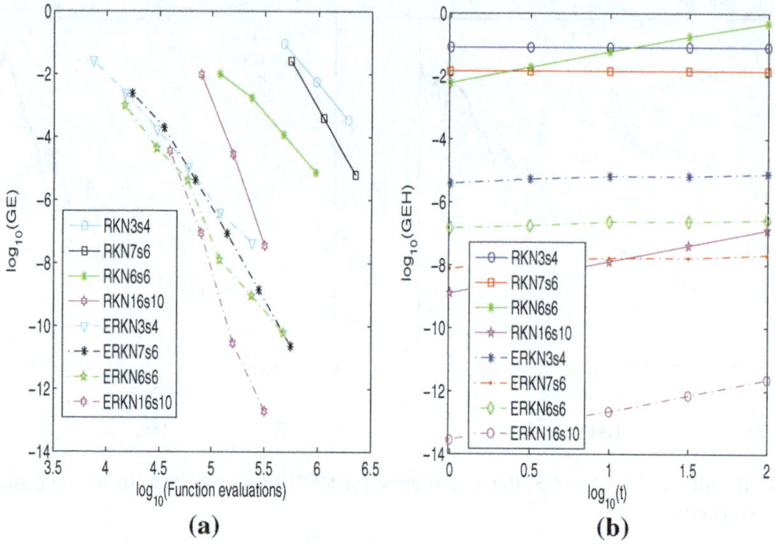

Fig. 6.12 Results for Problem 6.4: **a** The log-log plot of maximum global error GE against the number of function evaluations; **b** the logarithm of the maximum global error of Hamiltonian $GEH = max|H_n - H_0|$ against $\log_{10}(t)$ with the stepsize $h = 0.005$

In the experiment, we choose

$$m = 3,\ x_1(0) = 1,\ y_1(0) = 1,\ x_4(0) = \frac{1}{\omega},\ y_1(0) = 1,\ \omega = 200,$$

and choose zero for the remaining initial values. The numerical results are shown in Fig. 6.12. Similarly to Problem 6.3, we also integrate the equation over a short interval with $t_{end} = 20$ to decrease the influence of chaotic behavior. Both figures show good efficiency in the global error and energy error for the ERKN methods. In particular, symplecticity is also maintained by the map φ, such as ERKN3s4 and ERKN7s6 in Fig. 6.13 display a stable energy conservation in the sense of numerical computation.

6.6 Conclusions and Discussions

In this chapter, we studied in greater depth the ERKN methods for solving (6.1) based on the group structure of numerical methods. After the construction of the RKN group and the ERKN group, we first presented the inherent relationship between ERKN and RKN methods, that is, there exists an epimorphism η of the ERKN group onto the RKN group. This epimorphism gives a clear and exact meaning for the word *extension* from RKN methods to ERKN methods and describes the ERKN group in

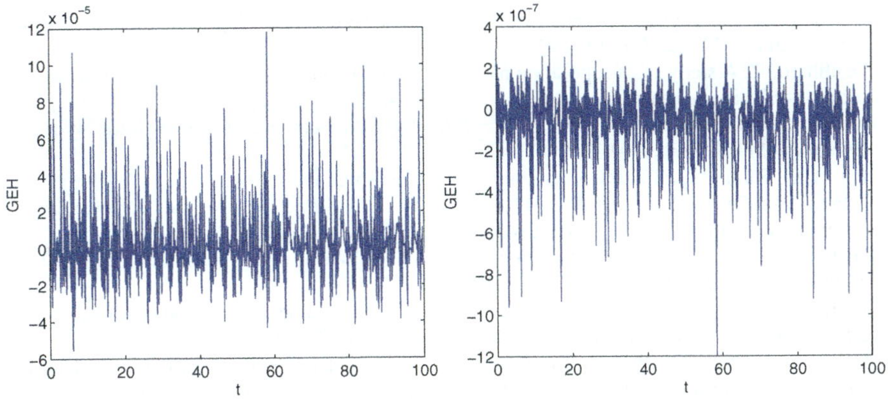

(a) ERKN3s4 **(b)** ERKN7s6

Fig. 6.13 Results for Problem 6.4: the global error for Hamiltonian of symplectic methods ERKN3s4 and ERKN7s6 with the stepsize $h = 0.01$

terms of the RKN group in the sense of structure preservation. Moreover, we established the particular mapping φ defined by (6.21), which maps an RKN method to an ERKN method. A series of theorems about the mapping show that the image element can be regarded as an ideal representative element for each congruence class of the ERKN group. That is, the image ERKN element almost preserves as many properties as the RKN element does. This mapping φ also provides us with an effective approach to constructing arbitrarily high order (symmetric or symplectic) ERKN methods, whereas the original way based directly on order conditions (symmetric or symplectic conditions) is more complicated. Furthermore, the numerical simulations in Sect. 6.5 strongly support our theoretical analysis in Sect. 6.4, and the numerical results are really promising. The high-order structure-preserving ERKN methods obtained in such a simple and effective way show better efficiency and accuracy than their corresponding reduced methods (letting $V = 0$), namely, the RKN methods.

Remember that the exponential Fourier collocation methods for first-order differential equations were derived and analysed in Sect. 6.3. Accordingly, the next chapter will present trigonometric collocation methods for multi-frequency and multidimensional second-order oscillatory systems.

The material of this chapter is based on the recent work by Mei and Wu [17].

References

1. Blanes, S., Casas, F., Ros, J.: Symplectic integrators with processing: a general study. SIAM J. Sci. Comput. **21**, 711–727 (1999)
2. Blanes, S., Casas, F., Ros, J.: New families of symplectic Runge-Kutta-Nyström integration methods. NAA 102–109 (2000)

3. Calvo, M.P., Sanz-Serna, J.M.: High-order symplectic Runge-Kutta-Nyström methods. SIAM J. Sci. Comput. **14**, 1237–1252 (1993)
4. Deuflhard, P.: A study of extrapolation methods based on multistep schemes without parasitic solutions. Z. angew. Math. Phys. **30**, 177–189 (1979)
5. Dormand, J.R., El-Mikkawy, M.E., Prince, P.J.: High order embedded Runge-Kutta-Nyström formulae. IMA J. Numer. Anal. **7**, 423–430 (1987)
6. Forest, E., Ruth, R.D.: Fourth-order symplectic integration. Phys. D **43**, 105–117 (1990)
7. García-Archilla, B., Sanz-Serna, J.M., Skeel, R.D.: Long-time-step methods for oscillatory differential equations. SIAM J. Sci. Comput. **20**, 930–963 (1998)
8. Gautschi, W.: Numerical integration of ordinary differential equations based on trigonometric polynomials. Numer. Math. **3**, 381–397 (1961)
9. Hairer, E., Lubich, C.: Long-time energy conservation of numerical methods for oscillatory differential equations. SIAM J. Numer. Anal. **38**, 414–441 (2000)
10. Hairer, E., Lubich, C., Wanner, G.: Geometric Numerical Integration: Structure-Preserving Algorithms for Ordinary Differential Equations, 2nd edn. Springer, Berlin (2006)
11. Hairer, E., Wanner, G.: On the Butcher group and general multi-value methods. Computing **13**, 1–15 (1974)
12. Hairer, E., Nørsett, S.P., Wanner, G.: Solving Ordinary Differential Equations I: Non-stiff Systems. Springer, Berlin (1987)
13. Hochbruck, M., Lubich, C.: A Gautschi-type method for oscillatory second-order differential equations. Numer. Math. **83**, 403–426 (1999)
14. Franco, J.M.: Runge-Kutta-Nyström methods adapted to the numerical integration of perturbed oscillators. Comput. Phys. Commun. **147**, 770–787 (2002)
15. Franco, J.M., Gómez, I.: Construction of explicit symmetric and symplectic methods of Runge-Kutta-Nyström type for solving perturbed oscillators. Appl. Math. Comput. **219**, 4637–4649 (2013)
16. Liu, K., Wu, X.Y.: High-order symplectic and symmetric composition methods for multi-frequency and multi-dimensional oscillatory Hamiltonian systems. J. Comput. Math. **33**, 355–377 (2015)
17. Mei, L.J., Wu, X.Y.: The construction of arbitrary order ERKN methods based on group theory for solving oscillatory Hamiltonian systems with applications. J. Comput. Phys. **323**, 171–190 (2016)
18. Nyström, E.J.: Ueber die numerische intrgration von differentialgleichungen. Acta Soc. Sci. Fenn. **50**, 1–54 (1925)
19. Okunbor, D.I., Skeel, R.D.: An explicit Runge-Kutta-Nyström method is canonical if and only if its adjoint is explicit. SIAM J. Numer. Anal. **29**, 521–527 (1992)
20. Okunbor, D.I., Skeel, R.D.: Explicit canonical methods for Hamiltonian systems. Math. Comput. **59**, 439–455 (1992)
21. Okunbor, D.I., Skeel, R.D.: Canonical Runge-Kutta-Nyström methods of orders five and six. J. Comput. Appl. Math. **51**, 375–382 (1994)
22. Papakostas, S.N., Tsitourasy, C.: High phase-lag-order Runge-Kutta-Nyström pairs. SIAM J. Sci. Comput. **21**, 747–763 (1999)
23. Sanz-Serna, J.M.: Symplectic integrators for Hamiltonian problems: an overview. Acta Numerica **1**, 243–286 (1992)
24. Sanz-Serna, J.M., Calvo, M.P.: Numerical Hamiltonian Problems. Chapman & Hall, London (1994)
25. Shi, W., Wu, X.Y., Xia, J.: Explicit multi-symplectic extended leap-frog methods for Hamiltonian wave equations. J. Comput. Phys. **231**, 7671–7694 (2012)
26. Suris, Y.B.: The canonicity of mappings generated by Runge-Kutta type methods when integrating the systems $\dot{x} = -\partial U/\partial x$, Zh. Vychisl. Mat. i Mat. Fiz. 29, 202.211 (in Russian); same as U.S.S.R. Comput. Maths. Phys. **29**, 138–144 (1989)
27. Wang, B., Wu, X.Y.: Explicit multi-frequency symmetric extended RKN integrators for solving multi-frequency and multidimensional oscillatory reversible systems. Calcolo **52**, 207–231 (2015)

28. Wu, X.Y., Liu, K., Shi, W.: Structure-Preserving Algorithms for Oscillatory Differential Equations II. Springer, Berlin (2015)
29. Wu, X.Y., Wang, B., Liu, K., Zhao, H.: ERKN methods for long-term integration of multidimensional orbital problems. Appl. Math. Model. **37**, 2327–2336 (2013)
30. Wu, X.Y., Wang, B., Xia, J.: Explicit symplectic multidimensional exponential fitting modified Runge-Kutta-Nyström methods. BIT Numer. Math. **52**, 773–795 (2012)
31. Wu, X.Y., You, X., Wang, B.: Structure-Preserving Algorithms for Oscillatory Differential Equations. Springer, Berlin (2013)
32. Wu, X.Y., You, X., Shi, W., Wang, B.: ERKN integrators for systems of oscillatory second-order differential equations. Comput. Phys. Commun. **181**, 1873–1887 (2010)
33. Wu, X.Y., You, X., Xia, J.: Order conditions for ARKN methods solving oscillatory systems. Comput. Phys. Commun. **180**, 2250–2257 (2009)
34. Yang, H., Wu, X.Y., You, X., Fang, Y.: Extended RKN-type methods for numerical integration of perturbed oscillators. Comput. Phys. Commun. **180**, 1777–1794 (2009)
35. Yang, H., Zeng, X., Wu, X.Y., Ru, Z.: A simplified Nyströ-tree theory for extended Runge-Kutta-Nyström integrators solving multi-frequency oscillatory systems. Comput. Phys. Commun. **185**, 2841–2850 (2014)
36. You, X., Zhao, J., Yang, H., Fang, Y., Wu, X.Y.: Order conditions for RKN methods solving general second-order oscillatory systems. Numer. Algorithms **66**, 147–176 (2014)

Chapter 7
Trigonometric Collocation Methods for Multi-frequency and Multidimensional Oscillatory Systems

This chapter presents a class of trigonometric collocation methods based on Lagrange basis polynomials for solving multi-frequency and multidimensional oscillatory systems $q''(t) + Mq(t) = f\big(q(t)\big)$. The properties of the collocation methods are investigated in detail. It is shown that the convergence condition of these methods is independent of $\|M\|$, which is crucial for solving multi-frequency oscillatory systems.

7.1 Introduction

The numerical treatment of multi-frequency oscillatory systems is a computational problem of overarching importance in a wide range of applications, such as quantum physics, circuit simulations, flexible body dynamics and mechanics (see, e.g. [3, 5, 6, 8, 9, 32, 33] and the references therein). The main purpose of this chapter is to construct and analyse a class of efficient collocation methods for solving multi-frequency and multidimensional oscillatory second-order differential equations of the form

$$q''(t) + Mq(t) = f\big(q(t)\big), \qquad q(0) = q_0, \ q'(0) = q_0', \qquad t \in [0, t_{\text{end}}], \tag{7.1}$$

where M is a $d \times d$ positive semi-definite matrix implicitly containing the dominant frequencies of the oscillatory problem and $f : \mathbb{R}^d \to \mathbb{R}^d$ is an analytic function. The solution of this system is a multi-frequency nonlinear oscillator because of the presence of the linear term Mq. The system (7.1) is a highly oscillatory problem when $\|M\| \gg 1$. In recent years, various numerical methods for approximating solutions of oscillatory systems have been developed by many researchers. Readers are referred to [12–14, 21–25, 31] and the references therein. Once it is further assumed that M is symmetric and f is the negative gradient of a real-valued function $U(q)$, the system (7.1) is identical to the following initial value Hamiltonian system

X. Wu and B. Wang, *Recent Developments in Structure-Preserving Algorithms for Oscillatory Differential Equations*, https://doi.org/10.1007/978-981-10-9004-2_7

$$\begin{cases} \dot{q}(t) = \nabla_p H(q(t), p(t)), & q(0) = q_0, \\ \dot{p}(t) = -\nabla_q H(q(t), p(t)), & p(0) = p_0 \equiv q_0', \end{cases} \tag{7.2}$$

with the Hamiltonian

$$H(q, p) = \frac{1}{2} p^\intercal p + \frac{1}{2} q^\intercal M q + U(q). \tag{7.3}$$

This is an important Hamiltonian problem which has seen studied by many authors (see, e.g. [3–5, 8, 9]).

In [26], the authors took advantage of shifted Legendre polynomials to obtain a local Fourier expansion of the system (7.1) and derived the so-called trigonometric Fourier collocation methods. Theoretical analysis and numerical experiments in [26] showed that the trigonometric Fourier collocation methods are more efficient than some earlier codes. Motivated by the work in [26], this chapter is devoted to the formulation and analysis of another trigonometric collocation method for solving multi-frequency oscillatory second-order systems (7.1). We will consider a classical approach and use Lagrange polynomials to derive a class of trigonometric collocation methods. Because of this different approach, compared with the methods in [26], the collocation methods have a simpler scheme and can be implemented at a lower cost in practical computations. These trigonometric collocation methods are designed by interpolating the function f of (7.1) by Lagrange basis polynomials, and incorporating the variation-of-constants formula and the idea of collocation. It is noted that these integrators are a class of collocation methods and they share all of the important features of collocation methods. We analyse the properties of trigonometric collocation methods and study the convergence of the fixed-point iteration for these methods. It is important to emphasize that for the trigonometric collocation methods, the convergence condition is independent of $\|M\|$, which is a crucial property for solving highly oscillatory systems.

This chapter is organized as follows. In Sect. 7.2, we formulate the scheme of trigonometric collocation methods based on Lagrange basis polynomials. The properties of the obtained methods are analysed in Sect. 7.3. In Sect. 7.4, a fourth-order scheme of the collocation methods is presented and numerical results confirm that the method proposed in this chapter yields a dramatic improvement. Conclusions are included in the last section.

7.2 Formulation of the Methods

We first restrict the multi-frequency oscillatory system (7.1) to the interval $[0, h]$ with any $h > 0$:

$$q''(t) + Mq(t) = f\big(q(t)\big), \quad q(0) = q_0, \ q'(0) = q_0', \quad t \in [0, h]. \tag{7.4}$$

With regard to the variation-of-constants formula for (7.1) given in [29], we have the following result on the exact solution $q(t)$ of the system (7.1) and its derivative $q'(t) = p(t)$:

$$\begin{cases} q(t) = \phi_0(t^2 M)q_0 + t\phi_1(t^2 M)p_0 + t^2 \int_0^1 (1-z)\phi_1\big((1-z)^2 t^2 M\big) f\big(q(tz)\big)dz, \\ p(t) = -tM\phi_1(t^2 M)q_0 + \phi_0(t^2 M)p_0 + t \int_0^1 \phi_0\big((1-z)^2 t^2 M\big) f\big(q(tz)\big)dz, \end{cases} \tag{7.5}$$

where $t \in [0, h]$ and

$$\phi_i(M) := \sum_{l=0}^{\infty} \frac{(-1)^l M^l}{(2l+i)!}, \qquad i = 0, 1. \tag{7.6}$$

It follows from (7.5) that

$$\begin{cases} q(h) = \phi_0(V)q_0 + h\phi_1(V)p_0 + h^2 \int_0^1 (1-z)\phi_1\big((1-z)^2 V\big) f\big(q(hz)\big)dz, \\ p(h) = -hM\phi_1(V)q_0 + \phi_0(V)p_0 + h \int_0^1 \phi_0\big((1-z)^2 V\big) f\big(q(hz)\big)dz, \end{cases} \tag{7.7}$$

where $V = h^2 M$.

The main idea in designing practical schemes to solve (7.1) is to approximate $f(q)$ in (7.7) by a quadrature. In this chapter, we interpolate $f(q)$ as

$$f\big(q(\xi h)\big) \sim \sum_{j=1}^{s} l_j(\xi) f\big(q(c_j h)\big), \qquad \xi \in [0, 1], \tag{7.8}$$

where

$$l_j(x) = \prod_{k=1, k\neq j}^{s} \frac{x - c_k}{c_j - c_k}, \tag{7.9}$$

for $j = 1, \ldots, s$, are the Lagrange basis polynomials, and $c_1, \ldots, c_s$ are distinct real numbers ($s \geq 1,\ 0 \leq c_i \leq 1$). Then replacing $f(q(\xi h))$ in (7.7) by the series (7.8) yields an approximation of $q(h)$, $p(h)$ as follows:

$$\begin{cases} \tilde{q}(h) = \phi_0(V)q_0 + h\phi_1(V)p_0 + h^2 \sum_{j=1}^{s} I_{1,j} f\big(\tilde{q}(c_j h)\big), \\ \tilde{p}(h) = -hM\phi_1(V)q_0 + \phi_0(V)p_0 + h \sum_{j=1}^{s} I_{2,j} f\big(\tilde{q}(c_j h)\big), \end{cases} \tag{7.10}$$

where

$$I_{1,j} := \int_0^1 l_j(z)(1-z)\phi_1\big((1-z)^2 V\big)dz, \quad I_{2,j} := \int_0^1 l_j(z)\phi_0\big((1-z)^2 V\big)dz. \tag{7.11}$$

From the variation-of-constants formula (7.5) for (7.4), the approximation (7.10) satisfies the following system

$$\begin{cases} \tilde{q}'(\xi h) = \tilde{p}(\xi h), & \tilde{q}(0) = q_0, \\ \tilde{p}'(\xi h) = -M\tilde{q}(\xi h) + \sum_{j=1}^{s} l_j(\xi) f\big(\tilde{q}(c_j h)\big), & \tilde{p}(0) = p_0. \end{cases} \tag{7.12}$$

In what follows we first approximate $f\big(\tilde{q}(c_j h)\big)$, $I_{1,j}$, $I_{2,j}$ in (7.10), and then formulate a class of trigonometric collocation methods.

7.2.1 The Computation of $f(\tilde{q}(c_j h))$

It follows from (7.12) that $\tilde{q}(c_i h)$ for $i = 1, 2, \ldots, s$, can be obtained by solving the following discrete problems:

$$\tilde{q}''(c_i h) + M\tilde{q}(c_i h) = \sum_{j=1}^{s} l_j(c_i) f\big(\tilde{q}(c_j h)\big), \quad \tilde{q}(0) = q_0, \ \tilde{q}'(0) = p_0. \tag{7.13}$$

Set $\tilde{q}_i = \tilde{q}(c_i h)$ for $i = 1, 2, \ldots, s$. Then (7.13) can be solved by the variation-of-constants formula (7.5) in the form:

$$\tilde{q}_i = \phi_0(c_i^2 V) q_0 + c_i h \phi_1(c_i^2 V) p_0 + (c_i h)^2 \sum_{j=1}^{s} \tilde{I}_{c_i,j} f(\tilde{q}_j), \quad i = 1, 2, \ldots, s,$$

where

$$\tilde{I}_{c_i,j} := \int_0^1 l_j(c_i z)(1-z)\phi_1\big((1-z)^2 c_i^2 V\big) dz, \qquad i, j = 1, \ldots, s. \tag{7.14}$$

7.2.2 The Computation of $I_{1,j}$, $I_{2,j}$, $\tilde{I}_{c_i,j}$

With the definition (7.9), the integrals $I_{1,j}$, $I_{2,j}$, $\tilde{I}_{c_i,j}$ appearing in (7.11) and (7.14) can be computed as follows:

$$\begin{aligned} I_{1,j} &= \int_0^1 l_j(z)(1-z)\phi_1\big((1-z)^2 V\big) dz \\ &= \prod_{k=1,k\neq j}^{s} \sum_{l=0}^{\infty} \int_0^1 \frac{z - c_k}{c_j - c_k} (1-z)^{2l+1} dz \frac{(-1)^l V^l}{(2l+1)!} \end{aligned}$$

$$= \sum_{l=0}^{\infty}\Big(\prod_{k=1,k\neq j}^{s}\frac{\frac{1}{2l+3}-c_k}{c_j-c_k}\Big)\frac{(-1)^l V^l}{(2l+2)!}=\sum_{l=0}^{\infty}l_j\Big(\frac{1}{2l+3}\Big)\frac{(-1)^l V^l}{(2l+2)!},$$

$$I_{2,j} = \int_0^1 l_j(z)\phi_0\big((1-z)^2V\big)dz = \prod_{k=1,k\neq j}^{s}\sum_{l=0}^{\infty}\int_0^1\frac{z-c_k}{c_j-c_k}(1-z)^{2l}dz\frac{(-1)^l V^l}{(2l)!}$$

$$= \sum_{l=0}^{\infty}\Big(\prod_{k=1,k\neq j}^{s}\frac{\frac{1}{2l+2}-c_k}{c_j-c_k}\Big)\frac{(-1)^l V^l}{(2l+1)!}=\sum_{l=0}^{\infty}l_j\Big(\frac{1}{2l+2}\Big)\frac{(-1)^l V^l}{(2l+1)!},$$

$$\tilde{I}_{c_i,j} = \int_0^1 l_j(c_iz)(1-z)\phi_1\big((1-z)^2c_i^2V\big)dz$$

$$= \prod_{k=1,k\neq j}^{s}\sum_{l=0}^{\infty}\int_0^1\frac{c_iz-c_k}{c_j-c_k}(1-z)^{2l+1}dz\frac{(-1)^l(c_i^2V)^l}{(2l+1)!}$$

$$= \sum_{l=0}^{\infty}\Big(\prod_{k=1,k\neq j}^{s}\frac{\frac{c_i}{2l+3}-c_k}{c_j-c_k}\Big)\frac{(-1)^l (c_i^2V)^l}{(2l+2)!}=\sum_{l=0}^{\infty}l_j\Big(\frac{c_i}{2l+3}\Big)\frac{(-1)^l (c_i^2V)^l}{(2l+2)!},$$

$$i,j=1,\ldots,s.$$

If M is symmetric and positive semi-definite, we have the decomposition of M as follows:

$$M = P^{\mathsf{T}}W^2P = \Omega_0^2 \text{ with } \Omega_0 = P^{\mathsf{T}}WP,$$

where P is an orthogonal matrix and $W = \mathrm{diag}(\lambda_k)$ with nonnegative diagonal entries which are the square roots of the eigenvalues of M. Hence the above integrals become

$$I_{1,j} = P^{\mathsf{T}}\int_0^1 l_j(z)W^{-1}\sin\big((1-z)W\big)dzP$$

$$= P^{\mathsf{T}}\mathrm{diag}\Big(\int_0^1 l_j(z)\lambda_k^{-1}\sin\big((1-z)\lambda_k\big)dz\Big)P,$$

$$I_{2,j} = P^{\mathsf{T}}\int_0^1 l_j(z)\cos\big((1-z)W\big)dzP = P^{\mathsf{T}}\mathrm{diag}\Big(\int_0^1 l_j(z)\cos\big((1-z)\lambda_k\big)dz\Big)P,$$

$$\tilde{I}_{c_i,j} = P^{\mathsf{T}}\int_0^1 l_j(c_iz)(c_iW)^{-1}\sin\big((1-z)c_iW\big)dzP$$

$$= P^{\mathsf{T}}\mathrm{diag}\Big(\int_0^1 l_j(c_iz)(c_i\lambda_k)^{-1}\sin\big((1-z)c_i\lambda_k\big)dz\Big)P,$$

$$i,j=1,\ldots,s.$$

Here, it is noted that $W^{-1}\sin\big((1-z)W\big)$, $(c_iW)^{-1}\sin\big((1-z)c_iW\big)$ are well defined also for singular W. The case $\lambda_k=0$ gives:

$$\int_0^1 l_j(z)\lambda_k^{-1}\sin\big((1-z)\lambda_k\big)dz=\int_0^1 l_j(z)(1-z)dz,$$
$$\int_0^1 l_j(z)\cos\big((1-z)\lambda_k\big)dz=\int_0^1 l_j(z)dz,$$
$$\int_0^1 l_j(c_iz)(c_i\lambda_k)^{-1}\sin\big((1-z)c_i\lambda_k\big)dz=\int_0^1 l_j(c_iz)(1-z)dz,$$

which can be evaluated easily since $l_j(z)$ is a polynomial function. If $\lambda_k\neq 0$, they can be evaluated as follows:

$$\begin{aligned}
&\int_0^1 l_j(z)\lambda_k^{-1}\sin\big((1-z)\lambda_k\big)dz\\
&=1/\lambda_k\int_0^1 l_j(z)\sin\big((1-z)\lambda_k\big)dz\\
&=1/\lambda_k^2\int_0^1 l_j(z)d\cos\big((1-z)\lambda_k\big)\\
&=1/\lambda_k^2 l_j(1)-1/\lambda_k^2 l_j(0)\cos(\lambda_k)-1/\lambda_k^2\int_0^1 l_j'(z)\cos\big((1-z)\lambda_k\big)dz\\
&=1/\lambda_k^2 l_j(1)-1/\lambda_k^2 l_j(0)\cos(\lambda_k)+1/\lambda_k^3\int_0^1 l_j'(z)d\sin\big((1-z)\lambda_k\big)\\
&=1/\lambda_k^2 l_j(1)-1/\lambda_k^2 l_j(0)\cos(\lambda_k)-1/\lambda_k^3 l_j'(0)\sin(\lambda_k)\\
&\quad-1/\lambda_k^3\int_0^1 l_j''(z)\sin\big((1-z)\lambda_k\big)dz\\
&=1/\lambda_k^2 l_j(1)-1/\lambda_k^2 l_j(0)\cos(\lambda_k)-1/\lambda_k^3 l_j'(0)\sin(\lambda_k)\\
&\quad-1/\lambda_k^4 l_j''(1)+1/\lambda_k^4 l_j''(0)\cos(\lambda_k)+1/\lambda_k^5 l_j^{(3)}(0)\sin(\lambda_k)\\
&\quad+1/\lambda_k^5\int_0^1 l_{l,j}^{(4)}(z)\sin\big((1-z)\lambda_k\big)dz\\
&=\cdots\\
&\quad\vdots\\
&=\sum_{k=0}^{\lfloor\deg(l_j)/2\rfloor}(-1)^k/\lambda_k^{2k+2}\Big(l_j^{(2k)}(1)-l_j^{(2k)}(0)\cos(\lambda_k)-1/\lambda_k l_j^{(2k+1)}(0)\sin(\lambda_k)\Big),
\end{aligned}$$

for $i=1,2,\ldots,s$, where $\deg(l_j)$ is the degree of l_j and $\lfloor\deg(l_j)/2\rfloor$ denotes the integral part of $\deg(l_j)/2$.

Likewise, we can obtain

$$\begin{aligned}
&\int_0^1 l_j(z)\cos\big((1-z)\lambda_k\big)dz\\
&=\sum_{k=0}^{\lfloor \deg(l_j)/2\rfloor}(-1)^k/\lambda_k^{2k+1}\Big(l_j^{(2k)}(0)\sin(\lambda_k)+1/\lambda_k l_j^{(2k+1)}(1)-1/\lambda_k^2 l_j^{(2k+1)}(0)\cos(\lambda_k)\Big),\\
&\int_0^1 l_j(c_i z)(c_i\lambda_k)^{-1}\sin\big((1-z)c_i\lambda_k\big)dz\\
&=\sum_{k=0}^{\lfloor \deg(l_j)/2\rfloor}(-1)^k/(c_i\lambda_k)^{2k+2}\Big(l_j^{(2k)}(c_i)-l_j^{(2k)}(0)\cos(c_i\lambda_k)-1/\lambda_k l_j^{(2k+1)}(0)\sin(c_i\lambda_k)\Big),
\end{aligned} \tag{7.15}$$

for $i,\ j=1,2,\ldots,s$.

7.2.3 The Scheme of Trigonometric Collocation Methods

We are now in a position to present a class of trigonometric collocation methods for the multi-frequency oscillatory second-order oscillatory system (7.1).

Definition 7.1 A trigonometric collocation method for integrating the multi-frequency oscillatory system (7.1) is defined as

$$\begin{cases}
\tilde{q}_i=\phi_0(c_i^2V)q_0+c_ih\phi_1(c_i^2V)p_0+(c_ih)^2\sum\limits_{j=1}^{s}\tilde{I}_{c_i,j}f(\tilde{q}_j), \quad i=1,2,\ldots,s,\\
\tilde{q}(h)=\phi_0(V)q_0+h\phi_1(V)p_0+h^2\sum\limits_{j=1}^{s}I_{1,j}f(\tilde{q}_j),\\
\tilde{p}(h)=-hM\phi_1(V)q_0+\phi_0(V)p_0+h\sum\limits_{j=1}^{s}I_{2,j}f(\tilde{q}_j),
\end{cases} \tag{7.16}$$

where h is the stepsize and $I_{1,j},\ I_{2,j},\ \tilde{I}_{c_i,j}$ can be computed as stated in Sect. 7.2.2.

Remark 7.1 In [26], the authors took advantage of shifted Legendre polynomials to obtain a local Fourier expansion of the system (7.1) and derived trigonometric Fourier collocation methods (TFCMs). TFCMs are a subclass of s-stage ERKN methods presented in [29] with the following Butcher tableau:

$$\begin{array}{c|ccc}
c_1 & \sum_{j=0}^{r-1} II_{1,j,c_1}(V)b_1\widehat{P}_j(c_1) & \dots & \sum_{j=0}^{r-1} II_{1,j,c_1}(V)b_s\widehat{P}_j(c_s) \\
\vdots & \vdots & \ddots & \vdots \\
c_s & \sum_{j=0}^{r-1} II_{1,j,c_s}(V)b_1\widehat{P}_j(c_1) & \cdots & \sum_{j=0}^{r-1} II_{1,j,c_s}(V)b_s\widehat{P}_j(c_s) \\
\hline
 & \sum_{j=0}^{r-1} II_{1,j}(V)b_1\widehat{P}_j(c_1) & \cdots & \sum_{j=0}^{r-1} II_{1,j}(V)b_s\widehat{P}_j(c_s) \\
\hline
 & \sum_{j=0}^{r-1} II_{2,j}(V)b_1\widehat{P}_j(c_1) & \cdots & \sum_{j=0}^{r-1} II_{2,j}(V)b_s\widehat{P}_j(c_s)
\end{array} \tag{7.17}$$

where

$$\begin{aligned}
II_{1,j}(V) &:= \int_0^1 \widehat{P}_j(z)(1-z)\phi_1\big((1-z)^2V\big)dz, \\
II_{2,j}(V) &:= \int_0^1 \widehat{P}_j(z)\phi_0\big((1-z)^2V\big)dz, \\
II_{1,j,c_i}(V) &:= \int_0^1 \widehat{P}_j(c_iz)(1-z)\phi_1\big((1-z)^2c_i^2V\big)dz,
\end{aligned}$$

r is an integer with the requirement: $2 \le r \le s$, all $\widehat{P}_j$ are shifted Legendre polynomials over the interval $[0, 1]$, and c_l, b_l for $l = 1, 2, \dots, s$ are the node points and the quadrature weights of a quadrature formula, respectively.

It is noted that the method (7.16) is also the subclass of s-stage ERKN methods with the following Butcher tableau:

$$\begin{array}{c|ccc}
c_1 & \tilde{I}_{c_1,1} & \dots & \tilde{I}_{c_1,s} \\
\vdots & \vdots & \ddots & \vdots \\
c_s & \tilde{I}_{c_s,1} & \cdots & \tilde{I}_{c_s,s} \\
\hline
 & I_{1,1} & \cdots & I_{1,s} \\
\hline
 & I_{2,1} & \cdots & I_{2,s}
\end{array}, \tag{7.18}$$

where

$$\begin{aligned}
I_{1,j} &:= \int_0^1 l_j(z)(1-z)\phi_1\big((1-z)^2V\big)dz, \\
I_{2,j} &:= \int_0^1 l_j(z)\phi_0\big((1-z)^2V\big)dz, \\
\tilde{I}_{c_i,j} &= \int_0^1 l_j(c_iz)(1-z)\phi_1\big((1-z)^2c_i^2V\big)dz.
\end{aligned}$$

From (7.17) and (7.18), it follows clearly that the coefficients of (7.18) are simpler than (7.17). Therefore, the scheme of the methods derived in this chapter is much simpler than that given in [26]. The obtained methods can be implemented at a lower cost in practical computations, which will be shown by the numerical experiments in Sect. 7.4. The reason for this better efficiency is that we use a classical approach and choose Lagrange polynomials to give a local Fourier expansion of the system (7.1).

Remark 7.2 We also note that in the recent monograph [2], it has been shown that the approach of constructing energy-preserving methods for Hamiltonian systems which are based upon the use of shifted Legendre polynomials (such as in [1]) and Lagrange polynomials constructed on Gauss–Legendre nodes (such as in [10]) leads to precisely the same methods. Therefore, by choosing special real numbers $c_1, \ldots, c_s$ for (7.18) and special quadrature formulae for (7.17), the methods given in this chapter may have some connections with those in [26], which need to be investigated.

Remark 7.3 It is noted that the method (7.16) can be applied to the system (7.1) with an arbitrary matrix M since trigonometric collocation methods do not need the symmetry of M. Moreover, the method (7.16) exactly integrates the linear system $q'' + Mq = 0$ and it has an additional advantage of energy preservation for linear systems while respecting structural invariants and geometry of the underlying problem. The method approximates the solution in the interval $[0, h]$. We then repeat this procedure with equal ease over the next interval. Namely, we can consider the obtained result as the initial condition for a new initial value problem in the interval $[h, 2h]$. In this way, the method (7.16) can approximate the solution in an arbitrary interval $[0, t_{\text{end}}]$ with $t_{\text{end}} = Nh$.

When $M = 0$, (7.1) reduces to a special and important class of systems of second-order ODEs expressed in the traditional form

$$q''(t) = f\big(q(t)\big), \qquad q(0) = q_0, \ \ q'(0) = q_0', \qquad t \in [0, t_{\text{end}}]. \tag{7.19}$$

For this case, with the definition (7.6) and the results of $I_{1,j}$, $I_{2,j}$, $\tilde{I}_{c_i,j}$ in Sect. 7.2.2, the trigonometric collocation method (7.16) reduces to the following RKN-type method.

Definition 7.2 An RKN-type collocation method for integrating the traditional second-order ODEs (7.19) is defined as

$$\begin{cases} \tilde{q}_i = q_0 + c_i h p_0 + (c_i h)^2 \displaystyle\sum_{j=1}^{s} \frac{1}{2} l_j\Big(\frac{c_i}{3}\Big) f(\tilde{q}_j), \quad i = 1, 2, \ldots, s, \\ \tilde{q}(h) = q_0 + h p_0 + h^2 \displaystyle\sum_{j=1}^{s} \frac{1}{2} l_j\Big(\frac{1}{3}\Big) f(\tilde{q}_j), \\ \tilde{p}(h) = p_0 + h \displaystyle\sum_{j=1}^{s} l_j\Big(\frac{1}{2}\Big) f(\tilde{q}_j), \end{cases} \tag{7.20}$$

where h is the stepsize.

Remark 7.4 The method (7.20) is the subclass of s-stage RKN methods with the following Butcher tableau:

$$\begin{array}{c|c} c & \bar{A}=(\bar{a}_{ij})_{s\times s} \\ \hline & \bar{b}^T \\ \hline & b^T \end{array} = \begin{array}{c|ccc} c_1 & l_1\left(\frac{c_1}{3}\right)/2 & \dots & l_s\left(\frac{c_1}{3}\right)/2 \\ \vdots & \vdots & \ddots & \vdots \\ c_s & l_1\left(\frac{c_s}{3}\right)/2 & \cdots & l_s\left(\frac{c_s}{3}\right)/2 \\ \hline & l_1\left(\frac{1}{3}\right)/2 & \cdots & l_s\left(\frac{1}{3}\right)/2 \\ \hline & l_1\left(\frac{1}{2}\right) & \cdots & l_s\left(\frac{1}{2}\right) \end{array} \tag{7.21}$$

Thus, by letting $M = 0$, the trigonometric collocation methods yield a subclass of RKN methods for solving traditional second-order ODEs, which demonstrates wide applications of the methods.

7.3 Properties of the Methods

For the exact solution of (7.2) at $t = h$, let $\mathbf{y}(h) = \left(q^{\intercal}(h), p^{\intercal}(h)\right)^{\intercal}$. Then the oscillatory Hamiltonian system (7.2) can be rewritten in the form

$$\mathbf{y}'(\xi h) = F(\mathbf{y}(\xi h)) := \begin{pmatrix} p(\xi h) \\ -Mq(\xi h) + f\left(q(\xi h)\right) \end{pmatrix}, \quad \mathbf{y}_0 = \begin{pmatrix} q_0 \\ p_0 \end{pmatrix}, \tag{7.22}$$

for $0 \le \xi \le 1$. The Hamiltonian is

$$H(\mathbf{y}) = \frac{1}{2}p^{\intercal}p + \frac{1}{2}q^{\intercal}Mq + U(q). \tag{7.23}$$

On the other hand, if we denote the updates of (7.16) by

$$\omega(h) = \left(\tilde{q}^{\intercal}(h), \tilde{p}^{\intercal}(h)\right)^{\intercal},$$

then we have

$$\omega'(\xi h) = \begin{pmatrix} \tilde{p}(\xi h) \\ -M\tilde{q}(\xi h) + \sum\limits_{j=1}^{s} l_j(\xi) f\left(\tilde{q}(c_j h)\right) \end{pmatrix}, \quad \omega_0 = \begin{pmatrix} q_0 \\ p_0 \end{pmatrix}. \tag{7.24}$$

The next lemma is useful for the subsequent analysis.

Lemma 7.1 *Let* $g : [0, h] \to \mathbb{R}^d$ *have* j *continuous derivatives. Then*

$$\int_0^1 P_j(\tau)g(\tau h)d\tau = \mathcal{O}(h^j),$$

where $P_j(\tau)$ is an orthogonal polynomial of degree j on the interval [0, 1].

Proof We assume that $g(\tau h)$ can be expanded in Taylor series at the origin for sake of simplicity. Then, for all $j \geq 0$, by considering that $P_j(\tau)$ is orthogonal to all polynomials of degree $n < j$:

$$\int_0^1 P_j(\tau)g(\tau h)d\tau = \sum_{n=1}^{\infty} \frac{g^{(n)}(0)}{n!}h^n \int_0^1 P_j(\tau)\tau^n d\tau = \mathcal{O}(h^j). \qquad \square$$

7.3.1 The Order of Energy Preservation

In this subsection we analyse the order of preservation of the Hamiltonian energy.

Theorem 7.1 *Assume that c_l for $l = 1, 2, \ldots, s$ are chosen as the node points of an s-point Gauss–Legendre's quadrature over the integral* [0, 1]. *Then we have*

$$H(\omega(h)) - H(\mathbf{y}_0) = \mathcal{O}(h^{2s+1}),$$

where the constant symbolized by $\mathcal{O}$ is independent of h.

Proof It follows from Lemma 7.1, (7.23) and (7.24) that

$$\begin{aligned}
H(\omega(h)) - H(\mathbf{y}_0) &= h\int_0^1 \nabla H(\omega(\xi h))^{\intercal}\omega'(\xi h)d\xi \\
&= h\int_0^1 \Big(\big(M\tilde{q}(\xi h) - f(\tilde{q}(\xi h))\big)^{\intercal}, \ \tilde{p}(\xi h)^{\intercal}\Big) \cdot \begin{pmatrix} \tilde{p}(\xi h) \\ -M\tilde{q}(\xi h) + \sum_{j=1}^{s} l_j(\xi) f(\tilde{q}(c_j h)) \end{pmatrix} d\xi \\
&= h\int_0^1 \tilde{p}(\xi h)^{\intercal}\Big(\sum_{j=1}^{s} l_j(\xi) f(\tilde{q}(c_j h)) - f\big(\tilde{q}(\xi h)\big)\Big)d\xi.
\end{aligned}$$

Moreover, we have

$$f\big(\tilde{q}(\xi h)\big) - \sum_{j=1}^{s} l_j(\xi) f\big(\tilde{q}(c_j h)\big) = \frac{f^{(s+1)}\big(\tilde{q}(\xi h)\big)|_{\xi=\zeta}}{(n+1)!}\prod_{i=1}^{s}(\xi h - c_i h).$$

Here $f^{(s+1)}\big(\tilde{q}(\xi h)\big)$ denotes the $(s+1)$th derivative of $f(\tilde{q}(t))$ with respect to t. We then obtain

$$
\begin{aligned}
H(\omega(h)) - H(\mathbf{y}_0) &= -h\int_0^1 \tilde{p}(\xi h)^\intercal \frac{f^{(s+1)}\big(\tilde{q}(\xi h)\big)|_{\xi=\zeta}}{(n+1)!}\prod_{i=1}^{s}(\xi h - c_i h)d\xi \\
&= -h^{s+1}\int_0^1 \tilde{p}(\xi h)^\intercal \frac{f^{(s+1)}\big(\tilde{q}(\xi h)\big)|_{\xi=\zeta}}{(n+1)!}\prod_{i=1}^{s}(\xi - c_i)d\xi.
\end{aligned}
$$

Since c_l for $l = 1, 2, \ldots, s$ are chosen as the node points of a s-point Gauss–Legendre's quadrature over the integral $[0, 1]$, $\prod\limits_{i=1}^{s}(\xi - c_i)$ is an orthogonal polynomial of degree s on the interval $[0, 1]$. Therefore, using Lemma 7.1 we obtain

$$
H(\omega(h)) - H(\mathbf{y}_0) = -h^{s+1}\mathscr{O}(h^s) = \mathscr{O}(h^{2s+1}).
$$

This gives the result of the theorem. □

7.3.2 The Order of Quadratic Invariant

We next turn to the quadratic invariant $Q(\mathbf{y}) = q^\intercal Dp$ of (7.1). The quadratic form Q is a first integral of (7.1) if and only if $p^\intercal Dp + q^\intercal D(f(q) - Mq) = 0$ for all $p, q \in \mathbb{R}^d$. This implies that D is a skew-symmetric matrix and that $q^\intercal D(f(q) - Mq) = 0$ for any $q \in \mathbb{R}^d$. The following result states the degree of accuracy of the method (7.16).

Theorem 7.2 *Under the condition in Theorem 7.1, we have*

$$
Q(\omega(h)) - Q(\mathbf{y}_0) = \mathscr{O}(h^{2s+1}),
$$

where the constant symbolized by $\mathscr{O}$ is independent of h.

Proof From $Q(\mathbf{y}) = q^\intercal Dp$ and $D^\intercal = -D$, it follows that

$$
Q(\omega(h)) - Q(\mathbf{y}_0) = h\int_0^1 \nabla Q(\omega(\xi h))^\intercal \omega'(\xi h)d\xi
$$

$$
= h\int_0^1 \Big(-\tilde{p}(\xi h)^\intercal D,\ \tilde{q}(\xi h)^\intercal D\Big)\begin{pmatrix} \tilde{p}(\xi h) \\ -M\tilde{q}(\xi h) + \sum\limits_{j=1}^{s} l_j(\xi) f\big(\tilde{q}(c_j h)\big)\end{pmatrix} d\xi.
$$

Since $q^\intercal D(f(q) - Mq) = 0$ for any $q \in \mathbb{R}^d$, we have

$$Q(\omega(h)) - Q(\mathbf{y}_0) = h \int_0^1 \tilde{q}(\xi h)^\intercal D\Big(- M\tilde{q}(\xi h) + \sum_{j=1}^{s} l_j(\xi) f\big(\tilde{q}(c_j h)\big)\Big) d\xi$$

$$= h \int_0^1 \tilde{q}(\xi h)^\intercal D \frac{f^{(s+1)}\big(\tilde{q}(\xi h)\big)|_{\xi=\zeta}}{(n+1)!} \prod_{i=1}^{s} (\xi h - c_i h) d\xi$$

$$= h^{s+1} \int_0^1 \tilde{q}(\xi h)^\intercal D \frac{f^{(s+1)}\big(\tilde{q}(\xi h)\big)|_{\xi=\zeta}}{(n+1)!} \prod_{i=1}^{s} (\xi - c_i) d\xi$$

$$= \mathscr{O}(h^{s+1})\mathscr{O}(h^s) = \mathscr{O}(h^{2s+1}).$$

This completes the proof. □

7.3.3 The Algebraic Order

To emphasize the dependence of the solutions of $\mathbf{y}'(t) = F(\mathbf{y}(t))$ on the initial values, for any given $\tilde{t} \in [0, h]$, we denote by $\mathbf{y}(\cdot, \tilde{t}, \tilde{\mathbf{y}})$ the solution satisfying the initial condition $\mathbf{y}(\tilde{t}, \tilde{t}, \tilde{\mathbf{y}}) = \tilde{\mathbf{y}}$ and set

$$\Phi(s, \tilde{t}, \tilde{\mathbf{y}}) = \frac{\partial \mathbf{y}(s, \tilde{t}, \tilde{\mathbf{y}})}{\partial \tilde{\mathbf{y}}}. \tag{7.25}$$

Recalling the elementary theory of ODEs, we have the following standard result (see, e.g. [11])

$$\frac{\partial \mathbf{y}(s, \tilde{t}, \tilde{\mathbf{y}})}{\partial \tilde{t}} = -\Phi(s, \tilde{t}, \tilde{\mathbf{y}}) F(\tilde{\mathbf{y}}). \tag{7.26}$$

The following theorem states the result on the order of the trigonometric collocation methods.

Theorem 7.3 *Under the condition in Theorem 7.1, the trigonometric collocation method* (7.16) *satisfies*

$$\mathbf{y}(h) - \omega(h) = \mathscr{O}(h^{2s+1}),$$

where the constant symbolized by $\mathscr{O}$ is independent of h.

Proof It follows from (7.25) and (7.26) that

$$\mathbf{y}(h)-\omega(h)=\mathbf{y}(h,0,\mathbf{y}_0)-\mathbf{y}\big(h,h,\omega(h)\big)=-\int_0^h\frac{d\mathbf{y}\big(h,\tau,\omega(\tau)\big)}{d\tau}d\tau$$

$$=-\int_0^h\Big[\frac{\partial\mathbf{y}\big(h,\tau,\omega(\tau)\big)}{\partial\tilde{t}}+\frac{\partial\mathbf{y}\big(h,\tau,\omega(\tau)\big)}{\partial\tilde{\mathbf{y}}}\omega'(\tau)\Big]d\tau$$

$$=h\int_0^1\Phi\big(h,\xi h,\omega(\xi h)\big)\Big[F\big(\omega(\xi h)\big)-\omega'(\xi h)\Big]d\xi$$

$$=h\int_0^1\Phi\big(h,\xi h,\omega(\xi h)\big)\begin{pmatrix}\mathbf{0}\\ f\big(\tilde{q}(\xi h)\big)-\sum\limits_{j=1}^{s}l_j(\xi)f\big(\tilde{q}(c_jh)\big)\end{pmatrix}d\xi.$$

We rewrite $\Phi\big(h,\xi h,\omega(\xi h)\big)$ as a block matrix:

$$\Phi\big(h,\xi h,\omega(\xi h)\big)=\begin{pmatrix}\Phi_{11}(\xi h)&\Phi_{12}(\xi h)\\ \Phi_{21}(\xi h)&\Phi_{22}(\xi h)\end{pmatrix},$$

where Φ_{ij} $(i,j=1,2)$ are $d\times d$ matrices.

We then obtain

$$\mathbf{y}(h)-\omega(h)=h\begin{pmatrix}\int_0^1\Phi_{12}(\xi h)\frac{f^{(s+1)}\big(\tilde{q}(\xi h)\big)|_{\xi=\zeta}}{(n+1)!}\prod\limits_{i=1}^{s}(\xi h-c_ih)d\xi\\ \int_0^1\Phi_{22}(\xi h)\frac{f^{(s+1)}\big(\tilde{q}(\xi h)\big)|_{\xi=\zeta}}{(n+1)!}\prod\limits_{i=1}^{s}(\xi h-c_ih)d\xi\end{pmatrix}$$

$$=h^{s+1}\begin{pmatrix}\int_0^1\Phi_{12}(\xi h)\frac{f^{(s+1)}\big(\tilde{q}(\xi h)\big)|_{\xi=\zeta}}{(n+1)!}\prod\limits_{i=1}^{s}(\xi-c_i)d\xi\\ \int_0^1\Phi_{22}(\xi h)\frac{f^{(s+1)}\big(\tilde{q}(\xi h)\big)|_{\xi=\zeta}}{(n+1)!}\prod\limits_{i=1}^{s}(\xi-c_i)d\xi\end{pmatrix}=h^{s+1}\mathcal{O}(h^s)=\mathcal{O}(h^{2s+1}).$$

The proof is complete. □

7.3.4 Convergence Analysis of the Iteration

Theorem 7.4 *Assume that M is symmetric and positive semi-definite and that f satisfies a Lipschitz condition in the variable q, i.e., there exists a constant L such that $\|f(q_1)-f(q_2)\|\leq L\,\|q_1-q_2\|$. If*

$$0<h<\frac{1}{\sqrt{L\max\limits_{i,j=1,\ldots,s}\int_0^1|l_j(c_iz)(1-z)|dz}},\tag{7.27}$$

then the fixed-point iteration for the method (7.16) *is convergent.*

Proof Following Definition 7.1, the first formula of (7.16) can be rewritten as

$$Q = \phi_0(c^2V)(e \otimes q_0) + hc\phi_1(c^2V)(e \otimes p_0) + h^2 A(V) f(Q), \tag{7.28}$$

where $c = (c_1, \dots, c_s)^{\intercal}$, $e = (1, \dots, 1)^{\intercal}$, $Q = (\tilde{q}_1, \dots, \tilde{q}_s)^{\intercal}$, $f(Q) = \big(f(\tilde{q}_1)^{\intercal}, \dots, f(\tilde{q}_s)^{\intercal}\big)^{\intercal}$, $A(V) = \big(a_{ij}(V)\big)_{s\times s}$ and $a_{ij}(V)$ are the block diagonal matrices defined by

$$\begin{aligned}
a_{ij}(V) &:= \int_0^1 l_j(c_i z)(1-z)\phi_1\big((1-z)^2 c_i^2 V\big)dz, \\
\phi_0(c^2V) &:= \operatorname{diag}\big(\phi_0(c_1^2V), \dots, \phi_0(c_s^2V)\big)^{\intercal}, \\
c\phi_1(c^2V) &:= \operatorname{diag}\big(c_1\phi_1(c_1^2V), \dots, c_s\phi_1(c_s^2V)\big)^{\intercal}.
\end{aligned}$$

It follows from Proposition 2.1 in [18] that $\big\|\phi_1\big((1-z)^2c_i^2V\big)\big\| \le 1$. We then obtain

$$\big\|a_{ij}(V)\big\| \le \int_0^1 |l_j(c_i z)(1-z)|dz.$$

Let

$$\varphi(x) = \phi_0(c^2V)(e \otimes q_0) + hc\phi_1(c^2V)(e \otimes p_0) + h^2 A(V) f(x).$$

Then,

$$\begin{aligned}
\|\varphi(x) - \varphi(y)\| &= \big\|h^2A(V)f(x) - h^2A(V)f(y)\big\| \le h^2 L\,\|A(V)\|\,\|x-y\| \\
&\le h^2 L \max_{i,j=1,\dots,s} \int_0^1 |l_j(c_i z)(1-z)|dz\,\|x-y\|,
\end{aligned}$$

which means that $\varphi(x)$ is a contraction from the assumption (7.27). The well-known Contraction Mapping Theorem then ensures the convergence of the fixed-point iteration. This proof is complete. □

Remark 7.5 We note that the convergence of the methods is independent of $\|M\|$. This point is of prime importance especially for highly oscillatory systems where $\|M\| \gg 1$, which will be shown by the numerical results of Problem 2 in Sect. 7.4.

7.3.5 Stability and Phase Properties

In this part we are concerned with the stability and phase properties. We consider the test equation:

$$q''(t) + \omega^2 q(t) = -\varepsilon q(t) \ \text{ with } \ \omega^2 + \varepsilon > 0, \tag{7.29}$$

where ω represents an estimation of the dominant frequency λ and $\varepsilon = \lambda^2 - \omega^2$ is the error of that estimation. Applying (7.16) to (7.29) produces

$$\begin{pmatrix} \tilde{q} \\ h\tilde{p} \end{pmatrix} = S(V, z) \begin{pmatrix} q_0 \\ hp_0 \end{pmatrix},$$

where the stability matrix $S(V, z)$ is given by

$$S(V, z) = \begin{pmatrix} \phi_0(V) - z\bar{b}^{\intercal}(V)N^{-1}\phi_0(c^2V) & \phi_1(V) - z\bar{b}^{\intercal}(V)N^{-1}(c \cdot \phi_1(c^2V)) \\ -V\phi_1(V) - zb^{\intercal}(V)N^{-1}\phi_0(c^2V) & \phi_0(V) - zb^{\intercal}(V)N^{-1}(c \cdot \phi_1(c^2V)) \end{pmatrix}$$

with $N = I + zA(V)$, $\bar{b}(V) = \left(I_{1,1}, \ldots, I_{1,s}\right)^{\intercal}$, $b(V) = \left(I_{2,1}, \ldots, I_{2,s}\right)^{\intercal}$.

Accordingly, we have the following definitions of stability and dispersion order and dissipation order for our method (7.16).

Definition 7.3 (*See* [30]) Let $\rho(S)$ be the spectral radius of S,

$$R_s = \{(V, z)|\ V > 0 \text{ and } \rho(S) < 1\}$$

and

$$R_p = \{(V, z)|\ V > 0,\ \rho(S) = 1 \text{ and } \operatorname{tr}(S)^2 < 4\det(S)\}.$$

Then R_s and and R_p are called the *stability region* and the *periodicity region* of the method (7.16) respectively. The quantities

$$\phi(\zeta) = \zeta - \arccos\left(\frac{\operatorname{tr}(S)}{2\sqrt{\det(S)}}\right), \quad d(\zeta) = 1 - \sqrt{\det(S)}$$

are called the dispersion error and the dissipation error of the method (7.16), respectively, where $\zeta = \sqrt{V + z}$. Then, a method is said to be dispersive of order r and dissipative of order s, if $\phi(\zeta) = \mathscr{O}(\zeta^{r+1})$ and $d(\zeta) = \mathscr{O}(\zeta^{s+1})$, respectively. If $\phi(\zeta) = 0$ and $d(\zeta) = 0$, then the corresponding method is said to be zero dispersive and zero dissipative, respectively.

7.4 Numerical Experiments

As an example of the trigonometric collocation methods (7.16), we choose the node points of a two-point Gauss–Legendre's quadrature over the integral [0, 1], as follows:

$$c_1 = \frac{3 - \sqrt{3}}{6}, \quad c_2 = \frac{3 + \sqrt{3}}{6}. \tag{7.30}$$

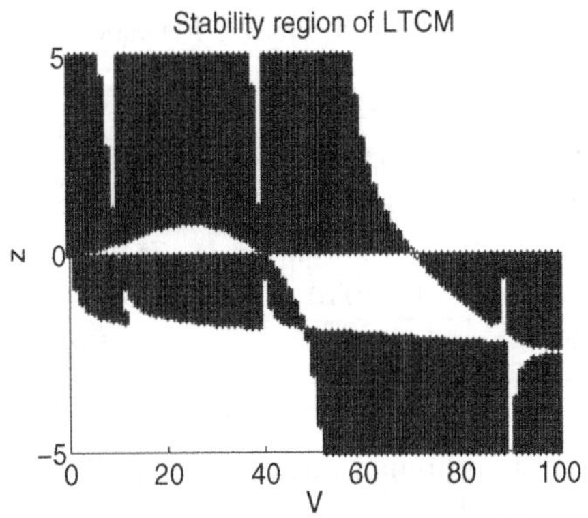

Fig. 7.1 Stability region (shaded area) of the method LTCM

Then we choose $s = 2$ in (7.16) and denote the corresponding fourth-order method as LTCM.

The stability region of this method is shown in Fig. 7.1. Here we choose the subset $V \in [0, 100]$, $z \in [-5, 5]$ and the region shown in Fig. 7.1 only gives an indication of the stability of this method.

The dissipative error and dispersion error are given respectively by

$$d(\zeta) = \frac{\varepsilon^2}{24(\varepsilon + \omega^2)^2}\zeta^4 + \mathscr{O}(\zeta^5), \quad \phi(\zeta) = \frac{\varepsilon^2}{6(\varepsilon + \omega^2)^2}\zeta^3 + \mathscr{O}(\zeta^4).$$

Note that when $M = 0$, the method LTCM reduces to a fourth-order RKN method given by the Butcher tableau (7.21) with nodes in (7.30).

In order to show the efficiency and robustness of the fourth-order method LTCM, several other integrators in the literature we select for comparison are:

- TFCM: a fourth-order trigonometric Fourier collocation method in [26] with $c_1 = \frac{3-\sqrt{3}}{6}$, $c_2 = \frac{3+\sqrt{3}}{6}$, $b_1 = b_2 = 1/2$, $r = 2$;
- SRKM1: the symplectic Runge–Kutta method of order five in [20] based on Radau quadrature;
- EPCM1: the "extended Lobatto IIIA method of order four" in [15], which is an energy-preserving collocation method (the case $s = 2$ in [10]);
- EPRKM1: the energy-preserving Runge–Kutta method of order four (formula (19) in [1]).

Since all of these methods are implicit, we use the classical waveform Picard algorithm. For each experiment, first we show the convergence rate of iterations for different error tolerances. Then, for different methods, we set the error tolerance as 10^{-16} and set the maximum number of iteration as 5. We display the global errors and the energy errors once the problem is a Hamiltonian system.

Table 7.1 Results for Problem 1: The total CPU time (s) of iterations for different error tolerances (tol)

Methods	$tol = 1.0e{-006}$	$tol = 1.0e{-008}$	$tol = 1.0e{-010}$	$tol = 1.0e{-012}$
LTCM	6.8215	8.8964	8.8500	10.5551
TFCM	9.7892	9.7553	9.9806	13.0105
SRKM1	67.0230	64.1777	75.9390	86.8317
EPCM1	104.4341	112.9710	126.4438	145.6188
EPRKM1	56.2409	64.3123	75.2503	84.9962

Problem 1 Consider the Hamiltonian equation which governs the motion of an artificial satellite (this problem has been considered in [19]) with the Hamiltonian

$$H(q,p)=\frac{1}{2}p^{\intercal}p+\frac{1}{2}\frac{\kappa}{2}q^{\intercal}q+\lambda\Big(\frac{(q_1q_3+q_2q_4)^2}{r^4}-\frac{1}{12r^2}\Big),$$

where $q=(q_1,q_2,q_3,q_4)^{\intercal}$ and $r=q^{\intercal}q$. The initial conditions are given on an elliptic equatorial orbit by

$$q_0=\sqrt{\frac{r_0}{2}}\Big(-1,-\frac{\sqrt{3}}{2},-\frac{1}{2},0\Big)^{\intercal},\quad p_0=\frac{1}{2}\sqrt{K^2\frac{1+e}{2}}\Big(1,\frac{\sqrt{3}}{2},\frac{1}{2},0\Big)^{\intercal}.$$

Here $M=\frac{\kappa}{2}$ and κ is the total energy of the elliptic motion which is defined by $\kappa=\frac{K^2-2|p_0|^2}{r_0}-V_0$ with $V_0=-\frac{\lambda}{12r_0^3}$. The parameters of this problem are chosen as $K^2=3.98601\times10^5$, $r_0=6.8\times10^3$, $e=0.1$, $\lambda=\frac{3}{2}K^2J_2R^2$, $J_2=1.08625\times10^{-3}$, $R=6.37122\times10^3$. First the problem is solved on the interval $[0,10^4]$ with the stepsize $h=\frac{1}{10}$ to show the convergence rate of iterations. Table 7.1 displays the CPU time of iterations for different error tolerances. Then this equation is integrated on $[0,1000]$ with the stepsizes $1/2^i$ for $i=2,3,4,5$. The global errors against CPU time are shown in Fig. 7.2i. We finally integrate this problem with the fixed stepsize $h=1/20$ on the interval $[0,t_{\text{end}}]$, and $t_{\text{end}}=10,100,10^3,10^4$. The maximum global errors of Hamiltonian energy against CPU time are presented in Fig. 7.2ii.

Problem 2 Consider the Fermi–Pasta–Ulam problem [9].

Fermi–Pasta–Ulam problem is a Hamiltonian system with the Hamiltonian

$$\begin{aligned}H(y,x)=&\frac{1}{2}\sum_{i=1}^{2m}y_i^2+\frac{\omega^2}{2}\sum_{i=1}^{m}x_{m+i}^2+\frac{1}{4}\Big[(x_1-x_{m+1})^4\\&+\sum_{i=1}^{m-1}(x_{i+1}-x_{m+i-1}-x_i-x_{m+i})^4+(x_m+x_{2m})^4\Big],\end{aligned}$$

where x_i is a scaled displacement of the ith stiff spring, x_{m+i} represents a scaled expansion (or compression) of the ith stiff spring, and y_i, y_{m+i} are their velocities (or momenta). This system can be rewritten as

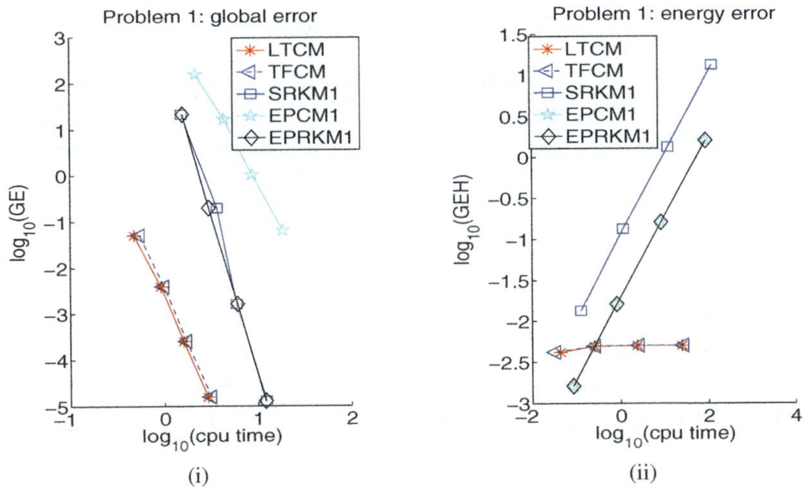

Fig. 7.2 Results for Problem 1. **i** The logarithm of the global error (GE) over the integration interval against the logarithm of CPU time. **ii** The logarithm of the maximum global error of Hamiltonian energy (GEH) against the logarithm of CPU time

$$x''(t) + Mx(t) = -\nabla U(x), \qquad t \in [t_0, t_{\text{end}}],$$

where

$$M = \begin{pmatrix} \mathbf{0}_{m\times m} \mathbf{0}_{m\times m} \\ \mathbf{0}_{m\times m} \omega^2 I_{m\times m} \end{pmatrix},$$

$$U(x) = \frac{1}{4}\Big[(x_1 - x_{m+1})^4 + \sum_{i=1}^{m-1}(x_{i+1} - x_{m+i-1} - x_i - x_{m+i})^4 + (x_m + x_{2m})^4\Big].$$

Following [9], we choose

$$m = 3,\ x_1(0) = 1,\ y_1(0) = 1,\ x_4(0) = \frac{1}{\omega},\ y_4(0) = 1,$$

with zero for the remaining initial values.

First, the problem is solved on the interval [0, 1000] with the stepsize $h = \frac{1}{100}$ and $\omega = 100,\ 200$ to show the convergence rate of iterations. See Table 7.2 for the total CPU time of iterations for different error tolerances. It can be observed that when ω increases, the convergence rates of LTCM and TFCM are almost unaffected. However, the convergence rates of the other methods vary greatly as ω becomes large.

We then integrate the system on the interval [0, 50] with $\omega = 50, 100, 150, 200$ and the stepsizes $h = 1/(20 \times 2^j)$ for $j = 1, \ldots, 4$. The global errors are shown in Fig. 7.4. Finally, we integrate this problem with a fixed stepsize $h = 1/100$ on

Table 7.2 Results for Problem 2: The total CPU time (s) of iterations for different error tolerances (tol)

Methods	$tol = 1.0e{-}006$	$tol = 1.0e{-}008$	$tol = 1.0e{-}010$	$tol = 1.0e{-}012$
LTCM ($\omega = 100$)	7.1570	9.7010	9.6435	12.2449
LTCM ($\omega = 200$)	7.5169	10.0160	9.2135	11.1672
TFCM ($\omega = 100$)	7.6434	10.3224	10.3341	12.7998
TFCM ($\omega = 200$)	7.8861	11.1322	10.0578	12.3621
SRKM1 ($\omega = 100$)	32.0491	39.4922	48.5822	57.0720
SRKM1 ($\omega = 200$)	58.2410	70.5585	86.1757	99.6403
EPCM1 ($\omega = 100$)	50.8899	70.5920	87.9782	102.9839
EPCM1 ($\omega = 200$)	121.2714	149.7104	189.4323	220.1096
EPRKM1 ($\omega = 100$)	31.0881	39.0050	47.6389	56.4456
EPRKM1 ($\omega = 200$)	55.2205	68.8459	82.5919	98.5277

the interval $[0, t_{\text{end}}]$ with $t_{\text{end}} = 1, 10, 100, 1000$. The maximum global errors of Hamiltonian energy are presented in Fig. 7.4. Here, it is noted that some results are too large, and hence we do not plot the corresponding points in Figs. 7.3 and 7.4. A similar situation occurs in the next two problems.

Problem 3 Consider the nonlinear Klein-Gordon equation [17]

$$\begin{cases} \frac{\partial^2 u}{\partial t^2} - \frac{\partial^2 u}{\partial x^2} = -u^3 - u, \quad 0 < x < L, \ t > 0, \\ u(x,0) = A(1 + \cos(\frac{2\pi}{L}x)), \quad u_t(x,0) = 0, \quad u(0,t) = u(L,t), \end{cases}$$

with $L = 1.28$, $A = 0.9$. Carrying out a semi-discretization on the spatial variable by using second-order symmetric differences yields

$$\frac{d^2U}{dt^2} + MU = F(U), \quad 0 < t \le t_{\text{end}},$$

where $U(t) = \big(u_1(t), \dots, u_N(t)\big)^\intercal$ with $u_i(t) \approx u(x_i, t)$ for $i = 1, \dots, N$,

$$M = \frac{1}{\Delta x^2} \begin{pmatrix} 2 & -1 & & & -1 \\ -1 & 2 & -1 & & \\ & \ddots & \ddots & \ddots & \\ & & -1 & 2 & -1 \\ -1 & & & -1 & 2 \end{pmatrix}_{N \times N}$$

with $\Delta x = L/N$, $x_i = i\Delta x$, $F(U) = \big(-u_1^3 - u_1, \dots, -u_N^3 - u_N\big)^\intercal$ and $N = 32$. The corresponding Hamiltonian of this system is

$$H(U', U) = \frac{1}{2}U'^\intercal U' + \frac{1}{2}U^\intercal MU + \frac{1}{2}u_1^2 + \frac{1}{4}u_1^4 + \dots + \frac{1}{2}u_N^2 + \frac{1}{4}u_N^4.$$

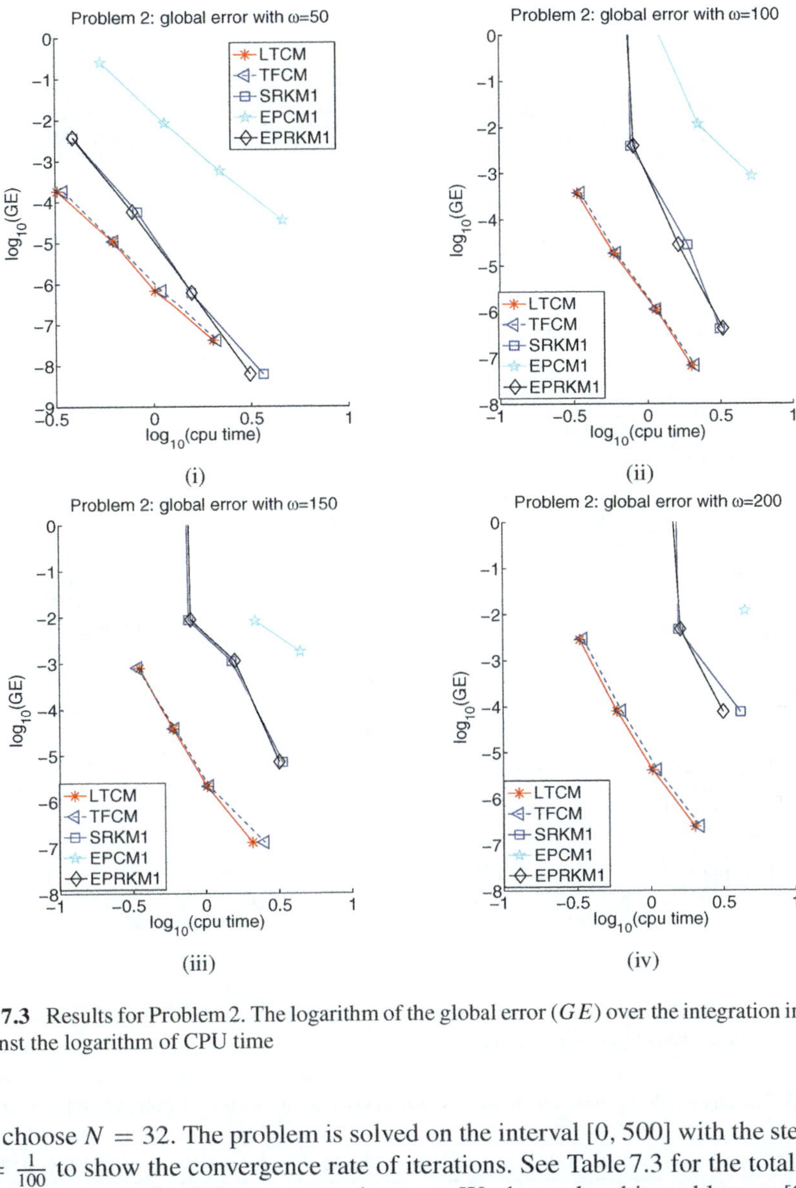

Fig. 7.3 Results for Problem 2. The logarithm of the global error (GE) over the integration interval against the logarithm of CPU time

We choose $N = 32$. The problem is solved on the interval $[0, 500]$ with the stepsize $h = \frac{1}{100}$ to show the convergence rate of iterations. See Table 7.3 for the total CPU time of iterations for different error tolerances. We then solve this problem on $[0, 20]$ with stepsizes $h = 1/(3 \times 2^j)$ for $j = 1, \ldots, 4$. Figure 7.5i shows the global errors. Finally this problem is integrated with a fixed stepsize $h = 0.002$ on the interval $[0, t_{\text{end}}]$ with $t_{\text{end}} = 10^i$ for $i = 0, 1, 2, 3$. The maximum global errors of Hamiltonian energy are presented in Fig. 7.5ii.

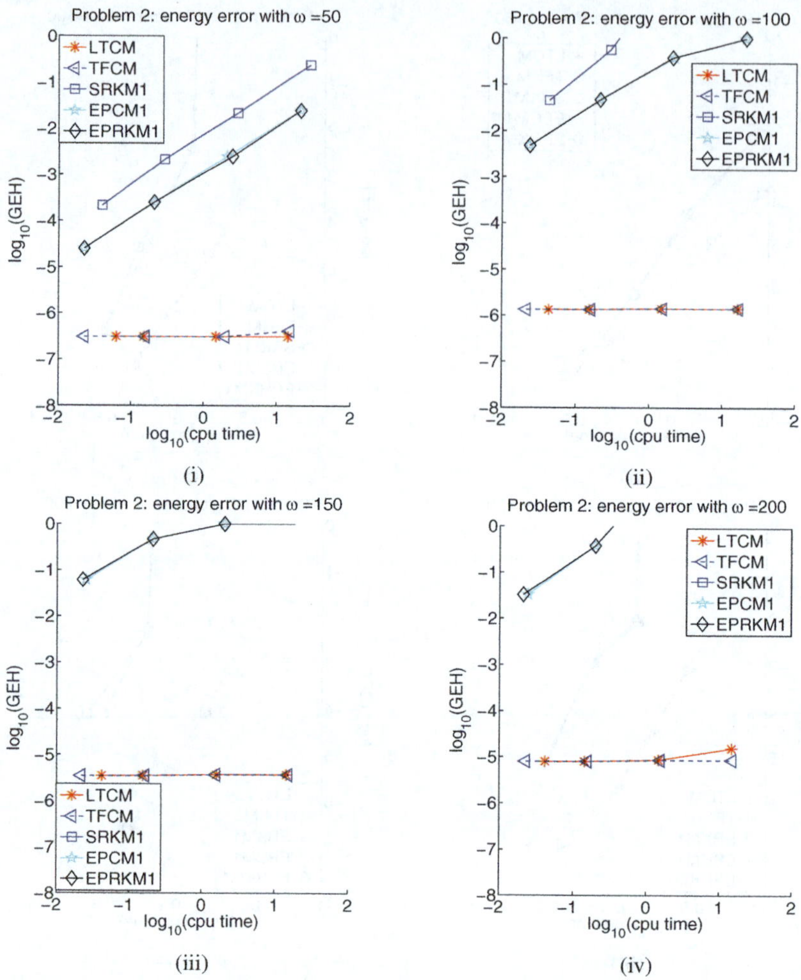

Fig. 7.4 Results for Problem 2. The logarithm of the maximum global error of Hamiltonian energy (GEH) against the logarithm of CPU timelabelfig

Table 7.3 Results for Problem 3: The total CPU time (s) of iterations for different error tolerances (tol)

Methods	$tol = 1.0e{-}006$	$tol = 1.0e{-}008$	$tol = 1.0e{-}010$	$tol = 1.0e{-}012$
LTCM	5.9325	7.9263	8.1816	10.0602
TFCM	6.5318	8.7008	8.8934	10.7489
SRKM1	24.1600	29.4173	34.5310	39.5161
EPCM1	37.2757	46.4011	53.1403	66.2339
EPRKM1	22.6571	27.8341	33.5435	39.4533

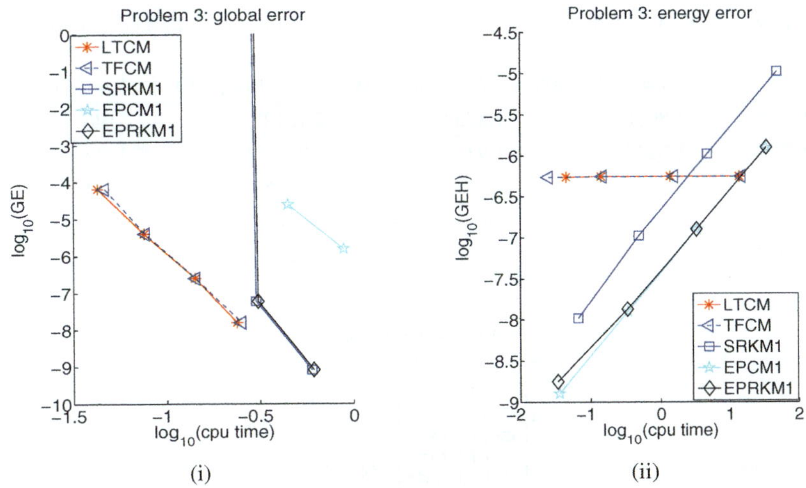

Fig. 7.5 Results for Problem 3. **i** The logarithm of the global error (GE) over the integration interval against the logarithm of CPU time. **ii** The logarithm of the maximum global error of Hamiltonian energy (GEH) against the logarithm of CPU time

Problem 4 Consider the wave equation

$$\frac{\partial^2 u}{\partial t^2} - a(x)\frac{\partial^2 u}{\partial x^2} + 92u = f(t,x,u), \quad 0 < x < 1, \quad t > 0,$$

$$u(0,t) = 0, \quad u(1,t) = 0, \quad u(x,0) = a(x), \quad u_t(x,0) = 0$$

with $a(x) = 4x(1-x)$, $f(t,x,u) = u^5 - a^2(x)u^3 + \frac{a^5(x)}{4}\sin^2(20t)\cos(10t)$. The exact solution of this problem is $u(x,t) = a(x)\cos(10t)$. Using semi-discretization on the spatial variable with second-order symmetric differences, we obtain

$$\frac{d^2U}{dt^2} + MU = F(t,U),\ U(0) = \big(a(x_1),\ldots,a(x_{N-1})\big)^{\intercal},\ U'(0) = \mathbf{0},\ 0 < t \le t_{\text{end}},$$

where $U(t) = \big(u_1(t),\ldots,u_{N-1}(t)\big)^{\intercal}$ with $u_i(t) \approx u(x_i,t)$, $x_i = i\Delta x$, $\Delta x = 1/N$ for $i = 1,\ldots,N-1$,

$$M = 92I_{N-1} + \frac{1}{\Delta x^2}\begin{pmatrix} 2a(x_1) & -a(x_1) & & & \\ -a(x_2) & 2a(x_2) & -a(x_2) & & \\ & \ddots & \ddots & \ddots & \\ & & -a(x_{N-2}) & 2a(x_{N-2}) & -a(x_{N-2}) \\ & & & -a(x_{N-1}) & 2a(x_{N-1}) \end{pmatrix},$$

and

$$F(t,U) = \big(f(t,x_1,u_1),\ldots,f(t,x_{N-1},u_{N-1})\big)^{\intercal}.$$

Table 7.4 Results for Problem 4: The total CPU time (s) of iterations for different error tolerances (tol)

Methods	$tol = 1.0e-006$	$tol = 1.0e-008$	$tol = 1.0e-010$	$tol = 1.0e-012$
LTCM	1.8980	1.8737	2.1212	2.3196
TFCM	1.9213	1.9345	2.2227	2.3736
SRKM1	13.8634	16.6963	19.0854	22.6142
EPCM1	23.5110	28.1288	32.2263	36.8443
EPRKM1	13.5526	17.2289	18.8744	23.0066

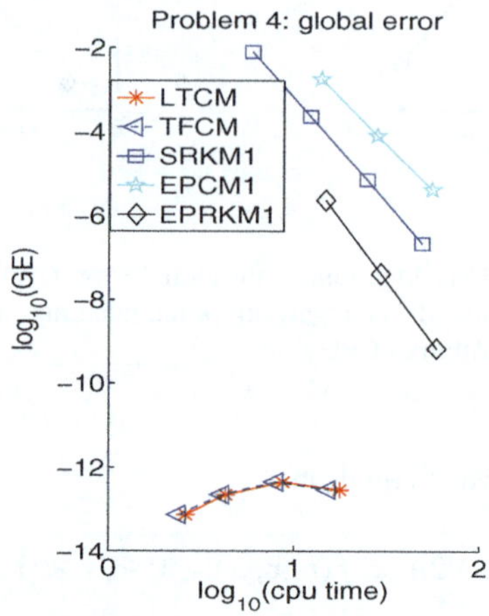

Fig. 7.6 Results for Problem 4: The logarithm of the global error (GE) over the integration interval against the logarithm of CPU time

The problem is solved on the interval [0, 100] with the stepsize $h = \frac{1}{40}$ to show the convergence rate of iterations. See Table 7.4 for the total CPU time of iterations for different error tolerances. Then, the system is integrated on the interval [0, 100] with $N = 40$ and $h = 1/2^j$ for $j = 5, \ldots, 8$. The global errors are shown in Fig. 7.6.

Remark 7.6 It follows from the numerical results that our method LTCM is very promising in comparison with the classical methods SRKM1, EPCM1 and EPRKM1. Although LTCM has a similar performance to TFCM in preserving the solution and the energy, it can be observed from Figs. 7.2i, 7.3 and 7.5i that LTCM performs a bit better than TFCM in presenting the solution. Moreover, it follows from Tables 7.1, 7.2, 7.3 and 7.4 that LTCM has a better convergence performance of iterations than TFCM. This means that LTCM can have a lower computational cost when the same error tolerance is required in the iteration procedure.

Remark 7.7 From Figs. 7.2ii, 7.4 and 7.5ii, it can be observed that the energy-preserving Runge–Kutta method EPRKM1 cannot preserve the Hamiltonian energy, and the errors seem to grow with the CPU time when the stepsize is reduced. The reason for this phenomenon may be that EPRKM1 does not take advantage of the special structure introduced by the linear term Mq of the oscillatory system (7.1) and its convergence depends on $\|M\|$. The method LTCM developed in this chapter makes good use of the matrix M appearing in the oscillatory systems (7.1) and its convergence condition is independent of $\|M\|$. This property enables LTCM to perform well in preserving Hamiltonian energy, although it is not an energy-preserving method.

7.5 Conclusions and Discussions

It is known that the trigonometric Fourier collocation method is a kind of collocation method for ODEs (see, e.g. [7, 9, 10, 16, 28]). In this chapter we have investigated a class of trigonometric collocation methods based on Lagrange basis polynomials, the variation-of-constants formula and the idea of collocation methods for solving multi-frequency oscillatory second-order differential equations (7.1) efficiently. It has been shown that the convergence condition of these trigonometric collocation methods is independent of $\|M\|$, which is crucial for solving highly oscillatory systems. This presents an approach to treating multi-frequency oscillatory systems. The numerical experiments were carried out, and the numerical results show that the trigonometric collocation methods based on Lagrange basis polynomials derived in this chapter have remarkable efficiency compared with standard methods in the literature. However, it is believed that other collocation methods based on suitable bases different from the Lagrange basis are also possible for the numerical simulation of ODEs.

The material of this chapter is based on the work by Wang et al. [27].

References

1. Brugnano, L., Iavernaro, F., Trigiante, D.: A simple framework for the derivation and analysis of effective one-step methods for ODEs. Appl. Math. Comput. **218**, 8475–8485 (2012)
2. Brugnano, L., Iavernaro, F.: Line Integral Methods for Conservative Problems. CRC Press, Boca Raton (2016)
3. Cohen, D., Hairer, E., Lubich, C.: Numerical energy conservation for multi-frequency oscillatory differential equations. BIT **45**, 287–305 (2005)
4. Cohen, D.: Conservation properties of numerical integrators for highly oscillatory Hamiltonian systems. IMA J. Numer. Anal. **26**, 34–59 (2006)
5. Cohen, D., Jahnke, T., Lorenz, K., Lubich, C.: Numerical integrators for highly oscillatory Hamiltonian systems: a review. In: Mielke, A. (ed.) Analysis, Modeling and Simulation of Multiscale Problems, pp. 553–576. Springer, Berlin (2006)
6. García-Archilla, B., Sanz-Serna, J.M., Skeel, R.D.: Long-time-step methods for oscillatory differential equations. SIAM J. Sci. Comput. **20**, 930–963 (1999)
7. Hairer, E., Nørsett, S.P., Wanner, G.: Solving Ordinary Differential Equations I, Nonstiff Problems. Springer Series in Computational Mathematics, 2nd edn. Springer, Berlin (1993)
8. Hairer, E., Lubich, C.: Long-time energy conservation of numerical methods for oscillatory differential equations. SIAM J. Numer. Anal. **38**, 414–441 (2000)
9. Hairer, E., Lubich, C., Wanner, G.: Geometric Numerical Integration: Structure-Preserving Algorithms for Ordinary Differential Equations, 2nd edn. Springer, Berlin (2006)

10. Hairer, E.: Energy-preserving variant of collocation methods. JNAIAM J. Numer. Anal. Ind. Appl. Math. **5**, 73–84 (2010)
11. Hale, J.K.: Ordinary Differential Equations. Roberte E. Krieger Publishing company, Huntington, New York (1980)
12. Hochbruck, M., Lubich, C.: A Gautschi-type method for oscillatory second-order differential equations. Numer. Math. **83**, 403–426 (1999)
13. Hochbruck, M., Ostermann, A.: Explicit exponential Runge–Kutta methods for semilineal parabolic problems. SIAM J. Numer. Anal. **43**, 1069–1090 (2005)
14. Hochbruck, M., Ostermann, A., Schweitzer, J.: Exponential Rosenbrock-type methods. SIAM J. Numer. Anal. **47**, 786–803 (2009)
15. Iavernaro, F., Trigiante, D.: High-order symmetric schemes for the energy conservation of polynomial Hamiltonian problems. NAIAM J. Numer. Anal. Ind. Appl. Math. **4**(1), 87–101 (2009)
16. Iserles, A.: A First Course in the Numerical Analysis of Differential Equations, 2nd edn. Cambridge University Press, Cambridge (2008)
17. Jiménez, S., Vázquez, L.: Analysis of four numerical schemes for a nonlinear Klein–Gordon equation. Appl. Math. Comput. **35**, 61–93 (1990)
18. Li, J., Wu, X.Y.: Adapted Falkner-type methods solving oscillatory second-order differential equations. Numer. Algorithm **62**, 355–381 (2013)
19. Stiefel, E.L., Scheifele, G.: Linear and regular celestial mechanics. Springer, New York (1971)
20. Sun, G.: Construction of high order symplectic Runge–Kutta methods. J. Comput. Math. **11**, 250–260 (1993)
21. Wang, B., Li, G.: Bounds on asymptotic-numerical solvers for ordinary differential equations with extrinsic oscillation. Appl. Math. Modell. **39**, 2528–2538 (2015)
22. Wang, B., Liu, K., Wu, X.Y.: A Filon-type asymptotic approach to solving highly oscillatory second-order initial value problems. J. Comput. Phys. **243**, 210–223 (2013)
23. Wang, B., Wu, X.Y.: A new high precision energy-preserving integrator for system of oscillatory second-order differential equations. Phys. Lett. A **376**, 1185–1190 (2012)
24. Wang, B., Wu, X.Y., Zhao, H.: Novel improved multidimensional Störmer-Verlet formulas with applications to four aspects in scientific computation. Math. Comput. Modell. **57**, 857–872 (2013)
25. Wang, B., Yang, H., Meng, F.: Sixth-order symplectic and symmetric explicit ERKN schemes for solving multi-frequency oscillatory nonlinear Hamiltonian equations. Calcolo (2016). https://doi.org/10.1007/s10092-016-0179-y
26. Wang, B., Iserles, A., Wu, X.Y.: Arbitrary-order trigonometric fourier collocation methods for multi-frequency oscillatory systems. Found. Comput. Math. **16**, 151–181 (2016)
27. Wang, B., Wu, X.Y., Meng, F.: Trigonometric collocation methods based on Lagrange basis polynomials for multi-frequency oscillatory second-order differential equations. J. Comput. Appl. Math. **313**, 185–201 (2017)
28. Wright, K.: Some relationships between implicit Runge-Kutta, collocation and Lanczos τ methods, and their stability properties. BIT **10**, 217–227 (1970)
29. Wu, X.Y., You, X., Shi, W., Wang, B.: ERKN integrators for systems of oscillatory second-order differential equations. Comput. Phys. Comm. **181**, 1873–1887 (2010)
30. Wu, X.Y.: A note on stability of multidimensional adapted Runge–Kutta–Nyström methods for oscillatory systems. Appl. Math. Modell. **36**, 6331–6337 (2012)
31. Wu, X.Y., Wang, B., Xia, J.: Explicit symplectic multidimensional exponential fitting modified Runge–Kutta–Nyström methods. BIT **52**, 773–795 (2012)
32. Wu, X.Y., Wang, B., Shi, W.: Efficient energy-preserving integrators for oscillatory Hamiltonian systems. J. Comput. Phys. **235**, 587–605 (2013)
33. Wu, X.Y., You, X., Wang, B.: Structure-Preserving Algorithms for Oscillatory Differential Equations. Springer, Berlin (2013)

Chapter 8
A Compact Tri-Colored Tree Theory for General ERKN Methods

This chapter develops a compact tri-colored rooted-tree theory for the order conditions for general ERKN methods. The bottleneck of the original tri-colored rooted-tree theory is the existence of numerous redundant trees. This chapter first introduces the extended elementary differential mappings. Then, the new compact tri-colored rooted tree theory is established based on a subset of the original tri-colored rooted-tree set. This new theory makes all redundant trees no longer appear, and hence the order conditions of ERKN methods for general multi-frequency and multidimensional second-order oscillatory systems are greatly simplified.

8.1 Introduction

Runge–Kutta–Nyström (RKN) methods (see [12]) are very popular for solving second-order differential equations. This chapter develops the rooted-tree theory and B-series for *extended Runge–Kutta–Nyström* (ERKN) methods solving general multi-frequency and multi-dimensional oscillatory second-order initial value problems (IVPs) of the form

$$\begin{cases} \boldsymbol{y}''(t) + M\boldsymbol{y}(t) = \boldsymbol{f}\big(\boldsymbol{y}(t), \boldsymbol{y}'(t)\big), \qquad t \in [t_0, T], \\ \boldsymbol{y}(t_0) = \boldsymbol{y}_0, \quad \boldsymbol{y}'(t_0) = \boldsymbol{y}'_0, \end{cases} \tag{8.1}$$

where M is a $d \times d$ constant matrix implicitly containing the dominant frequencies of the system, $\boldsymbol{y} \in \mathbb{R}^d$, and $\boldsymbol{f} : \mathbb{R}^d \times \mathbb{R}^d \to \mathbb{R}^d$, with position $\boldsymbol{y}$ and velocity $\boldsymbol{y}'$ as arguments. In the special case where the right-hand side function of (8.1) does not depend on velocity $\boldsymbol{y}'$, (8.1) reduces to the following special second-order oscillatory system

X. Wu and B. Wang, *Recent Developments in Structure-Preserving Algorithms for Oscillatory Differential Equations*, https://doi.org/10.1007/978-981-10-9004-2_8

$$\begin{cases} y''(t) + My(t) = f\big(y(t)\big), & t \in [t_0, T], \\ y(t_0) = y_0, \quad y'(t_0) = y_0'. \end{cases} \tag{8.2}$$

Furthermore, if M is symmetric and positive semi-definite and $f(q) = -\nabla U(q)$, then, with $q = y$, $p = y'$, (8.2) becomes identical to a multi-frequency and multidimensional oscillatory Hamiltonian system

$$\begin{cases} p'(t) = -\nabla_q H(p(t), q(t)), & p(t_0) = p_0, \\ q'(t) = \ \ \nabla_p H(p(t), q(t)), & q(t_0) = q_0, \end{cases} \tag{8.3}$$

with the Hamiltonian

$$H(p, q) = \frac{1}{2}p^{\mathsf{T}}p + \frac{1}{2}q^{\mathsf{T}}Mq + U(q),$$

where $U(q)$ is a smooth potential function. For solving the multi-frequency, multidimensional, oscillatory system (8.3), a large number of studies have been made (see, e.g. [6, 25, 28]). The methods for problems (8.1) and (8.2) are especially important when M has large positive eigenvalues, as in the case where the wave equations is semi-discretised in space (see, e.g. [11, 14, 26, 27, 29]). Such problems arise in a wide range of fields such as astronomy, molecular dynamics, classical mechanics, quantum mechanics, chemistry, biology and engineering.

ERKN methods were proposed originally in the papers [23, 32] to solve the special oscillatory system (8.2). ERKN methods work well in practical numerical simulation, since they are specially designed to be adapted to the structure of the underlying oscillatory system and do not depend on the decomposition of the matrix M. ERKN methods have been widely investigated and used in numerous applications in the fields of science and engineering. For example, the idea of ERKN methods has been extended to two-step hybrid methods (see, e.g. [8, 9]), to Falkner-type methods (see, e.g. [7]), to Störmer–Verlet methods (see, e.g. [17]), to energy-preserving methods (see, e.g. [11, 15, 24]), and to symplectic and multi-symplectic methods (see, e.g. [14, 16, 19, 20]). Meanwhile, further research on ARKN methods, including the symplectic conditions and symmetry, has been carried out in the following papers [10, 13, 21, 22, 31].

In a recent paper [33], ERKN methods were extended to the general oscillatory system (8.1), and a tri-colored tree theory called *extended Nyström tree theory* (EN-T theory) was analysed for the order conditions. Unfortunately, however, the EN-T theory is not completely satisfactory due to the existence of redundant trees. For example, there are 7 redundant trees out of 16 trees for third order ERKN methods. In practice, in order to gain the order conditions for a specific ERKN method of order r, one needs to draw all the trees of order up to r first, and then from them select and delete about half of the redundant trees. This will lead to inefficiency in the use of the EN-T theory to achieve the order conditions for ERKN methods.

Hence, in this chapter, we will present an improved theory to eliminate all such redundant trees. In a similar approach to the case of the special oscillatory system

(8.2) in [30], *extended elementary differentials* are required, and we will discuss this in detail in Sect. 8.4.

This chapter is organized as follows. We first summarise the ERKN method for the general oscillatory system (8.1) in Sect. 8.2, and then in Sect. 8.3 we illustrate drawbacks of the EN-T theory proposed in [33]. In Sect. 8.4, we introduce *the set of improved extended-Nyström trees* and show how this relates to other tree sets in the literature. Section 8.5 focuses on the B-series associated with the ERKN method for the general oscillatory system (8.1), and Sect. 8.6 analyses the corresponding order conditions for the ERKN methods, when applied to the general oscillatory system (8.1). In Sect. 8.7 we derive some ERKN methods of order up to four, exploiting the advantages of the new tree theory. The numerical experiments are made in Sect. 8.8. Conclusive remarks are included in Sect. 8.9.

8.2 General ERKN Methods

To begin with, we summarise the following general ERKN method based on the matrix-variation-of-constants formula (see [23]) and quadrature formulae.

Definition 8.1 (*See* [33]) An s-stage general extended Runge–Kutta–Nyström (ERKN) method for the numerical integration of the IVP (8.1) is defined by the following scheme

$$\begin{cases} Y_i = \phi_0(c_i^2 V)y_n + c_i\phi_1(c_i^2 V)hy_n' + h^2\displaystyle\sum_{j=1}^{s}\bar{a}_{ij}(V)f(Y_j, Y_j'), \qquad i = 1,\ldots,s, \\ hY_i' = -c_i V\phi_1(c_i^2 V)y_n + \phi_0(c_i^2 V)hy_n' + h^2\displaystyle\sum_{j=1}^{s}a_{ij}(V)f(Y_j, Y_j'), \qquad i = 1,\ldots,s, \\ y_{n+1} = \phi_0(V)y_n + \phi_1(V)hy_n' + h^2\displaystyle\sum_{i=1}^{s}\bar{b}_i(V)f(Y_i, Y_i'), \\ hy_{n+1}' = -V\phi_1(V)y_n + \phi_0(V)hy_n' + h^2\displaystyle\sum_{i=1}^{s}b_i(V)f(Y_i, Y_i'), \end{cases} \tag{8.4}$$

where $\phi_0(V), \phi_1(V), \bar{a}_{ij}(V), a_{ij}(V), \bar{b}_i(V)$ and $b_i(V)$ for $i, j = 1,\ldots,s$ are matrix-valued functions of $V = h^2M$, and are assumed to have the following series expansions

$$\bar{a}_{ij}(V) = \sum_{k=0}^{+\infty}\frac{\bar{a}_{ij}^{(2k)}}{(2k)!}V^k,\ a_{ij}(V) = \sum_{k=0}^{+\infty}\frac{a_{ij}^{(2k)}}{(2k)!}V^k,$$
$$\bar{b}_i(V) = \sum_{k=0}^{+\infty}\frac{\bar{b}_i^{(2k)}}{(2k)!}V^k,\ b_i(V) = \sum_{k=0}^{+\infty}\frac{b_i^{(2k)}}{(2k)!}V^k,\ \phi_i(V) = \sum_{k=0}^{+\infty}\frac{(-1)^k}{(2k+i)!}V^k$$

with real coefficients $\bar{a}_{ij}^{(2k)}$, $a_{ij}^{(2k)}$, $\bar{b}_i^{(2k)}$, $b_i^{(2k)}$ for $k = 0, 1, 2, \ldots$.

The ERKN method (8.4) in Definitions 8.1 can also be represented compactly in a Butcher tableau of the coefficients [4]:

$$\begin{array}{c|cccc|cccc} c_1 & \bar{a}_{11}(V) & \bar{a}_{12}(V) & \cdots & \bar{a}_{1s}(V) & a_{11}(V) & a_{12}(V) & \cdots & a_{1s}(V) \\ c_2 & \bar{a}_{21}(V) & \bar{a}_{22}(V) & \cdots & \bar{a}_{2s}(V) & a_{21}(V) & a_{22}(V) & \cdots & a_{2s}(V) \\ \vdots & \vdots & \vdots & \ddots & \vdots & \vdots & \vdots & \ddots & \vdots \\ c_s & \bar{a}_{s1}(V) & \bar{a}_{s2}(V) & \cdots & \bar{a}_{ss}(V) & a_{s1}(V) & a_{s2}(V) & \cdots & a_{ss}(V) \\ \hline & \bar{b}_1(V) & \bar{b}_2(V) & \cdots & \bar{b}_s(V) & b_1(V) & b_2(V) & \cdots & b_s(V) \end{array}. \tag{8.5}$$

In essence, ERKN methods incorporate the particular structure of the oscillatory system (8.1) into both the internal stages and the updates. Throughout this chapter, we call methods for the general oscillatory system (8.1) *general ERKN methods*, and *standard ERKN methods*, for the special case (8.2).

8.3 The Failure and the Reduction of the EN-T Theory

The EN-T theory for general ERKN methods was presented in the recent paper [33] in which some tri-colored trees are supplemented to *the classical Nyström trees* (N-Ts). The idea of the EN-T theory comes from the fact that the numbers of the N-Ts and of the elementary differentials are completely different. The paper [33] tried to eliminate the difference and then to make one elementary differential correspond to one tree uniquely. Unfortunately, however, the paper [33] did not succeed on this point. For example, the two different trees shown in Table 8.1 have the same elementary differentials $\mathscr{F}(\tau)(\boldsymbol{y}, \boldsymbol{y}')$.

Moreover, the great limitation of the EN-T theory is the existence of great numbers of redundant trees that cause trouble in applications. For example, in Table 8.2 (left), there are seven EN-Ts but five of them are redundant since their order $\rho(\tau)$, density $\gamma(\tau)$, weight $\Phi_i(\tau)$, and the consequent order conditions can be implied by others for the general ERKN methods (8.4).

Here, it should be pointed out that it is not necessary that one tree corresponds to one elementary differential. In other words, one tree may correspond to a set

Table 8.1 Two EN-Ts which have the same elementary differentials $\mathscr{F}(\tau)(\boldsymbol{y}, \boldsymbol{y}')$

EN-Ts	ρ	γ	Φ_i	α	$\mathscr{F}$
	4	4	$c_i^2 \sum_j a_{ij}^{(0)}$	3	$\boldsymbol{f}_{\boldsymbol{y}\boldsymbol{y}'}^{(2)}(-M\boldsymbol{y}, \boldsymbol{f})$
	4	8	$c_i \sum_j \bar{a}_{ij}^{(0)}$	3	$\boldsymbol{f}_{\boldsymbol{y}\boldsymbol{y}'}^{(2)}(-M\boldsymbol{y}, \boldsymbol{f})$

Table 8.2 Some EN-Ts and the redundance

EN-Ts	ρ	γ	Φ_i	α	$\mathscr{F}$
	2	2	c_i	1	$f^{(1)}_{y} y'$
	2	2	c_i	1	$f^{(1)}_{y'}(-My)$
	3	3	c_i^2	1	$f^{(2)}_{yy}(y', y')$
	3	3	c_i^2	2	$f^{(2)}_{yy'}(-My, y')$
	3	3	c_i^2	1	$f^{(2)}_{y'y'}(-My, -My)$
	3	3	c_i^2	1	$f^{(1)}_{y}(-My)$
	3	3	c_i^2	1	$f^{(1)}_{y'}(-My')$

EN-Ts	ρ	γ	Φ_i	α	$\mathscr{F}$
	2	2	c_i	1	$f^{(1)}_{y} y'$
					$+f^{(1)}_{y'}(-My)$
	3	3	c_i^2	1	$f^{(2)}_{yy}(y', y')$
					$+2f^{(2)}_{yy'}(-My, y')$
					$+ f^{(2)}_{y'y'}(-My, -My)$
					$+f^{(1)}_{y}(-My)$
					$+f^{(1)}_{y'}(-My')$

of elementary differentials. For example, just as shown in Table 8.2, the sum of the products of the coefficient $\alpha(\tau)$ and the elementary differentials $\mathscr{F}(\tau)(y, y')$ is meaningful. In fact, we have

$$f^{(1)}_{y} y' + f^{(1)}_{y'}(-My) = D^1_h f\Big(\phi_0(h^2M)y + \phi_1(h^2M)hy', \phi_0(h^2M)y' - hM\phi_1(h^2M)y\Big),$$

namely, $f^{(1)}_{y} y' + f^{(1)}_{y'}(-My)$ is the first-order derivative of function f with respect to h, at $h = 0$, where the function f is evaluated at point $(\hat{y}, \hat{y}')$ with

$$\hat{y} = \phi_0(h^2M)y + \phi_1(h^2M)hy', \tag{8.6}$$

$$\hat{y}' = \phi_0(h^2M)y' - hM\phi_1(h^2M)y. \tag{8.7}$$

Thus, in Table 8.2, we can choose these two bi-colored trees to respectively represent the sums, and omit all trees with meagre vertices. In this way, we can get rid of the redundance as shown in Table 8.2 (right).

On the other hand, although almost all tri-colored trees are redundant, there indeed exist tri-colored trees which are absolutely necessary in the research of order conditions for the general ERKN methods (8.4). For example, the fifth tree which is tri-colored in the fifth line in the Table 2 in [33] undoubtedly works for the order conditions. In a word, the theory for the general ERKN methods (8.4) is a tri-colored tree theory, but it is based on a subset of the EN-T set.

Hence, it is quite natural that this chapter starts from the Nth derivative of the function $\boldsymbol{f}^{(m+n)}_{\boldsymbol{y}^m \boldsymbol{y}'^n}\Big|_{(\hat{\boldsymbol{y}},\hat{\boldsymbol{y}}')}$ with respect to h, at $h = 0$. For details about multivariate Taylor series expansions and some related knowledge, readers are referred to [1, 30]. In what follows we will denote this derivative as $D_h^N \boldsymbol{f}^{(m+n)}_{\boldsymbol{y}^m \boldsymbol{y}'^n}$.

Remark 8.1 The dimension of the matrix $D_h^N \boldsymbol{f}^{(m+n)}_{\boldsymbol{y}^m \boldsymbol{y}'^n}$ is $d \times d^{m+n}$. If $\boldsymbol{z}$ is a $d^{m+n} \times 1$ matrix, the dimension of $D_h^N \boldsymbol{f}^{(m+n)}_{\boldsymbol{y}^m \boldsymbol{y}'^n}\boldsymbol{z}$ is $d \times 1$.

Remark 8.2 If the matrix M is null,

$$D_h^N \boldsymbol{f}^{(m+n)}_{\boldsymbol{y}^m \boldsymbol{y}'^n}\boldsymbol{z} = \boldsymbol{f}^{(m+n+N)}_{\boldsymbol{y}^{m+N} \boldsymbol{y}'^n}\Big(\underbrace{\boldsymbol{y}', \cdots, \boldsymbol{y}'}_{N\text{ fold}}, \boldsymbol{z}\Big),$$

where $\boldsymbol{f}^{(m+n+N)}_{\boldsymbol{y}^{m+N} \boldsymbol{y}'^n}$ is evaluated at the point $(\boldsymbol{y}, \boldsymbol{y}')$, and $(\cdot, \cdots, \cdot)$ is the Kronecker inner product (see [30]).

Remark 8.3 In the special case (8.2) where the function $\boldsymbol{f}$ is independent of $\boldsymbol{y}'$, $D_h^N \boldsymbol{f}^{(m+n)}_{\boldsymbol{y}^m \boldsymbol{y}'^n}\boldsymbol{z}$ is exactly $D_h^N \boldsymbol{f}^{(m)}\boldsymbol{z}$ in [30].

At the end of this section we give the following first three results of $D_h^N \boldsymbol{f}^{(m+n)}_{\boldsymbol{y}^m \boldsymbol{y}'^n}\boldsymbol{z}$, which contribute significantly to our understanding of the extended elementary differentials (see Definition 8.3 in Sect. 8.4).

$$\begin{aligned}
D_h^1 \boldsymbol{f}^{(m+n)}_{\boldsymbol{y}^m \boldsymbol{y}'^n}\boldsymbol{z} &= \boldsymbol{f}^{(m+n+1)}_{\boldsymbol{y}^{m+1} \boldsymbol{y}'^n}\Big(\boldsymbol{y}', \boldsymbol{z}\Big) + \boldsymbol{f}^{(m+n+1)}_{\boldsymbol{y}^m \boldsymbol{y}'^{n+1}}\Big(-M\boldsymbol{y}, \boldsymbol{z}\Big),\\
D_h^2 \boldsymbol{f}^{(m+n)}_{\boldsymbol{y}^m \boldsymbol{y}'^n}\boldsymbol{z} &= \boldsymbol{f}^{(m+n+2)}_{\boldsymbol{y}^{m+2} \boldsymbol{y}'^n}\Big(\boldsymbol{y}', \boldsymbol{y}', \boldsymbol{z}\Big) + \boldsymbol{f}^{(m+n+1)}_{\boldsymbol{y}^{m+1} \boldsymbol{y}'^n}\Big(-M\boldsymbol{y}, \boldsymbol{z}\Big) + 2\boldsymbol{f}^{(m+n+2)}_{\boldsymbol{y}^{m+1} \boldsymbol{y}'^{n+1}}\Big(\boldsymbol{y}', -M\boldsymbol{y}, \boldsymbol{z}\Big)\\
&\quad + \boldsymbol{f}^{(m+n+2)}_{\boldsymbol{y}^m \boldsymbol{y}'^{n+2}}\Big(-M\boldsymbol{y}, -M\boldsymbol{y}, \boldsymbol{z}\Big) + \boldsymbol{f}^{(m+n+1)}_{\boldsymbol{y}^m \boldsymbol{y}'^{n+1}}\Big(-M\boldsymbol{y}', \boldsymbol{z}\Big),\\
D_h^3 \boldsymbol{f}^{(m+n)}_{\boldsymbol{y}^m \boldsymbol{y}'^n}\boldsymbol{z} &= \boldsymbol{f}^{(m+n+3)}_{\boldsymbol{y}^{m+3} \boldsymbol{y}'^n}\Big(\boldsymbol{y}', \boldsymbol{y}', \boldsymbol{y}', \boldsymbol{z}\Big) + 3\boldsymbol{f}^{(m+n+3)}_{\boldsymbol{y}^{m+2} \boldsymbol{y}'^{n+1}}\Big(\boldsymbol{y}', \boldsymbol{y}', -M\boldsymbol{y}, \boldsymbol{z}\Big)\\
&\quad + 3\boldsymbol{f}^{(m+n+3)}_{\boldsymbol{y}^{m+1} \boldsymbol{y}'^{n+2}}\Big(\boldsymbol{y}', -M\boldsymbol{y}, -M\boldsymbol{y}, \boldsymbol{z}\Big) + \boldsymbol{f}^{(m+n+3)}_{\boldsymbol{y}^m \boldsymbol{y}'^{n+3}}\Big(-M\boldsymbol{y}, -M\boldsymbol{y}, -M\boldsymbol{y}, \boldsymbol{z}\Big)\\
&\quad + 3\boldsymbol{f}^{(m+n+2)}_{\boldsymbol{y}^{m+2} \boldsymbol{y}'^n}\Big(\boldsymbol{y}', -M\boldsymbol{y}, \boldsymbol{z}\Big) + 3\boldsymbol{f}^{(m+n+2)}_{\boldsymbol{y}^{m+1} \boldsymbol{y}'^{n+1}}\Big(-M\boldsymbol{y}, -M\boldsymbol{y}, \boldsymbol{z}\Big)\\
&\quad + 3\boldsymbol{f}^{(m+n+2)}_{\boldsymbol{y}^{m+1} \boldsymbol{y}'^{n+1}}\Big(\boldsymbol{y}', -M\boldsymbol{y}', \boldsymbol{z}\Big) + 3\boldsymbol{f}^{(m+n+2)}_{\boldsymbol{y}^m \boldsymbol{y}'^{n+2}}\Big(-M\boldsymbol{y}, -M\boldsymbol{y}', \boldsymbol{z}\Big)\\
&\quad + \boldsymbol{f}^{(m+n+1)}_{\boldsymbol{y}^{m+1} \boldsymbol{y}'^n}\Big(-M\boldsymbol{y}', \boldsymbol{z}\Big) + \boldsymbol{f}^{(m+n+1)}_{\boldsymbol{y}^m \boldsymbol{y}'^{n+1}}\Big((-M)^2\boldsymbol{y}, \boldsymbol{z}\Big).
\end{aligned}$$

Table 8.3 Four theory systems for second order differential equations

	IVPs	Methods	Trees (graphs)	Compact (T/F)
1	$y'' = f(y, y')$	General RKN methods	N-Ts	T
2	$y'' = f(y)$	Standard RKN methods	SN-Ts	T
3	$y'' + My = f(y, y')$	General ERKN methods	EN-Ts	F
4	$y'' + My = f(y)$	Standard ERKN methods	SSEN-Ts	T

8.4 The Set of Improved Extended-Nyström Trees

In the study of order conditions for second-order differential equations, there are four theory systems listed in Table 8.3, where the abbreviation "SSEN-T" is for *simplified special extended Nyström-tree* [30], and here the word "compact" should be interpreted as meaning that any order condition derived from a tree belonging the underlining rooted tree set cannot be obtained by another from the same rooted tree set.

The first two systems are very famous in the numerical analysis for ODEs, where the second is a special case of the first one. The rooted tree sets in these two systems are all bi-colored tree sets with the white vertex and the black vertex. The last two systems are constructed on tri-colored rooted tree sets by adding the meagre vertex to the graph of bi-colored trees. Similarly, the last system is the special case of the third.

Moreover, when the matrix M is null, the third is identical to the first, and the fourth to the second. In a word, the last two systems are the extensions of the first two systems respectively. However, the extension of the first system is not satisfied yet, since the last section in this chapter states that the third system is not compact. In order to make the extension better, a compact theory will be built to replace the third one, by introducing a completely new tri-colored rooted tree set and six mappings onto it. In this section, we will define the new tree set and study the relationships to the N-T set, the EN-T set, and the SSEN-T set.

8.4.1 *The IEN-T Set and the Related Mappings*

In what follows, we will recursively define a new set named *the improved extended-Nyström tree set*, and define six mappings on it.

Definition 8.2 The improved extended-Nyström tree (IEN-T) set, is recursively defined as follows:

Table 8.4 Tree $W_+B_+(b_+B_+)^p(\tau)$ (left), and tree $W_+(b_+B_+)^p(\tau)$ (right) in Definition 8.2

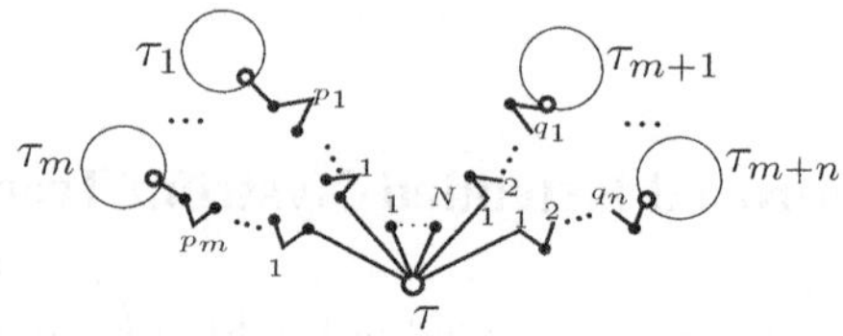

Fig. 8.1 The mode of the trees in the IEN-T set

(a) o,  belong to the IEN-T set.
(b) If τ belongs to the IEN-T set, then the graph obtained by grafting the root of tree τ to a new black fat node and then to a new meagre node, $\cdots$ (p times), and then to a new black fat node and then last to a new white node, denoted by $W_+B_+(b_+B_+)^p(\tau)$ (see Table 8.4), belongs to the IEN-T set for $\forall p = 0, 1, 2, \ldots$.
(c) If τ belongs to the IEN-T set, then the graph obtained by grafting the root of tree τ to a new black fat node and then to a new meagre node, $\cdots$ (p times), then last to a new white node, denoted by $W_+(b_+B_+)^p(\tau)$ (see Table 8.4), belongs to the IEN-T set for $\forall p = 0, 1, 2, \ldots$.
(d) If $\tau_1, \ldots, \tau_\mu$ belong to the IEN-T set, then $\tau_1 \times \cdots \times \tau_\mu$ belongs to the IEN-T set, where '$\times$' is the merging product [4].

Each tree τ in the IEN-T set can be denoted by

$$\tau := \underbrace{\tau_* \times \cdots \times \tau_*}_{N-\text{fold}} \times \Big(W_+B_+(b_+B_+)^{p_1}(\tau_1)\Big) \times \cdots \times \Big(W_+B_+(b_+B_+)^{p_m}(\tau_m)\Big)$$
$$\times \Big(W_+(b_+B_+)^{q_1}(\tau_{m+1})\Big) \times \cdots \times \Big(W_+(b_+B_+)^{q_n}(\tau_{m+n})\Big), \tag{8.8}$$

where $\tau_* =$ . Figure 8.1 gives the mode of the trees in the IEN-T set.

On the basis of Definition 8.2, the following rules for forming a tree τ in the IEN-T set can be obtained straightforwardly:

(i) The root of a tree is always a fat white vertex.
(ii) A white vertex has fat black children, or white children, or meagre children.

(iii) A fat black vertex has at most one child which can be white or meagre.
(iv) A meagre vertex must has one fat black vertex as its child, and must have a white vertex as its descendant.

Definition 8.3 The order $\rho(\tau)$, the extended elementary differential $\mathscr{F}(\tau)(\mathbf{y}, \mathbf{y}')$, the coefficient $\alpha(\tau)$, the weight $\Phi_i(\tau)$, the density $\gamma(\tau)$ and the sign $S(\tau)$ on the IEN-T set are recursively defined as follows.

1. $\rho(\circ) = 1$, $\mathscr{F}(\circ) = f$, $\alpha(\circ) = 1$, $\Phi_i(\circ) = 1$, $\gamma(\circ) = 1$ and $S(\circ) = 1$.
2. For $\tau \in$ IEN-T denoted by (8.8),
   - $\rho(\tau) = 1 + N + \sum\limits_{i=1}^{m}\Big(1 + 2p_i + \rho(\tau_i)\Big) + \sum\limits_{i=1}^{n}\Big(2q_i + \rho(\tau_{m+i})\Big)$,
   - $\mathscr{F}(\tau) = D_h^N f^{(m+n)}_{y^m y'^n}\Big((-M)^{p_1}\mathscr{F}(\tau_1), \cdots, (-M)^{p_{m+n}}\mathscr{F}(\tau_{m+n})\Big)$,

     where $p_{m+i} = q_i, i = 1, \cdots, n$, and $(\cdot, \cdots, \cdot)$ is the Kronecker inner product (see [30]),
   - $\alpha(\tau) = (\rho(\tau) - 1)! \cdot \frac{1}{N!} \cdot \prod\limits_{i=1}^{m}\left(\frac{\alpha(\tau_i)}{(1+2p_i+\rho(\tau_i))!}\right) \cdot \prod\limits_{i=1}^{n}\left(\frac{\alpha(\tau_{m+i})}{(2q_i+\rho(\tau_{m+i}))!}\right) \cdot \frac{1}{J_1!\ldots J_I!}$,

     where $J_1, \cdots, J_I$ count the same branches,
   - $\Phi_i(\tau) = c_i^N \cdot \prod\limits_{k=1}^{m}\Big(\sum\limits_{j=1}^{s} \bar{a}_{ij}^{(2p_k)}\Phi_j(\tau_k)\Big) \cdot \prod\limits_{k=1}^{n}\Big(\sum\limits_{j=1}^{s} a_{ij}^{(2q_k)}\Phi_j(\tau_{m+k})\Big)$,
   - $\gamma(\tau) = \rho(\tau) \cdot \prod\limits_{i=1}^{m}\left(\frac{(1+2p_i+\rho(\tau_i))!\gamma(\tau_i)}{(2p_i)!\rho(\tau_i)!}\right) \cdot \prod\limits_{i=1}^{n}\left(\frac{(2q_i+\rho(\tau_{m+i}))!\gamma(\tau_{m+i})}{(2q_i)!\rho(\tau_{m+i})!}\right)$,
   - $S(\tau) = \prod\limits_{i=1}^{m}\Big((-1)^{p_i}S(\tau_i)\Big) \cdot \prod\limits_{i=1}^{n}\Big((-1)^{q_i}S(\tau_{m+i})\Big)$,

where $\sum\limits_{k=1}^{0} = 0$ and $\prod\limits_{k=1}^{0} = 1$.

Definition 8.4 The set IEN-T$_m$ is defined as

$$\text{IEN-T}_m = \big\{\tau : \rho(\tau) = m, \ \tau \in \text{IEN-T}\big\}.$$

Remark 8.4 The order $\rho(\tau)$ is the number of the tree τ's vertices.

Remark 8.5 The extended elementary differential $\mathscr{F}(\tau)$ is a product of $(-M)^p$ (p is the number of meagre vertices between a white vertex and the next coming white vertex), and $D_h^N f^{(n+m)}_{y^m y'^n}$ (N is the number of end vertices from the white vertex, m is the number of the non-ending black vertices from the white vertex, and n is the number of the meagre vertices from the white vertex). We will see that the extended elementary differential is not only one function but a weighted sum of the traditional elementary differential.

Remark 8.6 One IEN-T corresponds to one extended elementary differential $\mathscr{F}(\tau)$.

Remark 8.7 The coefficient $\alpha(\tau)$ is the number of possible different monotonic labelings of τ.

Remark 8.8 The weight $\Phi_i(\tau)$ is a sum over the indices of all white vertices and of all end vertices. The general term of the sum is a product of $\bar{a}_{ij}^{(2p)}$ for $W_+B_+(b_+B_+)^p(\tau)$, of $a_{ij}^{(2p)}$ for $W_+(b_+B_+)^p(\tau)$ (p is the number of the meagre vertices between the white vertices i and j), and of c_i^m (m is the number of end vertices from the white vertex i).

Remark 8.9 One IEN-T corresponds to one weight $\Phi_i(\tau)$.

Remark 8.10 The density $\gamma(\tau)$ is the product of the density of a tree by overlooking the differences between vertices, and $\frac{1}{(2p)!}$, where p is the number of the meagre vertices between two white vertices.

Remark 8.11 The sign $S(\tau)$ is 1 if the number of the meagre vertices is even, and -1 if the number of the meagre vertices is odd.

Table 8.5 makes a list of the corresponding mappings: the order ρ, the sign S, the density γ, the weight Φ_i, the symmetry α and the extended elementary differential $\mathscr{F}$ for each τ in the IEN-T set of order up to 4.

8.4.2 *The IEN-T Set and the N–T Set*

In this subsection, we will see that with the disappearance of meagre vertices the IEN-T set is exactly the N-T set. In fact, in this case, each tree τ in the IEN-T set has the form shown in Fig 8.2, and the rules to form the tree set are straightforwardly reduced to:

(i) The root of a tree is always a fat white vertex.
(ii) A white vertex has fat black children, or white children.
(iii) A fat black vertex has at most one child which must be white.

In this case, from Remarks 8.4–8.10, the order $\rho(\tau)$, the coefficient' $\alpha(\tau)$ and the density $\gamma(\tau)$ are exactly the same as the ones on the N-T set respectively. If M is null, the weight $\Phi_i(\tau)$ and the extended elementary differential $\mathscr{F}(\tau)(\mathbf{y}, \mathbf{y}')$ on the IEN-T set are exactly the same as the ones on the N-T set respectively, too. In fact, from Definition 8.3, with the disappearance of meagre vertices, these two mappings are recursively defined respectively, for τ denoted by Fig. 8.2, as follows:

$$\Phi_i(\tau) = c_i^N \cdot \prod_{k=1}^{m}\Big(\sum_{j=1}^{s} \bar{a}_{ij}\Phi_j(\tau_k)\Big) \cdot \prod_{k=1}^{n}\Big(\sum_{j=1}^{s} a_{ij}\Phi_j(\tau_{m+k})\Big),$$

$$\mathscr{F}(\tau) = D_h^N \boldsymbol{f}^{(m+n)}_{\mathbf{y}^m\mathbf{y}'^n}\Big(\mathscr{F}(\tau_1), \cdots, \mathscr{F}(\tau_{m+n})\Big).$$

Table 8.5 IEN-Ts and mappings of order up to 4 and the corresponding elementary differentials on the N-T set

No.	IEN-Ts	ρ	S	γ	Φ_i	α	$\mathscr{F}$	$\mathscr{F}$ on the N-T set
1		1	1	1	1	1	$\boldsymbol{f}$	f
2		2	1	2	c_i	1	$D_h^1 \boldsymbol{f}$	$f'_y y'$
3		2	1	2	$\sum_j a_{ij}^{(0)}$	1	$\boldsymbol{f}_{y'}^{(1)} \boldsymbol{f}$	$f'_{y'} f$
4		3	1	3	c_i^2	1	$D_h^2 \boldsymbol{f}$	$f''_{yy}(y', y')$
5		3	1	3	$c_i \sum_j a_{ij}^{(0)}$	1	$D_h^1 \boldsymbol{f}_{y'} \boldsymbol{f}$	$f''_{yy'}(y', f)$
6		3	1	3	$\sum_{j,k} a_{ij}^{(0)} a_{ik}^{(0)}$	1	$\boldsymbol{f}_{y'y'}(\boldsymbol{f}, \boldsymbol{f})$	$f''_{y'y'}(f, f)$
7		3	1	6	$\sum_j \bar{a}_{ij}^{(0)}$	1	$\boldsymbol{f}_{y}^{(1)} \boldsymbol{f}$	$f'_y f$
8		3	1	6	$\sum_j a_{ij}^{(0)} c_j$	1	$\boldsymbol{f}_{y'}^{(1)} D_h^1 \boldsymbol{f}$	$f'_{y'} f_y y'$
9		3	1	6	$\sum_{j,k} a_{ij}^{(0)} a_{jk}^{(0)}$	1	$\boldsymbol{f}_{y'}^{(1)} \boldsymbol{f}_{y'}^{(1)} \boldsymbol{f}$	$f'_{y'} f'_{y'} f$
10		4	1	4	c_i^3	1	$D_h^3 \boldsymbol{f}$	$f_{yyy}^{(3)}(y', y', y')$
11		4	1	4	$c_i^2 \sum_j a_{ij}^{(0)}$	3	$D_h^2 \boldsymbol{f}_{y'}^{(1)} \boldsymbol{f}$	$f_{y'yy}^{(3)}(f, y', y')$
12		4	1	4	$c_i \sum_{j,k} a_{ij}^{(0)} a_{ik}^{(0)}$	3	$D_h^1 \boldsymbol{f}_{y'y'}^{(2)}(\boldsymbol{f}, \boldsymbol{f})$	$f_{yy'y'}^{(3)}(y', f, f)$
13		4	1	4	$\sum_{j,k,l} a_{ij}^{(0)} a_{ik}^{(0)} a_{il}^{(0)}$	1	$\boldsymbol{f}_{y'y'y'}^{(3)}(\boldsymbol{f}, \boldsymbol{f}, \boldsymbol{f})$	$f_{y'y'y'}^{(3)}(f, f, f)$
14		4	1	8	$c_i \sum_j \bar{a}_{ij}^{(0)}$	3	$D_h^1 \boldsymbol{f}_{y}^{(1)} \boldsymbol{f}$	$f''_{yy}(y', f)$
15		4	1	8	$\sum_{j,k} \bar{a}_{ij}^{(0)} a_{ik}^{(0)}$	3	$\boldsymbol{f}_{yy'}^{(2)}(\boldsymbol{f}, \boldsymbol{f})$	$f''_{yy'}(f, f)$
16		4	1	8	$c_i \sum_{j,k} a_{ij}^{(0)} a_{jk}^{(0)}$	3	$D_h^1 \boldsymbol{f}_{y'}^{(1)} \boldsymbol{f}_{y'} \boldsymbol{f}$	$f''_{yy'}(y', f_{y'} f)$
17		4	1	8	$\sum_{j,k,l} a_{ij}^{(0)} a_{ik}^{(0)} a_{kl}^{(0)}$	3	$\boldsymbol{f}_{y'y'}^{(2)}(\boldsymbol{f}, \boldsymbol{f}_{y'}^{(1)} \boldsymbol{f})$	$f''_{y'y'}(f_{y'} f, f)$
18		4	1	8	$c_i \sum_j a_{ij}^{(0)} c_j$	3	$D_h^1 \boldsymbol{f}_{y'}^{(1)} D_h^1 \boldsymbol{f}$	$f''_{yy'}(f_y y', y')$
19		4	1	8	$\sum_{j,k} a_{ij}^{(0)} a_{ik}^{(0)} c_k$	3	$\boldsymbol{f}_{y'y'}^{(2)}(\boldsymbol{f}, D_h^1 \boldsymbol{f})$	$f''_{y'y'}(f_y y', f)$

(continued)

Table 8.5 (continued)

No.	IEN-Ts	ρ	S	γ	Φ_i	α	$\mathscr{F}$	$\mathscr{F}$ on the N-T set
20		4	1	24	$\sum_j \bar{a}_{ij}^{(0)} c_j$	1	$f_y^{(1)} D_h^1 f$	$f_y' f_y' y'$
21		4	1	24	$\sum_{j,k} \bar{a}_{ij}^{(0)} a_{jk}^{(0)}$	1	$f_y^{(1)} f_{y'}^{(1)} f$	$f_y' f_{y'}' f$
22		4	1	24	$\sum_{j,k} a_{ij}^{(0)} \bar{a}_{jk}^{(0)}$	1	$f_{y'}^{(1)} f_y^{(1)} f$	$f_{y'}' f_y' f$
23		4	1	24	$\sum_{j,k} a_{ij}^{(0)} a_{jk}^{(0)} c_k$	1	$f_{y'}^{(1)} f_{y'}^{(1)} D_h^1 f$	$f_{y'}' f_{y'}' f_y' y'$
24		4	1	24	$\sum_{j,k,l} a_{ij}^{(0)} a_{jk}^{(0)} a_{kl}^{(0)}$	1	$f_{y'}^{(1)} f_{y'}^{(1)} f_{y'}^{(1)} f$	$f_{y'}' f_{y'}' f_{y'}' f$
25		4	-1	12	$\sum_j a_{ij}^{(2)}$	1	$f_{y'}^{(1)} (-M) f$	–
26		4	1	12	$\sum_j a_{ij}^{(0)} c_j^2$	1	$f_{y'}^{(1)} D_h^2 f$	$f_{y'}' f_{yy}''(y', y')$
27		4	1	12	c_i	2	$f_{y'}^{(1)} D_h^1 f_{y'}^{(1)} f$	$f_{y'}' f_{yy'}''(y', f)$
28		4	1	12	$\sum_{j,k,l} a_{ij}^{(0)} a_{jk}^{(0)} a_{jl}^{(0)}$	1	$f_{y'}^{(1)} f_{y'y'}^{(2)} (f, f)$	$f_{y'}' f_{y'y'}''(f, f)$

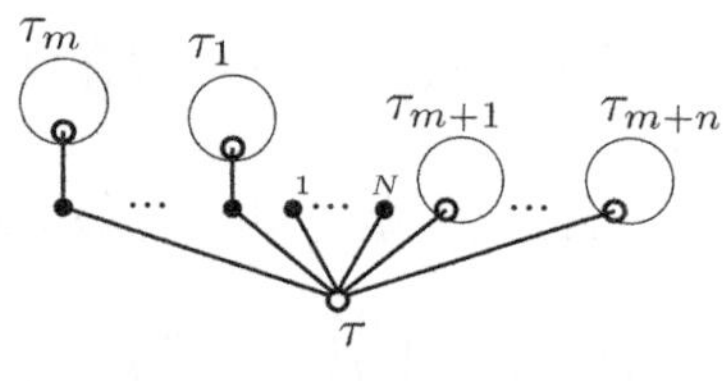

Fig. 8.2 The form of the trees with meagre vertices disappearing

Table 8.6 Tri-colored Trees which are appended to the set N-T_5 to form the set IEN-T_5

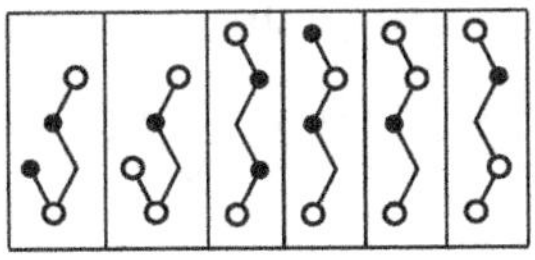

Clearly, the IEN-T set is really an extension of the N-T set (see, Table 14.3 on p. 292 in [4]). It can also be seen from Tables 8.5 and 8.6 that one 4th order tree, six 5th order trees are appended to the N-T set to form the IEN-T set. All these special and new appended trees have a meagre vertex (or some vertices) which correspond to nothing in the N-T set. In fact, the weights Φ_i in Table 8.6 are all the functions of $\bar{a}_{ij}^{(2k)}$ and $a_{ij}^{(2k)}$, high-order derivatives of $\bar{a}_{ij}(V)$ and $a_{ij}(V)$ with respect to h, at $h = 0$.

8.4.3 *The IEN-T Set and the EN-T Set*

First of all, we note that there are just five mappings defined on the EN-T set in the paper [33], while there are six mappings on the IEN-T set in this chapter. In the paper [33], the authors introduced *the signed density* $\tilde{\gamma}(\tau)$, but in this chapter we replace $\tilde{\gamma}(\tau)$ by the product of the two mappings, the density $\gamma(\tau)$ and the sign $S(\tau)$.

The IEN-T set is a subset of the EN-T set, once one overlooks the (extended) elementary differential $\mathscr{F}(\tau)$ on them.

8.4.4 *The IEN-T Set and the SSEN-T Set*

From the rules of the IEN-T set and of the SSEN-T set (see [30]), if the function f in the system (8.1) does not containing $\boldsymbol{y}'$ explicitly, the IEN-T set is exactly the SSEN-T set.

8.5 B-Series for the General ERKN Method

In Sect. 8.4 we presented the IEN-T set, on which six mappings are defined. With these preliminaries, motivated by the concept of B-series, we will describe a totally different approach from the one described in [33] to deriving the theory of order conditions for the general ERKN method.

The main results of the theory of B-series have their origins in the profound paper [2] of Butcher in 1972, and then are introduced in detail by Hairer and Wanner [5] in 1974. In what follows, we present the following two elementary theorems.

Theorem 8.1 *With Definition 8.3, $\boldsymbol{f}(\boldsymbol{y}(t+h), \boldsymbol{y}'(t+h))$ is a B-series*

$$\boldsymbol{f}(\boldsymbol{y}(t+h), \boldsymbol{y}'(t+h)) = \sum_{\tau\in IEN\text{-}T} \frac{h^{\rho(\tau)-1}}{(\rho(\tau)-1)!}\alpha(\tau)\mathscr{F}(\tau)(\boldsymbol{y}, \boldsymbol{y}').$$

Proof First, we expand $\boldsymbol{f}(\boldsymbol{y}(t+h), \boldsymbol{y}'(t+h))$ at point $(\hat{\boldsymbol{y}}, \hat{\boldsymbol{y}}')$, with the definitions of (8.6) and (8.7).

$$\boldsymbol{f}(\boldsymbol{y}(t+h), \boldsymbol{y}'(t+h)) = \sum_{m\geq 0, n\geq 0} \frac{1}{(m+n)!} \boldsymbol{f}^{(m+n)}_{\boldsymbol{y}^m\boldsymbol{y}'^n}\Big|_{(\hat{\boldsymbol{y}}, \hat{\boldsymbol{y}}')} \left(\boldsymbol{y}(t+h)-\hat{\boldsymbol{y}}\right)^{\otimes m} \otimes \left(\boldsymbol{y}'(t+h)-\hat{\boldsymbol{y}}'\right)^{\otimes n}, \tag{8.9}$$

where the second term $\boldsymbol{f}^{(m+n)}_{\boldsymbol{y}^m\boldsymbol{y}'^n}\Big|_{(\hat{\boldsymbol{y}}, \hat{\boldsymbol{y}}')}$ in this series is the matrix-valued function of h.

Definition 8.3 ensures that $\boldsymbol{f}(\boldsymbol{y}(t+h), \boldsymbol{y}'(t+h))$ is a B-series. In fact, if $\boldsymbol{f}(\boldsymbol{y}(t+h), \boldsymbol{y}'(t+h))$ is a B-series, from the matrix-variation-of-constants formula with $\mu = 1$, (see [33]), and from the properties of the ϕ-functions (see e.g. [25]), we have

$$\begin{aligned}
\boldsymbol{y}(t+h)-\hat{\boldsymbol{y}} &= h^2\textstyle\int_0^1 (1-z)\phi_1((1-z)^2V)\boldsymbol{f}(\boldsymbol{y}(t+hz), \boldsymbol{y}'(t+hz))\mathrm{d}z\\
&= \sum_{\tau\in \mathrm{IEN-T}} \textstyle\int_0^1 (1-z)\phi_1((1-z)^2V)\frac{z^{\rho(\tau)-1}}{(\rho(\tau)-1)!}\mathrm{d}z\cdot\left(h^{\rho(\tau)+1}\alpha(\tau)\mathscr{F}(\tau)(\boldsymbol{y}, \boldsymbol{y}')\right)\\
&= \sum_{\tau\in \mathrm{IEN-T}} \phi_{\rho(\tau)+1}(V)\cdot h^{\rho(\tau)+1}\alpha(\tau)\mathscr{F}(\tau)(\boldsymbol{y}, \boldsymbol{y}')\\
&= \sum_{\tau\in \mathrm{IEN-T}}\sum_{p\geq 0} \frac{(-1)^pV^p}{(\rho(\tau)+1+2p)!}h^{\rho(\tau)+1}\alpha(\tau)\mathscr{F}(\tau)(\boldsymbol{y}, \boldsymbol{y}'),
\end{aligned} \tag{8.10}$$

and

$$\boldsymbol{y}'(t+h)-\hat{\boldsymbol{y}}' = \sum_{\tau\in \mathrm{IEN-T}}\sum_{q\geq 0} \frac{(-1)^qV^q}{(\rho(\tau)+2q)!}h^{\rho(\tau)}\alpha(\tau)\mathscr{F}(\tau)(\boldsymbol{y}, \boldsymbol{y}'). \tag{8.11}$$

Taking the Taylor series of $\boldsymbol{f}^{(m+n)}_{\boldsymbol{y}^m\boldsymbol{y}'^n}\Big|_{(\hat{\boldsymbol{y}}, \hat{\boldsymbol{y}}')}$ at $h=0$, and from (8.10) and (8.11), the Eq. (8.9) becomes

$$\begin{aligned}
\boldsymbol{f}(\boldsymbol{y}(t+h), \boldsymbol{y}'(t+h)) = \sum_{N,n,m}\sum_{\tau\in \mathrm{IEN-T}} \frac{h^s}{N!(m+n)!}D_h^N\boldsymbol{f}^{(n+m)}_{\boldsymbol{y}^m\boldsymbol{y}'^n}\Big(\frac{(-\boldsymbol{M})^{p_1}\alpha(\tau_1)\mathscr{F}(\tau_1)(\boldsymbol{y})}{(\rho(\tau_1)+1+2p_1)!}, \ldots,\\
\frac{(-\boldsymbol{M})^{p_m}\alpha(\tau_m)\mathscr{F}(\tau_m)(\boldsymbol{y})}{(\rho(\tau_m)+1+2p_m)!}, \frac{(-\boldsymbol{M})^{q_1}\alpha(\tau_{m+1})\mathscr{F}(\tau_{m+1})(\boldsymbol{y})}{(\rho(\tau_{m+1})+2q_1)!}, \ldots, \frac{(-\boldsymbol{M})^{q_n}\alpha(\tau_{m+n})\mathscr{F}(\tau_{m+n})(\boldsymbol{y})}{(\rho(\tau_{m+n})+2q_n)!}\Big),
\end{aligned} \tag{8.12}$$

where

$$s = N + \sum_{k=1}^{m}(2p_k+\rho(\tau_k)+1) + \sum_{k=1}^{n}(2q_k+\rho(\tau_{m+k})).$$

By Definition 8.3, the proof is complete.

Theorem 8.2 *Given a general ERKN method (8.4), by Definition 8.3, each* $\boldsymbol{f}(Y_i, Y_i')$ *is a series of the form*

$$\boldsymbol{f}(Y_i, Y_i') = \sum_{\tau \in IEN\text{-}T} \frac{h^{\rho(\tau)-1}}{\rho(\tau)!} \boldsymbol{a}_i(\tau),$$

where $\boldsymbol{a}_i(\tau) = \Phi_i(\tau) \cdot \gamma(\tau) \cdot S(\tau) \cdot \alpha(\tau) \cdot \mathscr{F}(\tau)(\boldsymbol{y}_n, \boldsymbol{y}_n')$.

Proof In a similar way to the proof of Theorem 8.1, we expand $\boldsymbol{f}(Y_i, Y_i')$ at $(\tilde{\boldsymbol{y}}, \tilde{\boldsymbol{y}}')$ for the general ERKN method (8.4), where $\tilde{\boldsymbol{y}} = \phi_0(c_i^2 V)\boldsymbol{y}_n + \phi_1(c_i^2 V)c_i h \boldsymbol{y}_n'$ and $\tilde{\boldsymbol{y}}' = \phi_0(c_i^2 V)\boldsymbol{y}_n' - c_i h M \phi_1(c_i^2 V)\boldsymbol{y}_n$, and obtain the Taylor series expansion as follows:

$$\boldsymbol{f}(Y_i, Y_i') = \sum_{m,n \geq 0} \frac{1}{(m+n)!} \boldsymbol{f}_{\boldsymbol{y}^m \boldsymbol{y}'^n}^{(m+n)}\Big|_{\tilde{\boldsymbol{y}}, \tilde{\boldsymbol{y}}'} \Big(h^2 \sum_j \bar{a}_{ij}(V) \boldsymbol{f}(Y_j, Y_j')\Big)^{\otimes m} \otimes \Big(h \sum_j a_{ij}(V) \boldsymbol{f}(Y_j, Y_j')\Big)^{\otimes n}, \tag{8.13}$$

where the second term $\boldsymbol{f}_{\boldsymbol{y}^m \boldsymbol{y}'^n}^{(m+n)}\Big|_{\tilde{\boldsymbol{y}}, \tilde{\boldsymbol{y}}'}$ is a function of $c_i h$. Then the Taylor series expansion of $\boldsymbol{f}_{\boldsymbol{y}^m \boldsymbol{y}'^n}^{(m+n)}\Big|_{(\hat{\boldsymbol{y}}, \hat{\boldsymbol{y}}')}$ at $h = 0$ is given by

$$\boldsymbol{f}_{\boldsymbol{y}^m \boldsymbol{y}'^n}^{(m+n)}\Big|_{(\hat{\boldsymbol{y}}, \hat{\boldsymbol{y}}')} = \sum_{N \geq 0} \frac{c_i^N}{m!} h^N D_h^N \boldsymbol{f}_{\boldsymbol{y}^m \boldsymbol{y}'^n}^{(m+n)}. \tag{8.14}$$

Definition 8.3 ensures that each $\boldsymbol{f}(Y_i, Y_i')$ for $i = 1, \ldots, s$ is a B-series. In fact, the third and fourth terms in the Eq. (8.13) are given by

$$h^2 \sum_j \bar{a}_{ij}(V) \boldsymbol{f}(Y_j, Y_j') = \sum_{\tau \in \mathrm{IEN-T}} \sum_{p \geq 0} \frac{\sum_j \bar{a}_{ij}^{(2p)}}{\rho(\tau)!} \frac{V^p}{(2p)!} h^{\rho(\tau)+1} \boldsymbol{a}_j(\tau), \tag{8.15}$$

and

$$h \sum_j a_{ij}(V) \boldsymbol{f}(Y_j, Y_j') = \sum_{\tau \in \mathrm{IEN-T}} \sum_{q \geq 0} \frac{\sum_j a_{ij}^{(2q)}}{\rho(\tau)!} \frac{V^q}{(2q)!} h^{\rho(\tau)} \boldsymbol{a}_j(\tau). \tag{8.16}$$

We then obtain

$$\begin{aligned}\boldsymbol{f}(Y_i, Y_i') = \sum_{N,n,m} \sum_{\tau \in \mathrm{IEN-T}} & \frac{c_i^N h^s}{N!(n+m)!} D_h^N \boldsymbol{f}_{\boldsymbol{y}^m \boldsymbol{y}'^n}^{(m+n)} \Big(\frac{\sum_j \bar{a}_{ij}^{(2p_1)}}{\rho(\tau_1)!} \frac{M^{p_1}}{(2p_1)!} \boldsymbol{a}_j(\tau_1), \ldots, \frac{\sum_j \bar{a}_{ij}^{(2p_m)}}{\rho(\tau_m)!} \frac{M^{p_m}}{(2p_m)!} \boldsymbol{a}_j(\tau_m), \\ & \frac{\sum_j a_{ij}^{(2q_1)}}{\rho(\tau_{m+1})!} \frac{M^{q_1}}{(2q_1)!} \boldsymbol{a}_j(\tau_{m+1}), \ldots, \frac{\sum_j a_{ij}^{(2q_n)}}{\rho(\tau_{m+n})!} \frac{M^{q_n}}{(2q_n)!} \boldsymbol{a}_j(\tau_{m+n}) \Big),\end{aligned} \tag{8.17}$$

where $s = N + \sum_{k=1}^{m} (2p_k + \rho(\tau_k) + 1) + \sum_{k=1}^{n} (2q_k + \rho(\tau_{m+k}))$. Using Definition 8.3, we complete the proof.

8.6 The Order Conditions for the General ERKN Method

Theorem 8.3 *The scheme (8.4) for the general multi-frequency and multidimensional oscillatory second-order initial value problems* (8.1) *has order r if and only if the following conditions*

$$\sum_{i=1}^{s}\bar{b}_i(V)S(\tau)\gamma(\tau)\Phi_i(\tau)=\rho(\tau)!\phi_{\rho(\tau)+1}+O(h^{r-\rho(\tau)}),\quad \forall\tau\in IEN\text{-}T_m,\ m\leq r-1, \tag{8.18}$$

$$\sum_{i=1}^{s}b_i(V)S(\tau)\gamma(\tau)\Phi_i(\tau)=\rho(\tau)!\phi_{\rho(\tau)}+O(h^{r-\rho(\tau)+1}),\quad \forall\tau\in IEN\text{-}T_m,\ m\leq r, \tag{8.19}$$

are satisfied.

Proof It follows from the matrix-variation-of-constants formula, Theorems 8.1 and 8.2 that

$$\begin{aligned} y_{n+1}&=\phi_0(V)y_n+h\phi_1(V)y_n'\\ &\quad+\sum_{\tau\in\text{IEN}-\text{T}}\frac{h^{\rho(\tau)+1}}{\rho(\tau)!}\sum_{i=1}^{s}\bar{b}_i(V)\Phi_i(\tau)S(\tau)\gamma(\tau)\alpha(\tau)\mathscr{F}(\tau)(y_n,y_n'), \end{aligned} \tag{8.20}$$

$$\begin{aligned} y(t+h)&=\phi_0(V)y+h\phi_1(V)y'\\ &\quad+\sum_{\tau\in\text{IEN-T}}h^{\rho(\tau)+1}\alpha(\tau)\mathscr{F}(\tau)(y,y)\int_0^1(1-z)\frac{z^{\rho(\tau)-1}}{(\rho(\tau)-1)!}\phi_1((1-z)V)\,dz. \end{aligned} \tag{8.21}$$

Comparing the Eqs. (8.20) with (8.21) and using the properties of the ϕ-functions, we obtain the first result of Theorem 8.3. Likewise, we deduce the second part of the theorem.

Theorem 8.3 in this chapter and Theorem 4.1 in [33] share the same expression. However, it should be noted that there exist redundant order conditions in [33], while any order condition in this chapter cannot be replaced by others, provided the entries $\bar{a}_{ij}(V)$, $a_{ij}(V)$, $b_i(V)$ and $\bar{b}_i(V)$ in the general ERKN method (8.4) are independent. Obviously, the elimination of redundant order conditions makes the construction of high-order general ERKN methods (8.4) much clearer and simpler.

It is easy to see that Theorem 8.3 implies the order conditions for the standard ERKN methods in [23, 30] when the right-hand side function f does not depend on y'. It is noted that, if the matrix M is null, Theorem 8.3 reduces to the classical general RKN method when applied to $y''=f(y,y')$, since the IEN-T set is exactly the N-T set in this special case.

8.7 The Construction of General ERKN Methods

In this section, using Theorem 8.3, we present some general ERKN methods (8.4) of order up to 4. The approach to constructing new methods in this section is different from that described in [33].

8.7.1 Second-Order General ERKN Methods

From Theorem 8.3 and the three IEN-Ts with order no more than 2 which are listed in Table 8.5, for an s-stage general ERKN method (8.4) expressed in the Butcher tableau (8.5), we have the following second order conditions:

$$
\begin{aligned}
&\sum_{i=1}^{s} \bar{b}_i(V) = \phi_2(V) + O(h), \quad \sum_{i=1}^{s} b_i(V) = \phi_1(V) + O(h^2),\\
&\sum_{i=1}^{s} b_i(V)c_i = \phi_2(V) + O(h), \sum_{i=1}^{s} b_i(V)a_{ij}^{(0)} = \phi_2(V) + O(h).
\end{aligned}
$$

Comparing the coefficients of h^0 and h, we obtain 4 equations:

$$
\sum_{i=1}^{s} \bar{b}_i^{(0)} = \frac{1}{2}, \quad \sum_{i=1}^{s} b_i^{(0)} = 1, \quad \sum_{i=1}^{s} b_i^{(0)} c_i = \frac{1}{2}, \quad \sum_{i=1}^{s} b_i^{(0)} a_{ij}^{(0)} = \frac{1}{2}.
$$

It can be observed that these equations are exactly the second order conditions for the following traditional RKN method

$$
\begin{cases}
Y_i = y_n + c_i h y_n' + h^2 \sum\limits_{j=1}^{s} \bar{a}_{ij}^{(0)} \Big(f(Y_j, Y_j') - MY_j \Big), & i = 1, \cdots, s,\\
Y_i' = y_n' + h \sum\limits_{j=1}^{s} a_{ij}^{(0)} \Big(f(Y_j, Y_j') - MY_j \Big), & i = 1, \cdots, s,\\
y_{n+1} = y_n + h y_n' + h^2 \sum\limits_{i=1}^{s} \bar{b}_i^{(0)} \Big(f(Y_i, Y_i') - MY_i \Big),\\
y_{n+1}' = y_n' + h \sum\limits_{i=1}^{s} b_i^{(0)} \Big(f(Y_i, Y_i') - MY_i \Big),
\end{cases}
\tag{8.22}
$$

applied to the initial value problems (8.1), with the tableau

$$
\begin{array}{c|cccc|cccc}
c_1 & \bar{a}_{11}^{(0)} & \bar{a}_{12}^{(0)} & \cdots & \bar{a}_{1s}^{(0)} & a_{11}^{(0)} & a_{12}^{(0)} & \cdots & a_{1s}^{(0)} \\
c_2 & \bar{a}_{21}^{(0)} & \bar{a}_{22}^{(0)} & \cdots & \bar{a}_{2s}^{(0)} & a_{21}^{(0)} & a_{22}^{(0)} & \cdots & a_{2s}^{(0)} \\
\vdots & \vdots & \vdots & \ddots & \vdots & \vdots & \vdots & \ddots & \vdots \\
c_s & \bar{a}_{s1}^{(0)} & \bar{a}_{s2}^{(0)} & \cdots & \bar{a}_{s,s}^{(0)} & a_{s1}^{(0)} & a_{s2}^{(0)} & \cdots & a_{s,s}^{(0)} \\
\hline
 & \bar{b}_1^{(0)} & \bar{b}_2^{(0)} & \cdots & \bar{b}_s^{(0)} & b_1^{(0)} & b_2^{(0)} & \cdots & b_s^{(0)}
\end{array}
\quad . \tag{8.23}
$$

This means that we can easily solve $\left(c_i, \bar{a}_{ij}^{(0)}, a_{ij}^{(0)}, \bar{b}_i^{(0)}, b_i^{(0)}\right)$ in terms of a classical general RKN method. For example, from the explicit 2 stage second-order general RKN method with the Buthcher tableau

$$
\begin{array}{c|cc|cc}
0 & & & & \\
\frac{2}{3} & 0 & & \frac{2}{3} & \\
\hline
 & \frac{1}{4} & \frac{3}{4} & \frac{1}{4} & \frac{1}{4}
\end{array}
\quad , \tag{8.24}
$$

we can obtain 2 stage second-order explicit general ERKN methods. Two examples are given below.

Example 1 The first 2 stage second-order explicit general ERKN method (8.4) has Butcher tableau

$$
\begin{array}{c|cc|cc}
0 & & & & \\
\frac{2}{3} & 0 & & \frac{2}{3}I & \\
\hline
 & \frac{1}{4}I & \frac{3}{4}I & \frac{1}{4}I & \frac{1}{4}I
\end{array}
\quad . \tag{8.25}
$$

Example 2 The Butcher tableau of the second one is

$$
\begin{array}{c|cc|cc}
0 & & & & \\
\frac{2}{3} & 0 & & \frac{2}{3}\phi_0(\frac{4}{9}V) & \\
\hline
 & \frac{1}{4}\phi_1(V) & \frac{3}{4}\phi_1(\frac{1}{9}V) & \frac{1}{4}\phi_0(V) & \frac{1}{4}\phi_0(\frac{1}{9}V)
\end{array}
\quad . \tag{8.26}
$$

8.7.2 Third-Order General ERKN Methods

From Theorem 8.3 and 9 trees in the set of IEN-T$_m$, $(m \leq 3)$ in Table 8.5, for an s-stage general ERKN method (8.4) expressed in the Butcher tableau (8.5), we have the third order conditions as follows:

$$
\begin{array}{lll}
\sum\limits_{i=1}^{s} \bar{b}_i(V) = \phi_2(V) + O(h^2), & \sum\limits_{i=1}^{s} \bar{b}_i(V)c_i = \phi_3(V) + O(h), & \sum\limits_{i=1}^{s}\sum\limits_{j=1}^{s} \bar{b}_i(V)a_{ij}^{(0)} = \phi_3(V) + O(h), \\
\sum\limits_{i=1}^{s} b_i(V) = \phi_1(V) + O(h^3), & \sum\limits_{i=1}^{s} b_i(V)c_i = \phi_2(V) + O(h^2), & \sum\limits_{i=1}^{s}\sum\limits_{j=1}^{s} b_i(V)a_{ij}^{(0)} = \phi_2(V) + O(h^2), \\
\sum\limits_{i=1}^{s} b_i(V)c_i^2 = 2\phi_3(V) + O(h), & \sum\limits_{i=1}^{s}\sum\limits_{j=1}^{s} b_i(V)c_i a_{ij}^{(0)} = 2\phi_3(V) + O(h), & \sum\limits_{i=1}^{s}\sum\limits_{j=1}^{s}\sum\limits_{k=1}^{s} b_i(V)a_{ij}^{(0)}a_{ik}^{(0)} = 2\phi_3(V) + O(h), \\
\sum\limits_{i=1}^{s} b_i(V)\bar{a}_{ij}^{(0)} = \phi_3(V) + O(h), & \sum\limits_{i=1}^{s}\sum\limits_{j=1}^{s} b_i(V)a_{ij}^{(0)}c_j = \phi_3(V) + O(h), & \sum\limits_{i=1}^{s}\sum\limits_{j=1}^{s}\sum\limits_{k=1}^{s} b_i(V)a_{ij}^{(0)}a_{jk}^{(0)} = \phi_3(V) + O(h).
\end{array}
$$

Equating coefficients for each power of h, we obtain 13 equations, where 12 equations are exactly the third order conditions for the classical general RKN method (8.22) with the Butcher tableau (8.23)

$$\sum_{i=1}^{s} \bar{b}_i^{(0)} \gamma(\tau)\Phi_i(\tau) = \frac{1}{\rho(\tau)+1}, \quad \forall \tau \in \text{N-T}_m,\ m \le 2, \tag{8.27}$$

$$\sum_{i=1}^{s} b_i^{(0)} \gamma(\tau)\Phi_i(\tau) = 1, \quad \forall \tau \in \text{N-T}_m,\ m \le 3. \tag{8.28}$$

The extra equation is $\sum_{i=1}^{s} \bar{b}_i^{(2)} = -\frac{1}{3}$. We can solve $\left(c_i, \bar{a}_{ij}^{(0)}, a_{ij}^{(0)}, \bar{b}_i^{(0)}, b_i^{(0)}\right)$ from the Eqs. (8.27) and (8.28) via a classical general RKN method. We can then find $b_i^{(2)}$ from the extra equation. Using this approach, we can complete the construction of the general ERKN methods of order three. For example, from the explicit 3 stage third-order general RKN method with the Buthcher tableau

$$\begin{array}{c|ccc|ccc} 0 & & & & & & \\ \frac{1}{2} & 0 & & & \frac{1}{2} & & \\ 1 & 1 & 0 & & -1 & 2 & \\ \hline & \frac{1}{6} & \frac{2}{6} & 0 & \frac{1}{6} & \frac{4}{6} & \frac{1}{6} \end{array} \tag{8.29}$$

we can construct the 3 stage third-order explicit general ERKN methods straightforwardly. The three examples are listed below.

Example 3 The first 3 stage third-order explicit general ERKN method (8.4) is expressed in the Butcher tableau

$$\begin{array}{c|ccc|ccc} 0 & & & & & & \\ \frac{1}{2} & 0 & & & \frac{1}{2}I & & \\ 1 & I & 0 & & -I & 2I & \\ \hline & \frac{1}{6}I & \frac{2}{6}I & 0 & \frac{1}{6}(I-\frac{9}{20}V) & \frac{4}{6}(I-\frac{3}{20}V) & \frac{1}{6}(I+\frac{1}{20}V) \end{array}. \tag{8.30}$$

Example 4 The Butcher tableau of the second 3 stage third-order explicit general ERKN method (8.4) is given by

$$\begin{array}{c|ccc|ccc} 0 & & & & & & \\ \frac{1}{2} & 0 & & & \frac{1}{2}I & & \\ 1 & I & 0 & & -I & 2I & \\ \hline & \frac{1}{6}(I-\frac{1}{6}V) & \frac{2}{6}(I-\frac{1}{24}V) & 0 & \frac{1}{6}(I-\frac{1}{2}V) & \frac{4}{6}(I-\frac{1}{8}V) & \frac{1}{6}I \end{array}. \tag{8.31}$$

Example 5 The third 3 stage third-order explicit general ERKN method (8.4) is denoted by the Butcher tableau

$$\begin{array}{c|ccc|ccc} 0 & & & & & & \\ \frac{1}{2} & 0 & & & \frac{1}{2}\phi_0(\frac{1}{4}V) & & \\ 1 & \phi_1(V) & 0 & & -\phi_0(V) & 2\phi_0(\frac{1}{4}V) & \\ \hline & \frac{1}{6}\phi_1(V) & \frac{2}{6}\phi_1(\frac{1}{4}V) & 0 & \frac{1}{6}\phi_0(V) & \frac{4}{6}\phi_0(\frac{1}{4}V) & \frac{1}{6}I \end{array} \quad . \tag{8.32}$$

8.7.3 Fourth-Order General ERKN Methods

From Theorem 8.3 and Table 8.5, comparing the coefficients of the power of h of (8.18) and (8.19), for an s-stage general ERKN method (8.4) with the coefficient $\left(\bar{a}_{ij}(V), a_{ij}(V), \bar{b}_i(V), b_i(V)\right)$ displayed in the Butcher tableau (8.5), we can obtain 41 fourth order conditions, in which 36 conditions are as follows:

$$\sum_{i=1}^{s} \bar{b}_i^{(0)} \gamma(\tau)\Phi_i(\tau) = \frac{1}{\rho(\tau)+1}, \quad \forall \tau \in \text{N-T}_m,\ m \leq 3, \tag{8.33}$$

$$\sum_{i=1}^{s} b_i^{(0)} \gamma(\tau)\Phi_i(\tau) = 1, \quad \forall \tau \in \text{N-T}_m,\ m \leq 4. \tag{8.34}$$

The remaining 5 conditions are

$$\begin{aligned} &\sum_{i=1}^{s}\sum_{j=1}^{s} b_i^{(0)} a_{ij}^{(2)} = -\frac{1}{12}, \ \sum_{i=1}^{s} b_i^{(2)} = -\frac{1}{3}, \ \sum_{i=1}^{s} b_i^{(2)} c_i = -\frac{1}{12}, \\ &\sum_{i=1}^{s}\sum_{j=1}^{s} b_i^{(2)} a_{ij}^{(0)} = -\frac{1}{12}, \ \sum_{i=1}^{s} \bar{b}_i^{(2)} = -\frac{1}{12}. \end{aligned} \tag{8.35}$$

For each specific classical general RKN method of order four, we can solve for $\left(c_i, \bar{a}_{ij}^{(0)}, a_{ij}^{(0)}, \bar{b}_i^{(0)}, b_i^{(0)}\right)$ from (8.33) and (8.34), since these 36 conditions are exactly the order conditions for the classical general RKN method (8.22) with the Butcher tableau (8.23). Then we can find $\left(a_{ij}^{(2)}, \bar{b}_i^{(2)}, b_i^{(2)}\right)$ from conditions (8.35). In this way, we construct the general ERKN methods (8.4) of order four.

In what follows, we will construct explicit 4 stage fourth order general ERKN methods from the following explicit 4 stage fourth-order classical general RKN method (8.22) with the Butcher tableau

$$\begin{array}{c|cccc|cccc} 0 & & & & & & & & \\ \frac{1}{2} & \frac{1}{8} & & & & \frac{1}{2} & & & \\ \frac{1}{2} & \frac{1}{8} & 0 & & & 0 & \frac{1}{2} & & \\ 1 & 0 & 0 & \frac{1}{2} & & 0 & 0 & 1 & \\ \hline & \frac{1}{6} & \frac{1}{6} & \frac{1}{6} & 0 & \frac{1}{6} & \frac{2}{6} & \frac{2}{6} & \frac{1}{6} \end{array} \quad . \tag{8.36}$$

Some general ERKN methods of order four constructed in this approach are shown below.

Example 6 The Butcher tableau of the first explicit 4 stage fourth-order general ERKN method (8.4) is given by

$$\begin{array}{c|cccc|cccc}
0 & & & & & & & & \\
\frac{1}{2} & \frac{1}{8}I & & & & \frac{1}{2}I & & & \\
\frac{1}{2} & \frac{1}{8}I & 0 & & & 0 & \frac{1}{2}I & & \\
1 & 0 & 0 & \frac{1}{2}I & & 0 & 0 & I-\frac{1}{4}V & \\
\hline
 & \frac{1}{6}(I-\frac{1}{12}V) & \frac{1}{6}(I-\frac{1}{12}V) & \frac{1}{6}(I-\frac{1}{12}V) & 0 & \frac{1}{6}(I-\frac{1}{2}V) & \frac{2}{6}(I-\frac{1}{8}V) & \frac{2}{6}(I-\frac{1}{8}V) & \frac{1}{6}I
\end{array}. \quad (8.37)$$

Example 7 The second explicit 4 stage fourth-order general ERKN method is expressed in the Butcher tableau

$$\begin{array}{c|cccc|cccc}
0 & & & & & & & & \\
\frac{1}{2} & \frac{1}{8}I & & & & \frac{1}{2}(I-\frac{1}{8}V) & & & \\
\frac{1}{2} & \frac{1}{8}I & 0 & & & 0 & \frac{1}{2}I & & \\
1 & 0 & 0 & \frac{1}{2}I & & 0 & 0 & I-\frac{1}{8}V & \\
\hline
 & \frac{1}{6}(I-\frac{1}{6}V) & \frac{1}{6}(I-\frac{1}{24}V) & \frac{1}{6}(I-\frac{1}{24}V) & 0 & \frac{1}{6}(I-\frac{1}{2}V) & \frac{2}{6}(I-\frac{1}{8}V) & \frac{2}{6}(I-\frac{1}{8}V) & \frac{1}{6}I
\end{array}. \quad (8.38)$$

Example 8 The third explicit 4 stage fourth-order general ERKN method (8.4) has the Butcher tableau as follows:

$$\begin{array}{c|cccc|cccc}
0 & & & & & & & & \\
\frac{1}{2} & \frac{1}{8}\phi_1(\frac{1}{4}V) & & & & \frac{1}{2}\phi_0(\frac{1}{4}V) & & & \\
\frac{1}{2} & \frac{1}{8}\phi_1(\frac{1}{4}V) & 0 & & & 0 & \frac{1}{2}I & & \\
1 & 0 & 0 & \frac{1}{2}\phi_1(\frac{1}{4}V) & & 0 & 0 & \phi_0(\frac{1}{4}V) & \\
\hline
 & \frac{1}{6}\phi_1(V) & \frac{1}{6}\phi_1(\frac{1}{4}V) & \frac{1}{6}\phi_1(\frac{1}{4}V) & 0 & \frac{1}{6}\phi_0(V) & \frac{2}{6}\phi_0(\frac{1}{4}V) & \frac{2}{6}\phi_0(\frac{1}{4}V) & \frac{1}{6}I
\end{array}. \quad (8.39)$$

8.7.4 An Effective Approach to Constructing the General ERKN Methods

In the paper [33], in order to construct 4th order general ERKN methods for the systems (8.1), the authors first considered all 62 graphs of the EN-Ts (see Tables 1 and 2 in [33]), and then selected and deleted 34 redundant trees. Finally, they obtained 28 non-redundant EN-Ts (see Tables 3 and 4 in [33]). With these 28 EN-Ts, the authors in [33] achieved special 4th-order conditions, and then the authors derived a 4th-order ERKN method under two auxiliary simplifying assumptions.

Obviously, as shown in the paper [33] more than half of the construction effort was spent on drawing the redundant trees. In a word, the process described in the paper [33] is difficult to follow since the number of the redundant trees in the EN-T set is large.

However, in this chapter, these 28 trees can be directly obtained since 27 of them are exactly the classical N-Ts as shown in Sect. 8.4.2. In this way, it becomes quite easy to get the 4th-order conditions for the general ERKN method (8.4). Then using expansions of these order conditions, and equating each power of h, we can see that most are exactly the order conditions for the classical general RKN method (8.22). This approach to constructing the general ERKN integrators is very effective and efficient in practice, as shown in the previous sections where 2nd, 3rd and 4th order general ERKN methods are constructed as examples.

8.8 Numerical Experiments

In this section, some numerical experiments are implemented to illustrate the potential of the general ERKN methods (8.4) in comparison with the others in the literature. The criterion used in the numerical comparisons is the base-10 logarithm of the maximum global error ($\log_{10} \|MGE\|$) versus the base-2 logarithm of the stepsizes ($\log_2(h)$). The following 11 methods are used to solve the general system (8.1) for the comparison:

- RKN2: The 2 stage second-order general RKN method (8.24).
- ERKN2a: The first 2 stage second-order general ERKN method (8.25) given in Sect. 8.7 of this chapter.
- ERKN2b: The second 2 stage second-order general ERKN method (8.26) given in Sect. 8.7 of this chapter.
- RKN3: The 3 stage third-order general RKN method (8.29).
- ERKN3a: The first 3 stage third-order general ERKN method (8.30) given in Sect. 8.7 of this chapter.
- ERKN3b: The second 3 stage third-order general ERKN method (8.31) given in Sect. 8.7 of this chapter.
- ERKN3c: The third 3 stage third-order general ERKN method (8.32) given in Sect. 8.7 of this chapter.
- RKN4: The 4 stage fourth-order general RKN method (8.36).
- ERKN4a: The first 4 stage fourth-order general ERKN method (8.37) given in Sect. 8.7 of this chapter.
- ERKN4b: The second 4 stage fourth-order general ERKN method (8.38) given in Sect. 8.7 of this chapter.
- ERKN4c: The third 4 stage fourth-order general ERKN method (8.39) given in Sect. 8.7 of this chapter.

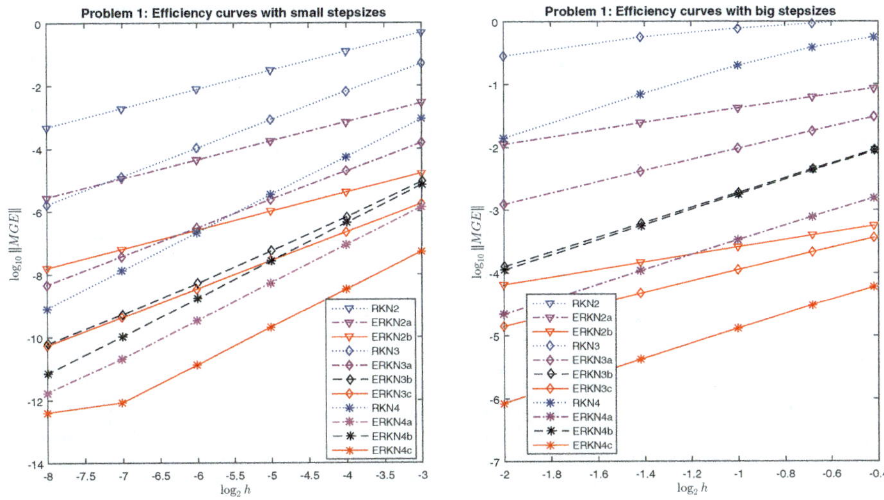

Fig. 8.3 Problem 1 integrated on $[0, 300]$

Problem 1 We consider the damped equation

$$my'' + by' + ky = 0,$$

as one of the test problems. When the damping constant b is small we would expect the system to still oscillate, but with decreasing amplitude as its energy is converted to heat. In this numerical test, the problem is integrated on the interval $[0, 300]$ with $m = 1$, $b = 0.01$, $k = 3$ and the initial conditions $\big(y(0), y'(0)\big) = \big(1, 0\big)$. The analytic solution to the problem is given by

$$y(t) = e^{-\frac{0.01}{2}t}\Big(\cos\Big(\frac{\sqrt{12-0.01^2}}{2}t\Big) + \frac{0.01}{\sqrt{12-0.01^2}}\sin\Big(\frac{\sqrt{12-0.01^2}}{2}t\Big)\Big).$$

The numerical results are displayed in Fig. 8.3, where the small stepsizes for the methods are $h = \frac{1}{2^j}$ for $j = 3, \ldots, 8$ and the big stepsizes are $h = \frac{j}{8}$ for $j = 2, \ldots, 6$.

Problem 2 We consider the initial value problem

$$y''(t) + \begin{pmatrix} 13 & -12 \\ -12 & 13 \end{pmatrix} y(t) = \frac{12\varepsilon}{5}\begin{pmatrix} 3 & 2 \\ -2 & -3 \end{pmatrix} y'(t) + \varepsilon^2 \begin{pmatrix} \frac{36}{5}\sin(t) + 24\sin(5t) \\ -\frac{24}{5}\sin(t) - 36\sin(5t) \end{pmatrix},$$

with the initial values $y(0) = (\varepsilon, \varepsilon)^{\mathsf{T}}$ and $y'(0) = (-4, 6)^{\mathsf{T}}$. The analytic solution is given by

$$y(t) = \begin{pmatrix} \sin(t) - \sin(5t) + \varepsilon\cos(t) \\ \sin(t) + \sin(5t) + \varepsilon\cos(5t) \end{pmatrix}.$$

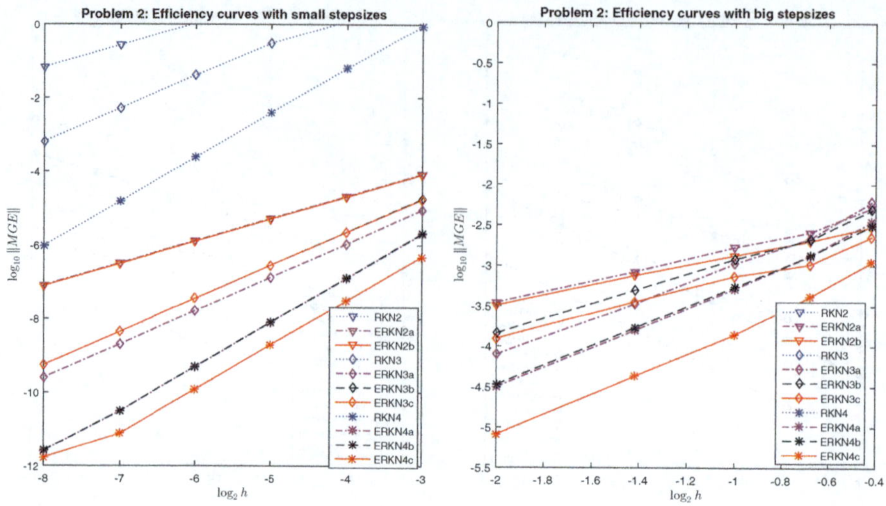

Fig. 8.4 Problem 2 integrated on [0, 300]

In the numerical experiment, we choose the parameter value $\varepsilon = 10^{-3}$ and integrate this problem on the interval [0, 300]. The numerical results are displayed in Fig. 8.4. The small stepsizes are $h = \frac{1}{2^j}$ for $j = 3, \ldots, 8$ and the big stepsizes are $h = \frac{j}{8}$ for $j = 2, \ldots, 6$. In this numerical test with the big stepsizes, the classical general RKN methods (RKN2, RKN3 and RKN4) give disappointing numerical results. Thus we do not depict the corresponding points in Fig. 8.4.

Problem 3 Consider the damped wave equation with periodic conditions (wave propagation in a medium, see e.g. Weinberger [18])

$$\begin{cases} \frac{\partial^2 u}{\partial t^2} + \delta \frac{\partial u}{\partial t} = \frac{\partial^2 u}{\partial x^2} - f(u), \quad -1 < x < 1, \ t > 0, \\ u(-1, t) = u(1, t), \end{cases}$$

where $f(u) = -\sin u$, (i.e., the damped sine Gordon equation) and $\delta = 1$. A semi-discretization in the spatial variable by second-order symmetric differences leads to the following system of second-order ODEs in time

$$\ddot{U} + MU = F(U, \dot{U}), \quad 0 < t \leq t_{end},$$

where $U(t) = \big(u_1(t), \cdots, u_N(t)\big)^{\intercal}$ with $u_i(t) \approx u(x_i, t)$ for $i = 1, \ldots, N$,

$$M = \frac{1}{\Delta x^2} \begin{pmatrix} 2 & -1 & & & -1 \\ -1 & 2 & -1 & & \\ & \ddots & \ddots & \ddots & \\ & & -1 & 2 & -1 \\ -1 & & & -1 & 2 \end{pmatrix},$$

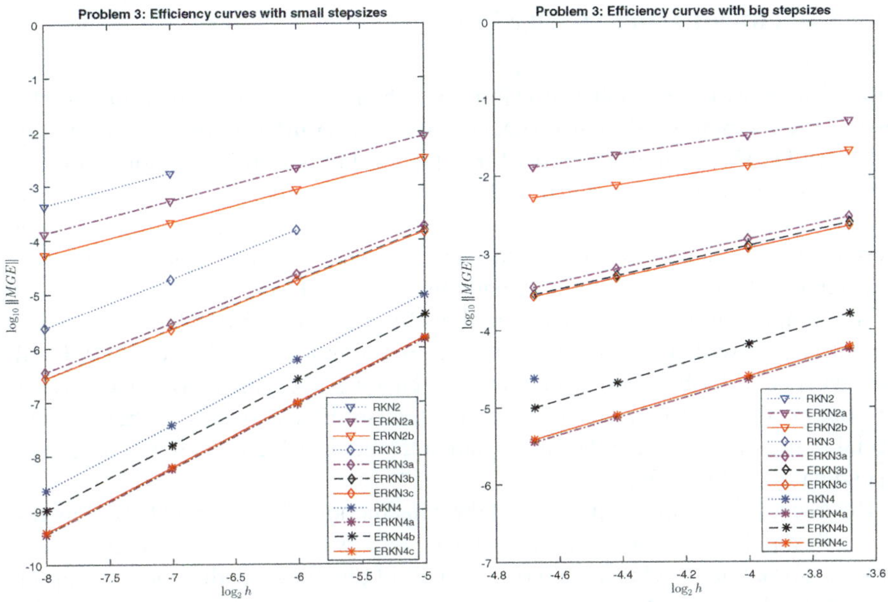

Fig. 8.5 Problem 3 integrated on [0, 300]

$\Delta x = 2/N$, $x_i = -1 + i\Delta x$ and $F(U, \dot{U}) = \big(f(u_1) - \delta\dot{u}_1, \cdots, f(u_N) - \delta\dot{u}_N\big)^{\mathsf{T}}$. Following the paper [3], we take the initial conditions as

$$U(0) = \Big(\pi, \cdots, \pi\Big)^{\mathsf{T}}, \quad U_t(0) = \sqrt{N}\Big(0.01 + \sin\big(\frac{2\pi}{N}\big), \cdots, 0.01 + \sin\big(\frac{2\pi N}{N}\big)\Big)^{\mathsf{T}},$$

with $N = 64$ and integrate the problem on the interval [0, 300] with small stepsizes $h = \frac{1}{2^j}$ for $j = 5, \ldots, 8$ and with big stepsizes $h = \frac{j}{128}$ for $j = 5, 6, 8, 10$. The numerical results are displayed in Fig. 8.5. In this numerical test for the big stepsizes, the classical general RKN methods (RKN2, RKN3 and RKN4) all behave badly, yielding large errors.

It can be observed from Figs. 8.3, 8.4 and 8.5 that

- The general ERKN methods perform more efficiently than the classical general RKN methods.
- The higher order general ERKN methods are more efficient than the lower ones.
- As the stepsize decreases, the difference among the general ERKN methods of the same order becomes negligible.
- The general ERKN methods behave perfectly for the large stepsizes.

8.9 Conclusions and Discussions

In this chapter, we have established an improved theory for the order conditions for the general ERKN methods designed specially for solving multi-frequency oscillatory system (8.1). The original tri-colored tree theory and the order conditions for the general ERKN methods presented in the paper [33] are not satisfied yet due to the existence of large numbers of redundant trees. This chapter has succeeded in making a simplification, by defining the IEN-T set on which some special mappings (especially the extended elementary differential mapping) are introduced.

This simplification of the order conditions for the general ERKN methods when applied to the oscillatory system (8.1) is of great importance. The new tri-colored tree theory and the B-series theory for the general ERKN methods when solving the general system (8.1) reduce to those for standard ERKN methods when solving special system (8.2), where the right-hand side vector-valued function $\boldsymbol{f}$ does not depend on $\boldsymbol{y}'$ (see [23, 30]).

This successful simplification makes the construction of the general ERKN methods much simpler and more efficient for the system (8.1). In light of the reduced tree theory analysed in this chapter, almost one half of algebraic conditions in the paper [33] can be eliminated. Furthermore, in this chapter, from the relation between the theories of order conditions for the general RKN method and for general ERKN method, we propose a simple approach to constructing new integrators. The numerical results show that the general ERKN methods are more suitable for long-term integration with a large stepsize, in comparison with the RKN methods in the literature.

The previous eight chapters concentrated on numerical integrators of oscillatory ordinary differential equations, although their applications to partial differential equations were implemented as well. However, in the next four chapters we will turn to structure-preserving schemes for partial differential equations.

The material of this chapter is based on the work by Zeng et al. [34].

References

1. Boik, R.J.: Lecture Notes: Statistics 550 Spring 2006, pp. 33–35 (2006). http://www.math.montana.edu/~rjboik/classes/550/notes.550.06.pdf
2. Butcher, J.C.: An algebraic theory of integration methods. Math. Comput. **26**, 79–106 (1972)
3. Franco, J.M.: New methods for oscillatory systems based on ARKN methods. Appl. Numer. Math. **56**, 1040C1053 (2006)
4. Hairer, E., Nørsett, S.P., Wanner, G.: Solving Ordinary Differnetial Equations I, Nonstiff Problems. Springer series in computational mathematics. Springer, Berlin (1993)
5. Hairer, E., Wanner, G.: On the Butcher group and general multi-value methods. Computing **13**, 1–15 (1974)
6. Hairer, E., Lubich, C., Wanner, G.: Geometric Numerical Integration, 2nd edn. Springer, Berlin (2006)
7. Li, J., Wu, X.Y.: Adapted Falkner-type methods solving oscillatory second-order differential equations. Numer. Algorithms **62**, 355–381 (2013)

8. Li, J., Wu, X.Y.: Error analysis of explicit TSERKN methods for highly oscillatory systems. Numer. Algorithms **65**, 465–483 (2014)
9. Li, J., Wang, B., You, X., Wu, X.Y.: Two-step extended RKN methods for oscillatory sysems. Comput. Phys. Commun. **182**, 2486–2507 (2011)
10. Liu, K., Wu, X.Y.: Multidimensional ARKN methods for general oscillatory second-order initial value problems. Comput. Phys. Commun. **185**, 1999–2007 (2014)
11. Liu, C., Wu, X.Y.: An energy-preserving and symmetric scheme for nonlinear Hamiltonian wave equations. J. Math. Anal. Appl. **440**, 167–182 (2016)
12. Nyström, E.J.: Numerische Integration von Differentialgleichungen. Acta. Soc. Sci. Fenn. **50**, 1–54 (1925)
13. Shi, W., Wu, X.Y.: A note on symplectic and symmetric ARKN methods. Comput. Phys. Commun. **184**, 2408–2411 (2013)
14. Shi, W., Wu, X.Y., Xia, J.: Explicit multi-symplectic extended leap-frog methods for Hamiltonian wave equations. J. Comput. Phys. **231**, 7671–7694 (2012)
15. Wang, B., Wu, X.Y.: A new high precision energy-preserving integrator for system of oscillatory second-order differential equations. Phys. Lett. A. **376**, 1185–1190 (2012)
16. Wang, B., Wu, X.Y.: A highly accurate explicit symplectic ERKN method for multi-frequency and multidimensional oscillatory Hamiltonian systems. Numer. Algorithms **65**, 705–721 (2014)
17. Wang, B., Wu, X.Y., Zhao, H.: Novel improved multifimensional Strömer-Verlet formulas with applications to four aspects in scientific computation. Math. Comput. Model. **57**, 857–872 (2013)
18. Weinberger, H.F.: A First Course in Partial Differential Equations with Complex Variables and Transform Methods. Dover Publications Inc., New York (1965)
19. Wu, X.Y., Wang, B., Xia, J.: Explicit symplectic multidimensional exponential fitting modified Runge-Kutta-Nyström methods. BIT Numer. Math. **52**, 773–795 (2012)
20. Wu, X.Y., Wang, B., Xia, J.: Extended symplectic Runge-Kutta-Nyström integrators for separable Hamiltonian systems. In: Proceedings of the 2010 International Conference on Computational and Mathematical Methods in Science and Engineering, vol. VIII, pp. 1016–1020. Spain (2010)
21. Wu, X.Y.: A note on stability of multidimensional adapted Runge-Kutta-Nyström methods for oscillatory systems. Appl. Math. Model. **36**, 6331–6337 (2012)
22. Wu, X.Y., You, X., Xia, J.: Order conditions for ARKN methods solving oscillatory system. Comput. Phys. Commun. **180**, 2250–2257 (2009)
23. Wu, X.Y., You, X., Shi, W., Wang, B.: ERKN integrators for systems of oscillatory second-order differential equations. Comput. Phys. Commun. **181**, 1873–1887 (2010)
24. Wu, X.Y., Wang, B., Shi, W.: Efficient energy-perserving integrators for oscillatory Hamiltonian systems. J. Comput. Phys. **235**, 587–605 (2013)
25. Wu, X.Y., You, X., Wang, B.: Structure-Preserving Algorithms for Oscillatory Differential Equations. Springer, Heidelberg (2013)
26. Wu, X.Y., Wang, B., Liu, K., Zhao, H.: ERKN methods for long-term integration of multidimensional orbital problems. Appl. Math. Model. **37**, 2327–2336 (2013)
27. Wu, X.Y., Wang, B., Shi, W.: Effective integrators for nonlinear second-order oscillatory systems with a time-dependent frequency matrix. Appl. Math. Model. **37**, 6505–6518 (2013)
28. Wu, X.Y., Liu, K., Shi, W.: Structure-Preserving Algorithms for Oscillatory Differential Equations II. Springer, Heidelberg (2015)
29. Wu, X.Y., Liu, C., Mei, L.J.: A new framework for solving partial differential equations using semi-analytical explicit RK(N)-type integrators. J. Comput. Appl. Math. **301**, 74–90 (2016)
30. Yang, H., Zeng, X., Wu, X.Y., Ru, Z.: A simplified Nyström-tree theory for extended Runge-Kutta-Nyström integrators solving multi-frequency oscillatory systems. Comput. Phys. Commun. **185**, 2841–2850 (2014)
31. Yang, H., Wu, X.Y.: Trigonometrically-fitted ARKN methods for perturbed oscillators. Appl. Numer. Math. **58**, 1375–1395 (2008)
32. Yang, H., Wu, X.Y., You, X., Fang, Y.: Extended RKN-type methods for numerical integration of perturbed oscillators. Comput. Phys. Commun. **180**, 1777–1794 (2009)

33. You, X., Zhao, J., Yang, H., Fang, Y., Wu, X.Y.: Order conditions for RKN methods solving general second-order oscillatory systems. Numer. Algorithms **66**, 147–176 (2014)
34. Zeng, X., Yang, H., Wu, X.Y.: An improved tri-colored rooted-tree theory and order conditions for ERKN methods for general multi-frequency oscillatory systems. Numer. Algorithms **75**, 909–935 (2017)

Chapter 9
An Integral Formula Adapted to Different Boundary Conditions for Arbitrarily High-Dimensional Nonlinear Klein–Gordon Equations

This chapter is concerned with the initial-boundary value problem for arbitrarily high-dimensional Klein–Gordon equations, posed on a bounded domain $\Omega \subset \mathbb{R}^d$ for $d \geq 1$ and subject to suitable boundary conditions. We derive and analyse an integral formula which proves to be adapted to different boundary conditions for general Klein–Gordon equations in arbitrarily high-dimensional spaces. The formula gives a closed-form solution to arbitrarily high-dimensional homogeneous linear Klein–Gordon equations, which is totally different from the well-known D'Alembert, Poisson and Kirchhoff formulas.

9.1 Introduction

Nonlinear phenomena appear in many areas of scientific and engineering applications such as solid state physics, plasma physics, fluid dynamics, gas dynamics, wave mechanics, mathematical biology and chemical kinetics, which can be modelled by partial differential equations (PDEs). For the past four decades, there has been broad interest in a class of nonlinear evolution equations that admits extremely stable solutions termed solitons (see, e.g. [1, 2, 5, 6, 8, 16, 21, 27]). An important and typical example of such equations is the Klein–Gordon equation which can be expressed in the form:

$$\begin{cases} U_{tt}(X,t) - a^2 \Delta U(X,t) = g\big(U(X,t)\big), \quad X \in \Omega,\ t_0 < t \leq T, \\ U(X,t_0) = U_0(X), \\ U_t(X,t_0) = U_1(X), \end{cases} \tag{9.1}$$

where g is a function of U, $U : \mathbb{R}^d \times \mathbb{R} \to \mathbb{R}$ with $d \geq 1$, representing the wave displacement at position $X \in \mathbb{R}^d$ and time t, and

X. Wu and B. Wang, *Recent Developments in Structure-Preserving Algorithms for Oscillatory Differential Equations*, https://doi.org/10.1007/978-981-10-9004-2_9

$$g\big(U(X,t)\big) = -G'(U) = -dG(U),$$

for some smooth function $G(U)$. The general Klein–Gordon equation can be written as

$$\begin{cases} U_{tt}(X,t) - a^2\Delta U(X,t) = -G'\big(U(X,t)\big), & X \in \Omega,\ t_0 < t \le T, \\ U(X,t_0) = U_0(X), \\ U_t(X,t_0) = U_1(X). \end{cases} \tag{9.2}$$

The Klein–Gordon equation was derived in 1928 as a relativistic version of the Schrödinger equation describing free particles. However, the Klein–Gordon equation was named after the physicists Oskar Klein and Walter Gordon, and proposed in 1926. The model describes relativistic electrons and correctly represents the spinless pion, a composite particle [17]. Here, it is assumed that (9.1) is subject to the given boundary conditions, such as Dirichlet boundary conditions, or Neumann boundary conditions, or Robin boundary conditions. Equation (9.1) is a natural generalization of the linear wave equation (see, e.g. [16]). A simple model of (9.1) with $d = 1$ and $g = 0$ is the homogeneous one-dimensional undamped wave equation,

$$\begin{cases} U_{tt} - a^2U_{xx} = 0, & x_l < x < x_r,\ t_0 < t \le T, \\ U(x,t_0) = u_0(x), \\ U_t(x,t_0) = u_1(x), \end{cases} \tag{9.3}$$

subject to the Dirichlet boundary conditions

$$U(x_l,t) = \alpha(t),\ U(x_r,t) = \beta(t),\ t_0 \le t \le T,$$

where a means the horizontal propagation speed of the wave motion.

In the numerical simulation it is well known that the method of lines is an effective approach to solving partial differential equations such as nonlinear wave equations. Using the method of lines [20], the semidiscretisation of (9.1) in space in the one-dimensional case suggests semi-discrete differential equations, namely, a system of second-order ordinary differential equations in time. Using this approach, each one-dimensional nonlinear wave equation can be converted into a system of second-order ordinary differential equations in time:

$$\begin{cases} q''(t) + Mq(t) = \tilde{g}\big(q(t), q'(t)\big), & t \in [t_0, t_{\text{end}}], \\ q(t_0) = q_0, \quad q'(t_0) = q_0', \end{cases} \tag{9.4}$$

where $\tilde{g} : \mathbb{R}^m \times \mathbb{R}^m \to \mathbb{R}^m$ is assumed to be continuous and M is a $m \times m$ positive semi-definite constant matrix. The solution of system (9.4) is a nonlinear multi-frequency oscillator. Such an oscillatory system has received a great deal of attention in the last few years (see, e.g. [3, 10, 12, 14, 24, 31]).

With regard to the exact solution of the system (9.4) and its derivative, the authors in [29, 32] established the following matrix-variation-of-constants formula which in fact is a semi-analytical expression of the solution of (9.4), or an integral formula for the oscillatory system (9.4).

Theorem 9.1 *If* $\tilde{g}: \mathbb{R}^m \times \mathbb{R}^m \to \mathbb{R}^m$ *is continuous in* (9.4), *then the solution of* (9.4) *and its derivative satisfy*

$$\begin{cases} q(t) = \phi_0\big((t-t_0)^2 M\big)q_0 + (t-t_0)\phi_1\big((t-t_0)^2 M\big)q_0' \\ \qquad + \displaystyle\int_{t_0}^{t} (t-\zeta)\phi_1\big((t-\zeta)^2 M\big)\tilde{g}\big(q(\zeta), q'(\zeta)\big)\mathrm{d}\zeta, \\ q'(t) = -(t-t_0)M\phi_1\big((t-t_0)^2 M\big)q_0 + \phi_0\big((t-t_0)^2 M\big)q_0' \\ \qquad + \displaystyle\int_{t_0}^{t} \phi_0\big((t-\zeta)^2 M\big)\tilde{g}\big(q(\zeta), q'(\zeta)\big)\mathrm{d}\zeta, \end{cases} \tag{9.5}$$

for $t_0, t \in (-\infty, +\infty)$, *where the unconditionally convergent matrix-valued functions are defined by*

$$\phi_j(M) := \sum_{k=0}^{\infty} \frac{(-1)^k M^k}{(2k+j)!}, \qquad j = 0, 1, \ldots. \tag{9.6}$$

Much attention has been paid to the matrix-variation-of-constants formula to develop new integrators such as ARKN (Adapted Runge–Kutta Nyström) methods, ERKN (Extended Runge–Kutta Nyström) methods, Gautschi-type methods, and trigonometric Fourier collocation methods for solving (9.4) (see, e.g. [9–13, 15, 22, 23, 25, 26, 29–32, 34]).

In practice, there exists a very small class of nonlinear PDEs that can be solved exactly by analytical methods. One such method is the well-known inverse scattering method (see, e.g. [4]), also called the inverse spectral transform, which is, for nonlinear PDEs, a direct generalization of the Fourier transform for linear PDEs. Regrettably, the inverse scattering method can solve the initial value problems for a very small class of nonlinear PDEs (see, e.g. [8]) with the requirement that U and various of its derivatives tend to zero as $\|X\| \to \infty$. For this reason, one therefore might think that the set of solvable nonlinear PDEs has "measure zero", and that linear PDEs and solvable nonlinear PDEs could be considered as belonging to a class in which solutions can be added in some function spaces (see, e.g. [16]). On the other hand, it is known that a formal solution to arbitrarily high-dimensional Klein–Gordon equations may be valuable in understanding new nonlinear physical phenomena and investigating novel numerical integrators for the simulation of nonlinear phenomena.

As stated above, nonlinear PDEs in general cannot be solved explicitly. Fortunately, however, we note that the mathematical structure of (9.1) is similar to (9.4), observing the fact that $-M$ in (9.4) can be regarded as a discrete operator of the Laplacian Δ in the one-dimensional case of the nonlinear wave equation based on the method of lines. This observation motivates us to derive and analyse an integral formula for the general Klein–Gordon equation (9.1) posed on a bounded domain $\Omega \subset \mathbb{R}^d$ for $d \geq 1$ equipped with the requirement of suitable boundary conditions.

The outline of this chapter is as follows. In Sect. 9.2, we analyse and derive an integral formula for (9.1). In Sect. 9.3, for the one-dimensional Klein–Gordon equations, we show in detail the consistency of the integral formula with the corresponding Dirichlet boundary conditions and Neumann boundary conditions, respectively. In Sect. 9.4, for arbitrarily high-dimensional Klein–Gordon equations, we prove the consistency of the formula with the underlying Dirichlet boundary conditions, and Neumann boundary conditions, respectively. To show the applications of the formula, illustrative examples are presented in Sect. 9.5. The last section is devoted to conclusions.

9.2 An Integral Formula for Arbitrarily High-Dimensional Klein–Gordon Equations

9.2.1 *General Case*

It is known that Δ is an unbounded operator which is not defined for all $v \in L^2(\Omega)$. In order to model boundary conditions, we restrict ourselves to the case where Δ is defined on a domain $D(\Delta) \subset L^2(\Omega)$, such that the underlying boundary condition is satisfied. For example, we will consider the one-dimensional Klein–Gordon equation of the form

$$\begin{cases} u_{tt} - a^2 \Delta u = f(u), \quad x \in [0, \Gamma], t \geq t_0, \\ u(x, t_0) = \varphi_1(x), \ u_t(x, t_0) = \varphi_2(x), x \in [0, \Gamma], \end{cases} \tag{9.7}$$

subject to the periodic boundary condition

$$u(0, t) = u(\Gamma, t),$$

where Γ is a fundamental period with respect to x, where $\Delta = \partial_x^2$ and $f(u) = -V'(u)$ is the negative derivative of a potential function $V(u)$. We then have

$$D(\Delta) = \{v(x) : \forall v \in L^2([0, \Gamma]) \text{ and } v(0) = v(\Gamma)\}.$$

The functions in $D(\Delta)$ are continuously differentiable and satisfy the underlying boundary condition.

In what follows, we will present an integral formula for the arbitrarily high-dimensional Klein–Gordon equation (9.1). To this end, we first define the formal operator-argument functions as follows:

$$\phi_j(\Delta) := \sum_{k=0}^{\infty} \frac{\Delta^k}{(2k+j)!}, \qquad j = 0, 1, \ldots, \tag{9.8}$$

where Δ is an operator defined on a normed space, such as the Laplacian defined on a subspace $D(\Delta)$ of $L^2(\Omega)$, and in this case, the operator-argument functions $\phi_j(\Delta)$ for $j = 0, 1, \ldots$ defined by (9.8) are bounded. Accordingly, $\phi_j(\Delta)$ in (9.8) can be called Laplacian-argument functions defined on $D(\Delta)$. Besides, Δ can also be a linear transformation such as a matrix and in the particular case of $\Delta = -M$, where M is a positive semi-definite constant matrix, (9.8) reduces to the matrix-valued functions (9.6) which have been widely used in the study of ARKN methods and ERKN methods for solving oscillatory or highly oscillatory differential equations (see, e.g. [32]).

It can be observed that (9.8) is obtained from replacing $-x$ by Δ in

$$\phi_j(x) = \sum_{k=0}^{\infty} \frac{(-1)^k x^k}{(2k+j)!}, \quad j = 0, 1, 2, \ldots,$$

and all $\phi_j(x)$ are bounded for any $x \geq 0$. Each of these operators has a complete system of orthogonal eigenfunctions in the complex Hilbert space $L^2(\Omega)$. Because of the isomorphism between L^2 and ℓ^2, the operator Δ on $L^2(\Omega)$ induces a corresponding operator on ℓ^2. *An elementary analysis which is similar to that for the exponential differential operator presented by Hochbruck and Ostermann in* [15] *can make sure that the Laplacian-argument functions defined on $D(\Delta)$ depending on different boundary conditions are bounded operators with respect to the norm* $\|\cdot\|_{L^2(\Omega)\leftarrow L^2(\Omega)}$, where Ω is the space region under consideration. The details can be found in [18]. It is noted that the exponential differential operator has the properties of a semigroup which are required for analysis. However, the operators defined by (9.8) do not have the semigroup property, but this is not needed in our analysis here.

Some useful properties of Laplacian-argument functions (9.8) are established in the next two theorems.

Theorem 9.2 *Suppose that Δ is the Laplacian defined on a subspace $D(\Delta)$ of $L^2(\Omega)$. The Laplacian-argument functions ϕ_0 and ϕ_1 defined by* (9.8) *satisfy:*

$$\begin{cases} \dfrac{d}{d\zeta}\big[\phi_0\big((t-\zeta)^2 a^2 \Delta\big)\big] = -(t-\zeta)a^2\Delta\phi_1\big((t-\zeta)^2 a^2\Delta\big), \\ \dfrac{d}{d\zeta}\big[(t-\zeta)\phi_1\big((t-\zeta)^2 a^2\Delta\big)\big] = -\phi_0\big((t-\zeta)^2 a^2\Delta\big), \qquad t,\ \zeta \in \mathbb{R}. \end{cases} \tag{9.9}$$

Proof

$$
\begin{aligned}
\frac{d}{d\zeta}\big[\phi_0\big((t-\zeta)^2a^2\Delta\big)\big] &= \frac{d}{d\zeta}\sum_{k=0}^{\infty}\frac{(t-\zeta)^{2k}a^{2k}\Delta^k}{(2k)!}\\
&= -\sum_{k=1}^{\infty}\frac{(t-\zeta)^{2k-1}a^{2k}\Delta^k}{(2k-1)!}\\
&= -\sum_{k=0}^{\infty}\frac{(t-\zeta)^{2k+1}a^{2k+2}\Delta^{k+1}}{(2k+1)!}\\
&= -(t-\zeta)a^2\Delta\phi_1\big((t-\zeta)^2a^2\Delta\big).
\end{aligned}
$$

The second formula of (9.9) can be proved in a similar way. □

Theorem 9.3 *For a symmetric negative (semi-) definite operator Δ, the ϕ-functions defined by* (9.8) *satisfy:*

(i)

$$
\begin{cases}
\phi_0(a^2\Delta) = \displaystyle\sum_{k=0}^{\infty}\frac{a^{2k}\Delta^k}{(2k)!} = \sum_{k=0}^{\infty}\frac{(a\sqrt{-\Delta})^{2k}(-1)^k}{(2k)!} = \cos(a\sqrt{-\Delta}),\\
\phi_1(a^2\Delta) = \displaystyle\sum_{k=0}^{\infty}\frac{a^{2k}\Delta^k}{(2k+1)!} = \sum_{k=0}^{\infty}\frac{(a\sqrt{-\Delta})^{2k}(-1)^k}{(2k+1)!} = \frac{1}{a\sqrt{-\Delta}}\sin(a\sqrt{-\Delta}),\ a\neq 0.
\end{cases}
\tag{9.10}
$$

(ii)

$$
\phi_0^2(a^2\Delta) - a^2\Delta\phi_1^2(a^2\Delta) = I, \tag{9.11}
$$

$$
\phi_0(a^2\Delta) - I = a^2\Delta\phi_2(a^2\Delta). \tag{9.12}
$$

(iii)

$$
\begin{cases}
\phi_1^2(a^2\Delta) - \phi_0(a^2\Delta)\phi_2(a^2\Delta) = \phi_2(a^2\Delta),\\
\phi_0(a^2\Delta)\phi_1(a^2\Delta) - a^2\Delta\phi_1(a^2\Delta)\phi_2(a^2\Delta) = \phi_1(a^2\Delta),\\
\dfrac{1}{2}\big(\phi_1^2(a^2\Delta) - a^2\Delta\phi_2^2(a^2\Delta)\big) = \phi_2(a^2\Delta).
\end{cases}
\tag{9.13}
$$

(iv)

$$
\begin{aligned}
&\int_0^1 \frac{(1-\xi)\phi_1\big(a^2(1-\xi)^2\Delta\big)\xi^j}{j!}\mathrm{d}\xi = \phi_{j+2}(a^2\Delta),\\
&\int_0^1 \frac{\phi_0\big(a^2(1-\xi)^2\Delta\big)\xi^j}{j!}\mathrm{d}\xi = \phi_{j+1}(a^2\Delta).
\end{aligned}
\tag{9.14}
$$

Proof These results can be derived straightforwardly and we omit the details of the proof for the sake of brevity. □

We are now in a position to present an integral formula for the initial-value problem of the general arbitrarily high-dimensional Klein–Gordon equation (9.1).

Theorem 9.4 *If Δ is a Laplacian defined on a subspace $D(\Delta)$ of $L^2(\Omega)$ and $g(U)$ in* (9.1) *is continuous, then the exact solution of* (9.1) *and its derivative satisfy*

$$\begin{cases} U(X,t) = \phi_0\big((t-t_0)^2a^2\Delta\big)U(X,t_0) + (t-t_0)\phi_1\big((t-t_0)^2a^2\Delta\big)U_t(X,t_0) \\ \qquad + \displaystyle\int_{t_0}^{t}(t-\xi)\phi_1\big((t-\xi)^2a^2\Delta\big)\tilde{f}(\xi)\mathrm{d}\xi, \\ U'(X,t) = (t-t_0)a^2\Delta\phi_1\big((t-t_0)^2a^2\Delta\big)U(X,t_0) + \phi_0\big((t-t_0)^2a^2\Delta\big)U_t(X,t_0) \\ \qquad + \displaystyle\int_{t_0}^{t}\phi_0\big((t-\xi)^2a^2\Delta\big)\tilde{f}(\xi)\mathrm{d}\xi \end{cases} \tag{9.15}$$

for $t_0, t \in (-\infty, +\infty)$, where $\tilde{f}(\xi) = g\big(U(X,\xi)\big)$, and the Laplacian-argument functions ϕ_0 and ϕ_1 are defined by (9.8).

Proof We first let

$$Y(X,t) = \big(U(X,t), U_t(X,t)\big)^{\mathsf{T}},$$
$$Y_0(X) = \big(U_0(X), U_1(X)\big)^{\mathsf{T}},$$
$$F\big(Y(X,t)\big) = \big(0, g(U(X,t))\big)^{\mathsf{T}},$$

and

$$W = \begin{pmatrix} 0 & I \\ a^2\Delta & 0 \end{pmatrix}.$$

Then the initial value problem (9.1) can be rewritten in a more compact form

$$\begin{cases} Y_t(X,t) = WY(X,t) + F\big(Y(X,t)\big), \\ Y(X,t_0) = Y_0(X), \quad t \geq t_0. \end{cases} \tag{9.16}$$

From the well-known result on inhomogeneous linear differential equations, the solution at $t \geq t_0$ of the system (9.16) has the form

$$Y(X,t) = \exp\big((t-t_0)W\big)Y_0(X) + \int_{t_0}^{t}\exp\big((t-\xi)W\big)F\big(Y(X,t-\xi)\big)\mathrm{d}\xi. \tag{9.17}$$

It follows from a careful calculation that

$$W^2 = \begin{pmatrix} a^2\Delta & 0 \\ 0 & a^2\Delta \end{pmatrix}, \quad W^3 = \begin{pmatrix} 0 & a^2\Delta \\ a^4\Delta^2 & 0 \end{pmatrix}, \quad W^4 = \begin{pmatrix} a^4\Delta^2 & 0 \\ 0 & a^4\Delta^2 \end{pmatrix},$$
$$W^5 = \begin{pmatrix} 0 & a^4\Delta^2 \\ a^6\Delta^3 & 0 \end{pmatrix}, \quad W^6 = \begin{pmatrix} a^6\Delta^3 & 0 \\ 0 & a^6\Delta^3 \end{pmatrix}, \quad W^7 = \begin{pmatrix} 0 & a^6\Delta^3 \\ a^8\Delta^4 & 0 \end{pmatrix},$$

$\ldots$.

An argument by induction leads to the result that, for each nonnegative integer k, we have

$$W^k = \begin{pmatrix} \frac{1+(-1)^k}{2}(a^2\Delta)^{\lfloor k/2\rfloor} & \frac{1-(-1)^k}{2}(a^2\Delta)^{\lfloor k/2\rfloor} \\ \frac{1-(-1)^k}{2}(a^2\Delta)^{\lfloor k/2\rfloor+1} & \frac{1+(-1)^k}{2}(a^2\Delta)^{\lfloor k/2\rfloor} \end{pmatrix},$$

where $\lfloor k/2\rfloor$ denotes the integer part of $k/2$, and then we have

$$\begin{aligned} \exp\big((t-t_0)W\big) &= \sum_{k=0}^{\infty}\frac{(t-t_0)^k}{k!}W^k \\ &= \begin{pmatrix} I+\frac{(t-t_0)^2}{2!}a^2\Delta+\frac{(t-t_0)^4}{4!}(a^2\Delta)^2+\ldots & (t-t_0)I+\frac{(t-t_0)^3}{3!}a^2\Delta+\ldots \\ (t-t_0)a^2\Delta+\frac{(t-t_0)^3}{3!}(a^2\Delta)^2+\ldots & I+\frac{(t-t_0)^2}{2!}a^2\Delta+\frac{(t-t_0)^4}{4!}(a^2\Delta)^2+\ldots \end{pmatrix} \\ &= \begin{pmatrix} \phi_0\big((t-t_0)^2a^2\Delta\big) & (t-t_0)\phi_1\big((t-t_0)^2a^2\Delta\big) \\ (t-t_0)a^2\Delta\phi_1\big((t-t_0)^2a^2\Delta\big) & \phi_0\big((t-t_0)^2a^2\Delta\big) \end{pmatrix}. \end{aligned} \tag{9.18}$$

Inserting the result of (9.18) into Eq. (9.17) yields

$$\begin{aligned} \begin{pmatrix} U(X,t) \\ U_t(X,t) \end{pmatrix} &= \begin{pmatrix} \phi_0\big((t-t_0)^2a^2\Delta\big) & (t-t_0)\phi_1\big((t-t_0)^2a^2\Delta\big) \\ (t-t_0)a^2\Delta\phi_1\big((t-t_0)^2a^2\Delta\big) & \phi_0\big((t-t_0)^2a^2\Delta\big) \end{pmatrix}\begin{pmatrix} U(X,t_0) \\ U_t(X,t_0) \end{pmatrix} \\ &+ \int_{t_0}^{t}\begin{pmatrix} \phi_0\big((t-\xi)^2a^2\Delta\big) & (t-\xi)\phi_1\big((t-\xi)^2a^2\Delta\big) \\ (t-\xi)a^2\Delta\phi_1\big((t-\xi)^2a^2\Delta\big) & \phi_0\big((t-\xi)^2a^2\Delta\big) \end{pmatrix}\begin{pmatrix} 0 \\ g(U(X,\xi)) \end{pmatrix}\mathrm{d}\xi \\ &= \begin{pmatrix} \phi_0\big((t-t_0)^2a^2\Delta\big)u(x,t_0)+(t-t_0)\phi_1\big((t-t_0)^2a^2\Delta\big)U_t(X,t_0) \\ (t-t_0)a^2\Delta\phi_1\big((t-t_0)^2a^2\Delta\big)U(X,t_0)+\phi_0\big((t-t_0)^2a^2\Delta\big)U_t(X,t_0) \end{pmatrix} \\ &+ \begin{pmatrix} \int_{t_0}^{t}(t-\xi)\phi_1\big((t-\xi)^2a^2\Delta\big)\tilde{f}(\xi)\mathrm{d}\xi \\ \int_{t_0}^{t}\phi_0\big((t-\xi)^2a^2\Delta\big)\tilde{f}(\xi)\mathrm{d}\xi \end{pmatrix}. \end{aligned}$$

This gives the form of (9.15) exactly and completes the proof. □

Let $U^n(X) = U(X,t_n)$ and $U_t^n(X) = U_t(X,t_n)$ represent the exact solution of (9.7) and its derivative with respect to t at $t = t_n$. It follows immediately from (9.15) with the change of variable $\xi = t_n + hz$ that

$$\begin{cases} U^{n+1}(X) = \phi_0\big(h^2a^2\Delta\big)U^n(X) + h\phi_1\big(h^2a^2\Delta\big)U_t^n(X) \\ \qquad + h^2\displaystyle\int_0^1 (1-z)\phi_1\big((1-z)^2h^2a^2\Delta\big)\tilde{f}(z)\mathrm{d}z, \\ U_t^{n+1}(X) = ha^2\Delta\phi_1\big(h^2a^2\Delta\big)U^n(X) + \phi_0\big(h^2a^2\Delta\big)U_t^n(X) \\ \qquad + h\displaystyle\int_0^1 \phi_0\big((1-z)^2h^2a^2\Delta\big)\tilde{f}(z)\mathrm{d}z, \end{cases} \tag{9.19}$$

where h is the temporal stepsize.

Remark 9.1 In comparison with the matrix-variation-of-constants formula (9.5) for (9.4) based on the method of lines for solving one-dimensional nonlinear wave equations, the formula (9.15) is a formal solution to the Klein–Gordon equation (9.1), whereas the matrix-variation-of-constants formula (9.5) is a formal solution to (9.4) but not a formal solution to (9.1). Thus, significant progress has been made on integral representations of solutions of the arbitrarily high-dimensional Klein–Gordon equation (9.1).

9.2.2 Homogeneous Case

We now turn to the special and important homogeneous case.

If $g(U) = 0$, then (9.1) reduces to the following homogeneous linear Klein–Gordon equation:

$$\begin{cases} U_{tt} - a^2 \Delta U = 0, \\ U(X, t_0) = U_0(X), \\ U_t(X, t_0) = U_1(X), \end{cases} \tag{9.20}$$

and then (9.15) becomes

$$\begin{cases} U(X,t) = \phi_0\big((t-t_0)^2 a^2 \Delta\big) U_0(X) + (t-t_0)\phi_1\big((t-t_0)^2 a^2 \Delta\big) U_1(X), \\ U'(X,t) = (t-t_0) a^2 \Delta \phi_1\big((t-t_0)^2 a^2 \Delta\big) U_0(X) + \phi_0\big((t-t_0)^2 a^2 \Delta\big) U_1(X), \end{cases} \tag{9.21}$$

which integrates (9.20) exactly. This means that (9.21) expresses a closed-form solution to the arbitrarily high-dimensional homogeneous linear Klein–Gordon equation (9.20). This fact shows that (9.21) possesses the additional advantage of energy preservation and quadratic invariant preservation for the homogeneous linear Klein–Gordon equation (9.20). Another key point is that, compared with the seminal D'Alembert, Poisson and Kirchhoff formulas, *the formula* (9.21) *doesn't depend on the evaluation of complicated integrals, whereas the evaluation of integrals is required by the D'Alembert, Poisson and Kirchhoff formulas.*

9.2.3 Towards Numerical Simulations

For the purpose of numerical simulations, we rewrite the Klein–Gordon equation (9.1) as

$$\begin{cases} U_{tt}(X,t) = g\big(U(X,t)\big) + a^2 \Delta U(X,t), & X \in \Omega \subseteq \mathbb{R}^d,\ t > t_0 \\ U(X,t_0) = \varphi_1(X),\ U_t(X,t_0) = \varphi_2(X), & X \in \Omega \cup \partial\Omega, \end{cases} \tag{9.22}$$

where

$$\Delta = \sum_{i=1}^{d} \frac{\partial^2}{\partial x_i^2}.$$

It follows from Theorem 9.4 that the solution to (9.22) is given by

$$\begin{cases} U(X,t) = U(X,t_0) + (t-t_0)U_t(X,t_0) + \int_{t_0}^{t} (t-\zeta)\hat{g}\big(U(X,\zeta)\big)\mathrm{d}\zeta, \\ U_t(X,t) = U_t(X,t_0) + \int_{t_0}^{t} \hat{g}\big(U(X,\zeta)\big)\mathrm{d}\zeta, \\ X \in \Omega \cup \partial\Omega, \end{cases} \tag{9.23}$$

i.e.,

$$\begin{cases} U(X,t) = \varphi_1(X) + (t-t_0)\varphi_2(x) + \int_{t_0}^{t} (t-\zeta)\hat{g}\big(U(X,\zeta)\big)\mathrm{d}\zeta, \\ U_t(X,t) = \varphi_2(X) + \int_{t_0}^{t} \hat{g}\big(U(X,\zeta)\big)\mathrm{d}\zeta, \\ X \in \Omega \cup \partial\Omega, \end{cases} \tag{9.24}$$

or

$$\begin{cases} U^{n+1}(X) = U^n(X) + hU_t^n(x) + h^2 \int_{0}^{1} (1-z)\hat{g}\big(U(X,t_n+zh)\big)\mathrm{d}z, \\ U_t^{n+1}(X) = U_t^n(X) + h \int_{0}^{1} \hat{g}\big(U(X,t_n+zh)\big)\mathrm{d}z, \\ X \in \Omega \cup \partial\Omega, \end{cases} \tag{9.25}$$

where $U^n(X) = U(X,t_n)$ and

$$\hat{g}\big(U(X,\zeta)\big) = g\big(U(X,\zeta)\big) + a^2\Delta U(X,\zeta).$$

Then, for each fixed $X \in \Omega \cup \partial\Omega$, approximating the integrals in (9.25) by using a quadrature formula yields a semi-analytical explicit RKN-type integrator.

Applying the modified midpoint rule (replacing $\hat{g}\big(U^{n+\frac{1}{2}}(X)\big)$ by $\hat{g}(\tilde{U}^n(X) + \frac{h}{2}\tilde{U}_t^n(X)))$ in the integrals in (9.25), we obtain

$$\begin{cases} \tilde{U}^{n+1}(X) = \tilde{U}^n(X) + h\tilde{U}_t^n(X) + \frac{h^2}{2}\hat{g}(\tilde{U}^n(X) + \frac{h}{2}\tilde{U}_t^n(X)), \\ \tilde{U}_t^{n+1}(X) = \tilde{U}_t^n(X) + h\hat{g}(\tilde{U}^n(X) + \frac{h}{2}\tilde{U}_t^n(X)), \end{cases} \tag{9.26}$$

where $\tilde{U}^n(X) \approx U^n(X) = U(X, t_n)$. This is the well-known Störmer–Verlet formula, and we call (9.26) the *SV-scheme* for (9.22). Hence, the SV-scheme is a symplectic integrator of order two.

In applications, (9.1) is defined on bounded domains on the boundary of which some physical conditions must be prescribed. These boundary conditions can be of different sorts. We will consider the most classical ones: Dirichlet boundary conditions, Neumann boundary conditions, and Robin boundary conditions. In what follows, we pay attention to the consistency of the formula (9.15) with the corresponding boundary conditions under suitable assumptions.

9.3 The Consistency of the Boundary Conditions for One-dimensional Klein–Gordon Equations

We now consider the initial problem in the one-dimensional case with $u : \mathbb{R} \times \mathbb{R} \to \mathbb{R}$ given by

$$\begin{cases} u_{tt} - a^2 \Delta u = f(u), \quad x_l < x < x_r, t > t_0, \\ u(x, t_0) = \varphi_1(x), u_t(x, t_0) = \varphi_2(x), \quad x_l \le x \le x_r, \end{cases} \tag{9.27}$$

where $f\big(u(x,t)\big) = -G'(u)$ for some smooth function $G(u)$. From Theorem 9.4, the solution of (9.27) satisfies

$$\begin{cases} u(x,t) = \phi_0\big((t-t_0)^2 a^2 \Delta\big)\varphi_1(x) + (t-t_0)\phi_1\big((t-t_0)^2 a^2 \Delta\big)\varphi_2(x) \\ \qquad + \displaystyle\int_{t_0}^{t} (t-\zeta)\phi_1((t-\zeta)^2 a^2 \Delta)\tilde{f}(\zeta)\mathrm{d}\zeta, \\ u'(x,t) = (t-t_0)a^2 \Delta \phi_1\big((t-t_0)^2 a^2 \Delta\big)\varphi_1(x) + \phi_0\big((t-t_0)^2 a^2 \Delta\big)\varphi_2(x) \\ \qquad + \displaystyle\int_{t_0}^{t} \phi_0\big((t-\zeta)^2 a^2 \Delta\big)\tilde{f}(\zeta)\mathrm{d}\zeta, \end{cases} \tag{9.28}$$

where $\Delta = \dfrac{\partial^2}{\partial x^2}$ and $\tilde{f}(\zeta) = f\big(u(x,\zeta)\big)$.

9.3.1 *Dirichlet Boundary Conditions*

Firstly, we consider the nonlinear wave equation (9.27) with the Dirichlet boundary conditions:

$$u(x_l, t) = \alpha(t), \quad u(x_r, t) = \beta(t), \quad t \ge t_0. \tag{9.29}$$

The next theorem shows the consistency of the formula (9.28) with the Dirichlet boundary conditions (9.29), i.e.,

$$\alpha(t) = \Big[\phi_0\big((t-t_0)^2a^2\Delta\big)\varphi_1(x) + (t-t_0)\phi_1\big((t-t_0)^2a^2\Delta\big)\varphi_2(x) \\ + \int_{t_0}^{t}(t-\zeta)\phi_1\big((t-\zeta)^2a^2\Delta\big)\tilde{f}(\zeta)\mathrm{d}\zeta\Big]\Big|_{x=x_l},$$

$$\beta(t) = \Big[\phi_0\big((t-t_0)^2a^2\Delta\big)\varphi_1(x) + (t-t_0)\phi_1\big((t-t_0)^2a^2\Delta\big)\varphi_2(x) \\ + \int_{t_0}^{t}(t-\zeta)\phi_1\big((t-\zeta)^2a^2\Delta\big)\tilde{f}(\zeta)\mathrm{d}\zeta\Big]\Big|_{x=x_r}.$$

Theorem 9.5 *Assume that $\alpha(t)$, $\beta(t)$, and $f\big(u(x,t)\big)$ are sufficiently differentiable with respect to t. Then the formula (9.28) is consistent with the Dirichlet boundary conditions (9.29).*

Proof Using the initial conditions, we obtain

$$\alpha(t_0) = \varphi_1(x_l),\quad \alpha'(t_0) = \varphi_2(x_l),\quad \beta(t_0) = \varphi_1(x_r),\quad \beta'(t_0) = \varphi_2(x_r).$$

It follows from (9.27) that

$$\begin{aligned}
u_{tt} &= a^2\Delta u + f(u) \\
&\Rightarrow \begin{cases} \alpha''(t_0) = a^2\Delta\varphi_1(x_l) + f\big(u(x_l,t_0)\big), \\ \beta''(t_0) = a^2\Delta\varphi_1(x_r) + f\big(u(x_r,t_0)\big), \end{cases} \\
u_t^{(3)} &= a^2\Delta u_t + f_t'(u) \\
&\Rightarrow \begin{cases} \alpha^{(3)}(t_0) = a^2\Delta\varphi_2(x_l) + f_t'\big(u(x_l,t_0)\big), \\ \beta^{(3)}(t_0) = a^2\Delta\varphi_2(x_r) + f_t'\big(u(x_r,t_0)\big), \end{cases} \\
u_t^{(4)} &= a^4\Delta^2 u + a^2\Delta f(u) + f_t^{(2)}(u) \\
&\Rightarrow \begin{cases} \alpha^{(4)}(t_0) = a^4\Delta^2\varphi_1(x_l) + a^2\Delta f\big(u(x_l,t_0)\big) + f_t^{(2)}\big(u(x_l,t_0)\big), \\ \beta^{(4)}(t_0) = a^4\Delta^2\varphi_1(x_r) + a^2\Delta f\big(u(x_r,t_0)\big) + f_t^{(2)}\big(u(x_r,t_0)\big), \end{cases} \\
u_t^{(5)} &= a^4\Delta^2 u_t + a^2\Delta f_t'(u) + f_t^{(3)}(u) \\
&\Rightarrow \begin{cases} \alpha^{(5)}(t_0) = a^4\Delta^2\varphi_2(x_l) + a^2\Delta f_t'\big(u(x_l,t_0)\big) + f_t^{(3)}\big(u(x_l,t_0)\big), \\ \beta^{(5)}(t_0) = a^4\Delta^2\varphi_2(x_r) + a^2\Delta f_t'\big(u(x_r,t_0)\big) + f_t^{(3)}\big(u(x_r,t_0)\big), \end{cases} \\
u_t^{(6)} &= a^6\Delta^3 u + a^4\Delta^2 f(u) + a^2\Delta f_t^{(2)}(u) + f_t^{(4)}(u) \\
&\Rightarrow \begin{cases} \alpha^{(6)}(t_0) = a^6\Delta^3\varphi_1(x_l) + a^4\Delta^2 f\big(u(x_l,t_0)\big) \\ \qquad + a^2\Delta f_t^{(2)}\big(u(x_l,t_0)\big) + f_t^{(4)}\big(u(x_l,t_0)\big), \\ \beta^{(6)}(t_0) = a^6\Delta^3\varphi_1(x_r) + a^4\Delta^2 f\big(u(x_r,t_0)\big) \\ \qquad + a^2\Delta f_t^{(2)}\big(u(x_r,t_0)\big) + f_t^{(4)}\big(u(x_r,t_0)\big), \end{cases}
\end{aligned}$$

$$u_t^{(7)} = a^6\Delta^3 u_t + a^4\Delta^2 f_t'(u) + a^2\Delta f_t^{(3)}(u) + f_t^{(5)}(u)$$
$$\Rightarrow \begin{cases} \alpha^{(7)}(t_0) = a^6\Delta^3\varphi_2(x_l) + a^4\Delta^2 f_t'\big(u(x_l,t_0)\big) \\ \qquad +a^2\Delta f_t^{(3)}\big(u(x_l,t_0)\big) + f_t^{(5)}\big(u(x_l,t_0)\big), \\ \beta^{(7)}(t_0) = a^6\Delta^3\varphi_2(x_r) + a^4\Delta^2 f_t'\big(u(x_r,t_0)\big) \\ \qquad +a^2\Delta f_t^{(3)}\big(u(x_r,t_0)\big) + f_t^{(5)}\big(u(x_r,t_0)\big). \end{cases}$$

$$\cdots.$$

An argument by induction leads to the results

$$\begin{aligned} \alpha^{(2k)}(t_0) &= a^{2k}\Delta^k\varphi_1(x_l) + \sum_{j=1}^{k} a^{2(k-j)}\Delta^{k-j} f_t^{(2j-2)}\big(u(x_l,t_0)\big), \\ \alpha^{(2k+1)}(t_0) &= a^{2k}\Delta^k\varphi_2(x_l) + \sum_{j=1}^{k} a^{2(k-j)}\Delta^{k-j} f_t^{(2j-1)}\big(u(x_l,t_0)\big), \end{aligned} \tag{9.30}$$

and

$$\begin{aligned} \beta^{(2k)}(t_0) &= a^{2k}\Delta^k\varphi_1(x_r) + \sum_{j=1}^{k} a^{2(k-j)}\Delta^{k-j} f_t^{(2j-2)}\big(u(x_r,t_0)\big) \\ \beta^{(2k+1)}(t_0) &= a^{2k}\Delta^k\varphi_2(x_r) + \sum_{j=1}^{k} a^{2(k-j)}\Delta^{k-j} f_t^{(2j-1)}\big(u(x_r,t_0)\big), \end{aligned} \tag{9.31}$$

for $k = 1, 2, \ldots$.

The Taylor expansion of $\alpha(t)$ and $\beta(t)$ at the point t_0 gives

$$\begin{aligned} \alpha(t) &= \sum_{k=0}^{\infty}\frac{(t-t_0)^k}{k!}\alpha^{(k)}(t_0) = \sum_{k=0}^{\infty}\frac{(t-t_0)^{2k}}{(2k)!}\alpha^{(2k)}(t_0) + \sum_{k=0}^{\infty}\frac{(t-t_0)^{2k+1}}{(2k+1)!}\alpha^{(2k+1)}(t_0), \\ \beta(t) &= \sum_{k=0}^{\infty}\frac{(t-t_0)^k}{k!}\beta^{(k)}(t_0) = \sum_{k=0}^{\infty}\frac{(t-t_0)^{2k}}{(2k)!}\beta^{(2k)}(t_0) + \sum_{k=0}^{\infty}\frac{(t-t_0)^{2k+1}}{(2k+1)!}\beta^{(2k+1)}(t_0). \end{aligned} \tag{9.32}$$

Inserting the results of (9.30) and (9.31) into (9.32) yields

$$\begin{aligned} \alpha(t) = \Big\{&\phi_0\big((t-t_0)^2a^2\Delta\big)\varphi_1(x) + (t-t_0)\phi_1\big((t-t_0)^2a^2\Delta\big)\varphi_2(x) \\ &+ \sum_{k=1}^{\infty}\Big[\frac{(t-t_0)^{2k}}{(2k)!}\sum_{j=1}^{k} a^{2(k-j)}\Delta^{k-j} f_t^{(2j-2)}\big(u(x,t_0)\big) \\ &+ \frac{(t-t_0)^{2k+1}}{(2k+1)!}\sum_{j=1}^{k} a^{2(k-j)}\Delta^{k-j} f_t^{(2j-1)}\big(u(x,t_0)\big)\Big]\Big\}\Big|_{x=x_l}, \end{aligned} \tag{9.33}$$

and

$$\begin{aligned}\beta(t) = \Big\{&\phi_0\big((t-t_0)^2a^2\Delta\big)\varphi_1(x) + (t-t_0)\phi_1\big((t-t_0)^2a^2\Delta\big)\varphi_2(x) \\ &+ \sum_{k=1}^{\infty}\Big[\frac{(t-t_0)^{2k}}{(2k)!}\sum_{j=1}^{k} a^{2(k-j)}\Delta^{k-j} f_t^{(2j-2)}\big(u(x,t_0)\big) \\ &+ \frac{(t-t_0)^{2k+1}}{(2k+1)!}\sum_{j=1}^{k} a^{2(k-j)}\Delta^{k-j} f_t^{(2j-1)}\big(u(x,t_0)\big)\Big]\Big\}\Big|_{x=x_r}.\end{aligned} \tag{9.34}$$

Let

$$F(x,t) \triangleq \int_{t_0}^{t}(t-\zeta)\phi_1\big((t-\zeta)^2a^2\Delta\big)\tilde{f}(\zeta)\mathrm{d}\zeta.$$

It is easy to see $F(x,t_0) = 0$, and a careful calculation gives

$$\begin{aligned}
F_t'(x,t) &= \int_{t_0}^{t}\phi_0\big((t-\zeta)^2a^2\Delta\big)\tilde{f}(\zeta)\mathrm{d}\zeta \\
&\Rightarrow F_t'(x,t_0) = 0, \\
F_t^{(2)}(x,t) &= f\big(u(x,t)\big) + \int_{t_0}^{t}(t-\zeta)a^2\Delta\phi_1\big((t-\zeta)^2a^2\Delta\big)\tilde{f}(\zeta)\mathrm{d}\zeta \\
&\Rightarrow F_t^{(2)}(x,t_0) = f\big(u(x,t_0)\big), \\
F_t^{(3)}(x,t) &= f_t'\big(u(x,t)\big) + \int_{t_0}^{t}a^2\Delta\phi_0\big((t-\zeta)^2a^2\Delta\big)\tilde{f}(\zeta)\mathrm{d}\zeta \\
&\Rightarrow F_t^{(3)}(x,t_0) = f_t'\big(u(x,t_0)\big), \\
F_t^{(4)}(x,t) &= f_t^{(2)}\big(u(x,t)\big) + a^2\Delta f\big(u(x,t)\big) + \int_{t_0}^{t}(t-\zeta)a^4\Delta^2\phi_1\big((t-\zeta)^2a^2\Delta\big)\tilde{f}(\zeta))\mathrm{d}\zeta \\
&\Rightarrow F_t^{(4)}(x,t_0) = f_t^{(2)}\big(u(x,t_0)\big) + a^2\Delta f\big(u(x,t_0)\big), \\
F_t^{(5)}(x,t) &= f_t^{(3)}\big(u(x,t)\big) + a^2\Delta f_t'\big(u(x,t)\big) + \int_{t_0}^{t}a^4\Delta^2\phi_0\big((t-\zeta)^2a^2\Delta\big)\tilde{f}(\zeta)\mathrm{d}\zeta \\
&\Rightarrow F_t^{(5)}(x,t_0) = f_t^{(3)}\big(u(x,t_0)\big) + a^2\Delta f_t'\big(u(x,t_0)\big), \\
F_t^{(6)}(x,t) &= f_t^{(4)}\big(u(x,t)\big) + a^2\Delta f_t^{(2)}\big(u(x,t)\big) + a^4\Delta^2 f\big(u(x,t)\big) \\
&\quad + \int_{t_0}^{t}(t-\zeta)a^6\Delta^3\phi_1\big((t-\zeta)^2a^2\Delta\big)\tilde{f}(\zeta)\mathrm{d}\zeta \\
&\Rightarrow F_t^{(6)}(x,t_0) = f_t^{(4)}\big(u(x,t_0)\big) + a^2\Delta f_t^{(2)}\big(u(x,t_0)\big) + a^4\Delta^2 f\big(u(x,t_0)\big)), \\
F_t^{(7)}(x,t) &= f_t^{(5)}\big(u(x,t)\big) + a^2\Delta f_t^{(3)}\big(u(x,t)\big) + a^4\Delta^2 f_t'\big(u(x,t)\big) \\
&\quad + \int_{t_0}^{t}a^6\Delta^3\phi_0\big((t-\zeta)^2a^2\Delta\big)\tilde{f}(\zeta)\mathrm{d}\zeta \\
&\Rightarrow F_t^{(7)}(x,t_0) = f_t^{(5)}\big(u(x,t_0)\big) + a^2\Delta f_t^{(3)}\big(u(x,t_0)\big) + a^4\Delta^2 f_t'\big(u(x,t_0)\big), \\
&\cdots.
\end{aligned}$$

An argument by induction then gives

$$F_t^{(2k)}(x,t_0)=\sum_{j=1}^{k}a^{2(k-j)}\Delta^{k-j}f_t^{(2j-2)}\big(u(x,t_0)\big)$$

$$F_t^{(2k+1)}(x,t_0)=\sum_{j=1}^{k}a^{2(k-j)}\Delta^{k-j}f_t^{(2j-1)}\big(u(x,t_0)\big),\quad k=1,2,\ldots.$$

The Taylor expansion of $F(x,t)$ at $t=t_0$ is

$$\begin{aligned}F(x,t)&=\sum_{k=0}^{\infty}\frac{(t-t_0)^k}{k!}F_t^{(k)}(x,t_0)=\sum_{k=2}^{\infty}\frac{(t-t_0)^k}{k!}F_t^{(k)}(x,t_0)\\&=\sum_{k=1}^{\infty}\frac{(t-t_0)^{2k}}{(2k)!}F_t^{(2k)}(x,t_0)+\sum_{k=1}^{\infty}\frac{(t-t_0)^{2k+1}}{(2k+1)!}F_t^{(2k+1)}(x,t_0)\\&=\sum_{k=1}^{\infty}\Big[\frac{(t-t_0)^{2k}}{(2k)!}\sum_{j=1}^{k}a^{2(k-j)}\Delta^{k-j}f_t^{(2j-2)}\big(u(x,t_0)\big)\\&\quad+\frac{(t-t_0)^{2k+1}}{(2k+1)!}\sum_{j=1}^{k}a^{2(k-j)}\Delta^{k-j}f_t^{(2j-1)}\big(u(x,t_0)\big)\Big].\end{aligned}\tag{9.35}$$

Inserting the result of (9.35) into (9.33) and (9.34) yields

$$\begin{aligned}\alpha(t)&=\Big[\phi_0\big((t-t_0)^2a^2\Delta\big)\varphi_1(x)+(t-t_0)\phi_1\big((t-t_0)^2a^2\Delta\big)\varphi_2(x)\\&+\int_{t_0}^{t}(t-\zeta)\phi_1\big((t-\zeta)^2a^2\Delta\big)\tilde f(\zeta)\mathrm{d}\zeta\Big]\Big|_{x=x_l}\\\beta(t)&=\Big[\phi_0\big((t-t_0)^2a^2\Delta\big)\varphi_1(x)+(t-t_0)\phi_1\big((t-t_0)^2a^2\Delta\big)\varphi_2(x)\\&+\int_{t_0}^{t}(t-\zeta)\phi_1\big((t-\zeta)^2a^2\Delta\big)\tilde f(\zeta)\mathrm{d}\zeta\Big]\Big|_{x=x_r}.\end{aligned}$$

The proof is complete. □

9.3.2 Neumann Boundary Conditions

We next consider the nonlinear wave equation (9.27) with the Neumann boundary conditions

$$\frac{\partial u}{\partial x}\Big|_{x_l}=\gamma(t),\quad \frac{\partial u}{\partial x}\Big|_{x_r}=\delta(t).\tag{9.36}$$

Theorem 9.6 *Assume that $\gamma(t)$, $\delta(t)$, and $f\big(u(x,t)\big)$ are sufficiently differentiable with respect to t. Then the formula* (9.28) *is consistent with the Neumann boundary conditions* (9.36).

Proof From the initial conditions,we have

$$\gamma(t_0)=\varphi_1'(x_l),\quad \gamma'(t_0)=\varphi_2'(x_l),\quad \delta(t_0)=\varphi_1'(x_r),\quad \delta'(t_0)=\varphi_2'(x_r).$$

Calculating the derivative of u with respect to x in (9.27) gives

$$\begin{cases} \left(\dfrac{\partial u}{\partial x}\right)_{tt}=a^2\Delta\left(\dfrac{\partial u}{\partial x}\right)+\dfrac{\partial}{\partial x}(f(u)), & x_l<x<x_r, t>t_0,\\ \dfrac{\partial u}{\partial x}(x,t_0)=\varphi_1'(x), \dfrac{\partial u_t}{\partial x}(x,t_0)=\varphi_2'(x), & x_l\le x\le x_r. \end{cases} \tag{9.37}$$

Let $v=\dfrac{\partial u}{\partial x}$. We then have the following initial-boundary problem

$$\begin{cases} v_{tt}=a^2\Delta v+\tilde{f}(u), & x_l<x<x_r,\ t\ge t_0,\\ v(x,t_0)=\varphi_1'(x),\ v_t(x,t_0)=\varphi_2'(x), & x_l\le x\le x_r,\\ v(x_l,t)=\gamma(t),\ v(x_r,t)=\delta(t), & t\ge t_0, \end{cases} \tag{9.38}$$

where

$$\tilde{f}(u)=f_x'\big(u(x,t)\big)=\frac{\partial}{\partial x}f\big(u(x,t)\big).$$

For the transformed initial-boundary value problem (9.38), after an analysis similarly to that in Sect. 9.3.1, we conclude that

$$\begin{aligned} \gamma(t)=&\Big[\phi_0\big((t-t_0)^2a^2\Delta\big)\varphi_1'(x)+(t-t_0)\phi_1\big((t-t_0)^2a^2\Delta\big)\varphi_2'(x)\\ &+\int_{t_0}^{t}(t-\zeta)\phi_1\big((t-\zeta)^2a^2\Delta\big)\hat{f}(\zeta)\mathrm{d}\zeta\Big]\Big|_{x=x_l},\\ \delta(t)=&\Big[\phi_0\big((t-t_0)^2a^2\Delta\big)\varphi_1'(x)+(t-t_0)\phi_1\big((t-t_0)^2a^2\Delta\big)\varphi_2'(x)\\ &+\int_{t_0}^{t}(t-\zeta)\phi_1\big((t-\zeta)^2a^2\Delta\big)\hat{f}(\zeta)\mathrm{d}\zeta\Big]\Big|_{x=x_r}, \end{aligned}$$

where $\hat{f}(\zeta)=f_x'\big(u(x,\zeta)\big)$.

The proof is complete. □

Another direct proof can be found in Appendix 1 of this chapter.

9.4 Towards Arbitrarily High-Dimensional Klein–Gordon Equations

Let Ω be a bounded Lipschitz domain in $\mathbb{R}^d$. We next consider the initial valued problem of the arbitrarily high-dimensional nonlinear Klein–Gordon equations

$$\begin{cases} U_{tt} - a^2 \Delta U = f\big(U\big), \quad X \in \Omega, \ t > t_0, \\ U(X, t_0) = \varphi_1(X), \ U_t(X, t_0) = \varphi_2(X), \quad X \in \Omega \cup \partial\Omega. \end{cases} \tag{9.39}$$

The integral formula for (9.39) is given by

$$\begin{cases} U(X,t) = \phi_0\big((t-t_0)^2 a^2 \Delta\big)\varphi_1(X) + (t-t_0)\phi_1\big((t-t_0)^2 a^2 \Delta\big)\varphi_2(X) \\ \qquad + \displaystyle\int_{t_0}^{t} (t-\zeta)\phi_1((t-\zeta)^2 a^2 \Delta)\tilde{f}(\zeta)\mathrm{d}\zeta, \\ U'(X,t) = (t-t_0)a^2 \Delta \phi_1\big((t-t_0)^2 a^2 \Delta\big)\varphi_1(X) + \phi_0\big((t-t_0)^2 a^2 \Delta\big)\varphi_2(X) \\ \qquad + \displaystyle\int_{t_0}^{t} \phi_0\big((t-\zeta)^2 a^2 \Delta\big)\tilde{f}(\zeta)\mathrm{d}\zeta, \end{cases} \tag{9.40}$$

where $\tilde{f}(\zeta) = f\big(U(X,\zeta)\big)$.

9.4.1 *Dirichlet Boundary Conditions*

Firstly, we consider the arbitrarily high-dimensional nonlinear Klein–Gordon equation (9.39) with the Dirichlet boundary condition:

$$U(X,t) = \alpha(X,t), \quad X \in \partial\Omega, \ t \geq t_0. \tag{9.41}$$

Theorem 9.7 *Assume that $\alpha(X,t)$ and $f\big(U(X,t)\big)$ are sufficiently differentiable with respect to t. Then, formula* (9.40) *is consistent with the Dirichlet boundary condition* (9.41).

Proof From the initial conditions, we obtain

$$\alpha(X,t_0) = \varphi_1(X), \quad \alpha'_t(X,t_0) = \varphi_2(X), \quad X \in \partial\Omega.$$

It follows from (9.39) that

$$
\begin{aligned}
U_{tt} &= a^2 \Delta U + f(U) \\
&\Rightarrow \alpha_t''(X, t_0) = \left[a^2 \Delta \varphi_1(X) + f\big(U(X, t_0)\big)\right]\Big|_{\partial\Omega}, \\
U_t^{(3)} &= a^2 \Delta U_t + f_t'(U) \\
&\Rightarrow \alpha_t^{(3)}(X, t_0) = \left[a^2 \Delta \varphi_2(X) + f_t'\big(U(X, t_0)\big)\right]\Big|_{\partial\Omega}, \\
U_t^{(4)} &= a^4 \Delta^2 U + a^2 \Delta f(U) + f_t^{(2)}(U) \\
&\Rightarrow \alpha_t^{(4)}(X, t_0) = \left[a^4 \Delta^2 \varphi_1(X) + a^2 \Delta f\big(U(X, t_0)\big) + f_t^{(2)}\big(U(X, t_0)\big)\right]\Big|_{\partial\Omega}, \\
U_t^{(5)} &= a^4 \Delta^2 U_t + a^2 \Delta f_t'(U) + f_t^{(3)}(U) \\
&\Rightarrow \alpha_t^{(5)}(X, t_0) = \left[a^4 \Delta^2 \varphi_2(X) + a^2 \Delta f_t'\big(U(X, t_0)\big) + f_t^{(3)}\big(U(X, t_0)\big)\right]\Big|_{\partial\Omega}, \\
U_t^{(6)} &= a^6 \Delta^3 U + a^4 \Delta^2 f(U) + a^2 \Delta f_t^{(2)}(U) + f_t^{(4)}(U) \\
&\Rightarrow \alpha_t^{(6)}(X, t_0) = \Big[a^6 \Delta^3 \varphi_1(X) + a^4 \Delta^2 f\big(U(X, t_0)\big) + a^2 \Delta f_t^{(2)}\big(U(X, t_0)\big) \\
&\qquad + f_t^{(4)}\big(U(X, t_0)\big)\Big]\Big|_{\partial\Omega}, \\
U_t^{(7)} &= a^6 \Delta^3 U_t + a^4 \Delta^2 f_t'(U) + a^2 \Delta f_t^{(3)}(U) + f_t^{(5)}(U) \\
&\Rightarrow \alpha_t^{(7)}(X, t_0) = \Big[a^6 \Delta^3 \varphi_2(X) + a^4 \Delta^2 f_t'\big(U(X, t_0)\big) + a^2 \Delta f_t^{(3)}\big(U(X, t_0)\big) \\
&\qquad + f_t^{(5)}\big(U(X, t_0)\big)\Big]\Big|_{\partial\Omega}, \\
&\dots .
\end{aligned}
$$

An argument by induction leads to the results

$$
\begin{aligned}
\alpha^{(2k)}(X, t_0) &= a^{2k} \Delta^k \varphi_1(X) + \sum_{j=1}^{k} a^{2(k-j)} \Delta^{k-j} f_t^{(2j-2)}\big(U(X, t_0)\big) \\
\alpha^{(2k+1)}(X, t_0) &= a^{2k} \Delta^k \varphi_2(X) + \sum_{j=1}^{k} a^{2(k-j)} \Delta^{k-j} f_t^{(2j-1)}\big(U(X, t_0)\big),
\end{aligned}
\tag{9.42}
$$

for $k = 1, 2, \dots,$ and $\forall X \in \partial\Omega$.

Inserting the results of (9.42) into the Taylor expansion of $\alpha(X, t)$ with respect to t at the point t_0 gives

$$
\begin{aligned}
\alpha(X, t) = \Big\{ &\phi_0\big((t - t_0)^2 a^2 \Delta\big)\varphi_1(X) + (t - t_0)\phi_1\big((t - t_0)^2 a^2 \Delta\big)\varphi_2(X) \\
&+ \sum_{k=1}^{\infty} \Big[\frac{(t - t_0)^{2k}}{(2k)!} \sum_{j=1}^{k} a^{2(k-j)} \Delta^{k-j} f_t^{(2j-2)}\big(U(X, t_0)\big) \\
&+ \frac{(t - t_0)^{2k+1}}{(2k+1)!} \sum_{j=1}^{k} a^{2(k-j)} \Delta^{k-j} f_t^{(2j-1)}\big(U(X, t_0)\big) \Big] \Big\} \Big|_{\partial\Omega}.
\end{aligned}
\tag{9.43}
$$

Let

$$F(X,t) \triangleq \int_{t_0}^{t} (t-\zeta)\phi_1\big((t-\zeta)^2 a^2 \Delta\big)\tilde{f}(\zeta)\mathrm{d}\zeta.$$

It is easy to see

$$F(X,t_0)=0,$$

and

$$\begin{aligned}
F_t'(X,t) &= \int_{t_0}^{t} \phi_0\big((t-\zeta)^2 a^2 \Delta\big)\tilde{f}(\zeta)\mathrm{d}\zeta \\
&\Rightarrow F_t'(X,t_0)=0, \\
F_t^{(2)}(X,t) &= f\big(U(X,t)\big) + \int_{t_0}^{t} (t-\zeta)a^2\Delta\phi_1\big((t-\zeta)^2 a^2 \Delta\big)\tilde{f}(\zeta)\mathrm{d}\zeta \\
&\Rightarrow F_t^{(2)}(X,t_0) = f\big(U(X,t_0)\big), \\
F_t^{(3)}(X,t) &= f_t'\big(U(X,t)\big) + \int_{t_0}^{t} a^2\Delta\phi_0\big((t-\zeta)^2 a^2 \Delta\big)\tilde{f}(\zeta)\mathrm{d}\zeta \\
&\Rightarrow F_t^{(3)}(X,t_0) = f_t'\big(U(X,t_0)\big), \\
F_t^{(4)}(X,t) &= f_t^{(2)}\big(U(X,t)\big) + a^2\Delta f\big(U(X,t)\big) + \int_{t_0}^{t} (t-\zeta)a^4\Delta^2\phi_1\big((t-\zeta)^2 a^2 \Delta\big)\tilde{f}(\zeta)\mathrm{d}\zeta \\
&\Rightarrow F_t^{(4)}(X,t_0) = f_t^{(2)}\big(U(X,t_0)\big) + a^2\Delta f\big(U(X,t_0)\big), \\
F_t^{(5)}(X,t) &= f_t^{(3)}\big(U(X,t)\big) + a^2\Delta f_t'\big(U(X,t)\big) + \int_{t_0}^{t} a^4\Delta^2\phi_0\big((t-\zeta)^2 a^2 \Delta\big)\tilde{f}(\zeta)\mathrm{d}\zeta \\
&\Rightarrow F_t^{(5)}(X,t_0) = f_t^{(3)}\big(U(X,t_0)\big) + a^2\Delta f_t'\big(U(X,t_0)\big), \\
F_t^{(6)}(X,t) &= f_t^{(4)}\big(U(X,t)\big) + a^2\Delta f_t^{(2)}\big(U(X,t)\big) + a^4\Delta^2 f\big(U(X,t)\big) \\
&\quad + \int_{t_0}^{t} (t-\zeta)a^6\Delta^3\phi_1\big((t-\zeta)^2 a^2 \Delta\big)\tilde{f}(\zeta)\mathrm{d}\zeta \\
&\Rightarrow F_t^{(6)}(X,t_0) = f_t^{(4)}\big(U(X,t_0)\big) + a^2\Delta f_t^{(2)}\big(U(X,t_0)\big) + a^4\Delta^2 f\big(U(X,t_0)\big), \\
F_t^{(7)}(X,t) &= f_t^{(5)}\big(U(X,t)\big) + a^2\Delta f_t^{(3)}\big(U(X,t)\big) + a^4\Delta^2 f_t'\big(U(X,t)\big) \\
&\quad + \int_{t_0}^{t} a^6\Delta^3\phi_0\big((t-\zeta)^2 a^2 \Delta\big)\tilde{f}(\zeta)\mathrm{d}\zeta \\
&\Rightarrow F_t^{(7)}(X,t_0) = f_t^{(5)}\big(U(X,t_0)\big) + a^2\Delta f_t^{(3)}\big(U(X,t_0)\big) + a^4\Delta^2 f_t'\big(U(X,t_0)\big),
\end{aligned}$$

$\dots$.

Likewise, an argument by induction yields the following results

$$F_t^{(2k)}(X,t_0)=\sum_{j=1}^{k}a^{2(k-j)}\Delta^{k-j}f_t^{(2j-2)}\big(U(X,t_0)\big)$$

$$F_t^{(2k+1)}(X,t_0)=\sum_{j=1}^{k}a^{2(k-j)}\Delta^{k-j}f_t^{(2j-1)}\big(U(X,t_0)\big),\quad k=1,2,\ldots.$$

The Taylor expansion of $F(X,t)$ at $t=t_0$ is

$$\begin{aligned}F(X,t)&=\sum_{k=1}^{\infty}\frac{(t-t_0)^{2k}}{(2k)!}F_t^{(2k)}(X,t_0)+\sum_{k=1}^{\infty}\frac{(t-t_0)^{2k+1}}{(2k+1)!}F_t^{(2k+1)}(X,t_0)\\&=\sum_{k=1}^{\infty}\Big[\frac{(t-t_0)^{2k}}{(2k)!}\sum_{j=1}^{k}a^{2(k-j)}\Delta^{k-j}f_t^{(2j-2)}\big(U(X,t_0)\big)\\&\quad+\frac{(t-t_0)^{2k+1}}{(2k+1)!}\sum_{j=1}^{k}a^{2(k-j)}\Delta^{k-j}f_t^{(2j-1)}\big(U(X,t_0)\big)\Big].\end{aligned}\tag{9.44}$$

Inserting the result of (9.44) into (9.43) gives

$$\begin{aligned}\alpha(X,t)&=\phi_0\big((t-t_0)^2a^2\Delta\big)\varphi_1(X)+(t-t_0)\phi_1\big((t-t_0)^2a^2\Delta\big)\varphi_2(X)\\&+\int_{t_0}^{t}(t-\zeta)\phi_1\big((t-\zeta)^2a^2\Delta\big)\tilde{f}(\zeta)\mathrm{d}\zeta,\ X\in\partial\Omega.\end{aligned}$$

The proof is complete. □

9.4.2 *Neumann Boundary Conditions*

We next consider the arbitrarily high-dimensional nonlinear wave equation (9.39) with the following Neumann boundary condition:

$$\nabla U\cdot\mathbf{n}=\gamma(X,t),\quad X\in\partial\Omega,\tag{9.45}$$

where $\mathbf{n}$ is the unit outward normal vectors on the boundary $\partial\Omega$.

Theorem 9.8 *Assume that* $\gamma(X,t)$ *and* $f\big(U(X,t)\big)$ *are sufficiently differentiable with respect to t. Then the formula* (9.40) *is consistent with the Neumann boundary conditions* (9.45).

Proof Using the initial condition, we have

$$\gamma(X,t_0)=\nabla\varphi_1(X)\cdot\mathbf{n}\triangleq\tilde{\varphi}_1(X),\quad \gamma_t'(X,t_0)=\nabla\varphi_2(X)\cdot\mathbf{n}\triangleq\tilde{\varphi}_2(X),\quad \forall X\in\partial\Omega.$$

Calculating the directional derivative of U with respect to X in (9.39) yields

$$\begin{cases} (\nabla U\cdot\mathbf{n})_{tt}=a^2\Delta(\nabla U\cdot\mathbf{n})+\tilde{f}(U(X,t)), & X\in\Omega,\ t>t_0,\\ (\nabla U\cdot\mathbf{n})(X,t_0)=\tilde{\varphi}_1(X),\ (\nabla U_t\cdot\mathbf{n})(X,t_0)=\tilde{\varphi}_2(X), & X\in\Omega\cup\partial\Omega,\end{cases} \tag{9.46}$$

where $\tilde{f}(U(X,t))=\nabla f(U(X,t))\cdot\mathbf{n}$.

It follows from (9.46) that

$$\begin{aligned}
&(\nabla U\cdot\mathbf{n})_{tt}=a^2\Delta(\nabla U\cdot\mathbf{n})+\tilde{f}(U)\\
&\quad\Rightarrow \gamma_t''(X,t_0)=\left[a^2\Delta\tilde{\varphi}_1(X)+\tilde{f}(U(X,t_0))\right]\big|_{\partial\Omega},\\
&(\nabla U\cdot\mathbf{n})_t^{(3)}=a^2\Delta(\nabla U\cdot\mathbf{n})_t'+\tilde{f}_t'(U)\\
&\quad\Rightarrow \gamma_t^{(3)}(X,t_0)=\left[a^2\Delta\tilde{\varphi}_2(X)+\tilde{f}_t'(U(X,t_0))\right]\big|_{\partial\Omega},\\
&(\nabla U\cdot\mathbf{n})_t^{(4)}=a^4\Delta^2(\nabla U\cdot\mathbf{n})+a^2\Delta\tilde{f}(U)+\tilde{f}_t^{(2)}(U)\\
&\quad\Rightarrow \gamma_t^{(4)}(X,t_0)=\left[a^4\Delta^2\tilde{\varphi}_1(X)+a^2\Delta\tilde{f}(U(X,t_0))+\tilde{f}_t^{(2)}(U(X,t_0))\right]\big|_{\partial\Omega},\\
&(\nabla U\cdot\mathbf{n})_t^{(5)}=a^4\Delta^2(\nabla U\cdot\mathbf{n})_t'+a^2\Delta\tilde{f}_t'(U)+\tilde{f}_t^{(3)}(U)\\
&\quad\Rightarrow \gamma_t^{(5)}(X,t_0)=\left[a^4\Delta^2\tilde{\varphi}_2(X)+a^2\Delta\tilde{f}_t'(U(X,t_0))+\tilde{f}_t^{(3)}(U(X,t_0))\right]\big|_{\partial\Omega},\\
&(\nabla U\cdot\mathbf{n})_t^{(6)}=a^6\Delta^3(\nabla U\cdot\mathbf{n})+a^4\Delta^2\tilde{f}(U)+a^2\Delta\tilde{f}_t^{(2)}(U)+\tilde{f}_t^{(4)}(U)\\
&\quad\Rightarrow \gamma_t^{(6)}(X,t_0)=\left[a^6\Delta^3\tilde{\varphi}_1(X)+a^4\Delta^2\tilde{f}(U(X,t_0))+a^2\Delta\tilde{f}_t^{(2)}(U(X,t_0))\right.\\
&\qquad\qquad\left.+\tilde{f}_t^{(4)}(U(X,t_0))\right]\big|_{\partial\Omega},\\
&(\nabla U\cdot\mathbf{n})_t^{(7)}=a^6\Delta^3(\nabla U\cdot\mathbf{n})_t'+a^4\Delta^2\tilde{f}_t'(U)+a^2\Delta\tilde{f}_t^{(3)}(U)+\tilde{f}_t^{(5)}(U)\\
&\quad\Rightarrow \gamma_t^{(7)}(X,t_0)=\left[a^6\Delta^3\tilde{\varphi}_2(X)+a^4\Delta^2\tilde{f}_t'(U(X,t_0))+a^2\Delta\tilde{f}_t^{(3)}(U(X,t_0))\right.\\
&\qquad\qquad\left.+\tilde{f}_t^{(5)}(U(X,t_0))\right]\big|_{\partial\Omega},\\
&\ldots.
\end{aligned}$$

This leads to the results

$$\begin{aligned}
\gamma_t^{(2k)}(X,t_0)&=a^{2k}\Delta^k\tilde{\varphi}_1(X)+\sum_{j=1}^{k}a^{2(k-j)}\Delta^{k-j}\tilde{f}_t^{(2j-2)}(U(X,t_0)),\\
\gamma_t^{(2k+1)}(X,t_0)&=a^{2k}\Delta^k\tilde{\varphi}_2(X)+\sum_{j=1}^{k}a^{2(k-j)}\Delta^{k-j}\tilde{f}_t^{(2j-1)}(U(X,t_0)),
\end{aligned}\tag{9.47}$$

for $k=1,2,\ldots,$ and $\forall X\in\partial\Omega$.

Inserting the results of (9.47) into the Taylor expansion of $\gamma(X,t)$ with respect to t at the point $t=t_0$ gives

$$
\begin{aligned}
\gamma(X,t) &= \sum_{k=0}^{\infty} \frac{(t-t_0)^k}{k!} \gamma_t^{(k)}(X,t_0) = \sum_{k=0}^{\infty} \frac{(t-t_0)^{(2k)}}{(2k)!} \gamma_t^{(2k)}(X,t_0) \\
&\quad + \sum_{k=0}^{\infty} \frac{(t-t_0)^{(2k+1)}}{(2k+1)!} \gamma_t^{(2k+1)}(X,t_0) \\
&= \Big\{ \phi_0\big((t-t_0)^2 a^2 \Delta\big)\tilde{\varphi}_1(x) + (t-t_0)\phi_1\big((t-t_0)^2 a^2 \Delta\big)\tilde{\varphi}_2(x) \\
&\quad + \sum_{k=1}^{\infty} \Big[\frac{(t-t_0)^{2k}}{(2k)!} \sum_{j=1}^{k} a^{2(k-j)} \Delta^{k-j} \tilde{f}_t^{(2j-2)}\big(U(x,t_0)\big) \\
&\quad + \frac{(t-t_0)^{2k+1}}{(2k+1)!} \sum_{j=1}^{k} a^{2(k-j)} \Delta^{k-j} \tilde{f}_t^{(2j-1)}\big(U(x,t_0)\big) \Big] \Big\} \Big|_{\partial\Omega}.
\end{aligned}
\tag{9.48}
$$

Let

$$
\tilde{F}(X,t) \triangleq \int_{t_0}^{t} (t-\zeta)\phi_1\big((t-\zeta)^2 a^2 \Delta\big) \tilde{f}\big(U(X,\zeta)\big) \mathrm{d}\zeta.
$$

Similarly to the case of Dirichlet boundary conditions, we can obtain

$$
\tilde{F}(X,t_0) = 0, \quad \tilde{F}_t'(X,t_0) = 0,
$$

and

$$
\begin{aligned}
\tilde{F}_t^{(2k)}(X,t_0) &= \sum_{j=1}^{k} a^{2(k-j)} \Delta^{k-j} \tilde{f}_t^{(2j-2)}\big(U(X,t_0)\big), \\
\tilde{F}_t^{(2k+1)}(X,t_0) &= \sum_{j=1}^{k} a^{2(k-j)} \Delta^{k-j} \tilde{f}_t^{(2j-1)}\big(U(X,t_0)\big), \quad k = 1, 2, \ldots .
\end{aligned}
$$

The Taylor expansion of $\tilde{F}(X,t)$ at the point t_0 with respect to t is

$$
\begin{aligned}
\tilde{F}(X,t) &= \sum_{k=1}^{\infty} \frac{(t-t_0)^{2k}}{(2k)!} \tilde{F}_t^{(2k)}(X,t_0) + \sum_{k=1}^{\infty} \frac{(t-t_0)^{2k+1}}{(2k+1)!} \tilde{F}_t^{(2k+1)}(X,t_0) \\
&= \sum_{k=1}^{\infty} \Big[\frac{(t-t_0)^{2k}}{(2k)!} \sum_{j=1}^{k} a^{2(k-j)} \Delta^{k-j} \tilde{f}_t^{(2j-2)}\big(U(x,t_0)\big) \\
&\quad + \frac{(t-t_0)^{2k+1}}{(2k+1)!} \sum_{j=1}^{k} a^{2(k-j)} \Delta^{k-j} \tilde{f}_t^{(2j-1)}\big(U(x,t_0)\big) \Big] \Big\}.
\end{aligned}
\tag{9.49}
$$

Comparing the result of (9.48) with (9.49), we obtain

$$
\begin{aligned}
\gamma(X,t) &= \phi_0\big((t-t_0)^2a^2\Delta\big)\tilde{\varphi}_1(X) + (t-t_0)\phi_1\big((t-t_0)^2a^2\Delta\big)\tilde{\varphi}_2(X) \\
&\quad + \int_{t_0}^{t}(t-\zeta)\phi_1\big((t-\zeta)^2a^2\Delta\big)\tilde{f}\big(U(X,\zeta)\big)\mathrm{d}\zeta \\
&= \phi_0\big((t-t_0)^2a^2\Delta\big)\big(\nabla\varphi_1(X)\cdot\mathbf{n}\big) + (t-t_0)\phi_1\big((t-t_0)^2a^2\Delta\big)\big(\nabla\varphi_2(X)\cdot\mathbf{n}\big) \\
&\quad + \int_{t_0}^{t}(t-\zeta)\phi_1\big((t-\zeta)^2a^2\Delta\big)\big(\nabla f\big(U(X,\zeta)\big)\cdot\mathbf{n}\big)\mathrm{d}\zeta \\
&= \nabla U(X,t)\cdot\mathbf{n}, \quad \forall X\in\partial\Omega.
\end{aligned}
$$

This proves the theorem. □

9.4.3 *Robin Boundary Condition*

In what follows we consider the arbitrarily high-dimensional nonlinear wave equation (9.39) with the following Robin boundary condition:

$$\nabla U\cdot\mathbf{n} + \lambda U = \beta(X,t), \quad X\in\partial\Omega, \tag{9.50}$$

where $\mathbf{n}$ is the unit outward normal vector on the boundary $\partial\Omega$ and λ is a constant.

Theorem 9.9 *Assume that $\beta(X,t)$ and $f\big(U(X,t)\big)$ are sufficiently differentiable with respect to t. Then, the formula* (9.40) *is consistent with the Robin boundary condition given by* (9.50).

The proof of Theorem 9.9 is similar to that in the recent paper [33] and we omit the details here.

Remark 9.2 As stated in Sects. 9.3 and 9.4, one need not care about the boundary conditions when the formula (9.15) is used directly since the formula is adapted to Dirichlet boundary conditions, Neumann boundary conditions, and Robin boundary conditions, respectively. In fact, (9.15) presents an exact analytical formal of the true solution to the initial-value problem of arbitrarily high-dimensional Klein–Gordon equations subject to the given boundary conditions, under the appropriate assumptions.

9.5 Illustrative Examples

To show applications of the integral formula presented in this chapter, we next present some illustrative examples.

Problem 9.1 We first consider the following two-dimensional equation

$$\begin{cases} u_{tt} - (u_{xx} + u_{yy}) = \omega^2 \sin(\omega(x-t))\sin(\omega y), \\ u|_{t=0} = \sin(\omega x)\sin(\omega y), \ u_t|_{t=0} = -\omega\cos(\omega x)\sin(\omega y). \end{cases} \tag{9.51}$$

Applying (9.15) to Problem 9.1 yields

$$\begin{cases} u(x,y,t) = \phi_0(t^2\Delta)\sin(\omega x)\sin(\omega y) - \omega t\phi_1(t^2\Delta)\cos(\omega x)\sin(\omega y) \\ \qquad + \omega^2 \int_0^t (t-\zeta)\phi_1((t-\zeta)^2\Delta)\sin(\omega(x-\zeta))\sin(\omega y)\mathrm{d}\zeta, \\ u_t(x,y,t) = t\Delta\phi_1(t^2\Delta)\sin(\omega x)\sin(\omega y) - \omega\phi_0(t^2\Delta)\cos(\omega x)\sin(\omega y) \\ \qquad + \omega^2 \int_0^t \phi_0((t-\zeta)^2\Delta)\sin(\omega(x-\zeta))\sin(\omega y)\mathrm{d}\zeta. \end{cases} \tag{9.52}$$

It follows from a careful calculation that

$$\begin{aligned} &\phi_0(t^2\Delta)\sin(\omega x)\sin(\omega y) = \sin(\omega x)\sin(\omega y)\cos(\sqrt{2}\omega t), \\ &-\omega t\phi_1(t^2\Delta)\cos(\omega x)\sin(\omega y) = -\frac{1}{\sqrt{2}}\cos(\omega x)\sin(\omega y)\sin(\sqrt{2}\omega t), \\ &\omega^2 \int_0^t (t-\zeta)\phi_1((t-\zeta)^2\Delta)\sin(\omega(x-\zeta))\sin(\omega y)\mathrm{d}\zeta \\ &= \frac{\omega}{\sqrt{2}}\int_0^t \sin(\sqrt{2}\omega(t-\zeta))\sin(\omega(x-\zeta))\sin(\omega y)\mathrm{d}\zeta. \end{aligned}$$

We then have

$$\begin{aligned} u(x,y,t) &= \sin(\omega x)\sin(\omega y)\cos(\sqrt{2}\omega t) - \frac{1}{\sqrt{2}}\cos(\omega x)\sin(\omega y)\sin(\sqrt{2}\omega t) \\ &\quad + \frac{\omega}{\sqrt{2}}\int_0^t \sin(\sqrt{2}\omega(t-\zeta))\sin(\omega(x-\zeta))\sin(\omega y)\mathrm{d}\zeta \\ &= \sin(\omega(x-t))\sin(\omega y). \end{aligned} \tag{9.53}$$

Likewise, we can obtain

$$u_t(x,y,t) = -\omega\cos(\omega(x-t))\sin(\omega y). \tag{9.54}$$

Problem 9.2 We next consider the three-dimensional linear homogeneous equation:

$$\begin{cases} u_{tt} = a^2(u_{xx} + u_{yy} + u_{zz}), \\ u|_{t=0} = x^3 + yz, \ u_t|_{t=0} = 0. \end{cases} \tag{9.55}$$

Applying (9.15) to Problem 9.2, we can obtain the analytical solution straightforwardly:

$$\begin{cases} u(x, y, z, t) = \phi_0(t^2a^2\Delta)(x^3 + yz) = x^3 + yz + 3a^2t^2x, \\ u_t(x, y, z, t) = ta^2\Delta\phi_1(t^2a^2\Delta)(x^3 + yz) = 6a^2tx. \end{cases} \tag{9.56}$$

We note that from Poisson's or Kirchhoff's formula (see, e.g. [7]) the solution to (9.55) can be expressed in the form

$$u(x, y, z, t) = \frac{1}{4\pi a^2 t}\frac{\partial}{\partial t}\iint_S (x^3 + yz)\mathrm{d}S, \tag{9.57}$$

where S is the sphere of radius a centered at (x_0, y_0, z_0). The calculation of the integral in (9.57) is quite complicated.

Problem 9.3 We next consider the following the initial valued problem of one dimensional linear Klein–Gordon equation (see, e.g. [19])

$$\begin{cases} u_{tt} - u_{xx} = -9u, \quad -\frac{5\pi}{8} < x < \frac{5\pi}{8}, \ t > 0, \\ u(x, 0) = \cos(4x), \quad u_t(x, 0) = 5\cos(4x), \quad -\frac{5\pi}{8} \le x \le \frac{5\pi}{8}, \end{cases} \tag{9.58}$$

subject to the Dirichlet boundary conditions

$$u(-\frac{5\pi}{8}, t) = 0, \quad u(\frac{5\pi}{8}, t) = 0. \tag{9.59}$$

The exact solution of the initial-boundary valued problem (9.58) and (9.59) is given by

$$u(x, t) = \cos(4x)\left(\cos(5t) + \sin(5t)\right). \tag{9.60}$$

Applying the *SV-scheme* (9.26) with the stepsize $h = 0.001$ to this initial-boundary valued problem, we obtain the numerical results, together with the true solution and the global error, which are shown in Fig. 9.1. It can be observed from Fig. 9.1 that

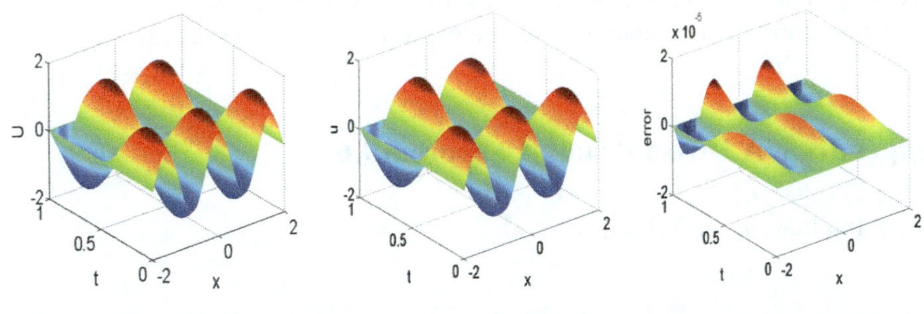

Fig. 9.1 The exact solution (left), the numerical solution (middle) and the global error (right) obtained by *SV-scheme* (9.26) with the stepsize $h = 0.001$, for Problem 9.3

the results show second-order behaviour of the *SV-scheme* (9.26). This indicates that the integral formula (9.15) is also helpful in gaining insight into developing efficient numerical integrators for Klein–Gordon equations.

9.6 Conclusions and Discussions

In this chapter, we considered the initial-boundary value problem of arbitrarily high-dimensional Klein–Gordon equations (9.1), posed on a bounded domain $\Omega \subset \mathbb{R}^d$ for $d \geq 1$ and equipped with various boundary conditions. We first defined the bounded operator-argument functions (9.8) which are restricted in a subspace $D(\Delta)$ of $L^2(\Omega)$, and then established an integral formula (9.15) for the Klein–Gordon equation in arbitrarily high-dimensional spaces. Thus, this chapter has made progress in research on integral representations of solutions of the arbitrarily high-dimensional Klein–Gordon equation (9.1). Another key aspect is that we showed in detail the consistency of the integral formula with Dirichlet boundary conditions, Neumann boundary conditions, and Robin boundary conditions, respectively. In other words, the integral formula (9.15) for (9.1) is completely adapted to the underlying boundary conditions under appropriate assumptions. If $g(U) = 0$, then (9.1) reduces to the arbitrarily high-dimensional homogeneous Klein–Gordon equation (9.20). Then, the integral formula (9.15) becomes (9.21), which integrates (9.20) exactly. In comparison with the seminal D'Alembert, Poisson and Kirchhoff formulas, formula (9.21) doesn't depend on the evaluation of complicated integrals, whereas the evaluation of integrals is required by the D'Alembert, Poisson and Kirchhoff formulas. To show the applications of the integral formula, some illustrative examples were also presented in Sect. 9.5.

Before the end of this chapter, we make a comment on the operator-variation-constants formula for PDEs. Once the operator-variation-constants formula is established for the underling PDEs, some structure-preserving schemes can be derived and analysed based on the formula. For example, Chaps. 10 and 11 will show the details for nonlinear wave equations. It is also believed that this approach to dealing with nonlinear wave PDEs can be extended to other PDEs, such as Maxwell's equations (see Yang et al. [35]). Further work on this research is in progress.

The material of this chapter is based on the work by Wu and Liu [28].

Appendix 1. A Direct Proof of Theorem 9.6

Proof It follows from (9.37) that

$$\left(\frac{\partial u}{\partial x}\right)_{tt} = a^2\Delta\left(\frac{\partial u}{\partial x}\right) + \tilde{f}(u)$$

$$\Rightarrow \begin{cases} \gamma''(t_0) = a^2\Delta\varphi_1'(x_l) + \tilde{f}(u), \\ \delta''(t_0) = a^2\Delta\varphi_1'(x_r) + \tilde{f}(u), \end{cases}$$

$$\left(\frac{\partial u}{\partial x}\right)_t^{(3)} = a^2\Delta\left(\frac{\partial u}{\partial x}\right)_t + \tilde{f}_t'(u)$$

$$\Rightarrow \begin{cases} \gamma^{(3)}(t_0) = a^2\Delta\varphi_2'(x_l) + \tilde{f}_t'(u), \\ \delta^{(3)}(t_0) = a^2\Delta\varphi_2'(x_r) + \tilde{f}_t'(u), \end{cases}$$

$$\left(\frac{\partial u}{\partial x}\right)_t^{(4)} = a^4\Delta^2\left(\frac{\partial u}{\partial x}\right) + a^2\Delta\tilde{f}(u) + \tilde{f}_t^{(2)}(u)$$

$$\Rightarrow \begin{cases} \gamma^{(4)}(t_0) = a^4\Delta^2\varphi_1'(x_l) + a^2\Delta\tilde{f}(u) + \tilde{f}_t^{(2)}(u), \\ \delta^{(4)}(t_0) = a^4\Delta^2\varphi_1'(x_r) + a^2\Delta\tilde{f}(u) + \tilde{f}_t^{(2)}(u), \end{cases}$$

$$\left(\frac{\partial u}{\partial x}\right)_t^{(5)} = a^4\Delta^2\left(\frac{\partial u}{\partial x}\right)_t + a^2\Delta\tilde{f}_t'(u) + \tilde{f}_t^{(3)}(u)$$

$$\Rightarrow \begin{cases} \gamma^{(5)}(t_0) = a^4\Delta^2\varphi_2'(x_l) + a^2\Delta\tilde{f}_t'(u) + \tilde{f}_t^{(3)}(u), \\ \delta^{(5)}(t_0) = a^4\Delta^2\varphi_2'(x_r) + a^2\Delta\tilde{f}_t'(u) + \tilde{f}_t^{(3)}(u), \end{cases}$$

$$\left(\frac{\partial u}{\partial x}\right)_t^{(6)} = a^6\Delta^3\left(\frac{\partial u}{\partial x}\right) + a^4\Delta^2\tilde{f}(u) + a^2\Delta\tilde{f}_t^{(2)}(u) + \tilde{f}_t^{(4)}(u)$$

$$\Rightarrow \begin{cases} \gamma^{(6)}(t_0) = a^6\Delta^3\varphi_1'(x_l) + a^4\Delta^2\tilde{f}(u) \\ \qquad\qquad +a^2\Delta\tilde{f}_t^{(2)}(u) + \tilde{f}_t^{(4)}(u), \\ \delta^{(6)}(t_0) = a^6\Delta^3\varphi_1'(x_r) + a^4\Delta^2\tilde{f}(u) \\ \qquad\qquad +a^2\Delta\tilde{f}_t^{(2)}(u) + \tilde{f}_t^{(4)}(u), \end{cases}$$

$$\left(\frac{\partial u}{\partial x}\right)_t^{(7)} = a^6\Delta^3\left(\frac{\partial u}{\partial x}\right)_t + a^4\Delta^2\tilde{f}_t'(u) + a^2\Delta\tilde{f}_t^{(3)}(u) + \tilde{f}_t^{(5)}(u)$$

$$\Rightarrow \begin{cases} \gamma^{(7)}(t_0) = a^6\Delta^3\varphi_2'(x_l) + a^4\Delta^2\tilde{f}_t'(u) \\ \qquad\qquad +a^2\Delta\tilde{f}_t^{(3)}(u) + \tilde{f}_t^{(5)}(u), \\ \delta^{(7)}(t_0) = a^6\Delta^3\varphi_2'(x_r) + a^4\Delta^2\tilde{f}_t'(u) \\ \qquad\qquad +a^2\Delta\tilde{f}_t^{(3)}(u) + \tilde{f}_t^{(5)}(u), \end{cases}$$

$\cdots$.

After an argument by induction we obtain the following results

$$\begin{aligned} \gamma^{(2k)} &= a^{2k}\Delta^k\varphi_1'(x_l) + \sum_{j=1}^{k} a^{2(k-j)}\Delta^{k-j}\tilde{f}_t^{(2j-2)}(u) \\ \gamma^{(2k+1)} &= a^{2k}\Delta^k\varphi_2'(x_l) + \sum_{j=1}^{k} a^{2(k-j)}\Delta^{k-j}\tilde{f}_t^{(2j-1)}(u), \quad k = 1, 2, \ldots. \end{aligned} \tag{9.61}$$

and

$$
\begin{aligned}
\delta^{(2k)}(t_0) &= a^{2k}\Delta^k\varphi_1'(x_r) + \sum_{j=1}^{k} a^{2(k-j)}\Delta^{k-j}\tilde{f}_t^{(2j-2)}(u) \\
\delta^{(2k+1)}(t_0) &= a^{2k}\Delta^k\varphi_2'(x_r) + \sum_{j=1}^{k} a^{2(k-j)}\Delta^{k-j}\tilde{f}_t^{(2j-1)}(u), \quad k = 1, 2, \ldots.
\end{aligned}
\tag{9.62}
$$

Inserting the results of (9.61) and (9.62) into the Taylor expansion of $\gamma(t)$ and $\delta(t)$ at point t_0 yields

$$
\begin{aligned}
\gamma(t) = \Big\{&\phi_0\big((t-t_0)^2a^2\Delta\big)\varphi_1'(x) + (t-t_0)\phi_1\big((t-t_0)^2a^2\Delta\big)\varphi_2'(x) \\
&+ \sum_{k=1}^{\infty}\Big[\frac{(t-t_0)^{2k}}{(2k)!}\sum_{j=1}^{k} a^{2(k-j)}\Delta^{k-j}\tilde{f}_t^{(2j-2)}(u) \\
&+ \frac{(t-t_0)^{2k+1}}{(2k+1)!}\sum_{j=1}^{k} a^{2(k-j)}\Delta^{k-j}\tilde{f}_t^{(2j-1)}(u)\Big]\Big\}\Big|_{x=x_l},
\end{aligned}
\tag{9.63}
$$

and

$$
\begin{aligned}
\delta(t) = \Big\{&\phi_0\big((t-t_0)^2a^2\Delta\big)\varphi_1'(x) + (t-t_0)\phi_1\big((t-t_0)^2a^2\Delta\big)\varphi_2'(x) \\
&+ \sum_{k=1}^{\infty}\Big[\frac{(t-t_0)^{2k}}{(2k)!}\sum_{j=1}^{k} a^{2(k-j)}\Delta^{k-j}\tilde{f}_t^{(2j-2)}(u) \\
&+ \frac{(t-t_0)^{2k+1}}{(2k+1)!}\sum_{j=1}^{k} a^{2(k-j)}\Delta^{k-j}\tilde{f}_t^{(2j-1)}(u)\Big]\Big\}\Big|_{x=x_r}.
\end{aligned}
\tag{9.64}
$$

Let

$$
\tilde{F}(x,t) \triangleq \int_{t_0}^{t}(t-\zeta)\phi_1\big((t-t_0)^2a^2\Delta\big)\hat{f}(\zeta)\mathrm{d}\zeta.
$$

As deduced for the Dirichlet boundary conditions in Sect. 9.4.1, it can be shown that

$$
\begin{aligned}
\tilde{F}_t^{(2k)} &= \sum_{j=1}^{k} a^{2(k-j)}\Delta^{k-j}\tilde{f}_t^{(2j-2)}(u) \\
\tilde{F}_t^{(2k+1)} &= \sum_{j=1}^{k} a^{2(k-j)}\Delta^{k-j}\tilde{f}_t^{(2j-1)}(u), \quad k = 1, 2, \ldots.
\end{aligned}
\tag{9.65}
$$

Inserting (9.65) into the Taylor expansion of $\tilde{F}(x,t)$ at the point $t = t_0$ gives

$$
\begin{aligned}
\tilde{F}(x,t) &= \sum_{k=0}^{\infty} \frac{(t-t_0)^k}{k!} \tilde{F}_t^{(k)} = \sum_{k=2}^{\infty} \frac{(t-t_0)^k}{k!} \tilde{F}_t^{(k)} \\
&= \sum_{k=1}^{\infty} \frac{(t-t_0)^{2k}}{(2k)!} \tilde{F}_t^{(2k)} + \sum_{k=1}^{\infty} \frac{(t-t_0)^{2k+1}}{(2k+1)!} \tilde{F}_t^{(2k+1)} \\
&= \sum_{k=1}^{\infty} \Big[\frac{(t-t_0)^{2k}}{(2k)!} \sum_{j=1}^{k} a^{2(k-j)} \Delta^{k-j} \tilde{f}_t^{(2j-2)}(u) \\
&\quad + \frac{(t-t_0)^{2k+1}}{(2k+1)!} \sum_{j=1}^{k} a^{2(k-j)} \Delta^{k-j} \tilde{f}_t^{(2j-1)}(u) \Big].
\end{aligned}
\tag{9.66}
$$

Comparing the results of (9.66) with (9.63) and (9.64) yields

$$
\begin{aligned}
\gamma(t) &= \Big[\phi_0\big((t-t_0)^2 a^2 \Delta\big)\varphi_1'(x) + (t-t_0)\phi_1\big((t-t_0)^2 a^2 \Delta\big)\varphi_2'(x) \\
&+ \int_{t_0}^{t} (t-\zeta)\phi_1\big((t-\zeta)^2 a^2 \Delta\big) \hat{f}(\zeta) \mathrm{d}\zeta \Big]\Big|_{x=x_l}, \\
\delta(t) &= \Big[\phi_0\big((t-t_0)^2 a^2 \Delta\big)\varphi_1'(x) + (t-t_0)\phi_1\big((t-t_0)^2 a^2 \Delta\big)\varphi_2'(x) \\
&+ \int_{t_0}^{t} (t-\zeta)\phi_1\big((t-\zeta)^2 a^2 \Delta\big) \hat{f}(\zeta) \mathrm{d}\zeta \Big]\Big|_{x=x_r}.
\end{aligned}
$$

This finishes the direct proof. □

References

1. Biswas, A.: Soliton perturbation theory for phi-four model and nonlinear Klein–Gordon equations. Commun. Nonlinear Sci. Numer. Simul. **14**, 3239–3249 (2009)
2. Bratsos, A.G.: On the numerical solution of the Klein–Gordon equation. Numer. Methods Partial Differ. Equ. **25**, 939–951 (2009)
3. Cohen, D., Jahnke, T., Lorenz, K., Lubich, C.: Numerical integrators for highly oscillatory Hamiltonian systems: a review. In: Mielke, A. (ed.) Analysis, Modeling and Simulation of Multiscale Problems, pp. 553–576. Springer, Berlin (2006)
4. Debnath, L.: Nonlinear Partial Differential Equations for Scientists and Engineers, 3rd edn. Birkhäuser, Springer, New York, Dordrecht, Heidelberg, London (2012)
5. Dodd, R.K., Eilbeck, I.C., Gibbon, J.D., Morris, H.C.: Solitons and Nonlinear Wave Equations. Academic, London (1982)
6. Eilbeck, J.C.: Numerical studies of solitons. In: Bishop, A.R., Schneider, T. (eds.) Solitons and Condensed Matter Physics, pp. 28–43. Springer, New York (1978)
7. Evans, L.C.: Partial Differential Equations. American Mathematical Society, Providence (1998)
8. Fordy, A.P.: Soliton Theory: A Survey of Results. Manchester University Press, Manchester (1990)
9. Franco, J.M.: New methods for oscillatory systems based on ARKN methods. Appl. Numer. Math. **56**, 1040–1053 (2006)

10. García-Archilla, B., Sanz-Serna, J.M., Skeel, R.D.: Long-time-step methods for oscillatory differential equations. SIAM J. Sci. Comput. **20**, 930–963 (1998)
11. Gautschi, W.: Numerical integration of ordinary differential equations based on trigonometric polynomials. Numer. Math. **3**, 381–397 (1961)
12. Grimm, V.: On error bounds for the Gautschi-type exponential integrator applied to oscillatory second-order differential equations. Numer. Math. **100**, 71–89 (2005)
13. Hairer, E., Lubich, C.: Long-time energy conservation of numerical methods for oscillatory differential equations. SIAM J. Numer. Anal. **38**, 414–441 (2000)
14. Hochbruck, M., Lubich, C.: A Gautschi-type method for oscillatory second-order differential equations. Numer. Math. **83**, 403–426 (1999)
15. Hochbruck, M., Ostermann, A.: Exponential integrators. Acta Numer. **19**, 209–286 (2010)
16. Infeld, E., Rowlands, G.: Nonlinear Waves, Solitons and Chaos. Cambridge University Press, New York (1990)
17. Kragh, H.: Equation with many fathers. Klein–Gordon equation in 1926. Am. J. Phys. **52**, 1024–1033 (1984)
18. Liu, C., Wu, X.Y.: The boundness of the operator-valued functions for multidimensional nonlinear wave equations with applications. Appl. Math. Lett. **74**, 60–67 (2017)
19. Polyanin, A.D.: Handbook of Linear Partial Differential Equations for Engineers and Scientists. Chapman & Hall/CRC, Boca Raton (2002)
20. Schiesser, W.: The Numerical Methods of Lines: Integration of Partial Differential Equation. Academic Press, San Diego (1991)
21. Shakeri, F., Dehghan, M.: Numerical solution of the Klein–Gordon equation via He's variational iteration method. Nonlinear Dyn. **51**, 89–97 (2008)
22. Shi, W., Wu, X.Y., Xia, J.: Explicit multi-symplectic extended leap-frog methods for Hamiltonian wave equations. J. Comput. Phys. **231**, 7671–7694 (2012)
23. Van de Vyver, H.: Scheifele two-step methods for perturbed oscillators. J. Comput. Appl. Math. **224**, 415–432 (2009)
24. Wang, B., Wu, X.Y.: A new high precision energy-preserving integrator for system of oscillatory second-order differential equations. Phys. Lett. A. **376**, 1185–1190 (2012)
25. Wang, B., Liu, K., Wu, X.Y.: A Filon-type asymptotic approach to solving highly oscillatory second-order initial value problems. J. Comput. Phys. **243**, 210–223 (2013)
26. Wang, B., Iserles, A., Wu, X.Y.: Arbitrary-order trigonometric Fourier collocation methods for multi-frequency oscillatory systems. Found. Comput. Math. (2014). https://doi.org/10.1007/s10208-014-9241-9
27. Wazwaz, A.M.: New travelling wave solutions to the Boussinesq and the Klein–Gordon equations. Commun. Nonlinear Sci. Numer. Simul. **13**, 889–901 (2008)
28. Wu, X.Y., Liu, C.: An integral formula adapted to different boundary conditions for arbitrarily high-dimensional nonlinear Klein–Gordon equations with its applications. J. Math. Phys. **57**, 021504 (2016)
29. Wu, X.Y., You, X., Xia, J.: Order conditions for ARKN methods solving oscillatory systems. Comput. Phys. Comm. **180**, 2250–2257 (2009)
30. Wu, X.Y., You, X., Shi, W., Wang, B.: ERKN integrators for systems of oscillatory second-order differential equations. Comput. Phys. Comm. **181**, 1873–1887 (2010)
31. Wu, X.Y., Wang, B., Shi, W.: Efficient energy-preserving integrators for oscillatory Hamiltonian systems. J. Comput. Phys. **235**, 587–605 (2013)
32. Wu, X.Y., You, X., Wang, B.: Structure-Preserving Algorithms for Oscillatory Differential Equations. Springer, Heidelberg (2013)
33. Wu, X.Y., Mei, L.J., Liu, C.: An analytical expression of solutions to nonlinear wave equations in higher dimensions with Robin boundary conditions. J. Math. Anal. Appl. **426**, 1164–1173 (2015)
34. Wu, X.Y., Liu, K., Shi, W.: Structure-Preserving Algorithms for Oscillatory Differential Equations II. Springer, Heidelberg (2015)
35. Yang, H., Zeng, X., Wu, X. Y.: Variation-of-constants formulae for Maxwell's equations in time domain, A seminar report at Nanjing University (2017)

Chapter 10
An Energy-Preserving and Symmetric Scheme for Nonlinear Hamiltonian Wave Equations

In this chapter, we derive and analyse an energy-preserving and symmetric scheme for nonlinear Hamiltonian wave equations, which can exactly preserve the energy of the underlying Hamiltonian wave equations. To this end, we first define and discuss the bounded operator-argument functions on the underlying domain. We then introduce an operator-variation-of-constants formula, based on which we present an energy-preserving scheme for nonlinear Hamiltonian wave equations. The scheme preserves the energy of the original continuous Hamiltonian system exactly. In comparison with the existing work on this topic, such as the well-known Average Vector Field (AVF) formula for Hamiltonian ordinary differential equations, *the energy-preserving scheme avoids the semi-discretisation of spatial derivatives and exactly preserves the Hamiltonian of the original continuous Hamiltonian wave equation*. This point is very significant in comparison with the AVF formula, since *the AVF formula can preserve only the energy of Hamiltonian ordinary differential equations*. Hence, the main theme of this chapter is to establish a *scheme* which can exactly preserve the energy of the nonlinear Hamiltonian wave equation. The chapter is also accompanied by some examples.

10.1 Introduction

Nonlinear wave or heat equations arise frequently in a wide variety of applications, which can usually be expressed in suitable nonlinear Hamiltonian forms, and partial differential equations with a Hamiltonian structure are important in the study of solitons. Hamiltonian systems have some characteristic properties such as inner symmetries and energy conservation. However, there are no general methods guaranteed to find closed form solutions to nonlinear Hamiltonian systems. Over the last 20 years, geometric numerical integration has become an important area of numerical analysis and scientific computing. Structure-preserving integrators have received much attention in recent years and have applications in many areas of physics, such

X. Wu and B. Wang, *Recent Developments in Structure-Preserving Algorithms for Oscillatory Differential Equations*, https://doi.org/10.1007/978-981-10-9004-2_10

as molecular dynamics, fluid dynamics, celestial mechanics, and particle accelerators. An integrator is said to be *structure-preserving* if it preserves one or more physical/geometric properties of the system exactly. In this chapter, we pay attention to an energy-preserving scheme for the nonlinear Hamiltonian wave equation of the form

$$\begin{cases} u_{tt} - a^2 \Delta u = f(u), \quad t \geq t_0, \\ u(x, t_0) = \varphi_1(x), \ u_t(x, t_0) = \varphi_2(x), \end{cases} \tag{10.1}$$

where $\Delta = \partial_x^2$, and $f(u) = -V'(u)$ is the negative derivative of a potential function $V(u)$ with respect to u.

The nonlinear wave equation (10.1) in this chapter is assumed to be subject to the following periodic boundary condition

$$u(x, t) = u(x + \Gamma, t), \tag{10.2}$$

where Γ is the fundamental period in x.

It is known that Δ is an unbounded operator which is not defined for every $v \in L^2([x, x + \Gamma])$. In order to model periodic boundary conditions, we restrict ourselves to the case where Δ is defined on the domain

$$D(\Delta) = \{v(x) : \forall v \in L^2([x, x + \Gamma]) \text{ and } v(x) = v(x + \Gamma)\}. \tag{10.3}$$

We consider (10.1) with independent variables $(x, t) \in [x_l, x_r] \times \{t \geq t_0\}$ and suppose that $\Gamma = x_r - x_l$ is the period. Let $v = (u, p)^\intercal$ with $p = u_t$. The nonlinear wave equation (10.1) can be thought of as an infinite dimensional Hamiltonian system of the form

$$\partial_t v = J \frac{\delta H}{\delta v}, \quad \forall v \in \mathscr{B}. \tag{10.4}$$

This is equivalent to

$$\begin{cases} u_t = p, \\ p_t = a^2 \Delta u + f(u), \\ u(x, t_0) = \varphi_1(x), \ p(x, t_0) = \varphi_2(x), \end{cases} \tag{10.5}$$

where the Hamiltonian

$$H(u, p) := \frac{1}{2} \int_{x_l}^{x_r} \left[p^2 + a^2 u_x^2 + 2V(u)\right] \mathrm{d}x \tag{10.6}$$

is defined on the infinite dimensional "phase-space" $\mathscr{B} := \mathscr{V} \times L^2([x_l, x_r])$, where $\mathscr{V} = \{u : u \in H^1([x_l, x_r]) \text{ and } u(x_l) = u(x_r)\}$, and the non-degenerate antisymmetric operator J represents a symplectic structure: the variables $v = (u, p)^\intercal$ are "Darboux coordinates" (see, e.g. [1]). The Hamiltonian system (10.4) preserves the energy (or the Hamiltonian) because J is skew-adjoint with respect to the L^2 inner product, i.e.,

$$\int_{x_l}^{x_r} uJu\mathrm{d}x = 0, \quad \forall u \in \mathscr{B}. \tag{10.7}$$

The conservation of the energy (or the Hamiltonian) is one of the most important properties of the nonlinear Hamiltonian wave equation (10.1), i.e.

$$E(t) = \frac{1}{2}\int_{x_l}^{x_r} \left[u_t^2(x,t) + a^2u_x^2(x,t) + 2V\big(u(x,t)\big)\right]\mathrm{d}x = E(t_0), \tag{10.8}$$

or

$$H(u,p) = H(u,p)\big|_{t=t_0}, \tag{10.9}$$

for the Hamiltonian system (10.5).

The nonlinear Hamiltonian wave equation (10.1) arises in a wide variety of application areas in science and engineering such as nonlinear optics, solid state physics and quantum field theory. Its description and understanding are of great importance both from the theoretical and practical point of view, and it has been investigated by many authors (see, e.g. [2, 8–11, 18, 26, 28, 35, 38]). This equation is the relativistic version of the Schrödinger equation [6, 7, 30]. Such a problem appears naturally in the study of some nonlinear dynamical problems of mathematical physics, including radiation theory, general relativity, the scattering and stability of kinks, vortices, and other coherent structures. This equation is the basis of much work in studying solitons and condensed matter physics, in investigating the interaction of solitons in collisionless plasma and the recurrence of initial states, in lattice dynamics, and in examining nonlinear phenomena.

Many authors have investigated energy preservation for semi-discrete Hamiltonian wave equations via classical spatial discretisation approximations. Usually, the semi-discrete systems are of the form

$$\begin{cases} y''(t) + My(t) = g(y(t)), & t \in [t_0, T], \\ y(t_0) = y_0, \quad y'(t_0) = y_0', \end{cases} \tag{10.10}$$

where $M \in \mathbb{R}^{m\times m}$ is a positive and semi-definite matrix (not necessarily diagonal or symmetric, in general). The solution of the system (10.10) exhibits nonlinear oscillations. When such oscillations occur, effective ODE solvers for (10.10) can be used, such as Gautschi-type methods (see, e.g. [16]), trigonometric Fourier collocation methods (see, e.g. [34]), and extended Runge–Kutta–Nyström (ERKN) methods (see, e.g. [42, 43]). With regard to the discrete energy-preserving method, the AVF formula is very popular (see, e.g. [3, 19, 21–25, 40]). However, the AVF formula cannot exactly preserve the energy of the original continuous Hamiltonian wave equation. In general, the discrete energy is different from the original energy of the continuous Hamiltonian wave equations. This means that the AVF formula based on classical discrete approximations cannot preserve the energy of the Hamiltonian wave equations exactly. This motivates an energy-preserving scheme for nonlinear Hamiltonian wave equations, which can exactly preserve the energy of the nonlinears Hamiltonian wave equation (10.1). *It should be noted that, in this chapter, all essential analytical features are present in the one-dimensional case* (10.1), *although the discussions are equally valid for high-dimensional nonlinear Hamiltonian wave equations.*

The outline of this chapter is as follows. Some preliminaries are given in Sect. 10.2. In Sect. 10.3, we introduce an operator-variation-of-constants formula for the nonlinear Hamiltonian wave equation (10.1), which is an analytical expression of the solution to the nonlinear Hamiltonian wave equation (10.1) expressed in a nonlinear integral equation. We then dicuss an energy-preserving scheme and analyse its properties in Sect. 10.4. Some illustrative examples are presented in Sect. 10.5. The last section is devoted to conclusions.

10.2 Preliminaries

This section presents some preliminaries in order to gain an insight into an exact energy-preserving scheme for the nonlinear Hamiltonian wave equation (10.1) subject to the periodic boundary condition (10.2).

To begin with, we define the following operator-argument functions:

$$\phi_j(\Delta) := \sum_{k=0}^{\infty} \frac{\Delta^k}{(2k+j)!}, \quad j = 0, 1, \ldots. \tag{10.11}$$

For example, Δ is the Laplace operator defined on $D(\Delta)$ in (10.3) and in this case, the operators defined by (10.11) is bounded on the subspace under the Sobolev norm $\|\cdot\|_{L^2 \leftarrow L^2}$ (see, e.g. [17, 20]). Accordingly, $\phi_j(\Delta)$ for $j = 0, 1, \ldots$ in (10.11) are called operator-argument functions. Besides, Δ can also be related to a linear mapping such as a positive semi-definite matrix $M \in \mathbb{R}^{m \times m}$ and in this particular case of $\Delta = -M$, (10.11) reduces to the matrix-valued functions which have been widely used in ARKN methods (Adapted Runge–Kutta–Nyström methods) and ERKN (Extended Runge–Kutta–Nyström methods) methods for the numerical solution of oscillatory or highly oscillatory differential equations (see, e.g. [12–14, 29, 31, 32, 40, 41, 43, 44]). These kinds of oscillatory problems have received a great deal of attention in the last few years (see, e.g. [4, 14–17, 34]).

In this chapter, some useful properties of these operator-argument functions are only sketched below for the sake of brevity.

Theorem 10.1 *For a symmetric negative (semi) definite operator Δ, the bounded ϕ-functions defined by (10.11) satisfy (9.10)–(9.14) in Chap. 9 and*

$$\begin{cases} \phi_0(m^2a^2\Delta)\phi_0(n^2a^2\Delta) + mna^2\Delta\phi_1(m^2a^2\Delta)\phi_1(n^2a^2\Delta) = \phi_0((m+n)^2a^2\Delta), \\ m\phi_0(n^2a^2\Delta)\phi_1(m^2a^2\Delta) + n\phi_0(m^2a^2\Delta)\phi_1(n^2a^2\Delta) = (m+n)\phi_1((m+n)^2a^2\Delta). \end{cases} \tag{10.12}$$

Proof These results are evident. □

Theorem 10.2 *Suppose that Δ is a Laplacian defined on the domain $D(\Delta)$. The bounded operator-argument functions ϕ_0 and ϕ_1 defined by (10.11) satisfy (9.9) in Chap. 9.*

Proof The results have been shown in Chap. 9. □

Some properties of the periodic functions are stated blow.

Theorem 10.3 *Assuming that $u(x,t)$, $v(x,t)$ are any sufficiently smooth periodic functions with respect to the variable x, i.e. $u(x+\Gamma,t) = u(x,t)$, and Γ is the fundamental period, then the following properties hold:*

(i) *For all $k, l = 0, 1, \ldots,$ we have*

$$\partial_x^k u(x+\Gamma,t) = \partial_x^k u(x,t), \quad \partial_t^l u(x+\Gamma,t) = \partial_t^l u(x,t). \tag{10.13}$$

(ii) *Applying integration by parts to the periodic functions $u(x,t)$, $v(x,t)$ yields*

$$\begin{aligned} &\int_{x_l}^{x_r} \partial_x^{2k} u(x,t)\cdot v(x,t)\mathrm{d}x = \int_{x_l}^{x_r} u(x,t)\cdot \partial_x^{2k} v(x,t)\mathrm{d}x, \\ &\int_{x_l}^{x_r} \partial_x^{2k+1} u(x,t)\cdot v(x,t)\mathrm{d}x = -\int_{x_l}^{x_r} u(x,t)\cdot \partial_x^{2k+1} v(x,t)\mathrm{d}x, \quad k = 0, 1, 2, \ldots, \end{aligned} \tag{10.14}$$

where the length $x_r - x_l$ of the interval $[x_l, x_r]$ is the period Γ or any nonnegative integer multiple of Γ.

(iii) *For any function $f(\cdot)$, the composite function $f\big(u(x,t)\big)$ is also a periodic function with respect to the variable x, and the fundamental period is Γ.*

10.3 Operator-Variation-of-Constants Formula for Nonlinear Hamiltonian Wave Equations

The next theorem presents the operator-variation-of-constants formula for the nonlinear Hamiltonian wave equation (10.1).

Theorem 10.4 *If $f(u)$ is continuous in (10.1) and Δ is the Laplace operator defined on the subspace $D(\Delta) \subset L^2$, then the exact solution of (10.1) and its derivative satisfy the following equations*

$$
\begin{cases}
u(x,t) = \phi_0\big((t-t_0)^2a^2\Delta\big)u(x,t_0) + (t-t_0)\phi_1\big((t-t_0)^2a^2\Delta\big)u_t(x,t_0) \\
\qquad + \displaystyle\int_{t_0}^{t} (t-\zeta)\phi_1\big((t-\zeta)^2a^2\Delta\big)\tilde{f}(\zeta)\mathrm{d}\zeta, \\
u_t(x,t) = (t-t_0)a^2\Delta\phi_1\big((t-t_0)^2a^2\Delta\big)u(x,t_0) + \phi_0\big((t-t_0)^2a^2\Delta\big)u_t(x,t_0) \\
\qquad + \displaystyle\int_{t_0}^{t} \phi_0\big((t-\zeta)^2a^2\Delta\big)\tilde{f}(\zeta)\mathrm{d}\zeta,
\end{cases}
\tag{10.15}
$$

for $x \in [x_l, x_r]$, t_0, $t \in (-\infty, +\infty)$, *where* $\tilde{f}(\zeta) = f\big(u(x,\zeta)\big)$, *and the bounded operator-argument functions* $\phi_0(\cdot)$ *and* $\phi_1(\cdot)$ *are defined by (10.11).*

Proof (10.15) solves the Eq. (10.1) exactly. In fact, this can be verified by directly inserting the formula (10.15) into the wave equation (10.1). The details of the proof for this theorem can be found in Appendix A of this chapter. □

Moreover, it can be proved that the operator-variation-of-constants formula (10.15) for the nonlinear Hamiltonian wave equation (10.1) is completely consistent with Dirichlet boundary conditions, Neumann boundary conditions, and Robin boundary conditions, respectively, under suitable assumptions. Readers are referred to the recent papers by Wu et al. [37, 39].

Let $u^n(x) = u(x, t_n)$ and $u_t^n(x) = u_t(x, t_n)$ represent the exact solution of (10.1) and its derivative with respect to t at $t = t_n$. It follows immediately from (10.15) with the change of variable $\xi = t_n + hz$ that

$$
\begin{cases}
u^{n+1}(x) = \phi_0\big(h^2a^2\Delta\big)u^n(x) + h\phi_1\big(h^2a^2\Delta\big)u_t^n(x) \\
\qquad + h^2\int_0^1 (1-z)\phi_1\big((1-z)^2h^2a^2\Delta\big)f\big(u(x,t_n+hz)\big)\mathrm{d}z, \\
u_t^{n+1}(x) = ha^2\Delta\phi_1\big(h^2a^2\Delta\big)u^n(x) + \phi_0\big(h^2a^2\Delta\big)u_t^n(x) \\
\qquad + h\int_0^1 \phi_0\big((1-z)^2h^2a^2\Delta\big)f\big(u(x,t_n+hz)\big)\mathrm{d}z,
\end{cases}
\tag{10.16}
$$

where h is a time stepsize.

It is noted that the Eq. (10.15) or (10.16) is not a closed-form solution to the nonlinear Hamiltonian wave equation (10.1), but a nonlinear integral equation. In order to gain an energy-preserving scheme for (10.1), a further analysis based on (10.16) is still required.

10.4 Exact Energy-Preserving Scheme for Nonlinear Hamiltonian Wave Equations

In this section we establish an exact energy-preserving scheme for nonlinear Hamiltonian wave equations.

In light of the operator-variation-of-constants formula (10.16), it is natural to consider the following scheme:

$$\begin{cases} u^{n+1}(x) = \phi_0(h^2a^2\Delta)u^n(x) + h\phi_1(h^2a^2\Delta)u_t^n(x) + h^2 I_1(x), \\ u_t^{n+1}(x) = ha^2\Delta\phi_1(h^2a^2\Delta)u^n(x) + \phi_0(h^2a^2\Delta)u_t^n(x) + hI_2(x), \end{cases} \tag{10.17}$$

where h is the stepsize, and $I_1(x)$, $I_2(x)$ are undetermined functions such that the following condition of energy preservation

$$E(t_{n+1}) = E(t_n) \quad \text{or} \quad H(u^{n+1}, p^{n+1}) = H(u^n, p^n),$$

is satisfied exactly, where $p = u_t$. It follows from (10.17) that

$$\begin{cases} u_x^{n+1}(x) = \phi_0(h^2a^2\Delta)u_x^n(x) + h\partial_x\phi_1(h^2a^2\Delta)u_t^n(x) + h^2\partial_x I_1(x), \\ u_t^{n+1}(x) = ha^2\partial_x\phi_1(h^2a^2\Delta)u_x^n(x) + \phi_0(h^2a^2\Delta)u_t^n(x) + hI_2(x). \end{cases} \tag{10.18}$$

To begin with, we compute

$$H(u^{n+1}, p^{n+1}) = H(u^{n+1}, u_t^{n+1}) = \frac{1}{2}\int_{x_l}^{x_r} \big[(u_t^{n+1}(x))^2 + a^2(u_x^{n+1}(x))^2 + 2V(u^{n+1}(x))\big]\mathrm{d}x. \tag{10.19}$$

Inserting (10.18) into (10.19), a careful calculation yields

$$\begin{aligned} H(u^{n+1}, u_t^{n+1}) &= \frac{1}{2}\int_{x_l}^{x_r} \big[\phi_0(h^2a^2\Delta)u_t^n \cdot \phi_0(h^2a^2\Delta)u_t^n + a^2h\partial_x\phi_1(h^2a^2\Delta)u_t^n \cdot h\partial_x\phi_1(h^2a^2\Delta)u_t^n\big]\mathrm{d}x \\ &+ \frac{a^2}{2}\int_{x_l}^{x_r} \big[ha^2\partial_x\phi_1(h^2a^2\Delta)u_x^n \cdot h\partial_x\phi_1(h^2a^2\Delta)u_x^n + \phi_0(h^2a^2\Delta)u_x^n \cdot \phi_0(h^2a^2\Delta)u_x^n\big]\mathrm{d}x \\ &+ \int_{x_l}^{x_r} \big[ha^2\partial_x\phi_1(h^2a^2\Delta)u_x^n \cdot \phi_0(h^2a^2\Delta)u_t^n + a^2\phi_0(h^2a^2\Delta)u_x^n \cdot h\partial_x\phi_1(h^2a^2\Delta)u_t^n\big]\mathrm{d}x \\ &+ \frac{1}{2}\int_{x_l}^{x_r} \big[\phi_0(h^2a^2\Delta)u_t^n \cdot hI_2(x) + a^2h\partial_x\phi_1(h^2a^2\Delta)u_t^n \cdot h^2\partial_x I_1(x)\big]\mathrm{d}x \\ &+ a^2\int_{x_l}^{x_r} \big[h\partial_x\phi_1(h^2a^2\Delta)u_x^n \cdot hI_2(x) + \phi_0(h^2a^2\Delta)u_x^n(x) \cdot h^2\partial_x I_1(x)\big]\mathrm{d}x \\ &+ \frac{1}{2}\int_{x_l}^{x_r} \big[h^2 I_2(x) \cdot I_2(x) + a^2h^4\partial_x I_1(x) \cdot \partial_x I_1(x)\big]\mathrm{d}x + \int_{x_l}^{x_r} V(u^{n+1}(x))\mathrm{d}x. \end{aligned} \tag{10.20}$$

Applying Theorem 10.3 to (10.20) gives

$$
\begin{aligned}
H(u^{n+1}, u_t^{n+1}) = &\frac{1}{2}\int_{x_l}^{x_r} [\phi_0^2(h^2a^2\Delta) - h^2a^2\Delta\phi_1^2(h^2a^2\Delta)]u_t^n \cdot u_t^n \mathrm{d}x \\
&+ \frac{a^2}{2}\int_{x_l}^{x_r} [\phi_0^2(h^2a^2\Delta) - h^2a^2\Delta\phi_1^2(h^2a^2\Delta)]u_x^n \cdot u_x^n \mathrm{d}x \\
&+ a^2h\int_{x_l}^{x_r} [\partial_x\phi_1(h^2a^2\Delta)\phi_0(h^2a^2\Delta) - \phi_0(h^2a^2\Delta)\partial_x\phi_1(h^2a^2\Delta)]u_x^n \cdot u_t^n \mathrm{d}x \\
&+ \int_{x_l}^{x_r} [\phi_0(h^2a^2\Delta)u_t^n \cdot hI_2(x) + a^2h\partial_x\phi_1(h^2a^2\Delta)u_t^n \cdot h^2\partial_x I_1(x)]\mathrm{d}x \\
&+ a^2\int_{x_l}^{x_r} [h\partial_x\phi_1(h^2a^2\Delta)u_x^n \cdot hI_2(x) + \phi_0(h^2a^2\Delta)u_x^n(x) \cdot h^2\partial_x I_1(x)]\mathrm{d}x \\
&+ \frac{1}{2}\int_{x_l}^{x_r} [h^2I_2(x) \cdot I_2(x) + a^2h^4\partial_x I_1(x) \cdot \partial_x I_1(x)]\mathrm{d}x + \int_{x_l}^{x_r} V(u^{n+1}(x))\mathrm{d}x.
\end{aligned} \tag{10.21}
$$

On noticing Theorem 10.1, (10.21) reduces to

$$
\begin{aligned}
&H(u^{n+1}, u_t^{n+1}) = \frac{1}{2}\int_{x_l}^{x_r} [(u_t^n(x))^2 + a^2(u_x^n(x))^2 + 2V(u^n(x))]\mathrm{d}x \\
&+ \int_{x_l}^{x_r} [\phi_0(h^2a^2\Delta)u_t^n \cdot hI_2(x) + a^2h\partial_x\phi_1(h^2a^2\Delta)u_t^n \cdot h^2\partial_x I_1(x)]\mathrm{d}x \\
&+ a^2\int_{x_l}^{x_r} [h\partial_x\phi_1(h^2a^2\Delta)u_x^n \cdot hI_2(x) + \phi_0(h^2a^2\Delta)u_x^n(x) \cdot h^2\partial_x I_1(x)]\mathrm{d}x \\
&+ \frac{1}{2}\int_{x_l}^{x_r} [h^2I_2(x) \cdot I_2(x) + a^2h^4\partial_x I_1(x) \cdot \partial_x I_1(x)]\mathrm{d}x + \int_{x_l}^{x_r} [V(u^{n+1}(x)) - V(u^n(x))]\mathrm{d}x.
\end{aligned} \tag{10.22}
$$

In what follows, we calculate

$$
\begin{aligned}
V(u^{n+1}) - V(u^n) = &\int_0^1 \mathrm{d}V((1-\tau)u^n + \tau u^{n+1}) = -\int_0^1 (u^{n+1} - u^n) \cdot f((1-\tau)u^n + \tau u^{n+1})\mathrm{d}\tau \\
&\triangleq -(u^{n+1} - u^n) \cdot I_f,
\end{aligned} \tag{10.23}
$$

where

$$
I_f = \int_0^1 f((1-\tau)u^n + \tau u^{n+1})\mathrm{d}\tau.
$$

The first equation of (10.17) gives

$$
\begin{aligned}
u^{n+1}(x) - u^n(x) &= [\phi_0(h^2a^2\Delta) - I]u^n(x) + h\phi_1(h^2a^2\Delta)u_t^n(x) + h^2I_1(x) \\
&= h^2a^2\partial_x\phi_2(h^2a^2\Delta)u_x^n(x) + h\phi_1(h^2a^2\Delta)u_t^n(x) + h^2I_1(x).
\end{aligned} \tag{10.24}
$$

Inserting (10.24) into (10.23) yields

$$
V(u^{n+1}) - V(u^n) = -h^2a^2\partial_x\phi_2(h^2a^2\Delta)u_x^n(x) \cdot I_f - h\phi_1(h^2a^2\Delta)u_t^n(x) \cdot I_f - h^2I_1(x) \cdot I_f. \tag{10.25}
$$

Then the scheme (10.22) can be rewritten by

$$\begin{aligned} H(u^{n+1}, u_t^{n+1}) &= \frac{1}{2}\int_{x_l}^{x_r} [(u_t^n(x))^2 + a^2(u_x^n(x))^2 + 2V(u^n(x))]\mathrm{d}x + \mathscr{R}^n \\ &= H(u^n, u_t^n) + \mathscr{R}^n, \end{aligned} \tag{10.26}$$

where

$$\begin{aligned} \mathscr{R}^n =& h\int_{x_l}^{x_r} [\phi_0(h^2a^2\Delta)u_t^n(x)\cdot I_2(x) + a^2h^2\partial_x\phi_1(h^2a^2\Delta)u_t^n\cdot\partial_x I_1(x) - \phi_1(h^2a^2\Delta)u_t^n(x)\cdot I_f]\mathrm{d}x \\ &+ a^2h^2\int_{x_l}^{x_r} [\partial_x\phi_1(h^2a^2\Delta)u_x^n\cdot I_2(x) + \phi_0(h^2a^2\Delta)u_x^n(x)\cdot\partial_x I_1(x) - \partial_x\phi_2(h^2a^2\Delta)u_x^n(x)\cdot I_f]\mathrm{d}x \\ &+ h^2\int_{x_l}^{x_r} [\frac{1}{2}(I_2(x)\cdot I_2(x) + h^2a^2\partial_x I_1(x)\cdot\partial_x I_1(x)) - I_1(x)\cdot I_f]\mathrm{d}x. \end{aligned} \tag{10.27}$$

It follows from the results of Theorem 10.3 that

$$\begin{aligned} \mathscr{R}^n =& h\int_{x_l}^{x_r} [\phi_0(h^2a^2\Delta)I_2(x) - h^2a^2\Delta\phi_1(h^2a^2\Delta)I_1(x) - \phi_1(h^2a^2\Delta)I_f]\cdot u_t^n(x)\mathrm{d}x \\ &+ a^2h^2\int_{x_l}^{x_r} [\Delta\phi_1(h^2a^2\Delta)I_2(x) - \Delta\phi_0(h^2a^2\Delta)I_1(x) - \Delta\phi_2(h^2a^2\Delta)I_f]\cdot u_x^n(x)\mathrm{d}x \\ &+ h^2\int_{x_l}^{x_r} [\frac{1}{2}(I_2(x)\cdot I_2(x) - h^2a^2\Delta I_1(x)\cdot I_1(x)) - I_1(x)\cdot I_f]\mathrm{d}x. \end{aligned} \tag{10.28}$$

The above analysis gives the following important theorem immediately.

Theorem 10.5 *The scheme (10.17) exactly preserves the energy (10.8) or the Hamiltonian (10.9), i.e.,*

$$E(t_{n+1}) = E(t_n) \quad or \quad H(u^{n+1}, p^{n+1}) = H(u^n, p^n), \quad n = 0, 1, \ldots, \tag{10.29}$$

if and only if $\mathscr{R}^n = 0$.

Based on Theorem 10.1, the following theorem gives a sufficient condition for the scheme (10.17) to be energy-preserving exactly.

Theorem 10.6 *If*

$$\begin{aligned} I_1(x) &= \phi_2(h^2a^2\Delta)\int_0^1 f((1-\tau)u^n(x) + \tau u^{n+1}(x))\mathrm{d}\tau, \\ I_2(x) &= \phi_1(h^2a^2\Delta)\int_0^1 f((1-\tau)u^n(x) + \tau u^{n+1}(x))\mathrm{d}\tau, \end{aligned} \tag{10.30}$$

then the scheme (10.17) exactly preserves the energy (10.8) or the Hamiltonian (10.9).

Proof It is clear from (10.28) that if the following three equations

$$\phi_0(h^2a^2\Delta)I_2(x) - h^2a^2\Delta\phi_1(h^2a^2\Delta)I_1(x) = \phi_1(h^2a^2\Delta)I_f, \tag{10.31}$$

$$\phi_1(h^2a^2\Delta)I_2(x) - \phi_0(h^2a^2\Delta)I_1(x) = \phi_2(h^2a^2\Delta)I_f, \tag{10.32}$$

$$\frac{1}{2}\int_{x_l}^{x_r}(I_2(x)\cdot I_2(x) - h^2a^2\Delta I_1(x)\cdot I_1(x))\mathrm{d}x = \int_{x_l}^{x_r} I_1(x)\cdot I_f\mathrm{d}x, \tag{10.33}$$

are satisfied, then $\mathscr{R}^n = 0$ for $n = 0, 1, \ldots$. Hence, we have

$$E(t_{n+1}) = E(t_n) \quad \text{or} \quad H(u^{n+1}, p^{n+1}) = H(u^n, p^n).$$

The difference of (10.31) times $\phi_1(h^2a^2\Delta)$ and (10.32) times $\phi_0(h^2a^2\Delta)$ is

$$[\phi_0^2(h^2a^2\Delta) - h^2a^2\Delta\phi_1^2(h^2a^2\Delta)]I_1(x) = [\phi_1^2(h^2a^2\Delta) - \phi_0(h^2a^2\Delta)\phi_2(h^2a^2\Delta)]I_f.$$

Likewise, the difference of (10.31) times $\phi_0(h^2a^2\Delta)$ and (10.32) times $h^2a^2\phi_1(h^2a^2\Delta)$ gives

$$\begin{aligned}&[\phi_0^2(h^2a^2\Delta) - h^2a^2\Delta\phi_1^2(h^2a^2\Delta)]I_2(x)\\ &= [\phi_0(h^2a^2\Delta)\phi_1(h^2a^2\Delta) - h^2a^2\Delta\phi_1(h^2a^2\Delta)\phi_2(h^2a^2\Delta)]I_f.\end{aligned}$$

Using Theorem 10.1, we obtain

$$I_1(x) = \phi_2(h^2a^2\Delta)I_f, \quad I_2(x) = \phi_1(h^2a^2\Delta)I_f. \tag{10.34}$$

It can be verified that under the condition (10.34) and Theorem 10.2, the equation (10.33) is also valid. Therefore, (10.30) are sufficient conditions for (10.17) to be an energy-preserving scheme. □

We are now in a position to present the following energy-preserving scheme for Hamiltonian PDEs.

Definition 10.1 The exact energy-preserving scheme for the nonlinear Hamiltonian wave equation (10.1) is defined by

$$\begin{cases} u^{n+1}(x) = \phi_0(h^2a^2\Delta)u^n(x) + h\phi_1(h^2a^2\Delta)u_t^n(x) \\ \quad + h^2\phi_2(h^2a^2\Delta)\displaystyle\int_0^1 f((1-\tau)u^n(x) + \tau u^{n+1}(x))\mathrm{d}\tau, \\ u_t^{n+1}(x) = ha^2\Delta\phi_1(h^2a^2\Delta)u^n(x) + \phi_0(h^2a^2\Delta)u_t^n(x) \\ \quad + h\phi_1(h^2a^2\Delta)\displaystyle\int_0^1 f((1-\tau)u^n(x) + \tau u^{n+1}(x))\mathrm{d}\tau, \end{cases} \tag{10.35}$$

where $h > 0$ is a time stepsize, $\phi_0(h^2a^2\Delta)$, $\phi_1(h^2a^2\Delta)$, and $\phi_2(h^2a^2\Delta)$ are bounded operator-argument functions defined by (10.11).

Since there is a very close similarity between the behaviour of solutions of reversible and Hamiltonian systems [27], in what follows, we show the symmetry of the energy-preserving scheme (10.35).

Theorem 10.7 *The energy-preserving scheme (10.35) is symmetric in time.*

Proof It follows from exchanging $u^{n+1}(x) \leftrightarrow u^n(x),\ u_t^{n+1}(x) \leftrightarrow u_t^n(x)$ and replacing h by $-h$ in (10.35) that

$$\begin{aligned} u^n(x) &= \phi_0(h^2a^2\Delta)u^{n+1}(x) - h\phi_1(h^2a^2\Delta)u_t^{n+1}(x) \\ &\quad + h^2\phi_2(h^2a^2\Delta)\int_0^1 f\big((1-\tau)u^{n+1}(x) + \tau u^n(x)\big)\mathrm{d}\tau, \\ u_t^n(x) &= -ha^2\Delta\phi_1(h^2a^2\Delta)u^{n+1}(x) + \phi_0(h^2a^2\Delta)u_t^{n+1}(x) \\ &\quad - h\phi_1(h^2a^2\Delta)\int_0^1 f\big((1-\tau)u^{n+1}(x) + \tau u^n(x)\big)\mathrm{d}\tau. \end{aligned} \tag{10.36}$$

From (10.36) and Theorem 10.1, it follows that

$$\begin{aligned} u^{n+1}(x) &= \phi_0(h^2a^2\Delta)u^n(x) + h\phi_1(h^2a^2\Delta)u_t^n(x) \\ &\quad + h^2\phi_2(h^2a^2\Delta)\int_0^1 f\big((1-\tau)u^{n+1}(x) + \tau u^n(x)\big)\mathrm{d}\tau, \\ u_t^{n+1}(x) &= ha^2\Delta\phi_1(h^2a^2\Delta)u^n(x) + \phi_0(h^2a^2\Delta)u_t^n(x) \\ &\quad + h\phi_1(h^2a^2\Delta)\int_0^1 f\big((1-\tau)u^{n+1}(x) + \tau u^n(x)\big)\mathrm{d}\tau. \end{aligned} \tag{10.37}$$

Letting $\xi = 1 - \tau$, we have

$$\begin{aligned} &\int_0^1 f\big((1-\tau)u^{n+1}(x) + \tau u^n(x)\big)\mathrm{d}\tau = \int_0^1 f\big(\xi u^{n+1}(x) + (1-\xi)u^n(x)\big)\mathrm{d}\xi \\ &= \int_0^1 f\big((1-\tau)u^n(x) + \tau u^{n+1}(x)\big)\mathrm{d}\tau, \end{aligned}$$

which shows (10.37) is exactly the same as (10.35).

Therefore, the conclusion of the theorem is proved. □

Remark 10.1 The extension of the scheme (10.35) to the general high-dimensional nonlinear Hamiltonian wave equation

$$\begin{cases} u_{tt}(X,t) - a^2\Delta u(X,t) = f\big(u(X,t)\big), & X \in \Omega \subseteq \mathbb{R}^d,\ t_0 < t \le T, \\ u(X,t_0) = \varphi_1(X),\ u_t(X,t_0) = \varphi_2(X), & x \in \Omega \cup \partial\Omega, \end{cases} \tag{10.38}$$

with the corresponding periodic boundary conditions, is straightforward (see [20]), where

$$\Delta = \sum_{i=1}^{d} \partial_{x_i}^2.$$

Remark 10.2 It is noted that the new approach described above for dealing with (10.1) is totally different from classical discrete approximations such as variational methods, and the method of lines (see, e.g. [26]), since the semidiscretisation of the spatial derivative is now avoided. Compared with classical discrete approximations, this approach to solving (10.1) is exact for the space variable x.

Remark 10.3 It can be observed that when the solution of the initial-boundary value problem (10.1) and (10.2) is independent of the spatial variable x, the system (10.1) becomes a Hamiltonian ordinary differential equation and, in this case, (10.35) reduces to the average vector field (AVF) method. Besides, when the spatial interval is divided into a set of finite points with a fixed spatial stepsize via the classical discrete approximations, then the $-\Delta$ is replaced by a symmetric semi-definite positive matrix which is from a discrete operator, such as the second-order central difference operator, (10.35) reduces to the adapted average vector field (AAVF) methods [44]. In other words, the exact energy-preserving scheme (10.35) is an essential extension of AVF to Hamiltonian wave equations based on the operator-variation-of-constants formula (10.15).

Theorem 10.8 *If* $V = V(\alpha u)$, *where* $\alpha \neq 0$, *then*

$$\int_0^1 f((1-\tau)u^n(x) + \tau u^{n+1}(x))\mathrm{d}\tau = \frac{-1}{u^{n+1}(x) - u^n(x)} \Big(V(\alpha u^{n+1}(x)) - V(\alpha u^n(x)) \Big).$$

Proof

$$\begin{aligned}
&\int_0^1 f((1-\tau)u^n(x) + \tau u^{n+1}(x))\mathrm{d}\tau = -\int_0^1 \alpha V'\Big(\alpha\big((1-\tau)u^n(x) + \tau u^{n+1}(x)\big)\Big)\mathrm{d}\tau \\
&= \frac{-\alpha}{\alpha\left(u^{n+1}(x) - u^n(x)\right)} \int_0^1 \frac{\mathrm{d}V\Big(\alpha\big((1-\tau)u^n(x) + \tau u^{n+1}(x)\big)\Big)}{\mathrm{d}\tau}\mathrm{d}\tau \\
&= \frac{-1}{u^{n+1}(x) - u^n(x)} \Big(V\left(\alpha u^{n+1}(x)\right) - V\left(\alpha u^n(x)\right) \Big).
\end{aligned}$$

□

From Theorem 10.8 we obtain the main result of this chapter.

Theorem 10.9 *An exact energy-preserving and symmetric scheme for the nonlinear Hamiltonian wave equation (10.1) is given by*

$$\begin{cases} u^{n+1}(x) = \phi_0(h^2a^2\Delta)u^n(x) + h\phi_1(h^2a^2\Delta)u_t^n(x) - h^2\phi_2(h^2a^2\Delta)J_n(x), \\ u_t^{n+1}(x) = ha^2\Delta\phi_1(h^2a^2\Delta)u^n(x) + \phi_0(h^2a^2\Delta)u_t^n(x) - h\phi_1(h^2a^2\Delta)J_n(x), \end{cases} \tag{10.39}$$

where $h > 0$ *is a time stepsize,* $\phi_0(h^2a^2\Delta)$, $\phi_1(h^2a^2\Delta)$, $\phi_2(h^2a^2\Delta)$ *are bounded operator-argument functions defined by* (10.11), *and*

$$J_n(x) = \frac{V\big(u^{n+1}(x)\big) - V\big(u^n(x)\big)}{u^{n+1}(x) - u^n(x)}. \tag{10.40}$$

Here, it can be observed that, if $u^{n+1}(x) - u^n(x) = 0$, *then* $J_n(x)$ *in* (10.39) *is* $\frac{0}{0}$, *which can be understood as*

$$J_n(x) = \frac{\mathrm{d}V\big(u^n(x)\big)}{\mathrm{d}u} = -f\big(u^n(x)\big). \tag{10.41}$$

Proof The conclusion of the theorem can be proved directly by applying Theorem 10.8 to the energy-preserving scheme (10.35). □

Theorem 10.9 establishes the exact energy-preserving scheme (10.39) for the nonlinear Hamiltonian wave equation (10.1) with the periodic boundary condition (10.2). In the special case $f(u) = \alpha(x)$, that is $V(u) = -\alpha(x)u + \beta(x)$ and $J_n(x) = -\alpha(x)$, the formula (10.39) yields the exact solution of the underlying problem.

If $f(u) = 0$, then (10.1) becomes the homogeneous linear wave equation:

$$\begin{cases} u_{tt} - a^2\Delta u = 0, \\ u(x, t_0) = \varphi_1(x), \ \ u_t(x, t_0) = \varphi_2(x), \end{cases} \tag{10.42}$$

and accordingly, from Theorem 10.1, (10.39) reduces to

$$\begin{cases} u^{n+1}(x) = \phi_0\big(h^2a^2\Delta\big)u^n(x) + h\phi_1\big(h^2a^2\Delta\big)u_t^n(x) \\ \qquad\qquad = \phi_0\big((n+1)^2h^2a^2\Delta\big)\varphi_1(x) + (n+1)h\phi_1\big((n+1)^2h^2a^2\Delta\big)\varphi_2(x), \\ u_t^{n+1}(x) = ha^2\Delta\phi_1\big(h^2a^2\Delta\big)u^n(x) + \phi_0\big(h^2a^2\Delta\big)u_t^n(x) \\ \qquad\qquad = (n+1)ha^2\Delta\phi_1\big((n+1)^2h^2a^2\Delta\big)\varphi_1(x) + \phi_0\big((n+1)^2h^2a^2\Delta\big)\varphi_2(x), \end{cases} \tag{10.43}$$

which exactly integrates the homogeneous linear wave equation (10.42). This implies that (10.43) possesses an additional advantage of energy preservation and quadratic invariant preservation for the homogeneous wave equation (10.42). Besides, compared with the well-known D'Alembert, Poisson and Kirchhoff formulas, *the formula* (10.43) *is independent of the computation of integrals and presents an exact closed-form solution to* (10.42).

10.5 Illustrative Examples

With regard to applications of the scheme (10.39) or (10.43), we now give some illustrative examples.

Problem 10.1 Consider the homogeneous linear wave equation

$$\begin{cases} u_{tt} = u_{xx}, \quad x \in (0, 2), \ t > 0, \\ u(x, 0) = \sin(\pi x), \ u_t(x, 0) = -\dfrac{1}{9}\sin(\pi x), \end{cases} \tag{10.44}$$

subject to the periodic boundary conditions $u(L, t) = u(0, t)$ where the period $L = 2$.

After a careful calculation, it is easy to see that (10.43) directly gives the analytic solution of Problem 10.1 and its derivative

$$\begin{cases} u(x, t) = \sin(\pi x)\cos(\pi t) - \dfrac{1}{9\pi}\sin(\pi x)\sin(\pi t), \\ u_t(x, t) = -\pi\sin(\pi x)\sin(\pi t) - \dfrac{1}{9}\sin(\pi x)\cos(\pi t), \end{cases} \tag{10.45}$$

on noticing that

$$\phi_0(t^2\frac{\partial^2}{\partial x^2})\sin(\pi x) = \sin(\pi x)\cos(\pi t),$$

$$t\phi_1(t^2\frac{\partial^2}{\partial x^2})\big(-\frac{1}{9}\sin(\pi x)\big) = -\frac{1}{9\pi}\sin(\pi x)\sin(\pi t),$$

and

$$t\frac{\partial^2}{\partial x^2}\phi_1(t^2\frac{\partial^2}{\partial x^2})\sin(\pi x) = -\pi\sin(\pi x)\sin(\pi t).$$

Problem 10.2 We consider the following two dimensional homogenous periodic wave equation

$$\begin{cases} u_{tt} - a^2(u_{xx} + u_{yy}) = 0, \ (x, y) \in (0, 2) \times (0, 2), t > 0, \\ u|_{t=0} = \sin(3\pi x)\sin(4\pi y), \ u_t|_{t=0} = 0. \end{cases} \tag{10.46}$$

Applying the formula (10.43) ($\Delta = \partial_x^2 + \partial_y^2$ in this case) to (10.46) leads to

$$\begin{cases} u(x, y, t) = \phi_0\big(t^2a^2\Delta\big)\sin(3\pi x)\sin(4\pi y), \\ u_t(x, y, t) = ta^2\Delta\phi_1\big(t^2a^2\Delta\big)\sin(3\pi x)\sin(4\pi y). \end{cases} \tag{10.47}$$

It follows from a simple calculation that

$$\begin{cases} u(x, y, t) = \sin(3\pi x)\sin(4\pi y)\cos(5t), \\ u_t(x, y, t) = -5\sin(3\pi x)\sin(4\pi y)\sin(5t). \end{cases} \tag{10.48}$$

Problem 10.3 Consider the following non-homogeneous linear periodic wave equation

$$\begin{cases} u_{tt} - u_{xx} = \cos x, \ x \in \left(\frac{\pi}{4}, 2\pi + \frac{\pi}{4}\right), \ t > 0, \\ u|_{t=0} = \sin x, \ u_t|_{t=0} = 0. \end{cases} \tag{10.49}$$

Applying (10.39) to (10.49) gives

$$\begin{cases} u(x,t) = \phi_0(t^2\Delta)\sin x + t^2\phi_2(t^2\Delta)\cos x, \\ u_t(x,t) = t\Delta\phi_1(t^2\Delta)\sin x + t\phi_1(t^2\Delta)\cos x. \end{cases} \tag{10.50}$$

Then a simple calculation yields

$$\begin{cases} u(x,t) = (\sin x - \cos x)\cos t + \cos x, \\ u_t(x,t) = -(\sin x - \cos x)\sin t, \end{cases} \tag{10.51}$$

which is exactly the solution of the problem (10.49).

Remark 10.4 The main purpose of this chapter is to establish a general framework for an exact energy-preserving scheme for nonlinear Hamiltonian wave equations, although we cannot achieve a closed-form solution for the nonlinear Hamiltonian wave equation (10.1). Consequently, we do not consider further computational issues in detail in this chapter.

10.6 Conclusions and Discussions

Energy-preserving schemes have a long history, and can date back to Courant, Friedrichs, and Lewy's work [5]. In this chapter, we considered the properties of energy-preserving schemes and presented an exact energy-preserving scheme for the nonlinear Hamiltonian wave equation (10.1) equipped with the periodic boundary condition (10.2), which is in fact identical to the infinite dimensional nonlinear Hamiltonian system (10.4) or (10.5). We first defined the bounded operator-argument functions (10.11) and analysed their properties, then established an operator-variation-of-constants formula for the nonlinear Hamiltonian wave equation (10.1). The proposed energy-preserving scheme is based on the operator-variation-of-constants formula which avoids the semidiscretisation of the spatial derivative and exactly preserves the energy of the original continuous Hamiltonian wave equation (10.1). This energy-preserving scheme (10.35) is a significant generalisation of the AVF formula and the AAVF formula (see, e.g. [33, 40]) as stated in Remark 10.3, since both the AVF formula and AAVF formula can preserve only the semi-discrete energy of the continuation Hamiltonian PDEs (10.1). In fact, both the AVF formula and AAVF formula are designed specially for Hamiltonian ordinary differential equations. In applications, such Hamiltonian ODEs in time can be obtained from Hamiltonian PDEs by the discretisation of the spatial derivative via classical discrete approximations such as variational methods, and the method of lines. Furthermore, we have also derived an exact energy-preserving and symmetric scheme (10.39) for the nonlinear Hamiltonian wave equation (10.1) with the periodic boundary condition (10.2), which avoids the evaluation of the integral

$$\int_0^1 f\big((1-\tau)u^n(x)+\tau u^{n+1}(x)\big)\mathrm{d}\tau$$

in the exact energy-preserving scheme (10.35). Therefore, *we have in fact derived an exact energy-preserving and symmetric scheme* (10.39) *for the nonlinear Hamiltonian wave equation* (10.1), although the closed form solution to (10.1) is not accessible (even though it exists).

Last but not least, the extension of scheme (10.35) to the general high-dimensional nonlinear wave equation (10.38) is straightforward, as stated in Remark 10.1. All essential analytical features presented for (10.1) are applicable to high-dimensional nonlinear Hamiltonian wave equations (10.38).

It should also be noted that the operator-variation-of-constants formula for wave equations makes it possible to systematically incorporate the inner structure properties of the original continuous system into numerical schemes in the design of structure-preserving integrators for nonlinear wave equations. Chapter 11 will try to demonstrate this point.

The material of this chapter is based on the work by Wu and Liu [36].

References

1. Berti, M.: Nonlinear Oscillations of Hamiltonian PDEs. Springer, Berlin (2007)
2. Biswas, A.: Soliton perturbation theory for phi-four model and nonlinear Klein–Gordon equations. Commun. Nonlinear Sci. Numer. Simul. **14**, 3239–3249 (2009)
3. Celledoni, E., Grimm, V., McLachlan, R.I., McLaren, D.I., O'Neale, D., Owren, B., Quispel, G.R.W.: Preserving energy resp. dissipation in numerical PDEs using the "Average Vector Field" method. J. Comput. Phys. **231**(20), 6770–6789 (2012)
4. Cohen, D., Jahnke, T., Lorenz, K., Lubich, C.: Numerical integrators for highly oscillatory Hamiltonian systems: a review. In: Mielke, A. (ed.) Analysis, Modeling and Simulation of Multiscale Problems, pp. 553–576. Springer, Berlin (2006)
5. Courant, R., Friedrichs, K., Lewy, H.: Über die partiellen Differenzengleichungen der mathematischen Physik, Math. Annal. **100**, 32–74 (1928); reprinted and translated. IBM J. Res. Dev. **11**, 215–234 (1967)
6. Dehghan, M.: Finite difference procedures for solving a problem arising in modeling and design of certain optoelectronic devices. Math. Comput. Simul. **71**, 16–30 (2006)
7. Dehghan, M., Mirezaei, D.: Numerical solution to the unsteady two-dimensional Schrödinger equation using meshless local boundary integral equation method. Int. J. Numer. Methods Eng. **76**, 501–520 (2008)
8. Dehghan, M., Shokri, A.: Numerical solution of the nonlinear KleinCGordon equation using radial basis functions. J. Comput. Appl. Math. **230**, 400–410 (2009)
9. Dodd, R.K., Eilbeck, I.C., Gibbon, J.D., Morris, H.C.: Solitons and Nonlinear Wave Equations. Academic, London (1982)
10. Eilbeck, J.C.: Numerical studies of solitons. In: Bishop, A.R., Schneider, T. (eds.) Solitons and Condensed Matter Physics, pp. 28–43. Springer, New York (1978)

11. Fordy, A.P.: Soliton Theory: A Survey of Results. Manchester University Press (1990)
12. Franco, J.M.: New methods for oscillatory systems based on ARKN methods. Appl. Numer. Math. **56**, 1040–1053 (2006)
13. García-Archilla, B., Sanz-Serna, J.M., Skeel, R.D.: Long-time-step methods for oscillatory differential equations. SIAM J. Sci. Comput. **20**, 930–963 (1998)
14. Hairer, E., Lubich, C.: Long-time energy conservation of numerical methods for oscillatory differential equations. SIAM J. Numer. Anal. **38**, 414–441 (2000)
15. Hairer, E., Lubich, C., Wanner, G.: Geometric Numerical Integration: Structure-Preserving Algorithms for Ordinary Differential Equations, 2nd edn. Springer, Berlin (2006)
16. Hochbruck, M., Lubich, C.: A Gautschi-type method for oscillatory second-order differential equations. Numer. Math. **83**, 403-426 (1999)
17. Hochbruck, M., Ostermann, A.: Exponential integrators. Acta Numer. **19**, 209–286 (2010)
18. Infeld, E., Rowlands, G.: Nonlinear Waves, Solitons and Chaos. Cambridge University Press, New York (1990)
19. Liu, K., Wu, X.Y.: An extended discrete gradient formula for oscillatory Hamiltonian systems. J. Phys. A: Math. Theor. **46**(165203), 19 (2013)
20. Liu, C., Wu, X.Y.: The boundness of the operator-valued functions for multidimensional nonlinear wave equations with applications. Appl. Math. Lett. **74**, 60–67 (2017)
21. Matsuo, T.: New conservative schemes with discrete variational derivatives for nonlinear wave equations. J. Comput. Appl. Math. **203**, 32–56 (2007)
22. Matsuo, T., Yamaguchi, H.: An energy-conserving Galerkin scheme for a class of nonlinear dispersive equations. J. Comput. Phys. **228**, 4346–4358 (2009)
23. McLachlan, R.I., Quispel, G.R.W., Robidoux, N.: Geometric integration using discrete gradients. Philos. Trans. R. Soc. A **357**, 1021–1045 (1999)
24. Quispel, G.R.W., McLaren, D.I.: A new class of energy-preserving numerical integration methods. J. Phys. A **41**(045206), 7 (2008)
25. Ringler, T.D., Thuburn, J., Klemp, J.B., Skamarock, W.C.: A unified approach to energy conservation and potential vorticity dynamics for arbitrarily structured C-grids. J. Comput. Phys. **229**, 3065–3090 (2010)
26. Schiesser, W.: The Numerical Methods Of Lines: Integration Of Partial Differential Equation. Academic Press, San Diego (1991)
27. Sevryuk, M. B., Lectures in Mathematics., 1211, Springer, Berlin (1986)
28. Shakeri, F., Dehghan, M.: Numerical solution of the Klein–Gordon equation via He's variational iteration method. Nonlinear Dyn. **51**, 89–97 (2008)
29. Shi, W., Wu, X.Y., Xia, J.: Explicit multi-symplectic extended leap-frog methods for Hamiltonian wave equations. J. Comput. Phys. **231**, 7671–7694 (2012)
30. Taleei, A., Dehghan, M.: Time-splitting pseudo-spectral domain decomposition method for the soliton solutions of the one and multi-dimensional nonlinear Schrödinger equations. Comput. Phys. Commun. **185**, 1515–1528 (2014)
31. Van de Vyver, H.: Scheifele two-step methods for perturbed oscillators. J. Comput. Appl. Math. **224**, 415–432 (2009)
32. Wang, B., Liu, K., Wu, X.Y.: A Filon-type asymptotic approach to solving highly oscillatory second-order initial value problems. J. Comput. Phys. **243**, 210–223 (2013)
33. Wang, B., Wu, X.Y.: A new high precision energy-preserving integrator for system of oscillatory second-order differential equations. Phys. Lett. A **376**, 1185–1190 (2012)
34. Wang, B., Iserles, A., Wu, X.Y.: Arbitrary-order trigonometric Fourier collocation methods for multi-frequency oscillatory systems. Found. Comput. Math. **16**, 151–181 (2016)
35. Wazwaz, A.M.: New travelling wave solutions to the Boussinesq and the Klein–Gordon equations. Commun. Nonlinear Sci. Numer. Simul. **13**, 889–901 (2008)
36. Wu, X.Y., Liu, C.: An energy-preserving and symmetric scheme for nonlinear Hamiltonian wave equations. J. Math. Anal. Appl. **440**, 167–182 (2016)
37. Wu, X.Y., Liu, C.: An integral formula adapted to different boundary conditions for arbitrarily high-dimensional nonlinear Klein–Gordon equations with its applications. J. Math. Phys. **57**, 021504 (2016)

38. Wu, X.Y., Liu, C., Mei, L.J.: A new framework for solving partial differential equations using semi-analytical explicit RK(N)-type integrators. J. Comput. Appl. Math. **301**, 74–90 (2016)
39. Wu, X.Y., Mei, L.J., Liu, C.: An analytical expression of solutions to nonlinear wave equations in higher dimensions with Robin boundary conditions. J. Math. Anal. Appl. **426**, 1164–1173 (2015)
40. Wu, X.Y., Wang, B., Shi, W.: Efficient energy-preserving integrators for oscillatory Hamiltonian systems. J. Comput. Phys. **235**, 587–605 (2013)
41. Wu, X.Y., You, X., Xia, J.: Order conditions for ARKN methods solving oscillatory systems. Comput. Phys. Commun. **180**, 2250–2257 (2009)
42. Wu, X.Y., Wang, B., Xia, J.: Explicit symplectic multidimensional exponential fitting modified Runge–Kutta–Nystrom methods. BIT Numer. Math. **52**, 773–795 (2012)
43. Wu, X.Y., You, X., Shi, W., Wang, B.: ERKN integrators for systems of oscillatory second-order differential equations. Comput. Phys. Commun. **181**, 1873–1887 (2010)
44. Wu, X.Y., You, X., Wang, B.: Structure-Preserving Algorithms for Oscillatory Differential Equations. Springer, Berlin (2013)

Chapter 11
Arbitrarily High-Order Time-Stepping Schemes for Nonlinear Klein–Gordon Equations

This chapter presents arbitrarily high-order time-stepping schemes for solving high-dimensional nonlinear Klein–Gordon equations with different boundary conditions. We first formulate an abstract ordinary differential equation (ODE) on a suitable infinite–dimensional function space based on the operator spectrum theory. We then introduce an operator-variation-of-constants formula for the nonlinear abstract ODE. The nonlinear stability and convergence are rigorously analysed once the spatial differential operator is approximated by an appropriate positive semi-definite matrix. With regard to the two dimensional Dirichlet or Neumann boundary problems, the time-stepping schemes coupled with *discrete Fast Sine/Cosine Transformation* can be applied to simulate the two-dimensional nonlinear Klein–Gordon equations effectively. The numerical results demonstrate the advantage of the schemes in comparison with the existing numerical methods for solving nonlinear Klein–Gordon equations in the literature.

11.1 Introduction

The computation of the Klein–Gordon equation which has a nonlinear potential function, is of great importance in a wide range of application areas in science and engineering. The nonlinear potential gives rise to major challenges. In this chapter, we begin with the following nonlinear Klein–Gordon equation in a single space variable:

$$\begin{cases} u_{tt} - a^2 \Delta u = f(u), \quad t_0 < t \leq T, \ x \in \Omega, \\ u(x, t_0) = \varphi_1(x), \ u_t(x, t_0) = \varphi_2(x), \ x \in \bar{\Omega}, \end{cases} \tag{11.1}$$

and suppose that the initial valued problem (11.1) is supplemented with the following periodic boundary condition on the domain $\Omega = (-\pi, \pi)$

X. Wu and B. Wang, *Recent Developments in Structure-Preserving Algorithms for Oscillatory Differential Equations*, https://doi.org/10.1007/978-981-10-9004-2_11

$$u(x,t) = u(x+2\pi, t), \tag{11.2}$$

where $u(x,t)$ represents the wave displacement at position x and time t, $\Delta = \frac{\partial^2}{\partial x^2}$, and $f(u)$ is a nonlinear function of u chosen as the negative derivative of a potential energy $V(u) \geq 0$. Generally, there are various choices of the potential function $f(u)$ to investigate solitons and nonlinear phenomena. For instance, the following sine–Gordon equation

$$u_{tt} - a^2 \Delta u + \sin(u) = 0, \tag{11.3}$$

is well known, and other nonlinear potential functions also appear in the literature such as $f(u) = \sinh u$ and polynomial $f(u)$. Moreover, if $u(\cdot,t) \in H^1(\Omega)$ and $u_t(\cdot,t) \in L^2(\Omega)$, energy conservation becomes another key feature of the Klein–Gordon equation, i.e.,

$$E(t) \equiv \frac{1}{2} \int_{\Omega} \big(u_t^2 + a^2|\nabla u|^2 + 2V(u)\big)\mathrm{d}x = E(t_0). \tag{11.4}$$

This is an essential property in the theory of solitons. Accordingly, it is also significant to test the effectiveness of a numerical method in preserving the corresponding discrete energy.

In a wide variety of application areas in science and engineering, such as nonlinear optics, solid state physics and quantum field theory [10, 23, 53], the nonlinear wave equation plays an important role and has been extensively investigated. In particular, the nonlinear Klein–Gordon equation (11.1) is used to model many different nonlinear phenomena, including the propagation of dislocations in crystals and the behavior of elementary particles and of Josephson junctions (see Chap. 8.2 in [24] for details). Its description and understanding are very important from both the analytical and numerical aspects, and have been investigated by many researchers. On the analytical front, the Cauchy problem was investigated (see, e.g. [7, 13, 26, 36]). If the potential function satisfies $V(u) \geq 0$ for $u \in \mathbb{R}$, the global existence of solutions for the defocusing case, was established in [13], whereas if the energy potential satisfies $V(u) \leq 0$ for $u \in \mathbb{R}$, the focusing case, possible finite time blow-up was shown in [7]. With regard to the numerical methods, there have been proposed and studied a variety of solution procedures for solving the nonlinear Klein–Gordon equation. For instance, the energy-preserving explicit, semi-implicit and symplectic conservative standard finite difference time domain (FDTD) discretisations were proposed and analysed in [1, 9, 25, 38, 44]. As far as the finite-difference method is concerned, on the basis of standard finite-difference approximations, a three-time-level scheme was derived by Strauss and Vázquez in [47]. Jiménez [35] derived conservative finite difference schemes with some analogous discretisations to that used in [47] for the nonlinear term. Other approaches, such as the finite element method and the spectral method, were also studied in [17, 18, 27, 50]. With respect to finite-element techniques, Tourigny [50] proved that the use of product approximations in Galerkin methods subject to Dirichlet boundary conditions does not affect the convergence rate of the method. Guo et al. [27] developed a conservative Legendre spectral method. Dehghan et al. used radial basis functions, the dual reciprocity boundary integral equation technique, the collocation and finite-difference collocation methods for solving the nonlinear Klein–Gordon equations, or coupled Klein–Gordon equations (see, e.g. [20–22, 37]). Although many numerical methods have been derived and investigated for solving the nonlinear Klein–Gordon equation in the literature, in general, the existing numerical methods have limited accuracy, and little attention was paid to the special structure brought by spatial discretisations. This motivates the main theme of this chapter, which is to consider arbitrarily high-order Lagrange collocation-type time-stepping schemes for efficiently solving nonlinear Klein–Gordon equations.

The plan of this chapter is as follows. In Sect. 11.2, based on the operator spectrum theory, we first formulate the one-dimensional nonlinear Klein–Gordon equation (11.1)–(11.2) as an abstract second-order ordinary differential equation on an infinite-dimensional Hilbert space $L^2(\Omega)$. Then, the operator-variation-of-constants formula for the abstract equation is introduced, which is in fact an integral equation of the solution for the nonlinear Klein–Gordon equation (11.1)–(11.2). In Sect. 11.3, using the derived operator-variation-of-constants formula, we calculate the nonlinear integrals appearing in this formula by Lagrange interpolation. This leads to a class of arbitrarily high-order Lagrange collocation-type time-stepping schemes. Furthermore, an investigation of the local error bounds is made, which arrives at the simplified order conditions in a much simpler form. Section 11.4 is devoted to semidiscretisation. This process enables us to take advantage of the properties of the undiscretised differential operator $\mathscr{A}$ and incorporate the special structure introduced by spatial discretisations with the new integrators. The main theoretical results of this work are presented in Sect. 11.5. We use the strategy of energy analysis to study the nonlinear stability and convergence of the fully discrete scheme. Since these fully discrete schemes are implicit, iterative solutions are required in practical computations. Therefore, we use fixed-point iteration and analyse its convergence in this section. In Sect. 11.6 we apply the Lagrange collocation-type time integrators to the two-dimensional nonlinear Klein–Gordon equations, equipped with homogenous Dirichlet or Neumann boundary conditions. In a similar way to the one-dimensional periodic boundary case, the abstract ordinary differential equations and the operator-variation-of-constants formula are established on the infinite-dimensional Hilbert space $L^2(\Omega)$. In Sect. 11.7, we are concerned with numerical experiments, and the numerical results show the advantage and effectiveness of our new schemes in comparison with the existing numerical methods in the literature. The last section is devoted to brief conclusions and discussions.

In this chapter, all essential features of the methodology are presented in the one-dimensional and two-dimensional cases, although the schemes to be analysed lend themselves with equal ease to higher dimensions.

11.2 Abstract Ordinary Differential Equation

Motivated by recent interest in exponential integrators for semilinear parabolic problems [30–32], and based on the operator spectrum theory (see, e.g. [6]), we will formulate the nonlinear problem (11.1)–(11.2) as an abstract ordinary differential equation on the Hilbert space $L^2(\Omega)$, and introduce an operator-variation-of-constants formula. To this end, some bounded operator-argument functions will be defined and analysed in advance, because these are essential to introducing the operator-variation-of-constants formula.

To begin with, we define the functions

$$\phi_j(x) := \sum_{k=0}^{\infty} \frac{(-1)^k x^k}{(2k+j)!}, \quad j = 0, 1, 2, \ldots, \quad \forall x \geq 0. \tag{11.5}$$

It can be observed that $\phi_j(x)$, $j = 0, 1, 2, \ldots$ are bounded functions for any $x \geq 0$. For example, we have

$$\phi_0(x) = \cos(\sqrt{x}), \quad \phi_1(x) = \frac{\sin(\sqrt{x})}{\sqrt{x}}, \tag{11.6}$$

with $\phi_1(0) = 1$, and it is obvious that $|\phi_j(x)| \leq 1$ for $j = 0, 1$ and $\forall x \geq 0$. For an abstract formulation of problem (11.1)–(11.2), we define the linear differential operator $\mathscr{A}$ by

$$(\mathscr{A}v)(x) = -a^2 v_{xx}(x). \tag{11.7}$$

It is known that the linear differential operator $\mathscr{A}$ is an unbounded operator and not defined for every $v \in L^2(\Omega)$. In order to model the periodic boundary condition (11.2), we consider $\mathscr{A}$ on the domain:

$$D(\mathscr{A}) := \left\{ v \in H^2(\Omega) : v(x) = v(x + 2\pi) \right\}. \tag{11.8}$$

Obviously, the defined operator $\mathscr{A}$ is positive semi definite, i.e.,

$$\Big(\mathscr{A}v(x), v(x)\Big) = \int_0^{2\pi} \mathscr{A}v(x) \cdot v(x)\mathrm{d}x = a^2 \int_0^{2\pi} v_x^2(x)\mathrm{d}x \geq 0, \quad \forall v(x) \in D(\mathscr{A}).$$

Here, $(\cdot, \cdot)$ denotes the inner product of $L^2(\Omega)$ and integration by parts, or Green's formula, has been used. Moreover, we note the important fact that the operator $\mathscr{A}$ has a complete system of orthogonal eigenfunctions $\left\{ e^{ikx} : k \in \mathbb{Z} \right\}$ in the Hilbert space $L^2(\Omega)$, and the corresponding eigenvalues are given by a^2k^2, $k = 0, \pm 1, \pm 2, \ldots$ (see, e.g. [49]). From the isomorphism between $L^2(\Omega)$ and $l^2 = \left\{ x = (x_i)_{i \in \mathbb{Z}} : \sum_{i \in \mathbb{Z}} |x_i|^2 < +\infty \right\}$, the operator $\mathscr{A}$ induces a corresponding operator on l^2

(see, e.g. [6, 32]). Then, it can be observed that the functions (11.5) imply the operator functions

$$\phi_j(t\mathscr{A}) : L^2(\Omega) \to L^2(\Omega),$$

for $j = 0, 1, 2, \dots$ and $t \geq t_0$ as follows:

$$\phi_j(t\mathscr{A})v(x) = \sum_{k=-\infty}^{\infty} \hat{v}_k \phi_j(ta^2k^2)e^{ikx}, \quad \text{for} \quad v(x) = \sum_{k=-\infty}^{\infty} \hat{v}_k e^{ikx}. \tag{11.9}$$

We next show that the defined operator functions are bounded. To do this, we need to clarify the norm of the function in $L^2(\Omega)$, which can be characterised in the frequency space by

$$\|v\|^2 = 2\pi \sum_{k=-\infty}^{\infty} |\hat{v}_k|^2. \tag{11.10}$$

The details can be found in [46]. Therefore, we have

$$\|\phi_j(t\mathscr{A})\|^2_{L^2(\Omega)\leftarrow L^2(\Omega)} = \sup_{\|v\|\neq 0} \frac{\|\phi_j(t\mathscr{A})v\|^2}{\|v\|^2} \leq \sup_{t\geq t_0} |\phi_j(ta^2k^2)| \leq \gamma_j, \tag{11.11}$$

where γ_j are bounds on the functions $|\phi_j(x)|$ for $j = 0, 1, 2, \dots$ and $x \geq 0$. For instance, we may choose $\gamma_0 = \gamma_1 = 1$ and then

$$\|\phi_0(t\mathscr{A})\|^2_{L^2(\Omega)\leftarrow L^2(\Omega)} \leq 1 \quad \text{and} \quad \|\phi_1(t\mathscr{A})\|^2_{L^2(\Omega)\leftarrow L^2(\Omega)} \leq 1. \tag{11.12}$$

By defining $u(t)$ as the function that maps x to $u(x, t)$:

$$u(t) = [x \mapsto u(x, t)],$$

we now can formulate the systems (11.1)–(11.2) as the following abstract ordinary differential equation on the Hilbert space $L^2(\Omega)$:

$$\begin{cases} u''(t) + \mathscr{A}u(t) = f\big(u(t)\big) \\ u(t_0) = \varphi_1(x), \; u'(t_0) = \varphi_2(x). \end{cases} \tag{11.13}$$

With this premise, we are now in a position to present an integral formula for the nonlinear Klein–Gordon equation (11.1)–(11.2). The solution of the abstract ordinary differential equations (11.13) can be given by the operator-variation-of-constants formula summarised in the following theorem.

Theorem 11.1 *The solution of* (11.13) *and its derivative satisfy*

$$\begin{cases} u(t) = \phi_0\big((t-t_0)^2\mathscr{A}\big)u(t_0) + (t-t_0)\phi_1\big((t-t_0)^2\mathscr{A}\big)u'(t_0) \\ \qquad + \displaystyle\int_{t_0}^{t} (t-\zeta)\phi_1\big((t-\zeta)^2\mathscr{A}\big)f\big(u(\zeta)\big)\mathrm{d}\zeta, \\ u'(t) = -(t-t_0)\mathscr{A}\phi_1\big((t-t_0)^2\mathscr{A}\big)u(t_0) + \phi_0\big((t-t_0)^2\mathscr{A}\big)u'(t_0) \\ \qquad + \displaystyle\int_{t_0}^{t} \phi_0\big((t-\zeta)^2\mathscr{A}\big)f\big(u(\zeta)\big)\mathrm{d}\zeta, \end{cases} \tag{11.14}$$

for $t \in [t_0, T]$, *where* $\phi_0\big((t-t_0)^2\mathscr{A}\big)$ *and* $\phi_1\big((t-t_0)^2\mathscr{A}\big)$ *are bounded functions of the operator* $\mathscr{A}$.

Proof Applying the *Duhamel Principle* to Eqs. (11.1) or (11.13), we have

$$\begin{pmatrix} u(t) \\ u'(t) \end{pmatrix} = e^{\mathscr{J}(t-t_0)} \begin{pmatrix} u(t_0) \\ u'(t_0) \end{pmatrix} + \int_{t_0}^{t} e^{\mathscr{J}(t-\zeta)} \begin{pmatrix} 0 \\ f\big(u(\zeta)\big) \end{pmatrix} \mathrm{d}\zeta, \tag{11.15}$$

where

$$\mathscr{J} = \begin{pmatrix} 0 & I \\ -\mathscr{A} & 0 \end{pmatrix}.$$

After expanding the exponential operator through its Taylor series, we obtain

$$e^{\mathscr{J}(t-t_0)} = \sum_{k=0}^{+\infty} \frac{\mathscr{J}^k (t-t_0)^k}{k!}.$$

An argument by induction leads to the following results

$$\mathscr{J}^k = (-1)^{\lfloor k/2 \rfloor} \begin{pmatrix} \frac{1+(-1)^k}{2}\mathscr{A}^{\lfloor k/2 \rfloor} & \frac{1-(-1)^k}{2}\mathscr{A}^{\lfloor k/2 \rfloor} \\ -\frac{1-(-1)^k}{2}\mathscr{A}^{\lfloor k/2 \rfloor + 1} & \frac{1+(-1)^k}{2}\mathscr{A}^{\lfloor k/2 \rfloor} \end{pmatrix}, \quad \forall k \in \mathbb{N},$$

where $\lfloor k/2 \rfloor$ denotes the integer part of $k/2$. According to the definition of $\phi_j(\mathscr{A})$ and a careful calculation, we obtain

$$e^{\mathscr{J}(t-t_0)} = \begin{pmatrix} \phi_0\big((t-t_0)^2\mathscr{A}\big) & (t-t_0)\phi_1\big((t-t_0)^2\mathscr{A}\big) \\ -(t-t_0)\mathscr{A}\phi_1\big((t-t_0)^2\mathscr{A}\big) & \phi_0\big((t-t_0)^2\mathscr{A}\big) \end{pmatrix}.$$

The conclusion of the theorem can be obtained straightforwardly by inserting the expansion into (11.15). □

Remark 11.1 Although equation (11.1) is one-dimensional in space, the method of analysis introduced in this section can be extended to the considerably more important high-dimensional Klein–Gordon equations

$$u_{tt} - a^2 \Delta u = f(u), \quad t \geq t_0, \quad \mathbf{x} \in [-\pi, \pi]^d, \tag{11.16}$$

where $u = u(\mathbf{x}, t)$ and $\Delta = \sum_{i=1}^{d} \frac{\partial^2}{\partial x_i^2}$, with periodic boundary conditions. In latter case, if we define the operator as the form $\mathscr{A} = -a^2\Delta$, the same operator-variation-of-constants formula as (11.14) for (11.16) can be achieved as well. An application of this approach can be found in a recent paper [56].

Remark 11.2 For the nonlinear Klein–Gordon equation, the formula (11.14) is a nonlinear integral equation which reflects the changes of the solution with time t. It will be helpful in deriving and analysing novel numerical integrators for the nonlinear Klein–Gordon equations. However, if the right-hand function f does not depend on u, i.e.,

$$\begin{cases} u_{tt} - a^2\Delta u = f(\mathbf{x}, t), \quad t_0 < t \le T, \ \mathbf{x} \in \Omega, \\ u(\mathbf{x}, t_0) = \varphi_1(\mathbf{x}), \ u_t(\mathbf{x}, t_0) = \varphi_2(\mathbf{x}), \end{cases} \tag{11.17}$$

this is a linear or homogenous ($f(\mathbf{x}, t) = 0$) wave equation. The closed-form solution to the linear problem (11.17) can be obtained by using the operator-variation-of-constants formula.

As an illustrative example, we consider the following two-dimensional homogenous periodic wave equation

$$\begin{cases} u_{tt} - a^2(u_{xx} + u_{yy}) = 0, \ (x, y) \in (0, 2) \times (0, 2), t > 0, \\ u|_{t=0} = \sin(3\pi x)\sin(4\pi y), \ u_t|_{t=0} = 0. \end{cases} \tag{11.18}$$

The homogeneous problem is equipped with periodic boundary conditions

$$u(x + L_x, y, t) = u(x, y + L_y, t) = u(x, y, t) \tag{11.19}$$

with the fundamental periods $L_x = \frac{2}{3}$ and $L_y = \frac{1}{2}$. Applying the formula (11.14) to (11.18) leads to

$$\begin{cases} u(x, y, t) = \phi_0(t^2\mathscr{A})\sin(3\pi x)\sin(4\pi y), \\ u_t(x, y, t) = ta^2\Delta\phi_1(t^2\mathscr{A})\sin(3\pi x)\sin(4\pi y). \end{cases} \tag{11.20}$$

It follows from a simple calculation that

$$\begin{cases} u(x, y, t) = \sin(3\pi x)\sin(4\pi y)\cos(5t), \\ u_t(x, y, t) = -5\sin(3\pi x)\sin(4\pi y)\sin(5t), \end{cases} \tag{11.21}$$

which is exactly the solution of problem (11.18) and its derivative.

We next consider the following nonhomogeneous linear wave equation

$$\begin{cases} u_{tt} - (u_{xx} + u_{yy}) = \pi^2\sin(\pi(x - t))\sin(\pi y), \\ u|_{t=0} = \sin(\pi x)\sin(\pi y), \ u_t|_{t=0} = -\pi\cos(\pi x)\sin(\pi y), \end{cases} \tag{11.22}$$

and suppose that the problem is subject to the periodic boundary conditions

$$u(x+L,y,t)=u(x,y+L,t)=u(x,y,t), \tag{11.23}$$

with the fundamental periods $L=2$. Applying formula (11.14) to (11.22) yields

$$\begin{cases} u(x,y,t) = \phi_0(t^2\Delta)\sin(\pi x)\sin(\pi y)-\pi t\phi_1(t^2\Delta)\cos(\pi x)\sin(\pi y) \\ \qquad +\pi^2\int_0^t (t-\zeta)\phi_1((t-\zeta)^2\Delta)\sin(\pi(x-\zeta))\sin(\pi y)d\zeta, \\ u_t(x,y,t) = t\Delta\phi_1(t^2\Delta)\sin(\pi x)\sin(\pi y)-\pi\phi_0(t^2\Delta)\cos(\pi x)\sin(\pi y) \\ \qquad +\pi^2\int_0^t \phi_0((t-\zeta)^2\Delta)\sin(\pi(x-\zeta))\sin(\pi y)d\zeta. \end{cases} \tag{11.24}$$

It follows from a careful calculation that

$$\begin{aligned} &\phi_0(t^2\Delta)\sin(\pi x)\sin(\pi y)=\sin(\pi x)\sin(\pi y)\cos(\sqrt{2}\pi t), \\ &-\pi t\phi_1(t^2\Delta)\cos(\pi x)\sin(\pi y)=-\frac{1}{\sqrt{2}}\cos(\pi x)\sin(\pi y)\sin(\sqrt{2}\pi t), \\ &\pi^2\int_0^t (t-\zeta)\phi_1((t-\zeta)^2\Delta)\sin(\pi(x-\zeta))\sin(\pi y)d\zeta \\ &=\frac{\pi}{\sqrt{2}}\int_0^t \sin(\sqrt{2}\pi(t-\zeta))\sin(\pi(x-\zeta))\sin(\pi y)d\zeta. \end{aligned}$$

We finally obtain the exact solution

$$\begin{aligned} u(x,y,t) &= \sin(\pi x)\sin(\pi y)\cos(\sqrt{2}\pi t)-\frac{1}{\sqrt{2}}\cos(\pi x)\sin(\pi y)\sin(\sqrt{2}\pi t) \\ &\quad +\frac{\pi}{\sqrt{2}}\int_0^t \sin(\sqrt{2}\pi(t-\zeta))\sin(\pi(x-\zeta))\sin(\pi y)d\zeta \\ &= \sin(\pi(x-t))\sin(\pi y), \end{aligned} \tag{11.25}$$

and its derivative

$$u_t(x,y,t)=-\pi\cos(\pi(x-t))\sin(\pi y). \tag{11.26}$$

11.3 Formulation of the Lagrange Collocation-Type Time Integrators

In light of the useful approach to dealing with the semiclassical Schrödinger equation (see [8]), this analysis will omit the standard steps of first semidiscretising in space and then approximating the semidiscretisation. In this section, based on the formula (11.14), we devote ourselves to constructing arbitrarily high-order Lagrange collocation-type time integrators for the nonlinear system (11.13) in the infinite-dimensional Hilbert space $L^2(\Omega)$. Furthermore, the local error bounds for the constructed time integrators will also be considered in detail.

11.3.1 Construction of the Time Integrators

From Theorem 11.1, the solution of (11.13) and its derivative at time $t_{n+1} = t_n + \Delta t$ for $n = 0, 1, 2, \ldots$ are given by

$$\begin{cases} u(t_{n+1}) = \phi_0(\mathscr{V})u(t_n) + \Delta t \phi_1(\mathscr{V})u'(t_n) \\ \qquad + \Delta t^2 \displaystyle\int_0^1 (1-z)\phi_1\big((1-z)^2\mathscr{V}\big)\tilde{f}(t_n + z\Delta t)\mathrm{d}z, \\ u'(t_{n+1}) = -\Delta t \mathscr{A}\phi_1(\mathscr{V})u(t_n) + \phi_0(\mathscr{V})u'(t_n) \\ \qquad + \Delta t \displaystyle\int_0^1 \phi_0\big((1-z)^2\mathscr{V}\big)\tilde{f}(t_n + z\Delta t)\mathrm{d}z, \end{cases} \tag{11.27}$$

where $\mathscr{V} = \Delta t^2 \mathscr{A}$ and $\tilde{f}(t_n + z\Delta t) = f\big(u(t_n + z\Delta t)\big)$.

In what follows, we pay our attention to deriving efficient methods for approximating the following two nonlinear integrals:

$$\begin{aligned} I_1 &:= \int_0^1 (1-z)\phi_1\big((1-z)\mathscr{V}\big)\tilde{f}(t_n + z\Delta t)\mathrm{d}z, \\ I_2 &:= \int_0^1 \phi_0\big((1-z)\mathscr{V}\big)\tilde{f}(t_n + z\Delta t)\mathrm{d}z. \end{aligned} \tag{11.28}$$

We choose non-confluent collocation nodes $c_1, \ldots, c_s$ and approximate the function $\tilde{f}(t_n + z\Delta t)$ involved in the integrals in (11.28) by its Lagrange interpolation polynomial at these quadrature nodes

$$\begin{aligned} \tilde{f}(t_n + z\Delta t) &= \sum_{i=1}^{s} l_i(z)\tilde{f}(t_n + c_i\Delta t) + R_s(t_n + z\Delta t) \\ &= \sum_{i=1}^{s} l_i(z) f\big(u(t_n + c_i\Delta t)\big) + R_s(t_n + z\Delta t). \end{aligned} \tag{11.29}$$

Here, $l_i(z)$ for $i = 1, 2, \ldots, s$ are the well-known Lagrange basis polynomials

$$l_i(z) = \prod_{\substack{j=1 \\ j\neq i}}^{s} \frac{z - c_j}{c_i - c_j}, \quad i = 1, 2, \ldots, s. \tag{11.30}$$

It is obvious that there exists a constant β satisfies $\max\limits_{1\le i\le s}\max\limits_{0\le z\le 1}|l_i(z)|\le\beta$. Moreover, the interpolation error on [0, 1] is given by

$$\begin{aligned} R_s(t_n+z\Delta t) &= \tilde{f}(t_n+z\Delta t)-\sum_{i=1}^{s} l_i(z)\tilde{f}(t_n+c_i\Delta t) \\ &= \frac{\Delta t^s}{s!}\tilde{f}_t^{(s)}(t_n+\xi^n\Delta t)w_s(z), \quad \xi^n\in(0,1), \end{aligned} \tag{11.31}$$

where $w_s(z)=\prod\limits_{i=1}^{s}(z-c_i)$ and $\tilde{f}_t^{(j)}(t)$ denotes the jth order derivative of $f\big(u(t)\big)$ with respect to t.

Suppose that the following approximations have been given:

$$u^n\approx u(t_n), \quad U^{ni}\approx u(t_n+c_i\Delta t).$$

Replacing $\tilde{f}(z)$ in (11.27) by the Lagrange interpolation (11.29) yields approximations to the exact solution and its derivative at time t_{n+1}

$$u^{n+1} = \phi_0(\mathscr{V})u^n+\Delta t\phi_1(\mathscr{V})u'^n+\Delta t^2\sum_{i=1}^{s} b_i(\mathscr{V})f(U^{ni}), \tag{11.32}$$

$$u'^{n+1} = -\Delta t\mathscr{A}\phi_1(\mathscr{V})u^n+\phi_0(\mathscr{V})u'^n+\Delta t\sum_{i=1}^{s}\bar{b}_i(\mathscr{V})f(U^{ni}), \tag{11.33}$$

where $b_i(\mathscr{V})$ and $\bar{b}_i(\mathscr{V})$ are determined by

$$b_i(\mathscr{V}) = \int_0^1 (1-z)\phi_1\big((1-z)^2\mathscr{V}\big)l_i(z)dz, \tag{11.34}$$

$$\bar{b}_i(\mathscr{V}) = \int_0^1 \phi_0\big((1-z)^2\mathscr{V}\big)l_i(z)dz. \tag{11.35}$$

It follows from (11.12) that

$$\|b_i(\mathscr{V})\|_{L^2(\Omega)\leftarrow L^2(\Omega)}\le\max_{0\le z\le 1}|l_i(z)|\le\beta \text{ and } \|\bar{b}_i(\mathscr{V})\|_{L^2(\Omega)\leftarrow L^2(\Omega)}\le\max_{0\le z\le 1}|l_i(z)|\le\beta,$$

and this means that $b_i(\mathscr{V})$ and $\bar{b}_i(\mathscr{V})$ are uniformly bounded.

Moreover, we note that the basis $l_i(z)$ for $i=1,\ldots,s$ are polynomials of degree at most $s-1$, the coefficients $b_i(\mathscr{V})$ are linear combinations of the functions

$$\int_0^1 (1-z)\phi_1\big((1-z)^2\mathscr{V}\big)z^j dz = \Gamma(j+1)\phi_{j+2}(\mathscr{V}), \tag{11.36}$$

and the coefficients $\bar{b}_i(\mathscr{V})$ are linear combinations of the functions

$$\int_0^1 \phi_0\big((1-z)^2\mathscr{V}\big)z^j \mathrm{d}z = \Gamma(j+1)\phi_{j+1}(\mathscr{V}), \tag{11.37}$$

where $\Gamma(j+1)$ is the Gamma function with $\Gamma(1) = 1$ (see, e.g. Abramowitz and Stegun [3]). Recall that, the Gamma function $\Gamma(j+1)$ satisfies the following recursion

$$\Gamma(j+1) = j\Gamma(j) = \cdots = j!.$$

It remains to determine the approximation U^{ni}. In a similar way to the formula (11.32), we replace Δt by $c_i \Delta t$ to define the internal stages:

$$U^{ni} = \phi_0\big(c_i^2\mathscr{V}\big)u^n + c_i\Delta t\phi_1\big(c_i^2\mathscr{V}\big)u'^n + c_i^2\Delta t^2\sum_{j=1}^{s} a_{ij}(\mathscr{V})f(U^{nj}), \tag{11.38}$$

where it is required that the weights $a_{ij}(\mathscr{V})$ are uniformly bounded. The weights $a_{ij}(\mathscr{V})$ will be determined by suitable order conditions, and we will derive these order conditions in Sect. 11.3.2.

On the basis of the above analysis and the formula (11.27), we present the following Lagrange collocation-type time-stepping integrators for the nonlinear system (11.13).

Definition 11.1 A Lagrange collocation-type time-stepping integrator for solving the nonlinear system (11.13) is defined by

$$\begin{cases} u^{n+1} = \phi_0(\mathscr{V})u^n + \Delta t\phi_1(\mathscr{V})u'^n + \Delta t^2\displaystyle\sum_{i=1}^{s} b_i(\mathscr{V})f(U^{ni}), \\ u'^{n+1} = -\Delta t\mathscr{A}\phi_1(\mathscr{V})u^n + \phi_0(\mathscr{V})u'^n + \Delta t\displaystyle\sum_{i=1}^{s} \bar{b}_i(\mathscr{V})f(U^{ni}), \\ U^{ni} = \phi_0(c_i^2\mathscr{V})u^n + c_i\Delta t\phi_1(c_i^2\mathscr{V})u'^n + c_i^2\Delta t^2\displaystyle\sum_{j=1}^{s} a_{ij}(\mathscr{V})f(U^{nj}), \quad i = 1, 2, \ldots, s, \end{cases} \tag{11.39}$$

where $b_i(\mathscr{V})$ and $\bar{b}_i(\mathscr{V})$ are defined by (11.34) and (11.35), respectively, and $a_{ij}(\mathscr{V})$ are uniformly bounded.

11.3.2 Error Analysis for the Lagrange Collocation-Type Time-Stepping Integrators

In this subsection, we will analyse the local error bounds of the Lagrange collocation-type time discretisations (11.39) for the nonlinear system (11.13). Our main hypothesis on the nonlinearity f is described below.

Assumption 1 It is assumed that (11.13) possesses a sufficiently smooth solution, and that $f : D(\mathscr{A}) \to \mathbb{R}$ is sufficiently often Fréchet differentiable in a strip along the exact solution.

Assumption 2 Let f be locally Lipschitz-continuous in a strip along the exact solution $u(t)$. Thus, there exists a real number L such that

$$\|f(v(t)) - f(w(t))\| \leq L\|v(t) - w(t)\| \tag{11.40}$$

for all $t \in [t_0, T]$ and $\max\big(\|v(t) - u(t)\|, \|w(t) - u(t)\|\big) \leq R$.

Inserting the exact solution into the time integrators (11.39) yields

$$\begin{cases} u(t_{n+1}) = \phi_0(\mathscr{V})u(t_n) + \Delta t\phi_1(\mathscr{V})u'(t_n) + \Delta t^2 \sum\limits_{i=1}^{s} b_i(\mathscr{V})\tilde{f}(t_n + c_i\Delta t) + \delta^{n+1}, \\ u'(t_{n+1}) = -\Delta t\mathscr{A}\phi_1(\mathscr{V})u(t_n) + \phi_0(\mathscr{V})u'(t_n) + \Delta t \sum\limits_{i=1}^{s} \bar{b}_i(\mathscr{V})\tilde{f}(t_n + c_i\Delta t) + \delta'^{n+1}, \\ u(t_n + c_i\Delta t) = \phi_0(c_i^2\mathscr{V})u(t_n) + c_i\Delta t\phi_1(c_i^2\mathscr{V})u'(t_n) + c_i^2\Delta t^2 \sum\limits_{j=1}^{s} a_{ij}(\mathscr{V})\tilde{f}(t_n + c_j\Delta t) + \Delta^{ni}, \\ i = 1, 2, \ldots, s, \end{cases} \tag{11.41}$$

where $b_i(\mathscr{V})$ and $\bar{b}_i(\mathscr{V})$ are defined by (11.34) and (11.35), respectively, and $a_{ij}(\mathscr{V})$ are uniformly bounded.

Applying the Lagrange interpolation polynomial (11.29) to the nonlinear integrals in the operator-variation-of-constants formula (11.27), and comparing with the first two equations of (11.41), we obtain the residuals δ^{n+1} and δ'^{n+1}:

$$\begin{aligned} \delta^{n+1} &= \frac{\Delta t^{s+2}}{s!} \int_0^1 (1-z)\phi_1\big((1-z)^2\mathscr{V}\big)w_s(z)\mathrm{d}z f_t^{(s)}\big(u(t_n + \xi^n\Delta t)\big), \\ \delta'^{n+1} &= \frac{\Delta t^{s+1}}{s!} \int_0^1 \phi_0\big((1-z)^2\mathscr{V}\big)w_s(z)\mathrm{d}z f_t^{(s)}\big(u(t_n + \xi^n\Delta t)\big). \end{aligned} \tag{11.42}$$

It follows from (11.42) that

$$\|\delta^{n+1}\| \leq C_1\Delta t^{s+2} \quad \text{and} \quad \|\delta'^{n+1}\| \leq C_1\Delta t^{s+1}, \tag{11.43}$$

where

$$C_1 = \frac{1}{s!}\max_{0\leq z\leq 1}|w_s(z)| \max_{t_0\leq t\leq T}\|f_t^{(s)}\big(u(t)\big)\| \tag{11.44}$$

is a constant.

In order to clarify the representation of the residuals Δ^{ni}, we expand $\tilde{f}(t_n + z\Delta t)$ into a Taylor series with remainder in integral form:

$$\tilde{f}(t_n+z\Delta t)=\sum_{k=1}^{s}\frac{z^{k-1}\Delta t^{k-1}}{(k-1)!}\tilde{f}_t^{(k-1)}(t_n)+\frac{\Delta t^s}{(s-1)!}\int_0^z\tilde{f}_t^{(s)}(t_n+\sigma\Delta t)(z-\sigma)^{s-1}\mathrm{d}\sigma. \tag{11.45}$$

On the one hand, inserting the Taylor formula (11.45) into the right-hand side of the operator-variation-of-constants formula gives

$$\begin{aligned}u(t_n+c_i\Delta t)&=\phi_0(c_i^2\mathscr{V})u(t_n)+c_i\Delta t\phi_1(c_i^2\mathscr{V})u'(t_n)+\sum_{k=1}^{s}c_i^{k+1}\Delta t^{k+1}\phi_{k+1}(c_i^2\mathscr{V})\tilde{f}_t^{(k-1)}(t_n)\\&+\frac{c_i^{s+2}\Delta t^{s+2}}{(s-1)!}\int_0^1(1-z)\phi_1\big((1-z)^2\mathscr{V}\big)\int_0^z\tilde{f}_t^{(s)}(t_n+\sigma c_i\Delta t)(z-\sigma)^{s-1}\mathrm{d}\sigma\mathrm{d}z.\end{aligned} \tag{11.46}$$

Substituting the Taylor formula (11.45) into the right-hand side of the last equations for $i=1,2,\ldots,s$ of (11.41) yields

$$\begin{aligned}u(t_n+c_i\Delta t)&=\phi_0(c_i^2\mathscr{V})u(t_n)+c_i\Delta t\phi_1\big(c_i^2\mathscr{V}\big)u'(t_n)\\&+\sum_{k=1}^{s}c_i^2\Delta t^{k+1}\sum_{j=1}^{s}a_{ij}(\mathscr{V})\frac{c_j^{k-1}}{(k-1)!}\tilde{f}_t^{(k-1)}(t_n)\\&+\frac{c_i^2\Delta t^{s+2}}{(s-1)!}\sum_{j=1}^{s}a_{ij}(\mathscr{V})\int_0^{c_j}\tilde{f}_t^{(s)}(t_n+\sigma\Delta t)(c_j-\sigma)^{s-1}\mathrm{d}\sigma+\Delta^{ni}.\end{aligned} \tag{11.47}$$

Subtracting (11.46) from (11.47), we obtain

$$\begin{aligned}\Delta^{ni}=&\sum_{k=1}^{s}c_i^2\Delta t^{k+1}\left(c_i^{k-1}\phi_{k+1}(c_i^2\mathscr{V})-\sum_{j=1}^{s}a_{ij}(\mathscr{V})\frac{c_j^{k-1}}{(k-1)!}\right)\tilde{f}_t^{(k-1)}(t_n)\\&+\frac{c_i^{s+2}\Delta t^{s+2}}{(s-1)!}\int_0^1(1-z)\phi_1\big((1-z)^2\mathscr{V}\big)\int_0^z\tilde{f}_t^{(s)}(t_n+\sigma c_i\Delta t)(z-\sigma)^{s-1}\mathrm{d}\sigma\mathrm{d}z\\&-\frac{c_i^2\Delta t^{s+2}}{(s-1)!}\sum_{j=1}^{s}a_{ij}(\mathscr{V})\int_0^{c_j}\tilde{f}_t^{(s)}(t_n+\sigma\Delta t)(c_j-\sigma)^{s-1}\mathrm{d}\sigma.\end{aligned}$$

By the following order conditions:

$$\sum_{j=1}^{s}a_{ij}(\mathscr{V})\frac{c_j^{k-1}}{(k-1)!}=c_i^{k-1}\phi_{k+1}(c_i^2\mathscr{V}),\quad k=1,2,\ldots,s,\quad i=1,2,\ldots,s, \tag{11.48}$$

the residuals Δ^{ni} can be explicitly expressed as:

$$\begin{aligned}\Delta^{ni}=&\frac{c_i^{s+2}\Delta t^{s+2}}{(s-1)!}\int_0^1(1-z)\phi_1\big((1-z)^2\mathscr{V}\big)\int_0^z\tilde{f}_t^{(s)}(t_n+\sigma c_i\Delta t)(z-\sigma)^{s-1}\mathrm{d}\sigma\mathrm{d}z\\&-\frac{c_i^2\Delta t^{s+2}}{(s-1)!}\sum_{j=1}^{s}a_{ij}(\mathscr{V})\int_0^{c_j}\tilde{f}_t^{(s)}(t_n+\sigma\Delta t)(c_j-\sigma)^{s-1}\mathrm{d}\sigma.\end{aligned} \tag{11.49}$$

Likewise, we can deduce the following results

$$\|\Delta^{ni}\| \leq \frac{c_i^2 \Delta t^{s+2}}{(s-1)!}\Big(c_i^s + \gamma \sum_{i=1}^{s} c_i^s\Big) \max_{t_0 \leq t \leq T} \|f_t^{(s)}\big(u(t)\big)\| \leq C_2 \Delta t^{s+2}, \quad i = 1, 2, \ldots, s, \tag{11.50}$$

where the constant C_2 is given by

$$C_2 = \frac{1 + s\gamma}{(s-1)!} \max_{t_0 \leq t \leq T} \|f_t^{(s)}\big(u(t)\big)\|, \tag{11.51}$$

and γ is the uniform bound on $a_{ij}(\mathscr{V})$ under the norm $\|\cdot\|_{L^2(\Omega) \leftarrow L^2(\Omega)}$.

Concerning the local error bounds of the Lagrange collocation-type time-stepping integrators (11.39), we have the following result.

Theorem 11.2 *Suppose that $f_t^{(s)} \in L^\infty(0, T; L^2(\Omega))$. Under the local assumptions of $u^n = u(t_n)$, $u'^n = u'(t_n)$, the local error bounds of the time integrators* (11.39) *satisfy the following inequalities*

$$\begin{aligned} \|u(t_{n+1}) - u^{n+1}\| &\leq 2\Delta t^2 \beta L \sum_{i=1}^{s} \|\Delta^{ni}\| + \|\delta^{n+1}\|, \\ \|u'(t_{n+1}) - u'^{n+1}\| &\leq 2\Delta t \beta L \sum_{i=1}^{s} \|\Delta^{ni}\| + \|\delta'^{n+1}\|, \end{aligned} \tag{11.52}$$

where the residuals δ^{n+1}, δ'^{n+1} and Δ^{ni} are explicitly represented by (11.42) *and* (11.49)*, respectively.*

Proof It follows on subtracting (11.39) from (11.41) that

$$\begin{cases} u(t_{n+1}) - u^{n+1} = \Delta t^2 \sum_{i=1}^{s} b_i(\mathscr{V})\Big(\tilde{f}(t_n + c_i \Delta t) - f(U^{ni})\Big) + \delta^{n+1}, \\ u'(t_{n+1}) - u'^{n+1} = \Delta t \sum_{i=1}^{s} \bar{b}_i(\mathscr{V})\Big(\tilde{f}(t_n + c_i \Delta t) - f(U^{ni})\Big) + \delta'^{n+1}, \\ u(t_n + c_i \Delta t) - U^{ni} = c_i^2 \Delta t^2 \sum_{j=1}^{s} a_{ij}(\mathscr{V})\big(\tilde{f}(t_n + c_j \Delta t) - f(U^{nj})\big) + \Delta^{ni}, \quad i = 1, 2, \ldots, s. \end{cases} \tag{11.53}$$

By taking norms on both sides of the Eq. (11.53) and using Assumption 2, the first two equations yield

$$\begin{aligned} \|u(t_{n+1}) - u^{n+1}\| &\leq \Delta t^2 \sum_{i=1}^{s} \|b_i(\mathscr{V})\|_{L^2(\Omega) \leftarrow L^2(\Omega)} \|\tilde{f}(t_n + c_i \Delta t) - f(U^{ni})\| + \|\delta^{n+1}\| \\ &\leq \Delta t^2 \beta L \sum_{i=1}^{s} \|u(t_n + c_i \Delta t) - U^{ni}\| + \|\delta^{n+1}\|, \end{aligned} \tag{11.54}$$

and

$$\begin{aligned}\|u'(t_{n+1})-u'^{n+1}\| \leq & \Delta t \sum_{i=1}^{s} \|\bar{b}_i(\mathscr{V})\|_{L^2(\Omega)\leftarrow L^2(\Omega)} \|\tilde{f}(t_n+c_i\Delta t)-f(U^{ni})\| + \|\delta'^{n+1}\| \\ \leq & \Delta t \beta L \sum_{i=1}^{s} \|u(t_n+c_i\Delta t)-U^{ni}\| + \|\delta'^{n+1}\|.\end{aligned} \tag{11.55}$$

The last equations of (11.53) give

$$\begin{aligned}\|u(t_n+c_i\Delta t)-U^{ni}\| \leq & c_i^2\Delta t^2 \sum_{j=1}^{s} \|\bar{a}_{ij}(\mathscr{V})\|_{L^2(\Omega)\leftarrow L^2(\Omega)} \|\tilde{f}(t_n+c_j\Delta t)-f(U^{nj})\| + \|\Delta^{ni}\| \\ \leq & c_i^2\Delta t^2 \gamma L \sum_{j=1}^{s} \|u(t_n+c_j\Delta t)-U^{nj}\| + \|\Delta^{ni}\|, \quad i=1,2,\ldots,s,\end{aligned} \tag{11.56}$$

where γ is the uniform bound on $a_{ij}(\mathscr{V})$ under the norm $\|\cdot\|_{L^2(\Omega)\leftarrow L^2(\Omega)}$. Summing up the results of (11.56) for i from 1 to s, we obtain

$$\sum_{i=1}^{s} \|u(t_n+c_i\Delta t)-U^{ni}\| \leq \Delta t^2\gamma L \sum_{i=1}^{s} c_i^2 \sum_{j=1}^{s} \|u(t_n+c_j\Delta t)-U^{nj}\| + \sum_{i=1}^{s} \|\Delta^{ni}\|. \tag{11.57}$$

If the sufficiently small time stepsize Δt satisfies $\Delta t^2\gamma L \sum_{i=1}^{s} c_i^2 \leq \frac{1}{2}$, namely,

$$\Delta t \leq \sqrt{\frac{1}{2\gamma L \sum_{i=1}^{s} c_i^2}}, \tag{11.58}$$

then we have

$$\sum_{i=1}^{s} \|u(t_n+c_i\Delta t)-U^{ni}\| \leq 2\sum_{i=1}^{s} \|\Delta^{ni}\|. \tag{11.59}$$

Inserting (11.59) into the right-hand sides of inequalities (11.54) and (11.55) yields the following results

$$\|u(t_{n+1})-u^{n+1}\| \leq 2\Delta t^2\beta L \sum_{i=1}^{s} \|\Delta^{ni}\| + \|\delta^{n+1}\|, \tag{11.60}$$

and

$$\|u'(t_{n+1})-u'^{n+1}\| \leq 2\Delta t\beta L \sum_{i=1}^{s} \|\Delta^{ni}\| + \|\delta'^{n+1}\|. \tag{11.61}$$

The statement of the theorem is proved. □

Using the estimate of the residuals δ^{n+1}, δ'^{n+1} and Δ^{ni} in (11.43) and (11.50), and inserting them into (11.52), the following corollary clarifies the local error bounds of the Lagrange collocation-type time integrators (11.39).

Corollary 11.1 *Under the condition of Theorem 11.2, the local error bounds of the time integrators* (11.39) *can be explicitly presented as*

$$\|u(t_{n+1}) - u^{n+1}\| \leq \tilde{C}_1 \Delta t^{s+2} \quad \text{and} \quad \|u'(t_{n+1}) - u'^{n+1}\| \leq \tilde{C}_1 \Delta t^{s+1}, \tag{11.62}$$

where $\tilde{C}_1 = (C_1 + 2C_2 s\beta L \Delta t^2)$, *and* C_1, C_2 *are defined as* (11.44) *and* (11.51), *respectively.*

Remark 11.3 The weights $b_i(\mathscr{V})$, $\bar{b}_i(\mathscr{V})$ and $a_{ij}(\mathscr{V})$ of the time integrators (11.39) are determined by (11.34), (11.35) and the order conditions (11.48) with appropriate nodes c_i for $i = 1, 2, \ldots, s$, respectively.

Remark 11.4 Furthermore, from the analysis of the local error bounds for the time integrators (11.39), it can be observed that there is a term $\max\limits_{0 \leq z \leq 1} |w_s(z)|$ appearing in the constant C_1. In order to minimise the constant C_1, it is wise to choose the nodes $\{c_i\}_{i=1}^s$ as the Gauss-Legendre nodes in this chapter.

Remark 11.5 Here, we should point out that the limitation on the time stepsize (11.58) is only a sufficient condition for our theoretical analysis. It is also important for the analysis of the stability and convergence for the proposed fully discrete schemes.

11.4 Spatial Discretisation

The proposed arbitrarily high-order Lagrange collocation-type time-stepping integrators (11.39) are expressed in terms of operator in the infinite dimensional Hilbert space $L^2(\Omega)$. In order to obtain proper numerical schemes, it remains to approximate the differential operator $\mathscr{A}$ with an appropriate differentiation matrix A acting on an M-dimensional space. Furthermore, it is our ideal choice to approximate the differential operator $\mathscr{A}$ by a positive semi-definite matrix A, in such a way that we can achieve a reasonable and rigorous nonlinear stability and convergence analysis. Fortunately, much research has been done on the spatial derivatives of nonlinear system (11.1) with periodic boundary conditions (11.2), from which it is easy to choose a suitable positive semi-definite differential matrix.

In this section, we mainly consider the following two types of spatial discretisations.

1. *Symmetric finite difference (SFD)* (see, e.g. R. Bank, R.L. Graham, J. Stoer, R. Varga, H. Yserentant [5])

As an option, we use the finite difference approximation to approximate the operator $\mathscr{A}$ by the following 9-diagonal differential matrix:

$$
A_{sfd} = \frac{-a^2}{\Delta x^2}
\begin{pmatrix}
\frac{-205}{72} & \frac{8}{5} & \frac{-1}{5} & \frac{8}{315} & \frac{-1}{560} & & & & \frac{-1}{560} & \frac{8}{315} & \frac{-1}{5} & \frac{8}{5} \\
\frac{8}{5} & \frac{-205}{72} & \frac{8}{5} & \frac{-1}{5} & \frac{8}{315} & \frac{-1}{560} & & & & \frac{-1}{560} & \frac{8}{315} & \frac{-1}{5} \\
\frac{-1}{5} & \frac{8}{5} & \frac{-205}{72} & \frac{8}{5} & \frac{-1}{5} & \frac{8}{315} & \frac{-1}{560} & & & & \frac{-1}{560} & \frac{8}{315} \\
\frac{8}{315} & \frac{-1}{5} & \frac{8}{5} & \frac{-205}{72} & \frac{8}{5} & \frac{-1}{5} & \frac{8}{315} & \frac{-1}{560} & & & & \frac{-1}{560} \\
\frac{-1}{560} & \frac{8}{315} & \frac{-1}{5} & \frac{8}{5} & \frac{-205}{72} & \frac{8}{5} & \frac{-1}{5} & \frac{8}{315} & \frac{-1}{560} & & & \\
 & & \ddots & \ddots & \ddots & \ddots & \ddots & \ddots & \ddots & \ddots & \ddots & \ddots \\
 & & & \frac{-1}{560} & \frac{8}{315} & \frac{-1}{5} & \frac{8}{5} & \frac{-205}{72} & \frac{8}{5} & \frac{-1}{5} & \frac{8}{315} & \frac{-1}{560} \\
\frac{-1}{560} & & & & \frac{-1}{560} & \frac{8}{315} & \frac{-1}{5} & \frac{8}{5} & \frac{-205}{72} & \frac{8}{5} & \frac{-1}{5} & \frac{8}{315} \\
\frac{8}{315} & \frac{-1}{560} & & & & \frac{-1}{560} & \frac{8}{315} & \frac{-1}{5} & \frac{8}{5} & \frac{-205}{72} & \frac{8}{5} & \frac{-1}{5} \\
\frac{-1}{5} & \frac{8}{315} & \frac{-1}{560} & & & & \frac{-1}{560} & \frac{8}{315} & \frac{-1}{5} & \frac{8}{5} & \frac{-205}{72} & \frac{8}{5} \\
\frac{8}{5} & \frac{-1}{5} & \frac{8}{315} & \frac{-1}{560} & & & & \frac{-1}{560} & \frac{8}{315} & \frac{-1}{5} & \frac{8}{5} & \frac{-205}{72}
\end{pmatrix}_{M\times M}.
$$

In general, the finite difference method is a local approximation and has limited accuracy. The accuracy of this approximation for the space derivative is of order eight with $O(\Delta x^8)$, and the differential matrix A_{sfd} is positive semi definite.

2. *Fourier spectral collocation (FSC)* (see, e.g. J. Shen, T. Tang, L. L. Wang [46] and J. S. Hesthaven, S. Gottlieb, D. Gottlieb [29])

Spectral methods are global in character, where the computation at any given point depends not only on the information at neighboring points, but on the information from the entire domain. The Fourier spectral collocation method is our other choice, which can be presented as a limit of local finite difference approximations of increasing orders of accuracy (see, e.g. J. S. Hesthaven, S. Gottlieb, D. Gottlieb [29]). The entries of the second-order Fourier differentiation matrix $A_{fsc} = (a_{kj})_{M\times M}$ are given by

$$
a_{kj} = \begin{cases} \frac{(-1)^{k+j}}{2}\sin^{-2}\left(\frac{(k-j)\pi}{M}\right), & k \neq j, \\[2ex] \frac{M^2}{12} + \frac{1}{6}, & k = j. \end{cases} \tag{11.63}
$$

The main appeal of spectral methods is that they exhibit spectral convergence to approximate $\mathscr{A}$: the error decays faster than $O(M^{-\alpha})$, $\forall \alpha > 0$ for sufficiently large M. In terms of classical concepts, the method is of an infinite order. Similarly, the differential matrix A_{fsc} is also a positive semi-definite matrix.

It has been noted that the energy conservation is a crucial property of the nonlinear Klein–Gordon equations (11.1)–(11.2). After approximating the operator $\mathscr{A}$ by a positive semi-definite differential matrix A, there is also a corresponding energy conservation law, which can be characterised in the following form:

$$
\tilde{E}(t) \equiv \frac{\Delta x}{2}\|u'(t)\|^2 + \frac{a^2 \Delta x}{2}\|Du(t)\|^2 + \Delta x \sum_{j=1}^{M} V\left(u_j(t)\right) = \cdots = \tilde{E}(t_0), \tag{11.64}
$$

where the norm $\|\cdot\|$ is the standard vector 2-norm and $\Delta x = \frac{2\pi}{M}$ is the spatial stepsize. Actually, this energy can be regarded as an approximate energy (a semi-discrete energy) of the original continuous system. Therefore, in the part of the numerical experiments, we will also test the effectiveness of our methods to preserve the semi-discrete energy (11.64).

11.5 The Analysis of Nonlinear Stability and Convergence for the Fully Discrete Scheme

The nonlinear stability and error analysis for the fully discrete scheme over a finite time interval $[t_0, T]$ will be investigated in this section. The main strategy used in this section is the popular energy analysis method. Here, it is noted that, throughout this section $\|\cdot\|$ presents the vector 2-norm or matrix 2-norm (spectral norm).

11.5.1 Analysis of the Nonlinear Stability

In this subsection, we will show the nonlinear stability of our time-steeping integrators (11.39), once the differential operator $\mathscr{A}$ is replaced by a differential matrix A.

Suppose that the perturbed problem of (11.13) is

$$\begin{cases} v''(t) + \mathscr{A}v(t) = f\big(v(t)\big), \quad t \in [t_0, T], \\ v(t_0) = \varphi_1(x) + \tilde{\varphi}_1(x), \ v'(t_0) = \varphi_2(x) + \tilde{\varphi}_2(x), \end{cases} \tag{11.65}$$

where $\tilde{\varphi}_1(x)$, $\tilde{\varphi}_2(x)$ are perturbation functions. Letting $\eta(t) = v(t) - u(t)$ and subtracting (11.13) from (11.65), we obtain

$$\begin{cases} \eta''(t) + \mathscr{A}\eta(t) = f\big(v(t)\big) - f\big(u(t)\big), \quad t \in [t_0, T], \\ \eta(t_0) = \tilde{\varphi}_1(x), \ \eta'(t_0) = \tilde{\varphi}_2(x). \end{cases} \tag{11.66}$$

In general, the operator $\mathscr{A}$ is approximated by a symmetric, positive semi-definite, differential matrix A in the sense of structure preservation. Then, there exists an orthogonal matrix P and a positive semi-definite diagonal matrix Λ such that

$$A = P^{\intercal}\Lambda P.$$

By defining the matrix $D = P^{\intercal}\Lambda^{\frac{1}{2}}P$, we obtain the decomposition of matrix A as $A = D^2$. The bounded operator functions $\phi_j(t^2\mathscr{A})$ are replaced by the matrix functions $\phi_j(t^2A)$. Similarly to the boundedness of the operator functions, we also have

$$\|\phi_j(t^2A)\| = \sqrt{\lambda_{\max}\left(\phi_j^2(t^2A)\right)} \leq \gamma_j, \quad j = 0, 1, 2 \dots. \tag{11.67}$$

Applying our time-stepping integrators (11.39) to (11.66), we obtain

$$\begin{cases} \eta^{n+1} = \phi_0(V)\eta^n + \Delta t\phi_1(V)\eta'^n + \Delta t^2 \sum_{i=1}^{s} b_i(V)(f(V^{ni}) - f(U^{ni})), \\ \eta'^{n+1} = -\Delta t A\phi_1(V)\eta^n + \phi_0(V)\eta'^n + \Delta t \sum_{i=1}^{s} \bar{b}_i(V)(f(V^{ni}) - f(U^{ni})), \\ V^{ni} - U^{ni} = \phi_0(c_i^2 V)\eta^n + c_i\Delta t\phi_1(c_i^2 V)\eta'^n + c_i^2\Delta t^2 \sum_{j=1}^{s} a_{ij}(V)(f(V^{nj}) - f(U^{nj})), \\ i = 1, 2, \dots, s, \end{cases} \tag{11.68}$$

where $V = \Delta t^2 A$ and $b_i(V)$ and $\bar{b}_i(V)$ are defined by (11.34) and (11.35), respectively. Likewise, we have

$$\|b_i(V)\| \leq \max_{0\leq z\leq 1} |l_i(z)| \leq \beta \quad \text{and} \quad \|\bar{b}_i(V)\| \leq \max_{0\leq z\leq 1} |l_i(z)| \leq \beta,$$

which are uniformly bounded.

We rewrite the first two formulae of (11.68) in the following matrix form:

$$\begin{pmatrix} D\eta^{n+1} \\ \eta'^{n+1} \end{pmatrix} = \Omega \begin{pmatrix} D\eta^n \\ \eta'^n \end{pmatrix} + \sum_{i=1}^{s} \Delta t \int_0^1 \Omega_i(z)\mathrm{d}z \begin{pmatrix} 0 \\ f(U^{ni}) - f(V^{ni}) \end{pmatrix}, \tag{11.69}$$

where

$$\Omega = \begin{pmatrix} \phi_0(V) & \Delta t D\phi_1(V) \\ -\Delta t D\phi_1(V) & \phi_0(V) \end{pmatrix} \tag{11.70}$$

and

$$\Omega_i(z) = l_i(z) \begin{pmatrix} \phi_0((1-z)^2V) & \Delta t(1-z)D\phi_1((1-z)^2V) \\ -\Delta t(1-z)D\phi_1((1-z)^2V) & \phi_0((1-z)^2V) \end{pmatrix}, \tag{11.71}$$

for $i = 1, \dots, s$. Before the stability analysis, we state a property of the operator-argument functions ϕ_0 and ϕ_1, and bound the spectral norm of matrices Ω and $\Omega_i(z)$ for $i = 1, 2, \dots, s$.

Lemma 11.1 *The bounded operator-argument functions $\phi_0(A)$ and $\phi_1(A)$ defined by* (11.5) *satisfy*

$$\phi_0^2(A) + A\phi_1^2(A) = I, \tag{11.72}$$

where A is any positive semi-definite operator or matrix.

Proof Lemma 11.1 can be obtained by a direct calculation based on (11.12). We omit the details of the proof. □

Lemma 11.2 *Assume that A is a symmetric positive semi-definite matrix and that $V = \Delta t^2 A$. Let the matrices Ω and $\Omega_i(z)$ for $i = 1, 2, \ldots, s$ be defined by* (11.70) *and* (11.71)*, respectively. Then, the spectral norms of matrices Ω and $\Omega_i(z)$ satisfy*

$$\|\Omega\| = 1 \quad and \quad \|\Omega_i(z)\| = |l_i(z)| \le \beta, \quad \forall z \in [0, 1], \quad i = 1, 2, \ldots, s, \tag{11.73}$$

where β is the uniform bound for the Lagrange basis $|l_i(z)|$.

Proof It is straightforward to verify that

$$\Omega^{\intercal}\Omega = \begin{pmatrix} \phi_0^2(V) + V\phi_1^2(V) & 0 \\ 0 & \phi_0^2(V) + V\phi_1^2(V) \end{pmatrix},$$

and

$$\Omega_i^{\intercal}(z)\Omega_i(z) = l_i^2(z)\begin{pmatrix} \Omega_i^{11} & 0 \\ 0 & \Omega_i^{22} \end{pmatrix},$$

where

$$\Omega_i^{11} = \phi_0^2((1-z)^2V) + (1-z)^2V\phi_1^2((1-z)^2V),$$
$$\Omega_i^{22} = \phi_0^2((1-z)^2V) + (1-z)^2V\phi_1^2((1-z)^2V).$$

It follows from Lemma 11.1 that

$$\Omega^{\intercal}\Omega = I_{2M\times 2M}, \quad \text{and} \quad \Omega_i(z)^{\intercal}\Omega_i(z) = l_i^2(z)I_{2M\times 2M}. \tag{11.74}$$

We then have

$$\|\Omega\| = 1 \quad \text{and} \quad \|\Omega_i(z)\| = |l_i(z)| \le \beta, \quad \forall z \in [0, 1], \quad i = 1, 2, \ldots, s.$$

The conclusion of the lemma is proved. □

Theorem 11.3 *Supposing that the nonlinear function f satisfies Assumption 2 and that the operator $\mathscr{A}$ is approximated by a symmetric positive semi-definite differential matrix A. Then, if the sufficiently small time stepsize Δt satisfies* (11.58)*, we have the following nonlinear stability results*

$$
\begin{aligned}
\|\eta^n\| &\le \exp\big((1+4s\beta L)T\big)\big(\|\tilde{\varphi}_1\| + \sqrt{\|D\tilde{\varphi}_1\|^2 + \|\tilde{\varphi}_2\|^2}\big),\\
\|\eta'^n\| &\le \exp\big((1+4s\beta L)T\big)\big(\|\tilde{\varphi}_1\| + \sqrt{\|D\tilde{\varphi}_1\|^2 + \|\tilde{\varphi}_2\|^2}\big),
\end{aligned}
\tag{11.75}
$$

where γ is a uniform bound for $\|a_{ij}(V)\|$.

Proof It follows from taking the l_2 norm on both sides of the first formula (11.68) and (11.69) that

$$
\|\eta^{n+1}\| \le \|\eta^n\| + \Delta t\|\eta'^n\| + \Delta t^2\beta\sum_{i=1}^{s}\big(\|f(V^{ni}) - f(U^{ni})\|\big),
\tag{11.76}
$$

$$
\sqrt{\|D\eta^{n+1}\|^2 + \|\eta'^{n+1}\|^2} \le \sqrt{\|D\eta^n\|^2 + \|\eta'^n\|^2} + \Delta t\beta\sum_{i=1}^{s}\big(\|f(V^{ni}) - f(U^{ni})\|\big).
\tag{11.77}
$$

Then summing up the results, we obtain

$$
\begin{aligned}
\|\eta^{n+1}\| + \sqrt{\|D\eta^{n+1}\|^2 + \|\eta'^{n+1}\|^2} \le& \|\eta^n\| + \sqrt{\|D\eta^n\|^2 + \|\eta'^n\|^2} + \Delta t\|\eta'^n\|\\
&+ \Delta t(1+\Delta t)\beta\sum_{i=1}^{s}\big(\|f(V^{ni}) - f(U^{ni})\|\big).
\end{aligned}
\tag{11.78}
$$

Applying Assumption 2 to the right-hand side of the inequality (11.78), we obtain

$$
\begin{aligned}
\|\eta^{n+1}\| + \sqrt{\|D\eta^{n+1}\|^2 + \|\eta'^{n+1}\|^2} \le& \|\eta^n\| + \sqrt{\|D\eta^n\|^2 + \|\eta'^n\|^2} + \Delta t\|\eta'^n\|\\
&+ \Delta t(1+\Delta t)\beta L\sum_{i=1}^{s}\big(\|V^{ni} - U^{ni}\|\big).
\end{aligned}
\tag{11.79}
$$

Likewise, it follows from the last equations in (11.68) that

$$
\begin{aligned}
\|V^{ni} - U^{ni}\| \le& \|\eta^n\| + c_i\Delta t\|\eta'^n\| + c_i^2\Delta t^2\sum_{j=1}^{s}\|a_{ij}(V)\|\cdot\|f(V^{nj}) - f(U^{nj})\|\\
\le& \|\eta^n\| + c_i\Delta t\|\eta'^n\| + c_i^2\Delta t^2\gamma L\sum_{j=1}^{s}\|V^{nj} - U^{nj}\|,\quad i = 1,\ldots,s.
\end{aligned}
\tag{11.80}
$$

Then, summing up the results of (11.80) between i from 1 to s, we obtain

$$
\sum_{i=1}^{s}\|V^{ni} - U^{ni}\| \le \sum_{i=1}^{s}\big(\|\eta^n\| + c_i\Delta t\|\eta'^n\|\big) + \Delta t^2\gamma L\sum_{i=1}^{s}c_i^2\sum_{j=1}^{s}\|V^{nj} - U^{nj}\|.
\tag{11.81}
$$

This gives

$$\Big(1 - \Delta t^2 \gamma L \sum_{i=1}^{s} c_i^2\Big) \sum_{i=1}^{s} \|V^{ni} - U^{ni}\| \le \sum_{i=1}^{s} \big(\|\eta^n\| + c_i \Delta t \|\eta'^n\|\big). \tag{11.82}$$

If the sufficiently small time stepsize Δt satisfies (11.58), then we have

$$\sum_{i=1}^{s} \|V^{ni} - U^{ni}\| \le 2 \sum_{i=1}^{s} \big(\|\eta^n\| + c_i \Delta t \|\eta'^n\|\big). \tag{11.83}$$

Inserting (11.83) into (11.79) yields

$$\begin{aligned} &\|\eta^{n+1}\| + \sqrt{\|D\eta^{n+1}\|^2 + \|\eta'^{n+1}\|^2} \\ \le &\|\eta^n\| + \sqrt{\|D\eta^n\|^2 + \|\eta'^n\|^2} + \Delta t \|\eta'^n\| + 2\Delta t(1+\Delta t)\beta L \sum_{i=1}^{s} \big(\|\eta^n\| + c_i \Delta t \|\eta'^n\|\big). \end{aligned} \tag{11.84}$$

An argument by induction leads to the following result

$$\begin{aligned} \|\eta^{n+1}\| + \sqrt{\|D\eta^{n+1}\|^2 + \|\eta'^{n+1}\|^2} \le &(1 + \Delta t(1 + 4s\beta L))^n \big(\|\eta^0\| + \sqrt{\|D\eta^0\|^2 + \|\eta'^0\|^2}\big) \\ \le &\exp\big(T(1 + 4s\beta L)\big)\big(\|\tilde{\varphi}_1\| + \sqrt{\|D\tilde{\varphi}_1\|^2 + \|\tilde{\varphi}_2\|^2}\big). \end{aligned} \tag{11.85}$$

Thus, the following inequalities are derived

$$\begin{aligned} \|\eta^n\| &\le \exp\big((1 + 4s\beta L)T\big)\big(\|\tilde{\varphi}_1\| + \sqrt{\|D\tilde{\varphi}_1\|^2 + \|\tilde{\varphi}_2\|^2}\big), \\ \|\eta'^n\| &\le \exp\big((1 + 4s\beta L)T\big)\big(\|\tilde{\varphi}_1\| + \sqrt{\|D\tilde{\varphi}_1\|^2 + \|\tilde{\varphi}_2\|^2}\big). \end{aligned} \tag{11.86}$$

The conclusions of the theorem are proved. □

11.5.2 Convergence of the Fully Discrete Scheme

As is known, the convergence of the classical methods for linear partial differential equations is governed by the Lax equivalence theorem: convergence equals consistency plus stability [33]. However, the Lax equivalence theorem does not directly apply to nonlinear problems.

In this subsection, the error analysis of the fully discrete scheme for nonlinear problems will be discussed. Based on some suitable assumptions of smoothness and spatial discretisation strategies, the original continuous system (11.1) or (11.13) can be discretised as follows:

$$\begin{cases} U''(t) + AU(t) = f\big(U(t)\big) + \delta(\Delta x), \quad t \in [t_0, T], \\ U(t_0) = \varphi_1, \; U'(t_0) = \varphi_2, \end{cases} \tag{11.87}$$

where A is a positive semi-definite differential matrix,

$$U(t) = \big(u(x_1,t), u(x_2,t), \ldots, u(x_M,t)\big)^{\mathsf{T}}$$

and

$$\varphi_l = \big(\varphi_l(x_1), \varphi_l(x_2), \ldots, \varphi_l(x_M)\big)^{\mathsf{T}},$$

for $l = 1, 2$.

Here, it should be noted that $\delta(\Delta x)$ is the truncation error produced by approximating the spatial differential operator $\mathscr{A}$ by a positive semi-definite matrix A. For example, if we were to approximate the spatial derivative by the classical fourth-order finite difference method (see, e.g. [5, 39]), then the truncation error $\delta(\Delta x)$ would be $\|\delta(\Delta x)\| = O(\Delta x^4)$.

Applying a time-stepping integrator (11.39) to the semi-discrete system (11.87) yields the following results

$$\begin{cases} U(t_{n+1}) = \phi_0(V)U(t_n) + \Delta t\phi_1(V)U'(t_n) + \Delta t^2 \displaystyle\sum_{i=1}^{s} b_i(V) f(U(t_n + c_i\Delta t)) + R^{n+1}, \\ U'(t_{n+1}) = -\Delta t A\phi_1(V)U(t_n) + \phi_0(V)U'(t_n) + \Delta t \displaystyle\sum_{i=1}^{s} \bar{b}_i(V) f(U(t_n + c_i\Delta t)) + R'^{n+1}, \\ U(t_n + c_i\Delta t) = \phi_0(c_i^2 V)U(t_n) + c_i\Delta t\phi_1(c_i^2 V)U'(t_n) + c_i^2\Delta t^2 \displaystyle\sum_{j=1}^{s} a_{ij}(V) f(U(t_n + c_j\Delta t)) + R^{ni}, \\ i = 1, 2, \ldots, s, \end{cases} \tag{11.88}$$

where the truncation errors R^{n+1}, R'^{n+1} and R^{ni} can be explicitly represented as

$$R^{n+1} = \frac{\Delta t^{s+2}}{s!}\int_0^1 (1-z)\phi_1\big((1-z)^2V\big)w_s(z)\mathrm{d}z f_t^{(s)}\big(U(t_n + \xi^n\Delta t)\big) \tag{11.89}$$

$$+ \Delta t^2 \int_0^1 (1-z)\phi_1\big((1-z)^2V\big)\delta(\Delta x)\mathrm{d}z, \tag{11.90}$$

$$R'^{n+1} = \frac{\Delta t^{s+1}}{s!}\int_0^1 \phi_0\big((1-z)^2V\big)w_s(z)\mathrm{d}z f_t^{(s)}\big(U(t_n + \xi^n\Delta t)\big) \tag{11.91}$$

$$+ \Delta t \int_0^1 \phi_0\big((1-z)^2V\big)\delta(\Delta x)\mathrm{d}z, \tag{11.92}$$

and

$$\begin{aligned} R^{ni} =& \frac{c_i^{s+2}\Delta t^{s+2}}{(s-1)!}\int_0^1 (1-z)\phi_1\big((1-z)^2V\big)\int_0^z f_t^{(s)}\big(U(t_n + \sigma c_i\Delta t)\big)(z-\sigma)^{s-1}\mathrm{d}\sigma\mathrm{d}z \\ &- \frac{c_i^2\Delta t^{s+2}}{(s-1)!}\sum_{j=1}^{s} a_{ij}(V)\int_0^{c_j} f_t^{(s)}\big(U(t_n + \sigma\Delta t)\big)(c_j - \sigma)^{s-1}\mathrm{d}\sigma \\ &+ c_i^2\Delta t^2 \int_0^1 (1-z)\phi_1\big((1-z)^2c_i^2V\big)\delta(\Delta x)\mathrm{d}z - c_i^2\Delta t^2 \sum_{j=1}^{s} a_{ij}(V)\delta(\Delta x). \end{aligned} \tag{11.93}$$

Under some suitable assumptions of smoothness, the truncation errors R^{n+1}, R'^{n+1} and R^{ni} satisfy

$$\|R^{n+1}\| \leq C_1 \Delta t^{s+2} + \frac{1}{2}\Delta t^2 \|\delta(\Delta x)\|, \quad \|R'^{n+1}\| \leq C_1 \Delta t^{s+1} + \Delta t \|\delta(\Delta x)\|, \tag{11.94}$$

and

$$\|R^{ni}\| \leq C_2 \Delta t^{s+2} + \Delta t^2 (1 + s\gamma) \|\delta(\Delta x)\|, \quad i = 1, 2, \ldots, s, \tag{11.95}$$

where the constants C_1 and C_2 are determined by (11.44) and (11.51), respectively.

Omitting the small terms R^{n+1}, R'^{n+1} and R^{ni} in (11.88) and using $u_j^n \approx u(x_j, t_n)$, we obtain the following fully discrete scheme

$$\begin{cases} u^{n+1} = \phi_0(V)u^n + \Delta t \phi_1(V) u'^n + \Delta t^2 \sum\limits_{i=1}^{s} b_i(V) f(U^{ni}), \\ u'^{n+1} = -\Delta t A \phi_1(V) u^n + \phi_0(V) u'^n + \Delta t \sum\limits_{i=1}^{s} \bar{b}_i(V) f(U^{ni}), \\ U^{ni} = \phi_0(c_i^2 V) u^n + c_i \Delta t \phi_1(c_i^2 V) u'^n + c_i^2 \Delta t^2 \sum\limits_{j=1}^{s} a_{ij}(V) f(U^{nj}), \quad i = 1, 2, \ldots, s. \end{cases} \tag{11.96}$$

We next consider the convergence of the fully discrete scheme (11.96) for nonlinear problems. To this end, we denote $e_j^n = u(x_j, t_n) - u_j^n$, $e_j'^n = u_t(x_j, t_n) - u_j'^n$ and $E_j^{ni} = u(x_j, t_n + c_i \Delta t) - U_j^{ni}$ for $j = 1, 2, \ldots, M$, i.e., $e^n = U(t_n) - u^n$, $e'^n = U'(t_n) - u'^n$ and $E^{ni} = U(t_n + c_i \Delta t) - U^{ni}$. Subtracting (11.96) from (11.88), and on noticing the exact initial conditions, we get a system of error equations expressed in the form

$$\begin{cases} e^{n+1} = \phi_0(V)e^n + \Delta t \phi_1(V) e'^n + \Delta t^2 \sum\limits_{i=1}^{s} b_i(V) \Big(f\big(U(t_n + c_i \Delta t)\big) - f(U^{ni}) \Big) + R^{n+1}, \\ e'^{n+1} = -\Delta t A \phi_1(V) e^n + \phi_0(V) e'^n + \Delta t \sum\limits_{i=1}^{s} \bar{b}_i(V) \Big(f\big(U(t_n + c_i \Delta t)\big) - f(U^{ni}) \Big) + R'^{n+1}, \\ E^{ni} = \phi_0(c_i^2 V) e^n + c_i \Delta t \phi_1(c_i^2 V) e'^n + c_i^2 \Delta t^2 \sum\limits_{j=1}^{s} a_{ij}(V) \Big(f\big(U(t_n + c_i \Delta t)\big) - f(U^{ni}) \Big) + R^{ni}, \\ i = 1, 2, \ldots, s, \end{cases} \tag{11.97}$$

with the initial conditions $e^0 = 0$, $e'^0 = 0$.

In what follows, we quote the following discrete Gronwall inequality, which plays an important role in the convergence analysis for the fully discrete scheme.

Lemma 11.3 (See, e.g. [48]) *Let μ be positive and a_k, b_k $(k = 0, 1, 2, \cdots)$ be nonnegative and satisfy*

$$a_k \leq (1 + \mu \Delta t) a_{k-1} + \Delta t b_k, \quad k = 1, 2, 3, \ldots,$$

then

$$a_k \leq \exp(\mu k \Delta t)\Big(a_0 + \Delta t \sum_{m=1}^{k} b_m\Big), \quad k = 1, 2, 3, \ldots.$$

Theorem 11.4 *Under the Assumptions 1 and 2, and suppose that $u(x, t)$ satisfies some suitable assumptions on smoothness. If the time stepsize Δt is sufficiently small and satisfies* (11.58), *then there exists a constant C such that*

$$\begin{aligned} \|e^n\| &\leq CT \exp\big((1 + 4s\beta L)T\big)\big(\Delta t^s + \|\delta(\Delta x)\|\big), \\ \|e'^n\| &\leq CT \exp\big((1 + 4s\beta L)T\big)\big(\Delta t^s + \|\delta(\Delta x)\|\big), \end{aligned} \tag{11.98}$$

where C is a constant independent of n, Δt and Δx.

Proof The first two equations of the error system (11.97) can be rewritten in the compact form

$$\begin{pmatrix} De^{n+1} \\ e'^{n+1} \end{pmatrix} = \Omega \begin{pmatrix} De^n \\ e'^n \end{pmatrix} + \Delta t \sum_{i=1}^{s} \int_0^1 \Omega_i(z) \mathrm{d}z \begin{pmatrix} 0 \\ f(U(t_n + c_i \Delta t)) - f(U^{ni}) \end{pmatrix} + \begin{pmatrix} DR^{n+1} \\ R'^{n+1} \end{pmatrix}, \tag{11.99}$$

where Ω and $\Omega_i(z)$ are defined by (11.70) and (11.71), respectively.

It follows from taking the l_2 norm on both sides of the first formula (11.97) and (11.99) that

$$\begin{aligned} &\|e^{n+1}\| \leq \|e^n\| + \Delta t \|e'^n\| + \Delta t^2 \beta \sum_{i=1}^{s} \|f(U(t_n + c_i \Delta t)) - f(U^{ni})\| + \|R^{n+1}\|, \\ &\sqrt{\|De^{n+1}\|^2 + \|e'^{n+1}\|^2} \leq \sqrt{\|De^n\|^2 + \|e'^n\|^2} + \Delta t \beta \sum_{i=1}^{s} \|f(U(t_n + c_i \Delta t)) - f(U^{ni})\| \\ &\qquad + \sqrt{\|DR^{n+1}\|^2 + \|R'^{n+1}\|^2}. \end{aligned} \tag{11.100}$$

Then, summing up the results of (11.100) and using the Assumption 2, we obtain

$$\begin{aligned} &\|e^{n+1}\| + \sqrt{\|De^{n+1}\|^2 + \|e'^{n+1}\|^2} \leq \|e^n\| + \Delta t \|e'^n\| + \sqrt{\|De^n\|^2 + \|e'^n\|^2} \\ &+ \Delta t (1 + \Delta t) \beta L \sum_{i=1}^{s} \|E^{ni}\| + \|R^{n+1}\| + \sqrt{\|DR^{n+1}\|^2 + \|R'^{n+1}\|^2}. \end{aligned} \tag{11.101}$$

Likewise, taking the l_2 norm on both sides of the last equations of the error system (11.97) yields

$$\|E^{ni}\| \leq \|e^n\| + c_i \Delta t \|e'^n\| + c_i^2 \Delta t^2 \gamma L \sum_{i=1}^{s} \|E^{ni}\| + \|R^{ni}\|, \quad i = 1, 2, \ldots, s. \tag{11.102}$$

Summing the results of (11.102) for i from 1 to s gives

$$\sum_{i=1}^{s}\|E^{ni}\| \le \sum_{i=1}^{s}(\|e^n\|+c_i\Delta t\|e'^n\|+\|R^{ni}\|)+\Delta t^2\gamma L\sum_{i=1}^{s}c_i^2\sum_{j=1}^{s}\|E^{nj}\|. \quad (11.103)$$

Under the condition (11.58), we obtain

$$\sum_{i=1}^{s}\|E^{ni}\| \le 2\sum_{i=1}^{s}(\|e^n\|+c_i\Delta t\|e'^n\|)+2\sum_{i=1}^{s}\|R^{ni}\|. \quad (11.104)$$

Inserting (11.104) into (11.101) yields

$$\begin{aligned}
&\|e^{n+1}\|+\sqrt{\|De^{n+1}\|^2+\|e'^{n+1}\|^2} \le \|e^n\|+\Delta t\|e'^n\|+\sqrt{\|De^n\|^2+\|e'^n\|^2}\\
&+2\Delta t(1+\Delta t)\beta L\sum_{i=1}^{s}(\|e^n\|+c_i\Delta t\|e'^n\|)+\|R^{n+1}\|+\sqrt{\|DR^{n+1}\|^2+\|R'^{n+1}\|^2}\\
&+2\Delta t(1+\Delta t)\beta L\sum_{i=1}^{s}\|R^{ni}\|.
\end{aligned} \quad (11.105)$$

It follows from the inequality (11.105) that

$$\begin{aligned}
&\|e^{n+1}\|+\sqrt{\|De^{n+1}\|^2+\|e'^{n+1}\|^2} \le \big(1+\Delta t(1+4s\beta L)\big)\big(\|e^n\|+\sqrt{\|De^n\|^2+\|e'^n\|^2}\big)\\
&+\|R^{n+1}\|+\sqrt{\|DR^{n+1}\|^2+\|R'^{n+1}\|^2}+2\Delta t(1+\Delta t)\beta L\sum_{i=1}^{s}\|R^{ni}\|.
\end{aligned} \quad (11.106)$$

We note that the truncation errors R^{n+1}, R'^{n+1} and R^{ni} satisfy (11.94) and (11.95), respectively. Then, there exists a constant C satisfying

$$\|R^{n+1}\|+\sqrt{\|DR^{n+1}\|^2+\|R'^{n+1}\|^2}+2\Delta t(1+\Delta t)\beta L\sum_{i=1}^{s}\|R^{ni}\| \le C\Delta t\big(\Delta t^s+\|\delta(\Delta x)\|\big). \quad (11.107)$$

Applying the discrete Gronwall inequality (Lemma 11.3) to (11.106) yields

$$\begin{aligned}
\|e^n\|+\sqrt{\|De^n\|^2+\|e'^n\|^2} \le &\exp\big(n\Delta t(1+4s\beta L)\big)\Big(\|e^0\|+\sqrt{\|De^0\|^2+\|e'^0\|^2}\\
&+Cn\Delta t\big(\Delta t^s+\|\delta(\Delta x)\|\big)\Big).
\end{aligned} \quad (11.108)$$

Therefore, we obtain the following estimates:

$$\begin{aligned}
&\|e^n\| \le CT\exp\big((1+4s\beta L)T\big)\big(\Delta t^s+\|\delta(\Delta x)\|\big),\\
&\|e'^n\| \le CT\exp\big((1+4s\beta L)T\big)\big(\Delta t^s+\|\delta(\Delta x)\|\big).
\end{aligned} \quad (11.109)$$

The conclusions of the theorem are confirmed. □

11.5.3 The Convergence of the Fixed-Point Iteration

The previous subsections derived and analysed the fully discrete scheme. However, the scheme (11.96) is implicit in general. Therefore, iteration is required in practical computations. Fortunately, a wide range of iterative methods (see, e.g. [34, 42, 52]) can be chosen for (11.96). Here, we will use the fixed-point iteration for the implicit scheme and analyse its convergence.

Actually, the iteration is needed only for the computation of the internal stages. The iterative procedure of the fixed-point iteration for (11.96) can be read as

$$\begin{cases} U^{ni}_{[0]} = \phi_0(c_i^2 V)u^n + c_i \Delta t \phi_1(c_i^2 V)u'^n, \\ U^{ni}_{[m+1]} = \phi_0(c_i^2 V)u^n + c_i \Delta t \phi_1(c_i^2 V)u'^n + c_i^2 \Delta t^2 \sum_{j=1}^{s} a_{ij}(V) f(U^{nj}_{[m]}), \\ i = 1, 2, \ldots, s, \quad m = 0, 1, 2, \ldots, \end{cases} \tag{11.110}$$

and

$$\begin{cases} u^{n+1} = \phi_0(V)u^n + \Delta t \phi_1(V)u'^n + \Delta t^2 \sum_{i=1}^{s} b_i(V) f(U^{ni}), \\ u'^{n+1} = -\Delta t A \phi_1(V)u^n + \phi_0(V)u'^n + \Delta t \sum_{i=1}^{s} \bar{b}_i(V) f(U^{ni}). \end{cases} \tag{11.111}$$

Theorem 11.5 *Let the nonlinear function f satisfy the Assumption 2. If the time stepsize Δt satisfies the condition* (11.58), *the iteration procedure determined by* (11.110) *and* (11.111) *is convergent.*

Proof According to Assumption 2 and (11.110), the following inequalities can be obtained

$$\begin{aligned} \|U^{ni}_{[m+1]} - U^{ni}_{[m]}\| \leq & c_i^2 \Delta t^2 \sum_{j=1}^{s} \|a_{ij}(V)\| \cdot \|f(U^{nj}_{[m]}) - f(U^{nj}_{[m-1]})\| \\ \leq & \Delta t^2 \gamma L c_i^2 \sum_{j=1}^{s} \|U^{nj}_{[m]} - U^{nj}_{[m-1]}\|, \quad i = 1, 2, \ldots, s. \end{aligned} \tag{11.112}$$

Then, summing over i in (11.112) yields

$$\sum_{i=1}^{s} \|U^{ni}_{[m]} - U^{ni}_{[m-1]}\| \leq \Delta t^2 \gamma L \sum_{i=1}^{s} c_i^2 \sum_{j=1}^{s} \|U^{nj}_{[m]} - U^{nj}_{[m-1]}\|. \tag{11.113}$$

An argument by induction then gives the following result:

$$\sum_{i=1}^{s} \|U_{[m]}^{ni} - U_{[m-1]}^{ni}\| \leq \left(\Delta t^2 \gamma L \sum_{i=1}^{s} c_i^2\right)^m \sum_{i=1}^{s} \|U_{[1]}^{ni} - U_{[0]}^{ni}\|. \tag{11.114}$$

The limitation of the time stepsize (11.58) leads to

$$\lim_{m\to+\infty} \left(\sum_{i=1}^{s} \|U_{[m]}^{ni} - U_{[m-1]}^{ni}\|\right) \leq \lim_{m\to+\infty} \frac{1}{2^m} \sum_{i=1}^{s} \|U_{[1]}^{ni} - U_{[0]}^{ni}\| = 0. \tag{11.115}$$

Therefore, the iterative procedure (11.110)–(11.111) is convergent.

11.6 The Application to Two-dimensional Dirichlet or Neumann Boundary Problems

The problem considered in (11.1) is the one-dimensional case, and is equipped with the special periodic boundary conditions (11.2). However, our approach can be extended to the considerably more important high-dimensional Klein–Gordon equations. The computational methodology developed in this chapter is very useful and has potential applications in solving more sophisticated multi-dimensional solitary wave equations. In this section, we mainly concentrate on discussing the application of our time-stepping schemes (11.39) to the two-dimensional nonlinear Klein–Gordon equations equipped with Dirichlet or Neumann boundary conditions. There has been a considerable amount of recent discussions on the computation of 2D sine–Gordon type solitons, in particular via different finite difference and finite element methods, splitting algorithms and predictor–corrector schemes (see, e.g. [2, 4, 11, 12, 19, 45]).

The two-dimensional nonlinear Klein–Gordon equation under consideration is expressed by

$$\begin{cases} u_{tt} - a^2(u_{xx} + u_{yy}) = f(u), & (x, y) \in \Omega,\ t_0 < t \leq T, \\ u(x, y, t_0) = \varphi_0(x, y),\ u_t(x, y, t_0) = \varphi_1(x, y), & (x, y) \in \bar{\Omega}, \end{cases} \tag{11.116}$$

where $f(u)$ is a nonlinear function of u chosen as the negative derivative of a potential energy $V(u)$. Here, we suppose that the 2D problem (11.116) is defined on the spatial domain $\Omega = (0, \pi)\times(0, \pi)$ and supplemented with homogenous *Dirichlet boundary conditions:*

$$u(0, y, t) = u(\pi, y, t) = 0, \quad u(x, 0, t) = u(x, \pi, t) = 0, \quad \forall t \in [t_0, T], \tag{11.117}$$

and homogenous *Neumann boundary conditions:*

$$\left.\frac{\partial u}{\partial x}\right|_{x=0,\pi} = 0, \quad \left.\frac{\partial u}{\partial y}\right|_{y=0,\pi} = 0, \quad \forall t \in [t_0, T]. \tag{11.118}$$

For an abstract formulation of the problem (11.116), the linear differential operator $\mathscr{A}$ now should be defined as

$$(\mathscr{A}v)(x, y) = -a^2\big(\partial_x^2 + \partial_y^2\big)v(x, y). \tag{11.119}$$

Likewise, $\mathscr{A}$ is an unbounded symmetric and positive semi-definite operator but not defined for every $v \in L^2(\Omega)$. For our further analysis, the inner product of the space $L^2(\Omega)$ is defined as

$$(u, v) = \int_0^{\pi} \int_0^{\pi} u(x, y)v(x, y)\mathrm{d}x\mathrm{d}y. \tag{11.120}$$

In order to model the homogenous Dirichlet and Neumann boundary conditions, the operator $\mathscr{A}$ should be defined on different function spaces respectively. In what follows, we will analyse the the two-dimensional case.

11.6.1 2D Klein–Gordon Equation with Dirichlet Boundary Conditions

The operator $\mathscr{A}$ is defined on the following domain

$$D(\mathscr{A}) = H^2(\Omega) \cap H_0^1(\Omega). \tag{11.121}$$

In this case, the functions $\sin(mx + ny)$ are orthogonal eigenfunctions of the operator $\mathscr{A}$ corresponding to the eigenvalues $a^2(m^2 + n^2), \ m, n = 1, 2, \ldots$. The functions of the operator $\mathscr{A}$ can be defined as:

$$\phi_j(t\mathscr{A})v(x, y) = \sum_{m=1}^{\infty}\sum_{n=1}^{\infty} \hat{v}_{m,n}\phi_j\big(a^2(m^2 + n^2)t\big)\sin(mx + ny) \tag{11.122}$$

for $v(x, y) = \sum_{m=1}^{\infty}\sum_{n=1}^{\infty} \hat{v}_{m,n}\sin(mx + ny) \in L^2(\Omega)$, where all $\hat{v}_{m,n}$ are the Fourier coefficients of $v(x, y)$. In order to show the operator functions $\phi_j(t\mathscr{A})$ for any $\forall t \in [t_0, T]$ are bounded, we will characterise the L^2 norm in the frequency space as

$$\|v\|^2 = \int_0^{\pi} \int_0^{\pi} |v(x, y)|^2\mathrm{d}x\mathrm{d}y = \frac{\pi^2}{4}\sum_{m=1}^{\infty}\sum_{n=1}^{\infty} |\hat{v}_{m,n}|^2. \tag{11.123}$$

Lemma 11.4 *The functions of the operator $\mathscr{A}$ defined by* (11.122) *are bounded operator under the norm* $\|\cdot\|_{L^2(\Omega)\leftarrow L^2(\Omega)}$, *i.e.*,

$$\|\phi_j(t\mathscr{A})\|_{L^2(\Omega)\leftarrow L^2(\Omega)} \le \gamma_j, \tag{11.124}$$

where γ_j are the bounds of the functions $\phi_j(x)$ for $j = 0, 1, 2, \ldots$ for $x \ge 0$, respectively.

Proof For any function $u(x, y) \in L^2(\Omega)$, its *Fourier series* can be expressed as

$$u(x, y) = \sum_{m=1}^{\infty}\sum_{n=1}^{\infty} \hat{u}_{m,n} \sin(mx + ny).$$

Considering the definition of the norm, we obtain

$$\|\phi_j(t\mathscr{A})u\|^2 = \frac{\pi^2}{4}\sum_{m=1}^{\infty}\sum_{n=1}^{\infty} |\hat{u}_{m,n}|^2 |\phi_j(a^2(m^2+n^2)t)|^2 \le \sup_{t\ge 0} |\phi_j(a^2(m^2+n^2)t|^2 \cdot \|u\|^2 \le \gamma_j^2\|u\|^2.$$

Thus, we deduce the following inequality

$$\|\phi_j(t\mathscr{A})\|^2_{L^2(\Omega)\leftarrow L^2(\Omega)} = \sup_{\|u\|\neq t_0} \frac{\|\phi_j(t\mathscr{A})u\|^2}{\|u\|^2} \le \gamma_j^2, \quad j = 0, 1, 2, \ldots.$$

The conclusion of the lemma is proved. □

The following lemma shows that the operator functions $\phi_j(t\mathscr{A})$ for $j = 0, 1, 2, \ldots$ are symmetric.

Lemma 11.5 *The bounded operator functions $\phi_j(t\mathscr{A})$ for $j = 0, 1, 2, \ldots$ are symmetric operators with respect to the inner product* (11.120).

Proof For any functions $u(x, y), v(x, y) \in L^2(\Omega)$, we have

$$\begin{aligned}(\phi_j(t\mathscr{A})u, v) &= \int_0^{\pi}\int_0^{\pi} \phi_j(t\mathscr{A})u(x, y)v(x, y)\mathrm{d}x\mathrm{d}y\\ &= \frac{\pi^2}{4}\sum_{m=1}^{\infty}\sum_{n=1}^{\infty} \hat{u}_{m,n}\hat{v}_{m,n}\phi_j(a^2(m^2+n^2)t),\end{aligned}$$

and

$$\begin{aligned}(u, \phi_j(t\mathscr{A})v) &= \int_0^{\pi}\int_0^{\pi} u(x, y)\phi_j(t\mathscr{A})v(x, y)\mathrm{d}x\mathrm{d}y\\ &= \frac{\pi^2}{4}\sum_{m=1}^{\infty}\sum_{n=1}^{\infty} \hat{u}_{m,n}\hat{v}_{m,n}\phi_j(a^2(m^2+n^2)t).\end{aligned}$$

Hence, we have

$$(\phi_j(t\mathscr{A})u, v) = (u, \phi_j(t\mathscr{A})v), \quad j = 0, 1, 2, \ldots.$$

The symmetry of the bounded operator functions is proved. □

11.6.2 2D Klein–Gordon Equation with Neumann Boundary Conditions

In this case, we define the operator $\mathscr{A}$ on the domain

$$D(\mathscr{A}) = \{v \in H^2(\Omega) : v_x = 0, \ v_y = 0, \ (x, y) \in \partial\Omega\}. \tag{11.125}$$

The orthogonal eigenfunctions of the operator $\mathscr{A}$ are $\cos(mx + ny)$, and the corresponding eigenvalues are $a^2(m^2 + n^2)$ for $m, \ n = 0, 1, 2, \ldots$. We define the operator functions of $\mathscr{A}$ as:

$$\phi_j(t\mathscr{A})v(x, y) = \sum_{m=0}^{\infty}\sum_{n=0}^{\infty} \hat{v}_{m,n}\phi_j\big(a^2(m^2 + n^2)t\big)\cos(mx + ny), \tag{11.126}$$

for $v(x, y) = \sum\limits_{m=0}^{\infty}\sum\limits_{n=0}^{\infty} \hat{v}_{m,n}\cos(mx + ny) \in L^2(\Omega)$, where $\hat{v}_{m,n}$ are the Fourier coefficients of $v(x, y)$. Similarly, the L^2 norm can be characterized in the frequency space by

$$\|v\|^2 = \int_0^{\pi}\int_0^{\pi} |v(x, y)|^2 \mathrm{d}x\mathrm{d}y = \frac{\pi^2}{4}\sum_{m=0}^{\infty}\sum_{n=0}^{\infty} |\hat{v}_{m,n}|^2. \tag{11.127}$$

In what follows, we show the boundedness of the operator functions $\phi_j(t\mathscr{A})$ for $j = 0, 1, 2, \ldots$ by the following Lemma.

Lemma 11.6 *The functions of the operator $\mathscr{A}$ defined by* (11.126) *are bounded operator under the norm $\|\cdot\|_{L^2(\Omega)\leftarrow L^2(\Omega)}$, i.e.,*

$$\|\phi_j(t\mathscr{A})\|_{L^2(\Omega)\leftarrow L^2(\Omega)} \le \gamma_j, \tag{11.128}$$

where γ_j are the bounds of the functions $\phi_j(x)$, $j = 0, 1, 2, \ldots$ for $x \ge 0$, respectively.

Proof For any function $u(x, y) \in L^2(\Omega)$, its *Fourier series* can be expressed by

$$u(x, y) = \sum_{m=0}^{\infty}\sum_{n=0}^{\infty} \hat{u}_{m,n}\cos(mx + ny).$$

Considering the definition of the norm, we obtain

$$\|\phi_j(t\mathscr{A})u\|^2 = \frac{\pi^2}{4}\sum_{m=0}^{\infty}\sum_{n=0}^{\infty}|\hat{u}_{m,n}|^2|\phi_j(a^2(m^2+n^2)t)|^2 \leq \sup_{t\geq 0}|\phi_j(a^2(m^2+n^2)t)|^2\cdot\|u\|^2 \leq \gamma_j^2\|u\|^2.$$

Hence, we deduce the following inequality

$$\|\phi_j\big(t\mathscr{A}\big)\|^2_{L^2(\Omega)\leftarrow L^2(\Omega)} = \sup_{\|u\|\neq 0}\frac{\|\phi_j\big(t\mathscr{A}\big)u\|^2}{\|u\|^2} \leq \gamma_j^2, \quad j=0,1,2,\ldots.$$

The conclusion of the lemma is proved. □

Similarly, the following lemma shows that the operator functions $\phi_j(t\mathscr{A})$ are symmetric for $j=0,1,2,\ldots$.

Lemma 11.7 *The bounded operator functions* $\phi_j(t\mathscr{A})$ *for* $j=0,1,2,\ldots$ *are symmetric operators with respect to the inner product* (11.120).

Proof For any functions $u(x,y), v(x,y)\in L^2(\Omega)$, we have

$$\begin{aligned}(\phi_j\big(t\mathscr{A}\big)u, v) &= \int_0^{\pi}\int_0^{\pi}\phi_j\big(t\mathscr{A}\big)u(x,y)v(x,y)\mathrm{d}x\mathrm{d}y\\ &=\frac{\pi^2}{4}\sum_{m=0}^{\infty}\sum_{n=0}^{\infty}\hat{u}_{m,n}\hat{v}_{m,n}\phi_j\big(a^2(m^2+n^2)t\big),\end{aligned}$$

and

$$\begin{aligned}(u, \phi_j\big(t\mathscr{A}\big)v) &= \int_0^{\pi}\int_0^{\pi}u(x,y)\phi_j\big(t\mathscr{A}\big)v(x,y)\mathrm{d}x\mathrm{d}y\\ &=\frac{\pi^2}{4}\sum_{m=0}^{\infty}\sum_{n=0}^{\infty}\hat{u}_{m,n}\hat{v}_{m,n}\phi_j\big(a^2(m^2+n^2)t\big).\end{aligned}$$

We then have

$$(\phi_j\big(t\mathscr{A}\big)u, v) = (u, \phi_j\big(t\mathscr{A}\big)v), \quad j=0,1,2,\ldots.$$

The statement of the theorem is confirmed. □

11.6.3 Abstract ODE Formulation and Spatial Discretisation

Similarly to the one dimensional periodic boundary problem (11.1)–(11.2), by defining $u(t)$ as the function that maps (x,y) to $u(x,y,t)$:

$$u(t) = [(x,y)\mapsto u(x,y,t)],$$

we can formulate the two-dimensional problem (11.116) equipped with the Dirichlet boundary conditions (11.121) or Neumann boundary conditions (11.125) as the following abstract ODE on the Hilbert space $L^2(\Omega)$:

$$\begin{cases} u''(t) + \mathscr{A}u(t) = f\big(u(t)\big), \\ u(t_0) = \varphi_1(x, y), \ u'(t_0) = \varphi_2(x, y). \end{cases} \tag{11.129}$$

Theorem 11.6 *The solution of the abstract ODE* (11.129) *and its derivative satisfy*

$$\begin{cases} u(t) = \phi_0\big((t-t_0)^2\mathscr{A}\big)u(t_0) + (t-t_0)\phi_1\big((t-t_0)^2\mathscr{A}\big)u'(t_0) \\ \qquad + \displaystyle\int_{t_0}^{t} (t-\zeta)\phi_1\big((t-\zeta)^2\mathscr{A}\big)f\big(u(\zeta)\big)\mathrm{d}\zeta, \\ u'(t) = -(t-t_0)\mathscr{A}\phi_1\big((t-t_0)^2\mathscr{A}\big)u(t_0) + \phi_0\big((t-t_0)^2\mathscr{A}\big)u'(t_0) \\ \qquad + \displaystyle\int_{t_0}^{t} \phi_0\big((t-\zeta)^2\mathscr{A}\big)f\big(u(\zeta)\big)\mathrm{d}\zeta, \end{cases} \tag{11.130}$$

where $\phi_0\big((t-t_0)^2\mathscr{A}\big)$, $\phi_1\big((t-t_0)^2\mathscr{A}\big)$ *are bounded functions of the operator* $\mathscr{A}$ *for* $\forall t \in [t_0, T]$.

Based on the above analysis, it is straightforward to extend our time-stepping integrators (11.39) to two-dimensional nonlinear Klein–Gordon equations with Dirichlet or Neumann boundary conditions. Moreover, we note that the orthogonal eigenfunctions of the operator $\mathscr{A}$ for Dirichlet and Neumann boundary problems are $\sin(mx)\sin(ny)$ and $\cos(mx)\cos(ny)$, respectively. In order to reduce the computation caused by the spatial discretisation, we focus much more on choosing Fourier spectral methods. The numerous related researches on the *discrete Fast Cosine / Sine Transformation* have been widely studied in the literature (see, e.g. [14–16, 43]). The corresponding spatial discretisation methods are the *discrete Fast Sine Transformation* for the underlying Dirichlet boundary problem, and the *discrete Fast Cosine Transformation* for the underlying Neumann boundary case.

11.7 Numerical Experiments

In this section, we derive three practical time integrators and illustrate the numerical results for one dimensional Klein–Gordon equation with periodic boundary conditions and two-dimensional sine–Gordon with homogenous Dirichlet or Neumann boundary conditions. It is clear that our time integrators (11.39) are determined by (11.34), (11.35) and (11.48) with appropriate nodes c_i for $i = 1, 2, \ldots, s$. Moreover, we note from our error analysis that there is a term $\max\limits_{0\le z\le 1} |w_s(z)|$ involved in the constant C_1. In order to minimise the constant C_1, here and in the following, we choose Gauss-Legendre nodes.

In the first example, we choose the two-point Gauss-Legendre nodes,

$$c_1 = \frac{3-\sqrt{3}}{6}, \quad c_2 = \frac{3+\sqrt{3}}{6}, \tag{11.131}$$

and the corresponding time integrator determined by (11.34), (11.35) and (11.48) is denoted by ***GLC2***.

For the second example, the following three-point Gauss-Legendre nodes

$$c_1 = \frac{5-\sqrt{15}}{10}, \quad c_2 = \frac{1}{2}, \quad c_3 = \frac{5+\sqrt{15}}{10}, \tag{11.132}$$

together with (11.34), (11.35) and (11.48) determine the three-point time integrator which is denoted by ***GLC3***.

For the third example, we take the four-point Gauss-Legendre nodes

$$\begin{aligned} c_1 &= \frac{1-\sqrt{\frac{15+2\sqrt{30}}{35}}}{2}, \quad c_2 = \frac{1-\sqrt{\frac{15-2\sqrt{30}}{35}}}{2}, \\ c_3 &= \frac{1+\sqrt{\frac{15-2\sqrt{30}}{35}}}{2}, \quad c_4 = \frac{1+\sqrt{\frac{15+2\sqrt{30}}{35}}}{2}, \end{aligned} \tag{11.133}$$

and denote the corresponding time integrator determined by (11.34), (11.35) and (11.48) by ***GLC4***.

For comparison, in what follows, we briefly describe a collection of classical finite difference and the method-of-lines approximations of the nonlinear Klein–Gordon equation. The methods are listed below:

1. The standard finite difference schemes (see, e.g. [9, 25, 48])

Let u_j^n be the approximation of $u(x_j, t_n)$ ($j = 0, 1, \ldots, M, n = 0, 1, \ldots, N$) and introduce the finite difference discretisation operators

$$\delta_t^2 u_j^n = \frac{u_j^{n+1} - 2u_j^n + u_j^{n-1}}{\Delta t^2} \quad \text{and} \quad \delta_x^2 u_j^n = \frac{u_{j+1}^n - 2u_j^n + u_{j-1}^n}{\Delta x^2}.$$

Here, we consider three frequently used *finite difference schemes* to discretise the problem (11.1)–(11.2) as follows:

- Explicit finite difference (*Expt-FD*) scheme

$$\delta_t^2 u_j^n - a^2 \delta_x^2 u_j^n = f(u_j^n);$$

- Semi-implicit finite difference (*Simpt-FD*) scheme

$$\delta_t^2 u_j^n - \frac{a^2}{2}\big(\delta_x^2 u_j^{n+1} + \delta_x^2 u_j^{n-1}\big) = f(u_j^n);$$

- Compact finite difference (*Compt-FD*) scheme

$$\big(I + \frac{\Delta x^2}{12}\delta_x^2\big)\delta_t^2 u_j^n - \frac{a^2}{2}\big(\delta_x^2 u_j^{n+1} + \delta_x^2 u_j^{n-1}\big) = \big(I + \frac{\Delta x^2}{12}\delta_x^2\big) f(u_j^n).$$

2. *The method-of-lines schemes*

Firstly, we approximate the spatial differential operator $\mathscr{A}$ to obtain a semi-discrete system of the form

$$u''(t) + Au(t) = f\big(u(t)\big), \tag{11.134}$$

where A is a symmetric and positive semi-definite matrix. Then, we use an ODE solver to deal with the semi-discrete system. There are many different ODE solvers for the semi-discrete system (11.134). Here, the time integrators selected for comparisons are:

- GAS2s4: the two-stage Gauss time integration method of order four presented in [28];
- LIIIB4s6: the Labatto IIIB method of order six presented in [28];
- IRKN2s4: the two-stage implicit symplectic Runge-Kutta-Nyström (IRKN) method of order four derived in [51];
- IRKN3s6: the three-stage implicit symplectic Runge-Kutta-Nyström (IRKN) method of order six derived in [51];
- ERKN3s4: the three-stage extended Runge-Kutta-Nyström (ERKN) time integration method of order four for second order ODEs proposed in [54];
- SMMERKN5s5: the five-stage explicit symplectic multi-frequency and multi-dimensional extended Runge–Kutta–Nyström (ERKN) method of order five with some small residuals for second order ODEs proposed in [55].

It is noted that we use fixed-point iteration for all of the implicit time integration methods in our numerical experiments. We set the error tolerance as 10^{-15}, and put the maximum iteration number $m = 1000$ in each iteration procedure. Here, it should be pointed out that, if the error produced by a method is too large for some time stepsize Δt, then the corresponding point will not be plotted in the figure.

All computations in the numerical experiments are carried out by using MATLAB 2011b on the the computer Lenovo ThinkCentre M8300t (CPU: Intel (R) Core (TM) i5-2400 CPU @ 3.10 GHz, Memory: 8 GB, Os: Microsoft Windows 7 with 64 bit).

11.7.1 One-dimensional Problem with Periodic Boundary Conditions

Problem 11.1 We consider the sine–Gordon equation

$$\frac{\partial^2 u}{\partial t^2}(x,t) - \frac{\partial^2 u}{\partial x^2}(x,t) = -\sin(u(x,t)), \tag{11.135}$$

on the region $-20 \le x \le 20$ and $0 \le t \le T$, subject to the initial conditions

$$u(x,0) = 0, \quad u_t(x,0) = 4\text{sech}\Big(x/\sqrt{1+c^2}\Big)/\sqrt{1+c^2},$$

where $\kappa = 1/\sqrt{1+c^2}$. The exact solution of this problem is given by

$$u(x,t) = 4\arctan\left(c^{-1}\sin(ct/\sqrt{1+c^2})\text{sech}(x/\sqrt{1+c^2})\right).$$

This problem is known as the breather solution of the sine–Gordon equation (see, e.g. [39]), and represents a pulse-type structure of a soliton. The parameter c is the velocity and we choose $c = 0.5$. The potential function is $V(u) = 1 - \cos(u)$.

In Figs. 11.1 and 11.2, we integrate the sine–Gordon equation (11.135) over the region $(x,t) \in [-20, 20] \times [0, 100]$ using the time integrator ***GLC4*** coupled with the eighth-order symmetric finite difference (SFD) method and the Fourier spectral collocation (FSC) method. The graphs of the errors are shown in Figs. 11.1 and 11.2 with the time stepsize $\Delta t = 0.01$ and several different values of M. The numerical results demonstrate the accuracy of the spatial discretisation, and also indicate that the Fourier spectral collocation method is much better to discretise the spatial derivative than the eighth-order finite difference method. Therefore, it is evident that the Fourier spectral collocation method is the best choice to discretise the spatial variable for this problem.

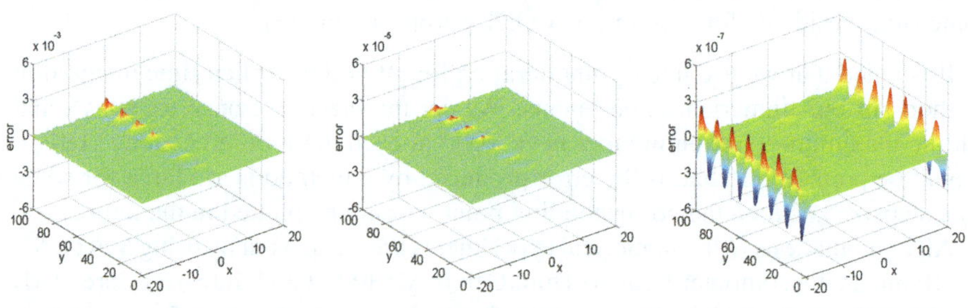

Fig. 11.1 The graphs of errors for the sine–Gorden equation obtained by combining the time integrator GLC4 with eighth-order finite difference spatial discretisation for the time stepsize $\Delta t = 0.01$ and several values of $M = 100$ (left), 200 (middle), and 400 (right)

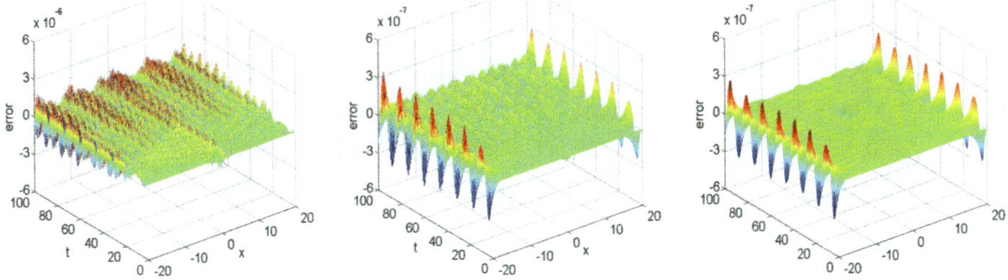

Fig. 11.2 The errors produced by combining the time integrator GLC4 with spatial discretisation by the Fourier spectral method for the time stepsize $\Delta t = 0.01$ and several values of $M = 100$ (left), 120 (middle), and 200 (right)

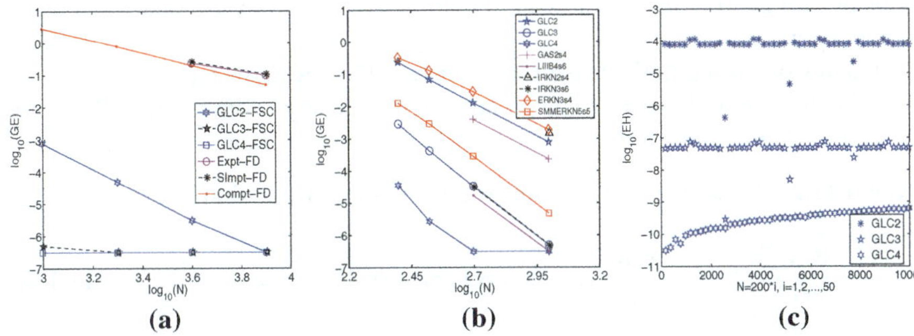

Fig. 11.3 The logarithms of the global error (GE) obtained by comparing our new schemes with the standard finite difference schemes (**a**) and the method-of-lines schemes (**b**) against different time integration stepsizes. **c** The conservation results of the GLCs with spatial discretisation by Fourier spectral collocation method (M=400). The time stepsize $\Delta t = 0.1$ for $T = 1000$

In Fig. 11.3a and b, the problem is integrated over the region $(x, t) \in [-20, 20] \times [0, 100]$ with different time stepsizes Δt and the spatial nodal values M. We compare our methods with the standard finite difference schemes in Fig. 11.3a. We choose $M = 1000$ for the finite difference schemes Expt-FD, SImpt-FD and Compt-FD and $M = 200$ for the time integrators GLCs coupled with the Fourier spectral collocation method (GLC-FSC). The logarithms of the global errors $GE = \|u(t_n) - u^n\|_\infty$ against different time stepsizes $\Delta t = 0.1/2^{j-1}$ for $j = 1, 2, 3, 4$ are displayed in Fig. 11.3a. In comparison with the method-of-lines schemes, we first discretise the spatial derivative by the Fourier spectral collocation method with fixed $M = 200$, and then integrate the semi-discrete system with different time stepsizes $\Delta t = 0.4, 0.3, 0.2$ and 0.1. The efficiency curves are shown in Fig. 11.3b.

Besides, in Fig. 11.3c, the problem is discretised by the Fourier spectral collocation method with the fixed $M = 400$. We then integrate the semi-discrete system over the time interval $t \in [0, 1000]$ by the derived time integrators GLCs with the time stepsize $\Delta t = 0.1$. The numerical results in Fig. 11.3c present the error of the semi-discrete energy conservation law as a function of the time stepsize calculated by

Table 11.1 The total numbers of iterations for different error tolerances with $M = 400$ and $\Delta t = 0.1$ for $T = 100$.

	IRKN2s4	IRKN3s6	GAS2s4	LIIIB4s6	GLC2	GLC3	GLC4
10^{-6}	2038	1996	4161	9309	1988	1988	1992
10^{-8}	2056	7967	7732	5745	2000	2000	2000
10^{-10}	2063	7579	8587	13519	2993	2993	2999
10^{-12}	2063	9508	45739	21820	3000	3000	3000

$\tilde{E}(t)$, where $\log_{10}(EH) = \log_{10}(|\tilde{E}(t_n) - \tilde{E}(t_0)|)$. We also display the total numbers of iterations in Table 11.1 when applying the different methods with different error tolerances to this problem for showing the efficiency of the fixed-point iteration in actual computations.

In conclusion, the numerical results demonstrate that the time-stepping integrators derived in this chapter have much better accuracy and energy conservation. They are more practical and efficient than existing methods in the literature.

Problem 11.2 We consider the nonlinear Klein–Gordon equation

$$\frac{\partial^2 u}{\partial t^2}(x,t) - a^2\frac{\partial^2 u}{\partial x^2}(x,t) + au(x,t) - bu^3(x,t) = 0, \tag{11.136}$$

on the region $(x, t) \in [-20, 20] \times [0, T]$, subject to the initial conditions

$$u(x,0) = \sqrt{\frac{2a}{b}}\operatorname{sech}(\lambda x), \quad u_t(x,0) = c\lambda\sqrt{\frac{2a}{b}}\operatorname{sech}(\lambda x)\tanh(\lambda x),$$

with $\lambda = \sqrt{\frac{a}{a^2-c^2}}$ and $a, b, a^2 - c^2 > 0$. The exact solution of Problem 11.2 is given by

$$u(x,t) = \sqrt{\frac{2a}{b}}\operatorname{sech}(\lambda(x - ct)). \tag{11.137}$$

The real parameter $\sqrt{2a/b}$ represents the amplitude of a soliton which travels with the velocity c. The potential function is $V(u) = \frac{a}{2}u^2 - \frac{b}{4}u^4$. The problem can be found in [39]. We consider the parameters $a = 0.3$, $b = 1$ and $c = 0.25$ which are similar to those in [39].

The Klein–Gordon equation 11.2 is solved by using the time integrator GLC4 coupled with the eighth-order symmetric finite difference method and the Fourier spectral collocation method. The graphs of errors are shown in Figs. 11.4 and 11.5 with the fixed time stepsize $\Delta t = 0.01$ and several values of M. The numerical results in Figs. 11.4 and 11.5 indicate that the Fourier spectral collocation method as a spatial discretisation method is much more accurate than the eighth-order finite difference method.

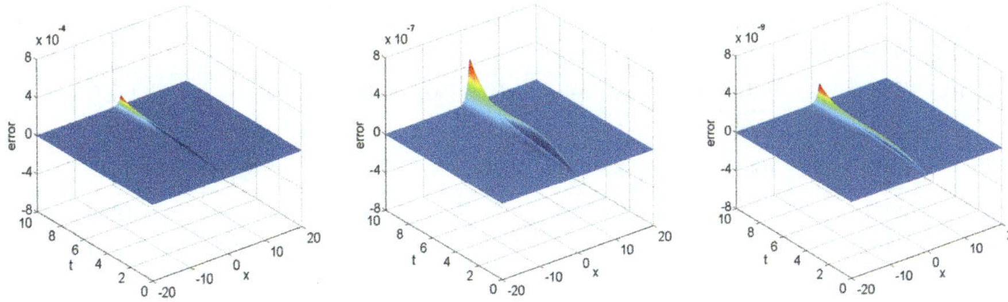

Fig. 11.4 The graphs of errors for the Klein-Gorden equation obtained by combing the time integrator GLC4 with the eighth-order finite difference spatial discretisation for the time stepsize $\Delta t = 0.01$ and several values of $M = 500$ (left), 1000 (middle) and 2000 (right)

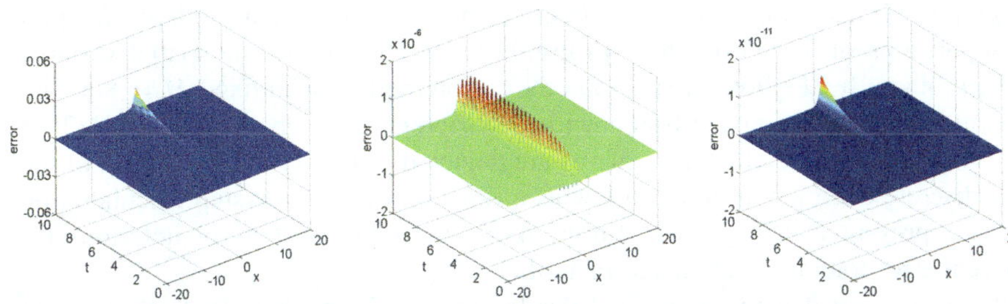

Fig. 11.5 The errors produced by combining the time integrator GLC4 with spatial discretisation by the Fourier spectral collocation method for the Klein-Gorden equation with the time stepsize $\Delta t = 0.01$ and $M = 200$ (left), 400 (middle) and 800 (right)

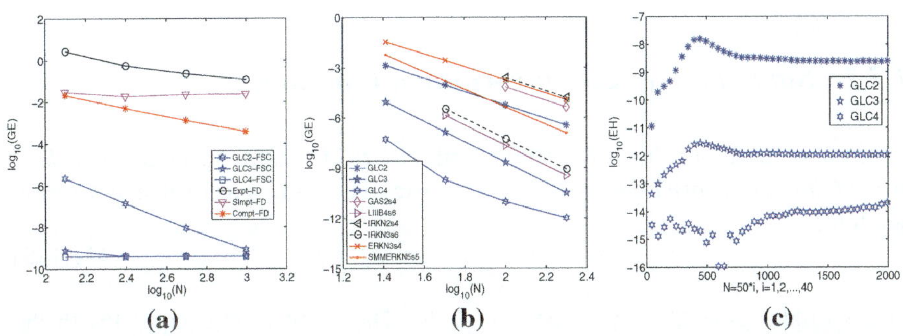

Fig. 11.6 The logarithms of the global error (GE) obtained by comparing our new schemes with **a** standard finite difference schemes and **b** the method-of-lines schemes against different time integration stepsizes. **c** The energy conservation results for the GLCs with spatial discretisation by the Fourier spectral collocation method (M=200). The time stepsize $\Delta t = 0.05$ for $T = 100$

In order to compare our methods with the classical finite difference schemes and the method-of-lines methods, we integrate the problem over the region $[-20, 20] \times [0, 10]$ with different time stepsizes Δt and spatial nodal values M. In Fig. 11.6a, we compare our methods with the classical finite difference schemes against different

Table 11.2 The total numbers of iterations for different error tolerances with $M = 800$ and $\Delta t = 0.1$ for $T = 100$

Tolerance	IRKN2s4	IRKN3s6	GAS2s4	LIIIB4s6	GLC2	GLC3	GLC4
10^{-6}	2396	2000	5207	5566	686	686	975
10^{-8}	3871	3000	8585	7551	707	707	994
10^{-10}	6460	4568	12600	10659	1038	1038	1464
10^{-12}	9462	6139	16327	13800	1085	1085	1491

time stepsizes $\Delta t = 0.08/2^{j-1}$ for $j = 1, 2, 3, 4$. We use $M = 1000$ for the finite difference schemes Expt-FD, SImpt-FD and Compt-FD and $M = 600$ for the time integrators GLCs coupled with the Fourier spectral collocation method. We plot the logarithms of the global error in Fig. 11.6a. We discretise the spatial variable of the problem by the Fourier spectral collocation method with fixed $M = 800$ and integrate the semi-discretised system with the different time stepsizes $\Delta t = 0.4/2^{j-1}$ for $j = 1, 2, 3, 4$. The efficiency curves are depicted in Fig. 11.6b. The errors of the semi-discrete energy conservation law as a function of the time-step calculated by $\tilde{E}(t)$ are presented in Fig. 11.6c. Furthermore, the total numbers of iterations for different error tolerances are listed in Table 11.2.

It can be seen that the numerical results again indicate that our time-stepping integrators have higher precision than existing methods in the literature, and the qualitative property of energy preservation is also quite promising.

11.7.2 Simulation of 2D Sine–Gordon Equation

In this subsection, our time integration method ***GLC4*** coupled with *Discrete Fast Cosine / Sine Transformation* is used to simulate the two-dimensional sine–Gordon equation:

$$u_{tt} - (u_{xx} + u_{yy}) = -\sin(u), \quad t > 0, \tag{11.138}$$

in the spatial region $\Omega = (-a, a) \times (-b, b)$. The problem is equipped with the following homogeneous Dirichlet or Neumann boundary conditions, namely,

- *Dirichlet boundary conditions:*

$$u(\pm a, y, t) = u(x, \pm b, t) = 0; \tag{11.139}$$

- *Neumann boundary conditions:*

$$u_x(\pm a, y, t) = u_y(x, \pm b, t) = 0. \tag{11.140}$$

The initial conditions are

$$u(x, y, 0) = f(x, y), \quad u_t(x, y, 0) = g(x, y). \tag{11.141}$$

It is known that different initial conditions lead to different numerical phenomena. In what follows, we will use our method to simulate three different types of circular ring solitons. The initial conditions and parameters are chosen similarly to those in [12, 45].

Problem 11.3 For the particular case of circular ring solitons (see, e.g. [12, 45]), we select the following initial conditions:

$$f(x, y) = 4 \arctan\left(\exp\left(3 - \sqrt{x^2 + y^2}\right)\right), \quad g(x, y) = 0, \tag{11.142}$$

over the two-dimensional domain $(x, y) \in [-14, 14] \times [-14, 14]$. The simulation results and the corresponding contour plots at the times $t = 0, 4, 8, 11.5, 13$ and 15 are presented in Figs. 11.7 and 11.8 in terms of $\sin(u/2)$ for the mesh region size 400×400 and time stepsize $\Delta t = 0.1$. It can be clearly observed from Fig. 11.7 that the ring soliton shrinks at the initial stage ($t = 0$), but oscillations and radiations begin to form and continue until time $t = 8$. Moreover, it can be seen from the graphs that a ring soliton is nearly formed again at time $t = 11.5$. In Fig. 11.8, the contour maps depict the movement of the soliton very clearly. The CPU time required to reach $t = 15$ is 668.056765 seconds.

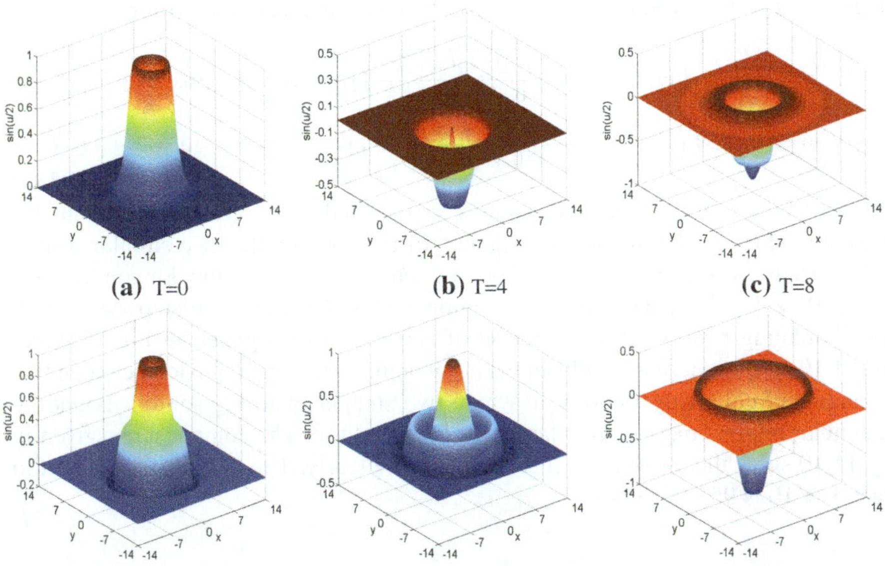

(a) T=0 **(b)** T=4 **(c)** T=8

(d) T=11.5 **(e)** T=13 **(f)** T=15

Fig. 11.7 Circular ring solitons: the function of $\sin(u/2)$ for the initial condition and numerical solutions at the times $t = 0, 4, 8, 11.5, 13$ and 15, successively

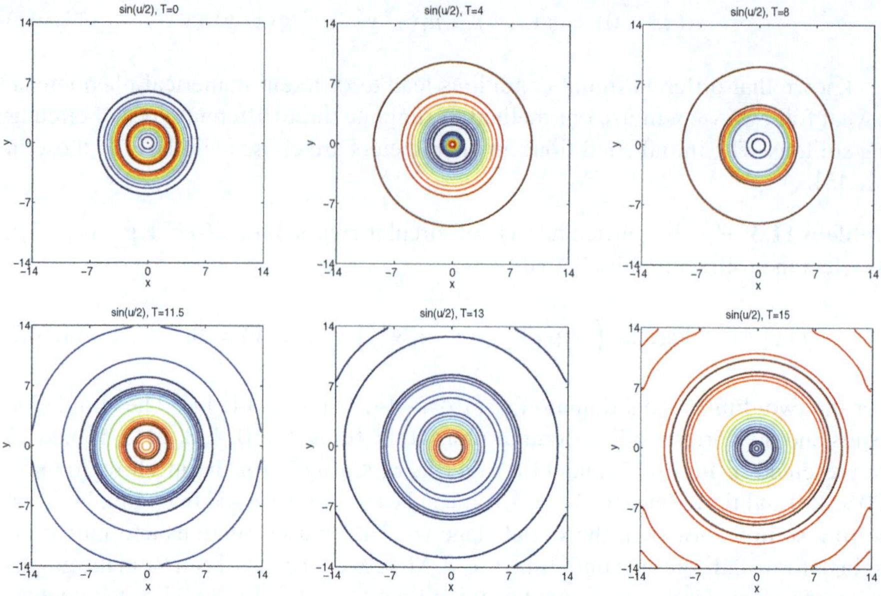

Fig. 11.8 Circular ring solitons: contours of $\sin(u/2)$ for the initial condition and numerical solutions at the times $t = 0, 4, 8, 11.5, 13$ and 15, successively

Problem 11.4 Furthermore, if we choose the following standard setting:

$$
\begin{aligned}
f(x, y) &= 4\arctan\Big(\exp\Big(\frac{4-\sqrt{(x+3)^2+(y+7)^2}}{0.436}\Big)\Big), \quad -10 \le x \le 10,\ -7 \le y \le 7,\\
g(x, y) &= 4.13\mathrm{sech}\Big(\exp\Big(\frac{4-\sqrt{(x+3)^2+(y+7)^2}}{0.436}\Big)\Big), \quad -10 \le x \le 10,\ -7 \le y \le 7,
\end{aligned}
\tag{11.143}
$$

and extend the solution across the sides $x = -10$ and $y = -7$ using the symmetry properties of the problem, the phenomenon of the collision for two circular soliton will be occurred (see, e.g. [12, 45]). We compute solutions over the domain $(x, y) \in [-30, 10] \times [-21, 7]$ with the mesh region 800×400 and time step $\Delta t = 0.1$. The simulating results, as the function of $\sin(u/2)$, are depicted in Fig. 11.9 and Fig. 11.10. The numerical results in Fig. 11.9 demonstrate the collision between two expanding circular ring solitons, in which two smaller oval ring solitons bounding an annular region emerge into a larger oval ring soliton. The contour maps given in Fig. 11.10 show the movement of solitons much clearly. The CPU time required to reach $t = 10$ is 953.263314 s.

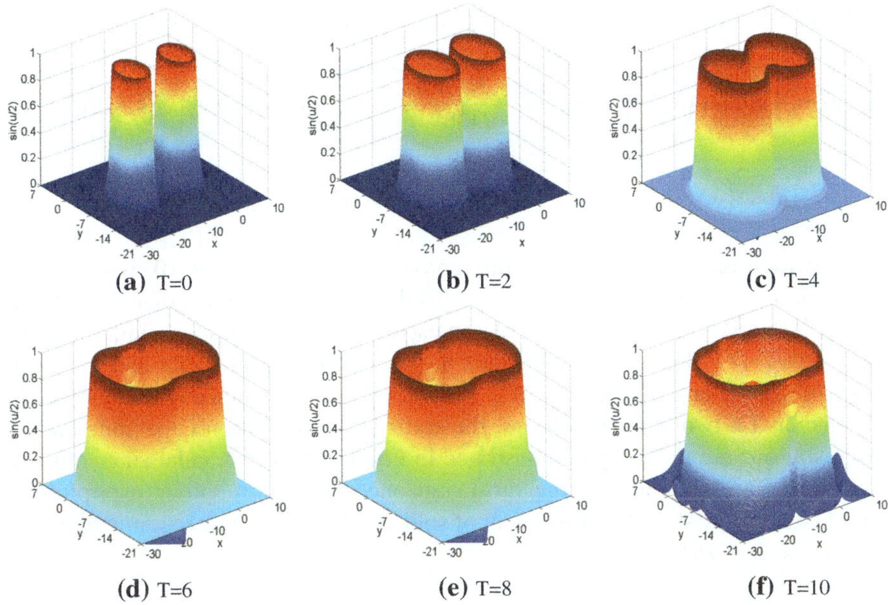

Fig. 11.9 Collision of two ring solitons: the function of $\sin(u/2)$ for the initial condition and numerical solutions at the times $t = 0, 2, 4, 6, 8$, and 10, successively

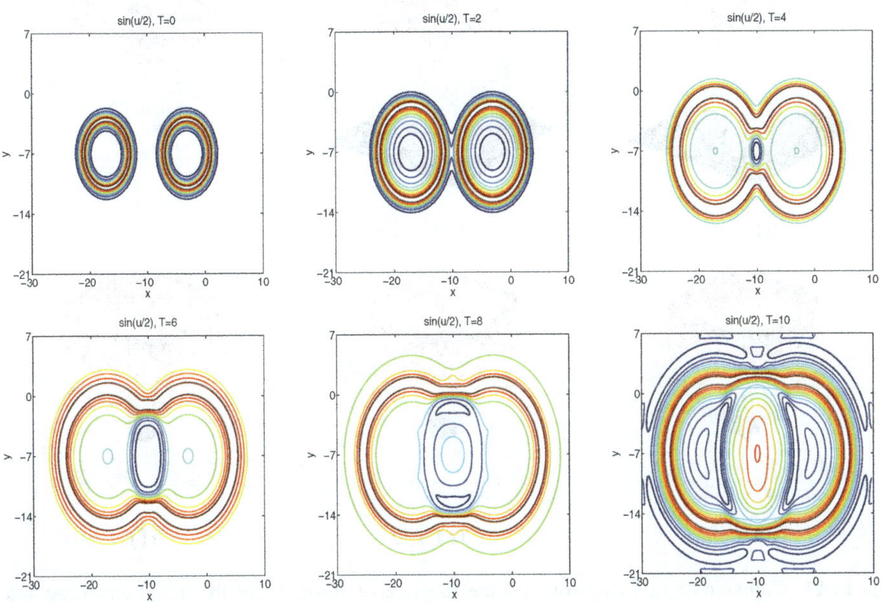

Fig. 11.10 Collision of two ring solitons: contours of $\sin(u/2)$ for the initial condition and numerical solutions at the times $t = 0, 2, 4, 6, 8$, and 10, successively

Problem 11.5 Finally, for collisions of four circular solitons, we take

$$
\begin{aligned}
f(x, y) &= 4 \arctan\Big(\exp\Big(\frac{4-\sqrt{(x+3)^2+(y+3)^2}}{0.436}\Big)\Big), \quad -10 \le x, y \le 10,\\
g(x, y) &= \frac{4.13}{\cosh\Big(\exp\Big(\big(4-\sqrt{(x+3)^2+(y+3)^2}\big)/0.436\Big)\Big)}, \quad -10 \le x, y \le 10.
\end{aligned}
\tag{11.144}
$$

The simulation of the problem over the region $[-30, 10] \times [-30, 10]$ is based on an extension across $x = -10$ and $y = -10$ due to the symmetry of the problem (see, e.g. [12, 45]). The size of mesh region used is 800×800 in space with the time stepsize $\Delta t = 0.1$. The numerical results are presented in Figs. 11.11 and 11.12 in terms of $\sin(u/2)$ at the times $t = 0, 2.5, 5, 7.5, 9$ and 10. Similarly to the case of the collisions for two circular solitons, the collision between four expanding circular ring solitons are precisely demonstrated in Fig. 11.11. The smaller ring solitons bounding an annular region emerge into a large one. Again, the contour maps plotted in Fig. 11.12 clearly show the movement of solitons. The CPU time required to reach $t = 10$ is 2492.677810 s.

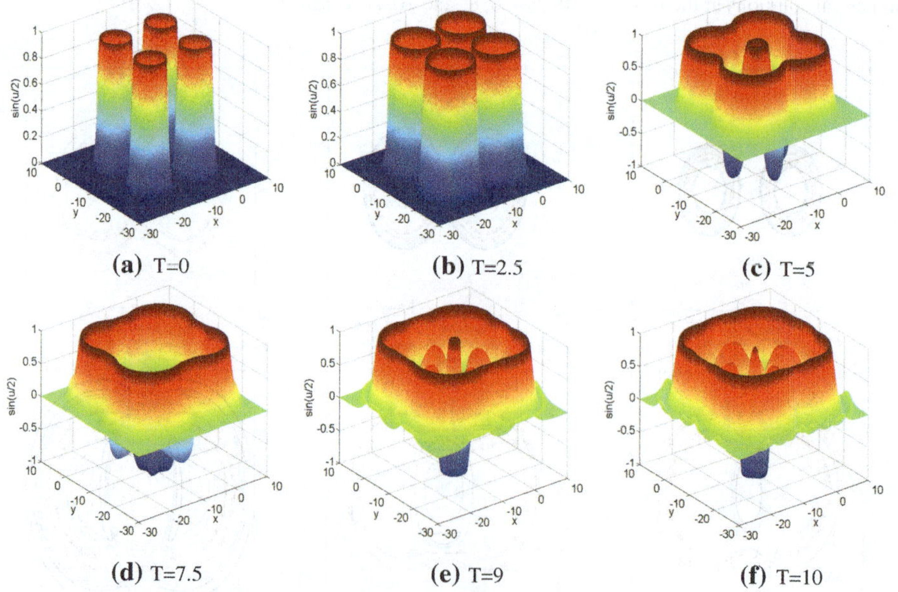

Fig. 11.11 Collision of four ring solitons: the function of $\sin(u/2)$ for the initial condition and numerical solutions at the times $t = 0, 2.5, 5, 7.5, 9$, and 10, successively

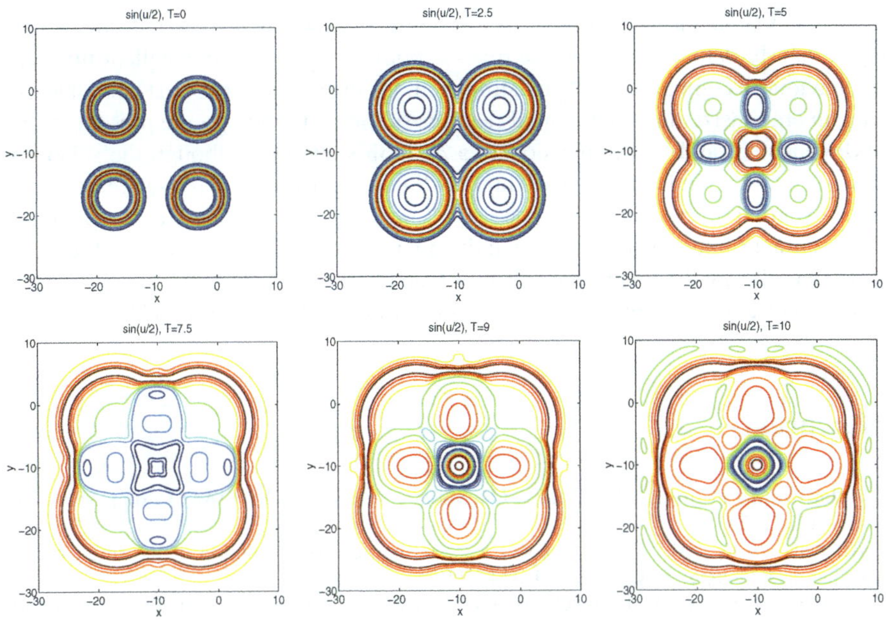

Fig. 11.12 Collision of four ring solitons: contours of $\sin(u/2)$ for the initial condition and numerical solutions at the times $t = 0, 2.5, 5, 7.5, 9$, and 10, successively

11.8 Conclusions and Discussions

In this chapter, the nonlinear Klein–Gordon equation (11.1)–(11.2) was firstly introduced as an abstract ODE on the Hilbert space $L^2(\Omega)$ on the basis of the operator spectral theory. Then, the operator-variation-of-constants formula (11.14) was derived based on the well-known *Duhamel Principle*, which is in fact an integral equation of the solution for the nonlinear Klein–Gordon equation. Using the formula (11.14) and keeping the eventual discretisation in mind, a novel class of time-stepping methods (11.39) has been derived and analysed. It has been shown that under the simplified order conditions (11.48) and chosen suitable collocation nodes the derived time-stepping integrator can have arbitrarily high-order. The spatial discretisation is implemented, following a Lagrange collocation-type time-stepping integrator. This allows us to consider a suitable spatial approximation and gives us a great degree of flexibility when handling nonlinear potentials. The stability and convergence for the fully discrete scheme were rigorously proved after spatial discretisation. Since the fully discrete scheme is implicit and iteration is required, we used the fixed-point iteration (11.110)–(11.111) in practical computation and analysed the convergence of the iteration. Moreover, we also showed that our time-stepping integrators coupled with *discrete Fast Sine / Cosine Transformation* can efficiently simulate the important two-dimensional Klein–Gordon equations, equipped with Dirichlet or Neumann boundary conditions. The numerical experiments carried out in this chapter clearly demonstrate that the time-stepping schemes have excellent numerical behaviour in comparison with existing methods in the literature. Last but not least, we again emphasize that all essential features of the methodology are present in the one-dimensional and two-dimensional cases in this chapter, although the schemes discussed equally lend themselves to higher-dimensional case. Moreover, remembering the eventual discretisation in space, applying a two–point Hermite interpolation to the nonlinear integrals that appear in the operator-variation-of-constants formula, we also can design different time schemes (see [40]).

The material of this chapter is based on the work by Liu and Wu [41].

References

1. Ablowitz, M.J., Kruskal, M.D., Ladik, J.F.: Solitary wave collisions. SIAM J. Appl. Math. **36**, 428–437 (1979)
2. Ablowitz, M.J., Herbst, B.M., Schober, C.: On the numerical solution of the sine-Gordon equation. J. Comput. Phys. **126**, 299–314 (1996)
3. Abramowitz, M., Stegun, I.: Handbook of Mathematical Functions with Formulas, Graphs, and Mathematical Tables. Dover publications, USA (1964)
4. Argyris, J., Haase, M., Heinrich, J.C.: Finite element approximation to two-dimensional sine-Gordon solitons. Comput. Methods Appl. Mech. Eng. **86**, 1–26 (1991)
5. Bank, R., Graham, R.L., Stoer, J., Varga, R., Yserentant, H.: Hight Order Difference Method for Time Dependent PDE. Springer, Berlin (2008)
6. Bátkai, A., Farkas, B., Csomós, P., Ostermann, A.: Operator semigroups for numerical analysis. In: 15th Internet Seminar (2011–12)
7. Baĭnov, D.D., Minchev, E.: Nonexistence of global solutions of the initial-boundary value problem for the nonlinear Klein–Gordon equation. J. Math. Phys. **36**, 756–762 (1995)
8. Bader, P., Iserles, A., Kropielnicka, K., Singh, P.: Effective approximation for the semiclassical Schrödinger equation. Found. Comput. Math. **14**, 689–720 (2014)
9. Bao, W.Z., Dong, X.C.: Analysis and comparison of numerical methods for the Klein-Gordon equation in the nonrelativistic limit regime. Numer. Math. **120**, 189–229 (2012)
10. Biswas, A.: Soliton perturbation theory for phi-four model and nonlinear Klein-Gordon equations. Commun. Nonlinear Sci. Numer. Simul. **14**, 3239–3249 (2009)
11. Bratsos, A.G.: A modified predictor-corrector scheme for the two-dimensional sine-Gordon equation. Numer. Algorithms **43**, 295–308 (2006)
12. Bratsos, A.G.: The solution of the two-dimensional sine-Gordon equation using the method of lines. J. Comput. Appl. Math. **206**, 251–277 (2007)
13. Brenner, P., van Wahl, W.: Global classical solutions of nonlinear wave equations. Math. Z. **176**, 87–121 (1981)
14. Briggs, W.L., Henson, V.E.: The DFT: An Owners Manual for the discrete Fourier Transform. SIAM, Philadelphia (2000)
15. Britanak, V., Yip, P.C., Rao, K.R.: Discrete Cosine and Sine transforms: General Properties. Fast Algorithms and Integer Approximations. Academic Press, Dublin (2006). ISBN 978-0-12-373624-6
16. Bueno-Orovio, A., Kay, D., Burrage, K.: Fourier spectral methods for fractional-in-space reaction-diffusion equations. BIT. Numer. Math. **54**, 937–957 (2014)
17. Cohen, D., Hairer, E., Lubich, C.: Conservation of energy, momentum and actions in numerical discretizations of non-linear wave equations. Numer. Math. **110**, 113–143 (2008)

18. Cao, W., Guo, B.: Fourier collocation method for solving nonlinear Klein-Gordon equation. J. Comput. Phys. **108**, 296–305 (1993)
19. Caputo, J.G., Flytzanis, N., Gaididei, Y.: Split mode method for the elliptic 2D sine-Gordon equation: application to Josephson junction in overlap geometry. Int. J. Mod. Phys. C **9**, 301–323 (1998)
20. Dehghan, M., Ghesmati, A.: Application of the dual reciprocity boundary integral equation technique to solve the nonlinear Klein-Gordon equation. Comput. Phys. Commun. **181**, 1410–1418 (2010)
21. Dehghan, M., Shokri, A.: Numerical solution of the nonlinear Klein-Gordon equation using radial basis functions. J. Comput. Appl. Math. **230**, 400–410 (2009)
22. Dehghan, M., Mohammadi, V.: Two numerical meshless techniques based on radial basis functions (RBFs) and the method of generalized moving least squares (GMLS) for simulation of coupled Klein-Gordon-Schrodinger (KGS) equations. Comput. Math. Appl. **71**, 892–921 (2016)
23. Dodd, R.K., Eilbeck, I.C., Gibbon, J.D., Morris, H.C.: Solitons and Nonlinear Wave Equations. Academic, London (1982)
24. Drazin, P.J., Johnson, R.S.: Solitons: An Introduction. Cambridge University Press, Cambridge (1989)
25. Duncan, D.B.: Symplectic finite difference approximations of the nonlinear Klein-Gordon equation. SIAM J. Numer. Anal. **34**, 1742–1760 (1997)
26. Ginibre, J., Velo, G.: The global Cauchy problem for the nonlinear Klein-Gordon equation. Math. Z. **189**, 487–505 (1985)
27. Guo, B.Y., Li, X.: V*á*zquez, L.: A Legendre spectral method for solving the nonlinear Klein-Gordon equation. Comput. Appl. Math. **15**, 19–36 (1996)
28. Hairer, E., Lubich, C., Wanner, G.: Geometric Numerical Integration: Structure-Preserving Algorithms for Ordinary Differential Equations, 2nd edn. Springer, Berlin (2006)
29. Hesthaven, J.S., Gottlieb, S., Gottlieb, D.: Spectral methods for Time-Dependent problems. Cambridge University Press, Cambridge Monographs on Applied and Computational Mathematics (2007)
30. Hochbruck, M., Ostermann, A.: Explicit exponential Runge-Kutta methods for semilinear parabolic problems. SIAM J. Numer. Anal. **43**, 1069–1090 (2005)
31. Hochbruck, M., Ostermann, A.: Exponential Runge-Kutta methods for parabolic problems. Appl. Numer. Math. **53**, 323–339 (2005)
32. Hochbruck, M., Ostermann, A.: Exponential integrators. Acta Numer. **19**, 209–286 (2010)
33. Iserles, A.: A First Course in the Numerical Analysis of Differential Equations, 2nd edn. Cambridge University Press, Cambridge (2008)
34. Janssen, J., Vandewalle, S.: On SOR waveform relaxation methods. SIAM J. Numer. Anal. **34**, 2456–2481 (1997)
35. Jim*é*nez, S.: Derivation of the discrete conservation laws for a family of finite difference schemes. Appl. Math. Comput. **64**, 13–45 (1994)
36. Kosecki, R.: The unit condition and global existence for a class of nonlinear Klein-Gordon equations. J. Differ. Equ. **100**, 257–268 (1992)
37. Lakestani, M., Dehghan, M.: Collocation and finite difference-collocation methods for the solution of nonlinear Klein-Gordon equation. Comput. Phys. Commun. **181**, 1392–1401 (2010)
38. Li, S., Vu-Quoc, L.: Finite difference calculus invariant structure of a class of algorithms for the nonlinear Klein-Gordon equation. SIAM J. Numer. Anal. **32**, 1839–1875 (1995)
39. Liu, C., Shi, W., Wu, X.Y.: An efficient high-order explicit scheme for solving Hamiltonian nonlinear wave equations. Appl. Math. Comput. **246**, 696–710 (2014)
40. Liu, C., Iserles, A., Wu, X.Y.: Symmetric and arbitrarily high-order Brikhoff-Hermite time integrators and their long-time behavior for solving nonlinear Klein–Gordon equations. J. Comput. Phys. https://doi.org/10.1016/j.jcp.2017.10.057
41. Liu, C., Wu, X.Y.: Arbitrarily high-order time-stepping schemes based on the operator spectrum theory for high-dimensional nonlinear Klein-Gordon equations. J. Comput. Phys. **340**, 243–275 (2017)

42. Lubich, C., Ostermann, A.: Multigrid dynamic iteration for parabolic equations. BIT **27**, 216–234 (1987)
43. Mulholland, L.S., Huang, W.Z., Sloan, D.M.: Pseudospectral solution of near-singular problems using numerical coordinate transformations based on adaptivity. SIAMJ. Sci. Comput. **19**, 1261–1289 (1998)
44. Pascual, P.J., Jiménez, S.: Vázquez, L. Numerical Simulations of a Nonlinear Klein-Gordon Model. Lecture Notes in Computational Physics (Granada, 1994), vol. 448, pp. 211–270. Springer, Berlin (1995)
45. Sheng, Q., Khaliq, A.Q.M., Voss, D.A.: Numerical simulation of two-dimensional sine-Gordon solitons via a split cosine scheme. Math. Comput. Simul. **68**, 355–373 (2005)
46. Shen, J., Tang, T., Wang, L.L.: Spectral Methods: Algorithms. Analysis, Applications. Springer, Berlin (2011)
47. Strauss, W.A., ázquez, L.V.: Numerical solution of a nonlinear Klein–Gordon equation. J. Comput. Phys. **28**, 271–278 (1978)
48. Sun, Z.Z.: Numerical Methods of Partial Differential Equations. Science Press, Beijing (2012). (2nd version, in Chinese)
49. Teman, R.: Applied Matematical Scinences. In: Infinite-dimensional dynamical systems in mechanics and physics. Springer, Berlin (2000)
50. Tourigny, Y.: Product approximation for nonlinear Klein-Gordon equations. IMA J. Numer. Anal. **9**, 449–462 (1990)
51. Tang, W.S., Ya, Y.J., Zhang, J.J.: High order symplectic integrators based on continuous-stage Runge–Kutta Nyström methods. arXiv:1510.04395
52. Vandewalle, S.: Parallel multigrid waveform relaxation for parabolic problems. In: Teubner Stuttgart, B.G. (ed.) Teubner Scripts on Numerical Mathematics (1993)
53. Wazwaz, A.M.: New travelling wave solutions to the Boussinesq and the Klein-Gordon equations. Commun. Nonlinear Sci. Numer. Simul. **13**, 889–901 (2008)
54. Wu, X.Y., You, X., Wang, B.: Structure-Preserving Algorithms for Oscillatory Differential Equations. Springer, Berlin (2013)
55. Wu, X.Y., Liu, K., Shi, W.: Structure-Preserving Algorithms for Oscillatory Differential Equations II. Springer, Heidelberg (2015)
56. Wu, X.Y., Liu, C., Mei, L.J.: A new framework for solving partial differential equations using semi-analytical explicit RK(N)-type integrators. J. Comput. Appl. Math. **301**, 74–90 (2016)

Chapter 12
An Essential Extension of the Finite-Energy Condition for ERKN Integrators Solving Nonlinear Wave Equations

This chapter is devoted to an essential extension of the finite-energy condition for extended Runge–Kutta–Nyström (ERKN) integrators when applied to nonlinear wave equations. We begin with an error analysis of ERKN integrators for multi-frequency highly oscillatory systems $y'' + My = f(y)$, where M is positive semi-definite, $\|M\| \gg max\{1, \|\frac{\partial f}{\partial y}\|\}$. These highly oscillatory problems arise from the semi-discretisation of conservative or dissipative nonlinear wave equations. The structure of M and the initial conditions are dependent on the particular spatial discretisation. A finite-energy condition for the semi-discretisation of nonlinear wave equations is introduced and analysed. This is similar to the error analysis for Gaustchi-type methods of order two, where a finite-energy condition bounding amplitudes of high oscillations is satisfied by the solution. These ensure that the error bound for ERKN methods is independent of $\|M\|$. Since stepsizes are not restricted by frequencies of M, large stepsizes can be employed by our ERKN integrators which may be of arbitrarily high order. The numerical experiments presented in this chapter demonstrate that our results are really promising, and consistent with our analysis and predictions.

12.1 Introduction

The study of numerical methods for highly oscillatory problems has become increasingly important in recent decades. A major source of such problems is the spatial discretisation of nonlinear wave equations, such as Klein–Gordon equations, which have received a great deal of attention in both their numerical and analytical aspects. In this chapter, we pay attention to an essential extension of the finite-energy condition for ERKN integrators and applications to nonlinear wave equations.

We commence with a system of multi-frequency highly oscillatory second-order differential equations

X. Wu and B. Wang, *Recent Developments in Structure-Preserving Algorithms for Oscillatory Differential Equations*, https://doi.org/10.1007/978-981-10-9004-2_12

$$\begin{cases} y''(t) + My(t) = f(y(t)), \quad t \in [t_0, T], \\ y(t_0) = y_0, \; y'(t_0) = y_0', \end{cases} \tag{12.1}$$

where $M \in \mathbb{R}^{d\times d}$ is a positive semi-definite matrix (not necessarily diagonal or symmetric, in general) with $\|M\| \gg max\{1, \|\frac{\partial f}{\partial y}\|\}$. This type of problem occurs in many aspects of science and engineering, among which the spatial discretisation of nonlinear wave equations by finite difference methods or spectral methods provides a large number of practical applications. In dealing with these oscillatory problems, the adapted Runge–Kutta–Nyström (ARKN) methods and ERKN integrators were respectively proposed by Franco [5] and Yang et al. [43] as developments of classical Runge–Kutta–Nyström (RKN) methods. As shown in the literature (see e.g. [5, 7, 22, 41]), based on the internal stages of traditional RKN methods, the ARKN methods adopt a modified form of updates given by

$$y_{n+1} = \phi_0(V)y_n + h\phi_1(V)y_n' + h^2\sum_{i=1}^{s}\bar{B}_i(V)f(Y_i),$$

$$y_{n+1}' = -hM\phi_1(V)y_n + \phi_0(V)y_n' + h\sum_{i=1}^{s}B_i(V)f(Y_i),$$

where $V = h^2M$, ϕ_0, ϕ_1, $\bar{B}_i$ and B_i are matrix-valued functions of V. However, as distinct from the ARKN methods, in light of the variation-of-constants formula for (12.1), the ERKN methods not only have a new form of updates, but also adopt a new form of internal stages given by

$$Y_i = \phi_0(C_i^2V)y_n + C_ih\phi_1(C_i^2V)y_n' + h^2\sum_{j=1}^{s}A_{ij}(V)f(Y_j),$$

to achieve a high level of harmony with the oscillatory structure of the problem (12.1). An ERKN method can be represented compactly in the Butcher tableau of coefficients

$$\begin{array}{c|c} c & \bar{A}(V) \\ \hline & \bar{b}^{(}V) \\ \hline & b^{(}V) \end{array} = \begin{array}{c|ccc} c_1 & \bar{A}_{11}(V) & \cdots & \bar{A}_{1s}(V) \\ \vdots & \vdots & \ddots & \vdots \\ c_s & \bar{A}_{s1}(V) & \cdots & \bar{A}_{ss}(V) \\ \hline & \bar{B}_1(V) & \cdots & \bar{B}_s(V) \\ \hline & B_1(V) & \cdots & B_s(V) \end{array} \,. \tag{12.2}$$

Well-known examples of explicit ERKN integrators are the Gautschi-type methods of order two [1, 8–12, 14]. As we will show in (12.12) in Sect. 12.2, Gautschi-type methods can be displayed in a Butcher tableau, which is exactly the form of ERKN methods. From this observation, ERKN integrators also can be thought of as generalized Gautschi-type methods.

Among other effective numerical methods for solving the oscillatory problem are the exponentially (or functionally) fitted methods, such as the the exponentially fitted Runge–Kutta (EFRK) method [28], the exponentially fitted Runge–Kutta–Nyström method (EFRKN) (see, e.g. [6]) and the functionally-fitted energy-preserving method [19]. As stated in the literature, the applications of these methods highly depend on the choice of a fitted frequency ω. For example, when solving the problem (12.1), both EFRK methods and EFRKN methods require symmetry of M and the preferred fitted frequencies may be inferred from the diagonalisation of the matrix M. Fortunately, however, the symmetry of M is not necessary for ERKN methods as we have mentioned above. This implies ERKN methods have a broader range of applications for solving oscillatory problems once M is not symmetric. An essential relation between ERKN methods and exponentially fitted methods is that the ERKN methods are consistent with the exponentially fitted Runge–Kutta–Nyström methods (EFRKN), provided that M is symmetric. This point has been definitely proved by Wu et al. [36]. It is noted that the energy-preserving methods have also been developed for solving oscillatory Hamiltonian PDEs (see, e.g. [3, 18, 19, 42]) in recent years.

As a class of structure-preserving algorithms, ERKN integrators display some important properties in dealing with the multi-frequency and highly oscillatory system (12.1). The most notable one is that the multi-frequency and highly oscillatory homogeneous equation $y'' + My = 0$ can be exactly integrated by *both the updates and the internal stages* of ERKN integrators. Another one to merit our attention is their superiority over the classical RKN methods in numerical behaviour, such as the local truncation error, the global error, dispersion and dissipation. However, this superiority in errors for ERKN integrators is usually supported by a variety of numerical experiments in applications, noting that only a few studies of theoretical analysis (see, e.g. [9, 11, 14, 17, 32]) have been conducted, in which some further restrictions are needed on the variable y, y'', and the right-hand side function f of the system (12.1).

Meanwhile, it should be noted that Gautschi-type methods of order two have been successfully applied to oscillatory second-order differential equations [8, 14], among which the Ferm–Pasta–Ulam problem [13] is the most notable one. An observation from the Ferm–Pasta–Ulam problem leads to the finite-energy condition, which plays a very important role in effectively solving oscillatory second-order differential equations (see, e.g. [14]). For example, it is shown that a large time-step can be used for these explicit methods when applied to oscillatory differential equations (see e.g. [8]). Long-time energy conservation for these methods also can be achieved with this condition (see e.g. [12]). The most important result is that, under the finite-energy condition, the error bounds of Gautschi-type exponential methods of order two are proved to be independent of $\|M\|$ (see, e.g. [9, 11]). This point is crucial to the underlying multi-frequency highly oscillatory problem. The so-called Gautschi-type exponential integrators also have been applied to the Klein-Gordon equation in the nonrelativistic limit regime (see, e.g. [1]) and the references therein.

With this observation, and the fact that Gautschi-type methods are special ERKN methods of order two, we further investigate an extension of the finite-energy condition in this chapter. Using an analogical methodology, we pay attention to nonlinear wave equations, for which the appropriate spatial discretisation plays a similar role to the finite-energy condition for ERKN methods. Furthermore, the finite-energy condition for ERKN integrators could be naturally derived for nonlinear wave equations by a suitable spatial discretisation, whose differentiation matrix M is symmetric and positive semi-definite. We then prove that the error bounds for ERKN methods are entirely independent of $\|M\|$, when applied to the semi-discrete wave equations. This result is completely consistent with those stated in [9, 11]. Moreover, another promising result is that large stepsizes are allowed for explicit ERKN methods. This point is supported by the numerical experiments in this chapter, where corresponding classical RKN methods could hardly be employed with the same stepsizes.

This chapter is organised as follows. In Sect. 12.2, we briefly summarise the ERKN integrators in dealing with the multi-frequency highly oscillatory problems, and some results on error analysis are included as well. In Sect. 12.3, we obtain error bounds independent of $\|M\|$ for ERKN integrators when applied to conservative or dissipative nonlinear wave equations. We conduct numerical experiments in Sect. 12.4, and the numerical results support our theoretical analysis presented in this chapter. The last section is concerned with conclusions and discussions.

Throughout this chapter, the dimension of the system (12.1) is d, i.e., $y, y' \in \mathbb{R}^d$.

12.2 Preliminaries

In this section, we first summarise the results on ERKN integrators for the second-order oscillatory autonomous system (12.1). To this end, we introduce the following unconditionally convergent matrix-valued functions

$$\phi_j(V) := \sum_{k=0}^{\infty} \frac{(-1)^k V^k}{(2k+j)!}, \quad j = 0, 1, \ldots. \tag{12.3}$$

The definition of ERKN integrators is given below.

Definition 12.1 (*See, e.g.* [39]) An ERKN integrator for the second-order oscillatory initial value problem (12.1) is defined by

$$\begin{cases} Y_i = \phi_0(C_i^2 V)y_n + C_i h\phi_1(C_i^2 V)y_n' + h^2 \sum\limits_{j=1}^{s} A_{ij}(V) f(Y_j), \quad i = 1, \ldots, s, \\ y_{n+1} = \phi_0(V)y_n + h\phi_1(V)y_n' + h^2 \sum\limits_{i=1}^{s} \bar{B}_i(V) f(Y_i), \\ y_{n+1}' = -hM\phi_1(V)y_n + \phi_0(V)y_n' + h \sum\limits_{i=1}^{s} B_i(V) f(Y_i), \end{cases} \tag{12.4}$$

where $C_1, ..., C_s$ are real constants, $B_i(V)$, $\bar{B}_i(V)$ for $i = 1, ..., s$, and $A_{ij}(V)$ for $i, j = 1, ..., s$ are matrix-valued functions of $V \equiv h^2M$.

ERKN integrators can be expressed in Butcher tableaux (12.2). However, for convenience, in the remainder of this chapter, we denote the coefficients of an ERKN integrator in upper-case $(C, B, \bar{B}, A)$. Some useful properties related to the unconditionally convergent matrix-valued functions $\phi_j(V)$ for $j = 0, 1, \ldots$ are established in [39], and summarised below.

Proposition 12.1 (See [39]) *The matrix-valued functions $\phi_j(M)$ defined by (12.3) satisfy:*

1. $\lim_{M\to 0}\phi_j(M) = \frac{1}{j!}I$ *for $j = 0, 1, \ldots$, where I is the identity matrix;*
2. *All $\phi_j(M)$ for $j = 0, 1, \ldots$, are bounded when M is positive semi-definite, i.e. $\|\phi_j(M)\| \le \tilde{c}$, where $\tilde{c}$ is a constant depending on $\|M\|$ in general. However, an important and special case is that $\tilde{c}$ is independent of $\|M\|$ provided M is symmetric and positive semi-definite;*
3. $$\begin{cases} \displaystyle\int_0^1 \frac{(1-\zeta)\phi_1\big(x^2(1-\zeta)^2M\big)\zeta^j}{j!}d\zeta = \phi_{j+2}(x^2M), & x \in \mathbb{R}, \\ \displaystyle\int_0^1 \frac{\phi_0\big(x^2(1-\zeta)^2M\big)\zeta^j}{j!}d\zeta = \phi_{j+1}(x^2M), & x \in \mathbb{R}, \end{cases} \tag{12.5}$$

   for j=0,1,....

We then quote the following theorem, which is related to the order conditions for ERKN integrators.

Theorem 12.1 (See [44]) *An s-stage ERKN integrator (12.4) is of order p if and only if the conditions*

$$\begin{cases} \displaystyle\sum_{i=1}^{s} \bar{B}_i\Phi_i(\tau) = \frac{\rho(\tau)!}{\gamma(\tau)s(\tau)}\phi_{\rho(\tau)+1} + \mathcal{O}(h^{p-\rho(\tau)}), & \forall\tau \in SSENT_m, \quad m \le p-1, \\ \displaystyle\sum_{i=1}^{s} B_i\Phi_i(\tau) = \frac{\rho(\tau)!}{\gamma(\tau)s(\tau)}\phi_{\rho(\tau)} + \mathcal{O}(h^{p-\rho(\tau)+1}), & \forall\tau \in SSENT_m, \quad m \le p, \end{cases} \tag{12.6}$$

are satisfied.

Here, the definitions and properties associated with the order $\rho(\tau)$, the sign $s(\tau)$, the density $\gamma(\tau)$, and the weight $\Phi_i(\tau)$ can be found in [44].

As stated in [15, 32], we also admit the following two assumptions throughout the error analysis in this chapter, but restrict them to a more relaxed setting.

Assumption 3 *It is assumed that the solution $y(t)$ of (12.1) and its derivative $y'(t)$ are sufficiently smooth and uniformly bounded with respect to t.*

Assumption 4 *It is assumed that all the occurring derivatives $f^{(k)}(y)$ (with respect to y) of $f(y)$ are uniformly bounded.*

Remark 12.1 We note that the uniform bound is established with a particular norm. For the vectors $y(t)$, $y'(t)$ and the vector-valued function $f(y)$, we naturally use the Euclidean norm $\|\cdot\|_2$. Remember that each derivative $f^{(k)}(y)$ can be regarded as a k-linear mapping [13]

$$f^{(k)}(y): \quad \underbrace{\mathbb{R}^d \times \cdots \times \mathbb{R}^d}_{k-fold} \longrightarrow \mathbb{R}^d, \tag{12.7}$$

where d is the dimension of the problem (12.1). In this sense, we can take the induced norm $|||\cdot|||$:

$$|||f^{(k)}(y)||| = \sup\left\{\|f^{(k)}(y)(v_1,\ldots,v_k)\|_2 : \|v_i\|_2 = 1, i = 1,\ldots,k\right\}, \tag{12.8}$$

subordinated to the Euclidean norm $\|\cdot\|_2$ for $f^{(k)}(y)$. In the special case that $f'(y)$ is a negative semi-definite matrix, it is known that the induced norm $|||\cdot|||$ is just the spectral norm. In the remainder of this chapter, we will denote all norms by the uniform symbol $\|\cdot\|$ for convenience, if there is no confusion.

An earlier error analysis has been made in part by Wang et al. [32] for explicit ERKN methods. We briefly summarise the results below.

Theorem 12.2 (See [32]) *Under suitable assumptions, the explicit ERKN methods converge for $0 \le nh \le T - t_0$ when applied to the problem (12.1). In particular, the numerical solution and its derivative satisfy the following error bounds*

$$\|e_n\| \le \widehat{C}_1 h^p,$$
$$\|e_n'\| \le \widehat{C}_2 h^p,$$

where $\widehat{C}_1$ and $\widehat{C}_2$ depend on T and $\|M\|$, and they are independent of h and n. However, if M is symmetric and positive semi-definite, then $\widehat{C}_1$ is also independent of $\|M\|$.

Even though this result seems promising, a stronger restriction on the function f has been implicitly used. That is, all occurring derivatives of f (especially the total derivatives of $f(y(t))$ with respect to t) are assumed to be uniformly bounded. This stronger restriction assumption exactly leads to the conclusion that the norms of the high-order derivatives (with respective to t) of the exact solution $y(y)$ are independent of $\|M\|$. Hence, the global error bounds are independent of $\|M\|$. Another consequence of the result is that the s-stage explicit ERKN integrators of order p are restricted to those with $s \le 3$ and $p \le 3$, and hence the accuracy of the numerical methods is limited.

If the m-th derivatives of the exact solution $y(t)$ to the problem (12.1) are bounded by

$$y^{(m)}(t) = \mathcal{O}(\|M\|^k), \quad m, k \in \mathbb{N}^+, \tag{12.9}$$

then both the local truncation error bound and the global error bound must be dependent on $\|M\|$. In the case of multi-frequency highly oscillatory ODEs, (12.9) always holds with the given initial conditions (y_0, y_0'). In this sense, the global error bound for ERKN integrators depends on $\|M\|$ since the derivatives of $f(y)$ are involved with M. Fortunately, however, when M is symmetric positive semi-definite, the earlier seminal work on error analysis had discussed the so-called *finite-energy condition* (see, e.g. [8, 9, 11, 12, 14]):

$$\frac{1}{2}\|y'(t)\|^2 + \frac{1}{2}\|\Omega y(t)\|^2 \leq \frac{1}{2}K^2, \quad (\Omega^2 = M). \tag{12.10}$$

With the condition (12.10), it has been proved that the global error bound for the Gaustchi-type exponential integrators of order two is independent of $\|M\|$. The next theorem confirms this point.

Theorem 12.3 (See [9]) In (12.1), *let $M = \Omega^2$ be an arbitrary symmetric positive semi-definite matrix. Suppose that f, f_y and f_{yy} are bounded in the Euclidean norm or the norms induced by the operator Euclidean norm, respectively. Moreover, it is assumed that the solution $y(t)$ satisfies the finite-energy condition (12.10). Then, under suitable assumptions for the even analytic functions Ψ, Ψ_0 and Ψ_1, we have the following estimation of global errors*

$$\|y(t_n) - y_n\| \leq Ch^2, \quad t_n \in [t_0, T], \tag{12.11}$$

for the Gaustchi-type exponential integrators, where the constant C only depends on T, K, $\tilde{C}$, $\|f\|$, $\|f_y\|$ and $\|f_{yy}\|$, and $\tilde{C}$ is a small constant independent of $\|M\|$.

Since the Gaustchi-type integrators of order two described therein are exactly explicit ERKN integrators of order two with the Butcher tableau

$$\begin{array}{c|c} c & A(V) \\ \hline & \bar{b}(V) \\ \hline & b(V) \end{array} = \begin{array}{c|cc} 0 & 0 & 0 \\ 1 & \frac{1}{2}\Psi(V) & 0 \\ \hline & \frac{1}{2}\Psi(V) & 0 \\ \hline & \frac{1}{2}\Psi_0(V) & \frac{1}{2}\Psi_1(V) \end{array}, \tag{12.12}$$

where Ψ, Ψ_0 and Ψ_1 are matrix-valued functions respectively fixed for different Gaustchi-type integrators, we also make an attempt to find suitable conditions such that the error bound for ERKN integrators is independent of $\|M\|$, when applied to nonlinear wave equations. With this idea and observation, in what follows we will concentrate our attention on nonlinear wave equations, which can be converted to the form of (12.1) by an appropriate spatial discretisation.

12.3 Error Analysis for ERKN Integrators Applied to Nonlinear Wave Equations

We now consider the wave equation

$$\begin{cases} \dfrac{\partial^2 u(x,t)}{\partial t^2} - a^2 \Delta u(x,t) = f(u), \quad x \in D,\ t_0 \le t \le T, \\ \qquad u(x,t_0) = \varphi(x), \quad u_t(x,t_0) = \psi(x), \quad x \in D, \\ B(x)u(x,t) = 0, \quad x \in \partial D, \end{cases} \tag{12.13}$$

where a means the horizontal propagation speed of the wave motion, D is a spatial domain with boundary ∂D, Δ is the Laplacian operator and $B(x)$ is a linear boundary operator. Here, we suppose that the problem (12.13) is well-posed and satisfies the conditions described in the recent papers [34, 35]. As shown in [34, 35], the exact solution $u(x,t)$ and its derivative $u_t(x,t)$ satisfy

$$\begin{cases} u(x,t) = \phi_0\big((t-t_0)^2 a^2 \Delta\big)\varphi(x) + (t-t_0)\phi_1\big((t-t_0)^2 a^2 \Delta\big)\psi(x) \\ \qquad + \displaystyle\int_{t_0}^{t} (t-\zeta)\phi_1\big((t-\zeta)^2 a^2 \Delta\big)\tilde{f}(\zeta)d\zeta, \\ u_t(x,t) = (t-t_0)a^2\Delta\phi_1\big((t-t_0)^2 a^2 \Delta\big)\varphi(x) + \phi_0\big((t-t_0)^2 a^2 \Delta\big)\psi(x) \\ \qquad + \displaystyle\int_{t_0}^{t} \phi_0\big((t-\zeta)^2 a^2 \Delta\big)\tilde{f}(\zeta)d\zeta, \end{cases} \tag{12.14}$$

under some regularity conditions, where $\tilde{f}(\zeta) = f(u(x,\zeta))$, and the operator determined by the Laplacian-argument functions $\phi_j(\Delta)$ are defined by

$$\phi_j(\Delta) := \sum_{k=0}^{\infty} \frac{\Delta^k}{(2k+j)!}, \quad j = 0, 1, \ldots. \tag{12.15}$$

It can be seen that (12.15) is obtained from replacing z by Δ in the functions

$$\phi_j(z) = \sum_{k=0}^{\infty} \frac{z^k}{(2k+j)!}, \quad j = 0, 1, 2, \ldots, \tag{12.16}$$

and all $\phi_j(z)$ for $j = 0, 1, 2, \ldots$ are bounded for any $z \le 0$ (note that, here, the definition of $\phi_j(z)$ for (12.14) is different from (12.3) for (12.4) and there are minus signs in (12.3)).

It is known that Δ is not defined for every $v \in L^2(D)$. In order to model the boundary conditions, we restrict ourselves to the case where Δ is defined on the domain $\Omega(\Delta) \subset L^2(D)$, and the underlying boundary condition is satisfied. We then consider the linear differential operator $\mathscr{A}$ defined by $(\mathscr{A}v)(x) = a^2\Delta v(x)$. The operator has a complete system of orthogonal eigenfunctions in the Hilbert space $L^2(D)$. The operator on $L^2(D)$ induces a corresponding operator on ℓ^2 due to the isomorphism between L^2 and ℓ^2. An elementary analysis which is similar to that for the *exponential operator* presented by Hochbruck and Ostermann [16], implies that the Laplacian-argument functions $\phi_j(\Delta)$ are bounded operators with respect to the norm $\|\cdot\|_{L^2(D)\leftarrow L^2(D)}$.

In general, the domain $\Omega(\Delta)$, the eigenvalues and the eigenfunctions of the operator Δ will depend on the specified boundary conditions. As an example, we prove that the operators $\phi_j(\Delta)$ for $j = 0, 1, \ldots$ defined by (12.15) are bounded for periodic boundary problems with $D = [0, 2\pi]$. In this case, the operator Δ is defined on the domain

$$\Omega(\Delta) = \left\{v \in H^2(D) : v(x) = v(x + 2\pi)\right\}.$$

The operator has a complete system of orthogonal eigenfunctions $\{e^{ikx} : k \in \mathbb{Z}\}$ in the complex Hilbert space $L^2(D)$, and the corresponding eigenvalues are $-k^2,\ k \in \mathbb{Z}$. The functions $\phi_j(z)$ for $j = 0, 1, \ldots,$ defined by (12.16) allow us to define the operators:

$$\phi_j(t\Delta) : L^2(D) \to L^2(D) \tag{12.17}$$

given by

$$\phi_j(t\Delta)v(x) = \sum_{k=-\infty}^{\infty} \hat{v}_k \phi_j(-k^2 t)e^{ikx} \quad \text{for} \quad v(x) = \sum_{k=-\infty}^{\infty} \hat{v}_k e^{ikx}.$$

Clearly, the functions $\phi_j(z)$ for $j = 0, 1, \ldots$ are bounded for any $z \le 0$, i.e. $|\phi_j(z)| \le 1$ for $j = 0, 1, \ldots$. We then show that the functions $\phi_j(z)$ define the bounded operators $\phi_j(t\Delta)$ for any $t \ge 0$.

Theorem 12.4 *The operators $\phi_j(\Delta)$ defined by (12.15) and (12.3) are bounded operators under the norm $\|\cdot\|_{L^2(D)\leftarrow L^2(D)}$, i.e.,*

$$\|\phi_j(t\Delta)\|_{L^2(D)\leftarrow L^2(D)} \le \gamma_j, \quad t \ge 0, \tag{12.18}$$

where γ_j are the bounds of the functions $\phi_j(x)$ with $x \le 0$ for $j = 0, 1, 2, \ldots$, respectively.

Proof Before presenting the proof for the boundedness of the operators $\phi_j(t\Delta)$ for $j = 0, 1, 2, \ldots$, we first clarify the inner product of the space $L^2(D)$ expressed by

$$(u, v) = \int_D u(x)\overline{v(x)}\mathrm{d}x, \quad \forall u, v \in L^2(D). \tag{12.19}$$

For any function $v(x) \in L^2(D)$, its *Fourier series* can be represented as

$$v(x) = \sum_{k=-\infty}^{\infty} \hat{v}_k e^{ikx}.$$

The norm of the function in $L^2(D)$ can be characterized in the frequency space by

$$\|v\|^2 = \int_D |v(x)|^2 \mathrm{d}x = \sum_{k=-\infty}^{\infty} |\hat{v}_k|^2. \tag{12.20}$$

Therefore, we have

$$\|\phi_j(t\Delta)v\|^2 = \sum_{k=-\infty}^{\infty} |\hat{v}_k \phi_j(-k^2 t)|^2 \leq \sup_{t\geq 0} |\phi_j(-k^2 t)|^2 \cdot \|v\|^2 \leq \gamma_j^2 \|v\|^2, \tag{12.21}$$

where γ_j are the bounds of the functions $\phi_j(z)$ with $z \leq 0$ for $j = 0, 1, 2, \ldots$. Hence, it follows from the definition of operator norm and (12.21) that

$$\|\phi_j(t\Delta)\|_{L^2(D)\leftarrow L^2(D)} = \sup_{\|v\|\neq 0} \frac{\|\phi_j(t\Delta)v\|}{\|v\|} \leq \sup_{t\geq 0} |\phi_j(-k^2 t)| \leq \gamma_j, \quad j = 0, 1, 2, \ldots, t \geq 0. \tag{12.22}$$

The conclusion of the theorem is proved. □

The boundness of $\phi_j(\Delta)$ for other boundary conditions can be proved in a similar way.

Remark 12.2 Here it should be noted that the operators $\phi_j(\Delta)$ for $j = 0, 1, \ldots$ are well-defined and bounded on the Hilbert space $L^2(D)$. Each bounded operator $\phi_j(\Delta)$ is an indivisible whole and on no account can one think that it can be thought of as any finite sum of the Laplacian Δ successively operated one by one. In fact, all the operators defined by (12.15) are bounded operators, and completely different from the Laplacian. Therefore, it is of great importance to keep it in mind that each operator $\phi_j(\Delta)$ defined by (12.15) is well-defined and bounded on the space $L^2(D)$.

On the basis of the above analysis, we define $u(t)$ as the function that maps x to $u(x, t)$:

$$u(t) := [x \mapsto u(x, t)].$$

Then the system (12.13) can be formulated by an abstract second-order ordinary differential equation on the infinity-dimensional function space $L^2(D)$,

$$\begin{cases} u''(t) - \mathscr{A}u(t) = f\big(u(t)\big), \ t_0 < t \leq T, \\ \qquad u(t_0) = \varphi(x), \ u'(t_0) = \psi(x). \end{cases} \tag{12.23}$$

Approximating the operator $-\mathscr{A}$ in (12.23) by the differentiation matrix M (with some manipulation if necessary), we obtain an initial value problem of ODEs:

$$\begin{cases} U''(t) + MU = f(U), \quad t \in [t_0, T], \\ \qquad U(t_0) = U_0, \ U'(t_0) = U_0'. \end{cases} \tag{12.24}$$

For the applications of ERKN integrators to (12.24), it is required that the differentiation matrix M is positive semi-definite. It is easy to fulfill this requirment, since the suitable difference discretisation (see, e.g. [21]) and the well-known Fourier or Chebyshev pseudospectral spatial discretisation yield positive semi-definite differentiation matrices M. In what follows, we assume that the spatial approximation used here is consistent with the original PDEs, and that $U(t)$ is convergent to $u(t) \equiv u(x,t)$. Note that the matrix M and the initial values (U_0, U_0') of (12.24) are simultaneously generated by the underlying spatial discretisation.

Another point which should be especially emphasized is that if we replace the operator $-\mathscr{A}$ by the differentiation matrix M in (12.14), then the solution $U(t)$ of (12.24) and its derivative $U_t(t)$ can be exactly expressed as follows:

$$\begin{cases} U(t) = \phi_0\big((t-t_0)^2 M\big)U_0 + (t-t_0)\phi_1\big((t-t_0)^2 M\big)U_0' \\ \qquad + \displaystyle\int_{t_0}^{t} (t-\zeta)\phi_1\big((t-\zeta)^2 M\big)\tilde{f}(\zeta)d\zeta, \\ U_t(t) = -(t-t_0)M\phi_1\big((t-t_0)^2 M\big)U_0 + \phi_0\big((t-t_0)^2 M\big)U_0' \\ \qquad + \displaystyle\int_{t_0}^{t} \phi_0\big((t-\zeta)^2 M\big)\tilde{f}(\zeta)d\zeta, \end{cases} \tag{12.25}$$

where $\tilde{f}(\zeta) = f(U(x,\zeta))$.

We now turn to an important property of the nonlinear wave equations (12.23). By denoting

$$H(u) = -\int_0^u f(s)ds$$

(the function $H(u)$ is assumed to be positive since $H(u)$ is only used through its gradient here), we obtain that

$$E = E(t) = \frac{1}{2}\int_D \big((u_t)^2 + a^2(u_x)^2 + 2H(u)\big)dx, \tag{12.26}$$

is conserved during the evolution of the wave equation, when suitable boundary conditions, such as periodic or homogeneous Dirichlet or Neumann boundary conditions are prescribed. For the case where f explicitly depends on t, i.e., $f = f(t,u)$, it is noted that $E(t)$ is no longer conserved and the wave equation is then dissipative. Thus, after a suitable spatial discretisation for (12.13) we can give an approximate relation between the continuous energy and the discrete energy for (12.24):

$$\frac{1}{2}\|U'(t)\|^2 + \frac{1}{2}\|\Omega U(t)\|^2 \approx \frac{1}{2}\int_D \big((u_t)^2 + a^2(u_x)^2\big)dx, \quad (\Omega^2 = M). \tag{12.27}$$

Then, the *finite-energy condition* (12.10) is easily satisfied, since the constant K in (12.10) can now be roughly chosen as

$$K = \sqrt{2(|E(t_0)| + |\int_D H(u(t_0))dx|)},$$

for both the conservative and dissipative nonlinear wave equations. Here, it should be noted that, though the constant K depends also on the boundary condition according to its formula, this could be neglected by virtue of Assumption 3. This point is the essential idea behind our fundamental error analysis of ERKN methods for efficiently solving conservative or dissipative nonlinear wave equations. The fact that the finite-energy condition (12.10) can be satisfied by our ERKN methods motivates the study of global errors for ERKN methods. This implies that we are hopeful of obtaining the same results of global errors for ERKN methods of arbitrary order as those for Gaustchi-type exponential integrators of order two.

Theorem 12.5 *Under Assumptions 3 and 4, if the matrix M in* (12.24) *is symmetric positive semi-definite and the spatial discretisation is consistent with the original equation and convergent to the exact solution, then the spatial discretisation error satisfies*

$$\|U(t) - u(t)\| \le C_2 \tau^k, \tag{12.28}$$

where τ is the maximal spatial stepsize, C_2 is a constant depending on T but independent of $\|M\|$, and k is a positive integer number depending on the spatial discretisation.

Proof Let r_τ be the natural restriction operator on the space grids and $\rho(\tau)$ stand for the maximal distance in the grids. We denote $u_\tau(t) = r_\tau u(x,t)$. By denoting

$$u''_\tau(t) = \frac{d^2 u_\tau(t)}{dt^2} = r_\tau u_{tt}(x,t),$$

we introduce the space truncation error

$$\delta(t) = -Mu_\tau(t) + f(u_\tau(t)) - u''_\tau(t) = a^2 r_\tau \Delta u(x,t) - Mu_\tau(t). \tag{12.29}$$

It is trivial to obtain that $\|\delta(t)\| \to 0$ as $\varrho(\tau) \to 0$ uniformly in t due to the consistency of the spatial discretisation with the wave equation. It follows from (12.29) that $\delta(t)$ is only involved with the spatial discretisation. Hence it can be estimated by

$$\|\delta(\tau)\| \le C\tau^k, \tag{12.30}$$

where k is some positive integer depending on the spatial discretisation, and C is a constant independent of $\|M\|$.

If we denote

$$\eta(t) = U(t) - u_\tau(t),$$

then the spatial discretisation error $\eta(t)$ is the solution of the system

$$\eta''(t) = -MU(t) + f(U) + Mu_\tau(t) - f(u_\tau(t)) + \delta(t),$$

i.e.,

$$\eta''(t) + M\eta(t) = G(t)\eta(t) + \delta(t), \tag{12.31}$$

where

$$G(t) = \int_0^1 F'(u_\tau(t) + \theta\eta(t))d\theta$$

and $F'(\cdot)$ is the Jacobian of $f(\cdot)$. The application of the variation-of-constants formula to (12.31) yields

$$\begin{aligned}\eta(t) &= \phi_0\big((t-t_0)^2M\big)\eta_0 + (t-t_0)\phi_1\big((t-t_0)^2M\big)\eta_0' \\ &\quad + \int_{t_0}^t (t-\zeta)\phi_1\big((t-\zeta)^2M\big)\widetilde{G}(\zeta)d\zeta,\end{aligned} \tag{12.32}$$

where $\eta_0 = \eta(t_0)$, $\eta_0' = \eta'(t_0)$ and $\widetilde{G}(\zeta) = G(\zeta)\eta(\zeta) + \delta(\zeta)$.

It follows from Assumption 4 that $\|F'\| \leq L$, where L is a constant independent of $\|M\|$. On noting that $\eta(t_0) = U(t_0) - u_\tau(t_0) = 0$ and $\eta'(t_0) = U'(t_0) - u_\tau'(t_0) = 0$, we then obtain

$$\begin{aligned}\|\eta(t)\| &= \Big\| \int_{t_0}^t (t-\zeta)\phi_1\big((t-\zeta)^2M\big)\widetilde{G}(\zeta)d\zeta \Big\| \\ &\leq L\|\eta(t)\| \cdot \Big\|(t-t_0)^2 \cdot \int_0^1 (1-\zeta)\phi_1\big((1-\zeta)^2(t-t_0)^2M\big)d\zeta \Big\| \\ &\quad + \|\delta(t)\| \cdot \Big\|(t-t_0)^2 \cdot \int_0^1 (1-\zeta)\phi_1\big((1-\zeta)^2(t-t_0)^2M\big)d\zeta \Big\| \\ &\leq L|T-t_0|^2 \cdot \|\eta(t)\| \cdot \|\phi_2((t-t_0)^2M)\| \\ &\quad + C|T-t_0|^2 \cdot \|\phi_2((t-t_0)^2M)\|\tau^k.\end{aligned} \tag{12.33}$$

The last inequality follows from Proposition 12.1 (4) and (12.30). Since M is symmetric and positive semi-definite, $\|\phi_2((t-t_0)^2M)\|$ is uniformly bounded and independent of $\|M\|$ on the basis of Proposition 12.1. Consequently, (12.33) gives $\|\eta(t)\| \leq C_2\tau^k$, where C_2 is dependent on T but independent of $\|M\|$. This completes the proof. □

As the mesh partition in the space discretisation increases for (12.23), $\|M\|$ will increase in (12.24). The larger $\|M\|$ is, the higher the accuracy will be in the spatial approximations. This argument implies that

$$U(t) \to u_\tau(t) \qquad as \quad \|M\| \to \infty. \tag{12.34}$$

Since the derivatives of $u_\tau(t)$ are entirely independent of $\|M\|$, the exact solution $U(t)$ to (12.24) and its high-order derivatives (with respect to t) are also independent of $\|M\|$ due to the convergence stated by (12.34). This leads to the uniform boundness and the independence of $\|M\|$ for all occurring derivatives of $\tilde{f}(t) = f(U)$, which has been stated by *Assumption 1* in [32]. With this insight into the underlying problem, and following the analysis in [32], we present the following important result on the global error bounds for ERKN integrators when applied to nonlinear wave equations (12.13).

Theorem 12.6 *Let M be real symmetric and positive semi-definite. Suppose that initial-boundary value problems (12.13) are well-posed and conservative (or dissipative), and a suitable spatial discretisation for (12.13) leads to (12.24). Then, applying a p-th order ERKN integrator to the semi-discrete problem (12.24), we have the global error bound*

$$\|U_n - u(t_n)\| \leq \|U_n - U(t_n)\| + \|U(t_n) - u(t_n)\| \leq C_1 h^p + C_2 \tau^k, \tag{12.35}$$

where C_1 and C_2 are dependent on T but independent of $\|M\|$, h and τ are time and spatial stepsizes respectively, and k is a positive integer determined by the spatial discretisation.

Proof We begin with the fact that the high-order derivatives of $U(t)$ are independent of $\|M\|$, as (12.34) claims. We denote $\tilde{f}(t) = f(U(t))$. Since the high-order total derivatives $\tilde{f}^{(j)}(t) = \frac{d^j}{dt^j} f(U(t))$ are the combinations of $f^{(i)}(U) = \frac{\partial^i}{\partial U^i} f(U)$ and $U^{(i)}(t)$, it can be deduced that all $\tilde{f}^{(j)}(t)$ are bounded and independent of $\|M\|$ based on Assumption 4. From the variation-of-constants formula (12.25) for (12.24) and the definition of the ERKN integrator (12.4), under the local assumptions $U_n = U(t_n)$ and $U'_n = U'(t_n)$, the local truncation error T^1_{ERKN} satisfies

$$\begin{aligned} T^1_{ERKN} &= U(t_n + h) - U_{n+1} \\ &= h^2 \int_0^1 (1-\zeta)\phi_1\big((1-\zeta)V\big)\tilde{f}(\zeta)d\zeta - h^2 \sum_{i=1}^{s} \bar{B}_i(V) f(U^i) \\ &= \sum_{j=0}^{\infty} h^{j+2}\Big(\phi_{j+2}(V) - \sum_{k=1}^{s} \bar{B}_k(V)\frac{c_k^j}{j!}\Big)\tilde{f}^{(j)}(t_n) \\ &= \tilde{C}h^{p+1} + \sum_{j=p}^{\infty} h^{j+2}\Big(\phi_{j+2}(V) - \sum_{k=1}^{s} \bar{B}_k(V)\frac{c_k^j}{j!}\Big)\tilde{f}^{(j)}(t_n), \end{aligned} \tag{12.36}$$

where $\tilde{C}$ is independent of $\|M\|$. The second identity of (12.36) follows from [32] and third identity holds because the ERKN integrator is of order p. Thus, (12.36) gives

$$\|T^1_{ERKN}\| \leq K_1 h^{p+1},$$

where K_1 is independent of $\|M\|$. Likewise, we can obtain

$$\|T^2_{ERKN}\| = \|U'(t_n + h) - U'_{n+1}\| \leq K_2 h^{p+1},$$

where K_2 is independent of $\|M\|$.

We next analyse the global error of ERKN methods when applied to the nonlinear wave equation (12.13). To this end, we denote

$$E^i = u(t_n + c_i h) - U^i, \quad e_n = u(t_n) - U_n, \quad e'_n = u'(t_n) - U'_n.$$

It follows from a Taylor expansion in time at t_n that

$$\begin{cases} e_{n+1} = \phi_0(V)e_n + h\phi_1(V)e'_n + h^2 \sum\limits_{i=1}^{s} \bar{B}_i(V)\Big(f\big(u(t_n + c_i h)\big) - f(U^i)\Big) \\ \qquad + K_1 h^{p+1} + \hat{C}h^2 \displaystyle\int_0^1 (1-z)\phi_1\big((1-z)^2 V\big)\mathrm{d}z\tau^q, \\ e'_{n+1} = -hM\phi_1(V)e_n + \phi_0(V)e'_n + h\sum\limits_{i=1}^{s} B_i(V)\Big(f\big(u(t_n + c_i h)\big) - f(U^i)\Big) \\ \qquad + K_2 h^{p+1} + \hat{C}h \displaystyle\int_0^1 \phi_0\big((1-z)^2 V\big)\mathrm{d}z\tau^q, \\ E^i = \phi_0(c_i^2 V)e_n + c_i h\phi_1(c_i^2 V)e'_n + h^2 \sum\limits_{j=1}^{s} A_{ij}(V)\Big(f\big(u(t_n + c_j h)\big) - f(U^j)\Big) \\ \qquad + \tilde{K}_1 h^{p+1} + \hat{C}c_i^2 h^2 \displaystyle\int_0^1 (1-z)\phi_1\big((1-z)^2 c_i^2 V\big)\mathrm{d}z\tau^q, \\ \qquad i = 1, 2. \ldots, s. \end{cases} \tag{12.37}$$

Furthermore, we rewrite the first two equations in (12.37) as the matrix-vector form:

$$\begin{aligned} \begin{bmatrix} \Omega e_{n+1} \\ e'_{n+1} \end{bmatrix} &= \begin{bmatrix} \phi_0(V) & h\Omega\phi_1(V) \\ -h\Omega\phi_1(V) & \phi_0(V) \end{bmatrix} \begin{bmatrix} \Omega e_n \\ e'_n \end{bmatrix} \\ &\quad + h\sum_{i=1}^{s} \begin{bmatrix} h\Omega \bar{B}_i(V)\Big(f\big(u(t_n + c_i h)\big) - f(U^i)\Big) \\ B_i(V)\Big(f\big(u(t_n + c_i h)\big) - f(U^i)\Big) \end{bmatrix} \\ &\quad + \hat{C}h \int_0^1 \begin{bmatrix} h(1-z)\Omega\phi_1\big((1-z)^2 V\big) \\ \phi_0\big((1-z)^2 V\big) \end{bmatrix} \mathrm{d}z \cdot \tau^q + Kh^{p+1}. \end{aligned} \tag{12.38}$$

Due to the finite-energy condition (12.10), taking the l^2-norm of both sides of (12.38) and the first equation in (12.37) and summing up the results leads to

$$\|e_{n+1}\| + \sqrt{\|e'_{n+1}\|^2 + \|\Omega e_{n+1}\|^2} \le \|e_n\| + \sqrt{\|e'_n\|^2 + \|\Omega e_n\|^2} + h\|e'_n\| \\ + hL(\bar{B} + \hat{B} + B)\sum_{i=1}^{s}\|E^i\| + C_1 h\Big(h^p + \tau^q\Big). \tag{12.39}$$

Likewise, by taking the l^2-norm of both sides of the last equations in (12.37) yields

$$\|E^i\| \le \|e_n\| + c_i h\|e'_n\| + h^2 LA\sum_{j=1}^{s}\|E^j\| + C_2 h\Big(h^p + \tau^q\Big), \tag{12.40}$$

If the time stepsize h satisfies $h \le \sqrt{\frac{1}{2LA}}$, the inequality (12.40) implies

$$\sum_{i=1}^{s}\|E^i\| \le 2s\big(\|e_n\| + h\|e'_n\|\big) + 2C_2 sh\Big(h^p + \tau^q\Big). \tag{12.41}$$

Inserting (12.41) into (12.39), we obtain

$$\|e_{n+1}\| + \sqrt{\|e'_{n+1}\|^2 + \|\Omega e_{n+1}\|^2} \le (1 + C_4 h)\left(\|e_n\| + \sqrt{\|e'_n\|^2 + \|\Omega e_n\|^2}\right) \\ + C_3 h\Big(h^p + \tau^q\Big). \tag{12.42}$$

It then follows from the well-known discrete Gronwall inequality that

$$\begin{aligned} &\|e_n\| + \sqrt{\|e'_n\|^2 + \|\Omega e_n\|^2} \\ &\le \exp(C_4 nh)\left(\|e_0\| + \sqrt{\|e'_0\|^2 + \|\Omega e_0\|^2} + C_3 nh\Big(h^p + \tau^q\Big)\right) \\ &\le C_3 T\exp(C_4 T)\Big(h^p + \tau^q\Big). \end{aligned} \tag{12.43}$$

This means that

$$\begin{cases} \|e_n\| \le C\Big(h^p + \tau^q\Big), \\ \|e'_n\| \le C\Big(h^p + \tau^q\Big). \end{cases} \tag{12.44}$$

The constant C is only involved with the spatial discretisation, and hence independent of $\|M\|$.

This completes the proof. □

The discussion about the convergence of the discretisation in space for (12.34) and that of the full discretisation for (12.35) can be found in [29], in which the authors discussed the convergence of *Method of Lines* (MOL) when applied to PDEs. Here we assume the convergence and give our main attention to the analysis of global error bounds. This theorem reveals that the bound of the global error is independent of $\|M\|$ when the semi-discrete wave equation (12.24) is numerically solved by ERKN integrators. It is noted that (12.24) is a multi-frequency oscillatory system. The result is a natural extension of that for Gaustchi-type exponential integrators of order two [9, 11], though the analysis for Gaustchi-type exponential integrators is carried out by a different approach. Furthermore, it gives a physical description of the *finite-energy condition*, namely, this condition essentially originates from an appropriate spatial discretisation for the conservative or dissipative wave equation. Another point which should be noted is that although one has a similar result for classical RKN methods when applied to (12.24), it is not recommended to use them in practice because of the larger global error bound for RKN methods compared with ERKN integrators. Moreover, totally different from Gaustchi-type exponential integrators and the generalized Gaustchi-type exponential integrators, i.e., ERKN integrators, classical RKN methods are not designed specially for solving multi-frequency highly oscillatory problems, and the original idea of classical RKN methods takes no account of the oscillatory structure introduced by the term MU in (12.24). Finally, as stated in [39], ERKN methods always have a larger stability region than the corresponding RKN methods in dealing with the oscillatory problem (12.24). Hence, large stepsizes are also allowed by ERKN methods when solving nonlinear wave equations, whereas the corresponding RKN methods may behave badly with the same stepsizes as those for ERKN methods. This point will be demonstrated by the numerical experiments in the next section.

12.4 Numerical Experiments

In our numerical simulations, we prefer to use explicit ERKN methods rather than implicit ones due to the easy manipulation of the former, even though both of them have the same properties as stated in Theorem 12.6. Here, the key point is that implicit ERKN integrators require iterative solutions, whereas explicit ERKN integrators avoid the complexity brought by the iterative procedure. The explicit ERKN integrators are selected as follows:

- ERKN3s4: the three-stage ERKN method of order four given in [40];
- ERKN7s6: the seven-stage symplectic ERKN method of order six given in [23].

It is known that an ERKN method reduces to an RKN method when $M \to \mathbf{0}$. Thus, the degenerate ERKN methods are assigned as the corresponding RKN methods, which are denoted by RKN3s4 and RKN7s6, respectively. It will be observed from the numerical experiments that the numerical behaviour of ERKN methods is much better than that of the corresponding RKN methods. Although the Gaustchi-type exponential methods of order two mentioned in previous section are well known, we do not consider them in the numerical comparison due to their limited accuracy.

Problem 12.1 Consider the Duffing equation (see e.g., [20])

$$\begin{cases} \ddot{q} + \omega^2 q = k^2(2q^3 - q), \\ q(0) = 0, \quad \dot{q}(0) = \omega, \end{cases}$$

with $0 \leq k < \omega$. It is a Hamiltonian system with the Hamiltonian

$$H(p,q) = \frac{1}{2}p^2 + \frac{1}{2}\omega^2 q^2 + \frac{k^2}{2}(q^2 - q^4).$$

The analytic solution is given by

$$q(t) = sn(\omega t, k/\omega),$$

where sn is the Jacobian elliptic function.

The problem is solved on the interval [0,1000] with $k = 0.03$. In this problem, we investigate the influence of $\|M\|$ on the global errors for ERKN integrators in the case of traditional ODEs. Figure 12.1 gives the numerical results obtained by ERKN3s4 and ERKN7s6 with the different stepsizes when applied to Problem 12.1. As shown in Fig. 12.1, the larger $\|M\|$ $(= \omega^2)$ becomes, the larger the global errors are for both the two ERKN integrators. It is very much in accordance with the conclusion that the global error bounds of ERKN methods are usually dependent on $\|M\|$, when applied to the general oscillatory problem (12.1), where M is independent of the initial conditions y_0, y_0'. A similar result is given in Fig. 12.2, where the same stepsizes are respectively used for the corresponding RKN methods. Note that the case of $\omega = 20$ is not plotted in Fig. 12.2b for RKN7s6, since the scheme is unstable with the stepsize $h = 1/10$. It is shown that the RKN methods give disappointing numerical solutions in the high-frequency case ($\omega = 20$), whereas the ERKN methods are robust and give satisfactory numerical solutions of high accuracy with the same stepsizes. Moreover, the results of ERKN methods are also much more accurate than those of classical RKN methods in the low-frequency case of $\omega = 5$. Finally, we plot the efficiency curves in Fig. 12.3a for all the four methods, where the ERKN methods show higher efficiency than the corresponding RKN methods. As shown in Fig. 12.3b, the numerical convergence orders of the selected numerical methods are 4.00, 5.99, 4.21 and 5.93 respectively for RKN3s4, RKN7s6, ERKN3s4 and ERKN7s6, which demonstrate the high consistency of their numerical convergence order with the theoretical convergence order.

Problem 12.2 (*Breather soliton*) We consider the well-known sine–Gordon equation (see e.g. [25])

$$\frac{\partial^2 u}{\partial t^2} = \frac{\partial^2 u}{\partial x^2} - \sin u, \tag{12.45}$$

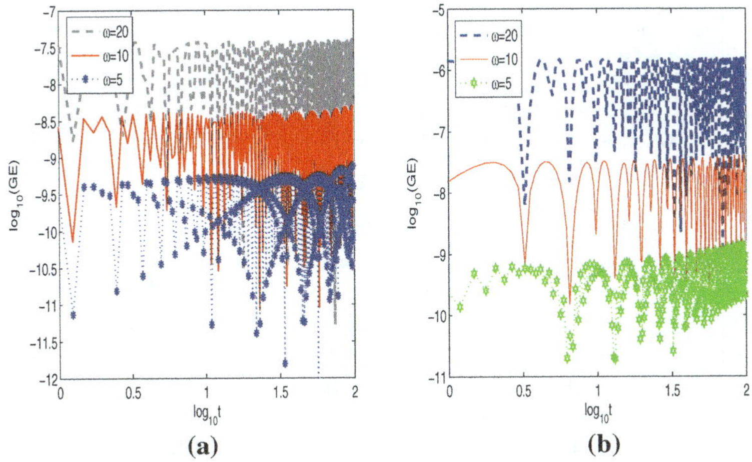

Fig. 12.1 The log–log plot of maximum global error GE against t with different ω for Problem 12.1: **a** the result for ERKN3s4 with the stepsize $h = 1/40$ (left); **b** the result for ERKN7s6 with the stepsize $h = 1/10$ (right)

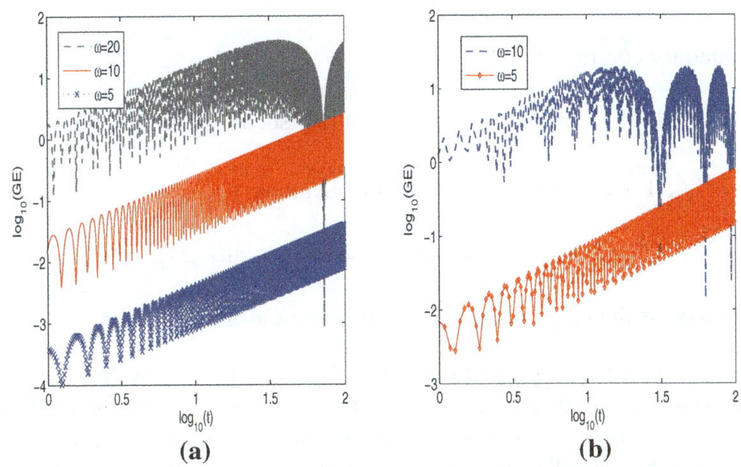

Fig. 12.2 The log–log plot of maximum global error GE against t with different ω for Problem 12.1: **a** the result for RKN3s4 with the stepsize $h = 1/40$ (left); **b** the result for RKN7s6 with the stepsize $h = 1/10$ (right)

on the region $-10 \le x \le 10$ and $-20 \le t \le 20$, with the initial conditions

$$u(x, -20) = -4 \arctan\left(c^{-1}\text{sech}(\kappa x)\sin(20c\kappa)\right),$$
$$u_t(x, -20) = \frac{4\kappa\cos(20c\kappa)\text{sech}(\kappa x)}{1 + c^{-2}sech^2(\kappa x)\sin^2(20c\kappa)},$$

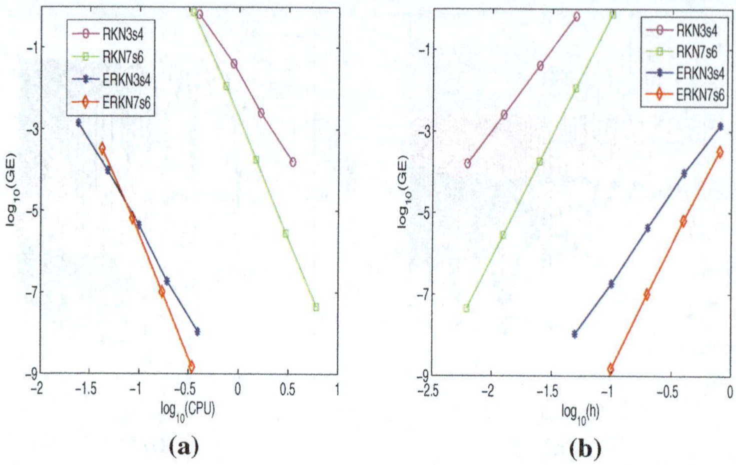

Fig. 12.3 Numerical results for Problem 12.1 with $\omega = 5$: **a** The efficiency curves, i.e., the log–log plot of maximum global error GE against the consumed CPU time (left); **b** The numerical convergence order, i.e., the log–log plot of maximum global error GE against the stepsize h (right)

and the boundary conditions

$$u(-10, t) = u(10, t) = -4 \arctan \left(c^{-1}\text{sech}(10\kappa) \sin(c\kappa t)\right),$$

where $\kappa = 1/\sqrt{1 + c^2}$. The exact solution is given by

$$u(x, t) = -4 \arctan \left(c^{-1}\text{sech}(\kappa x) \sin(c\kappa t)\right),$$

which is known as the breather solution of the sine-Gordon equation.

For the spatial discretisation of Problem 12.2, we use the Chebyshev pseudospectral discretisation [27]. The parameter c is selected as $c = 0.5$, and the total number of spatial mesh grids is denoted by N. During the experiment, we choose various values of N for the two ERKN integrators, and hence the value of $\|M\|$ varies with N. As shown in Fig. 12.4, the larger N is, the larger $\|M\|$ becomes. Fortunately, however, the large $\|M\|$ is, the better accuracy becomes for the global error. This result strongly supports our analysis in Sect. 12.3, which shows that the global error bounds of ERKN integrators are independent of $\|M\|$ when applied to the underlying nonlinear wave equations. For the stepsize $h = 1/64$, the global errors of the two RKN methods are too large to plot here. This means that the two corresponding RKN methods are unstable with the stepsize $h = 1/64$. To exclude the possible influence of spatial discretisation on the accuracy of numerical solutions, we select $N = 200$ and plot the efficiency curves in Fig. 12.5a by varying the stepsize h, where the higher order ERKN method gives higher efficiency. Note that only the numerical results of the two ERKN methods are plotted in Fig. 12.5a, since the corresponding

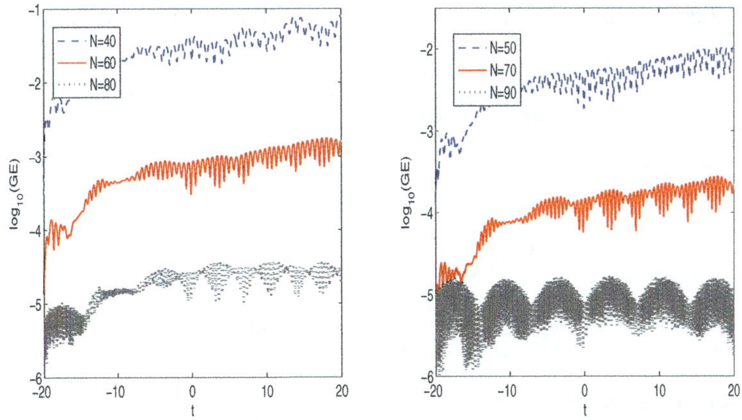

(a) (b)

Fig. 12.4 The log–log plot of maximum global error GE against t with the time stepsize $h = 1/64$ for Problem 12.2: **a** the results for ERKN3s4 for different pairs $(N, \|M\|)$: $(40, 1.4677 \times 10^3)$, $(60, 7.4169 \times 10^3)$, $(80, 2.3426 \times 10^4)$ (left); **b** the results for ERKN7s6 for different pairs $(N, \|M\|)$: $(50, 3.5791 \times 10^3)$, $(70, 1.3736 \times 10^4)$, $(90, 3.7519 \times 10^4)$ (right)

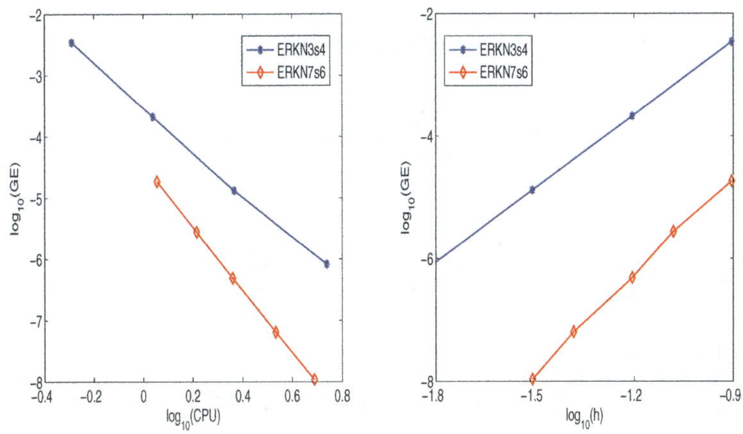

(a) (b)

Fig. 12.5 Numerical results for for Problem 12.2 with $N = 200$: **a** The efficiency curves, i.e., the log–log plot of maximum global error GE against the consumed CPU time (left); **b** The numerical convergence order, i.e., the log–log plot of maximum global error GE against the stepsize h (right)

RKN methods give unstable numerical solutions for such large N. The numerical convergence orders of the two ERKN methods are respectively 4.01 and 5.89 for ERKN3s4 and ERKN7s6, as shown in Fig. 12.5b.

Problem 12.3 (*Single soliton*) We consider the nonlinear Klein–Gordon equation (see e.g., [2])

$$\frac{\partial^2 u}{\partial t^2} - a^2 \frac{\partial^2 u}{\partial x^2} = -au + bu^3, \tag{12.46}$$

on the region $-10 \leq x \leq 10$ and $0 \leq t \leq 10$, with the initial conditions

$$u(x, 0) = \sqrt{\frac{2a}{b}} \operatorname{sech}(\lambda x),$$
$$u_t(x, 0) = c\lambda \sqrt{\frac{2a}{b}} \operatorname{sech}(\lambda x) \tanh(\lambda x),$$

and the boundary conditions

$$u(-10, t) = \sqrt{\frac{2a}{b}} \operatorname{sech}(\lambda(-10 - ct)),$$
$$u(10, t) = \sqrt{\frac{2a}{b}} \operatorname{sech}(\lambda(10 - ct)),$$

where $\lambda = \sqrt{\frac{a}{a^2 - c^2}}$. The exact solution is given by

$$u(x, t) = \sqrt{\frac{2a}{b}} \operatorname{sech}(\lambda(x - ct)).$$

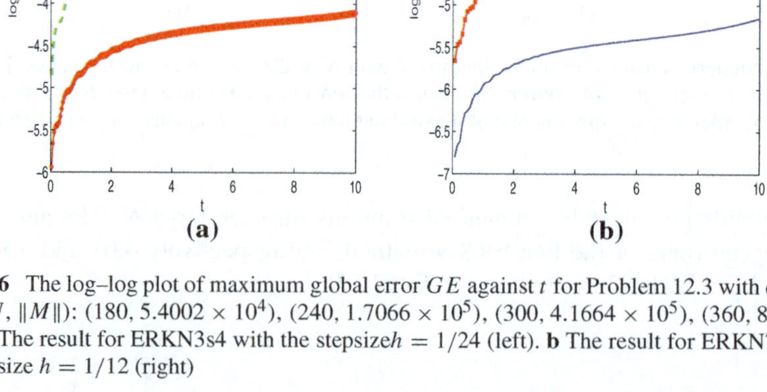

Fig. 12.6 The log–log plot of maximum global error GE against t for Problem 12.3 with different pairs $(N, \|M\|)$: $(180, 5.4002 \times 10^4)$, $(240, 1.7066 \times 10^5)$, $(300, 4.1664 \times 10^5)$, $(360, 8.6392 \times 10^5)$. **a** The result for ERKN3s4 with the stepsize$h = 1/24$ (left). **b** The result for ERKN7s6 with the stepsize $h = 1/12$ (right)

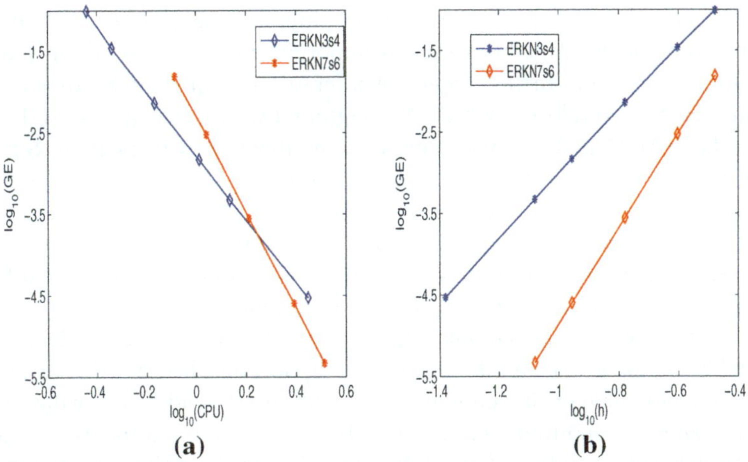

Fig. 12.7 Numerical results for for Problem 12.3 with $N = 400$: **a** the log–log plot of maximum global error GE against the consumed CPU time (left); **b** the log–log plot of maximum global error GE against the stepsize h (right)

With regard to the spatial discretisation of this problem, we use the Chebyshev pseudospectral discretisation. In this experiment, we take the parameters $a = 0.3$, $b = 1$ and $c = 0.25$, as in [2]. Various values of N are selected for the two ERKN methods. As shown in Fig. 12.6, a large value of $\|M\|$ always implies higher accuracy for the global error, which is entirely similar to the result obtained in Problem 12.2. The result obtained here also supports our analysis in Sect. 12.3 that the global error bounds of ERKN integrators are independent of $\|M\|$ when applied to wave equations. Similarly to the case in Problem 12.2, the global errors for RKN3s4 and RKN7s6 are not plotted here, since instability and nonconvergence occur for the two RKN methods with the corresponding stepsizes $h = 1/12$ and $h = 1/24$, respectively. Similarly to Problem. 12.2, we also plot the efficiency curves in Fig. 12.7a for the fixed $N = 400$, from which it can be observed that ERKN3s4 gives better performance when the accuracy is lower than 10^{-4}, while the higher order method ERKN7s6 is preferred for the high accuracy case. The two corresponding RKN methods are omitted in this figure due to their instability for $N = 400$. Fig. 12.7b shows the numerical convergence order of the two ERKN methods, which are respectively 3.91 and 5.86 for ERKN3s4 and ERKN7s6.

12.5 Conclusions and Discussions

Highly oscillatory problems have a very wide range of applications such as celestial mechanics, nonlinear dynamical systems, quantum chemistry and molecular modelling. The computation of highly oscillatory problems has generated a large number of different numerical approaches and algorithms. High accuracy and structure preservation for numerical methods are very important when applied to highly oscillatory Hamiltonian systems. In this chapter, in order to gain an insight into the extension of the finite-energy condition for ERKN integrators we further discussed the error estimation for ERKN methods when applied to nonlinear wave equations. Since the solutions of (12.1) are high-frequency oscillators, due to the existence of the linear term My, the error bounds for numerical methods depend on $\|M\|$ in general. However, we have shown the error bounds for ERKN integrators are independent of $\|M\|$ when applied to conservative or dissipative wave equations. This result is of great importance for ERKN integrators, i.e., the so-called generalised Gautschi-type methods. Furthermore, we conducted numerical experiments which clearly demonstrate this point. These numerical simulations firmly support our theoretical analysis.

To sum up, motivated by the seminal work on the error bound based on the analysis of the finite-energy condition (see, e.g. [8, 9, 11, 12, 14]), the intensive study of this chapter has successfully achieved the extension of the finite-energy condition for ERKN integrators when applied to nonlinear wave equations. Meanwhile we have obtained a new result on the error analysis of ERKN integrators. It is believed that the new investigation of error analysis has led to further insight into ERKN integrators when applied to the initial-boundary value problems of nonlinear wave equations (12.13). In particular, it is noted that the multi-frequency highly oscillatory Hamiltonian system with the Hamiltonian

$$H(q, p) = \frac{1}{2}pp + \frac{1}{2}qMq + U(q)$$

has been efficiently solved by the ERKN integrators in the literature (see, e.g. [4, 11–14, 23, 26, 30, 31, 33, 36–39]) from the accuracy to the qualitative behaviour of the symplecticity preservation, energy preservation, symmetry preservation etc. in long-term numerical simulation of nonlinear wave equations (12.13).

To conclude, we emphasise that all essential features of the analysis of this chapter were presented in the case of one-dimensional nonlinear wave equations, but the arguments naturally extend to the higher-dimensional case.

The material of this chapter is based on the work by Mei et al. [24].

References

1. Bao, W., Dong, X.: Analysis and comparison of numerical methods for the Klein–Gordon equation in the nonrelativistic limit regime. Numer. Math. **120**, 189–229 (2012)
2. Bratsos, A.G.: On the numerical solution of the Klein–Gordon equation. Numer. Methods Partial Differ. Equ. **25**, 939–951 (2009)
3. Brugnano, L., Frasca Caccia, G., Iavernaro, F.: Energy conservation issues in the numerical solution of the semilinear wave equation. Appl. Math. Comput. **270**, 842–870 (2015)
4. Cohen, D., Hairer, E., Lubich, C.: Numerical energy conservation for multi-frequency oscillatory differential equations. BIT **45**, 287–305 (2005)

5. Franco, J.M.: Runge–Kutta–Nyström methods adapted to the numerical integration of perturbed oscillators. Comput. Phys. Commun. **147**, 770–787 (2002)
6. Franco, J.M.: Exponentially fitted explicit Runge–Kutta–Nyström methods. J. Comput. Appl. Math. **167**, 1–19 (2004)
7. Franco, J.M.: New methods for oscillatory systems based on ARKN methods. Appl. Numer. Math. **56**, 1040–1053 (2006)
8. García-Archillay, B., Sanz-Serna, J.M., Skeel, R.D.: Long-time-step methods for oscillatory differential equations. SIAM J. Sci. Comput. **20**, 930–963 (1998)
9. Grimm, V.: On error bounds for the Gautschi-type exponential integrator applied to oscillatory second-order differential equations. Numer. Math. **100**, 71–89 (2005)
10. Grimm, V.: On the use of the Gautschi-type exponential integrator for wave equations. In: The 6th European Conference on Numerical Mathematics and Advanced Applications, Santiago de Compostela, Spain (2005)
11. Grimm, V., Hochbruck, M.: Error analysis of exponential integrators for oscillatory second-order differential equations. J. Phys. A. Math. Gen. **39**, 5495–5507 (2006)
12. Hairer, E., Lubich, C.: Long-time energy conservation of numerical methods for oscillatory differential equations. SIAM J. Numer. Anal. **38**, 414–441 (2000)
13. Hairer, E., Lubich, C., Wanner, G.: Geometric Numerical Integration: Structure-Preserving Algorithms for Ordinary Differential Equations, 2nd edn. Springer, Berlin (2006)
14. Hochbruck, M., Lubich, C.: A Gautschi-type method for oscillatory second-order differential equations. Numer. Math. **83**, 403–426 (1999)
15. Hochbruck, M., Ostermann, A.: Explicit exponential Runge–Kutta methods for semilinear parabolic problems. SIAM J. Numer. Anal. **43**, 1069–1090 (2005)
16. Hochbruck, M., Ostermann, A.: Exponential integrators. Acta Numer. **19**, 209–286 (2010)
17. Li, J., Wu, X.Y.: Error analysis of explicit TSERKN methods for highly oscillatory systems. Numer. Algorithms **65**, 465–483 (2014)
18. Li, Y.W., Wu, X.Y.: Exponential integrators preserving first integrals or Lyapunov functions for conservative or dissipative systems. SIAM J. Sci. Comput. **38**, A1876–A1895 (2016)
19. Li, Y.W., Wu, X.Y.: Functionally-fitted energy-preserving methods for solving oscillatory nonlinear Hamiltonian systems. SIAM J. Numer. Anal. **54**, 2036–2059 (2016)
20. Liu, K., Shi, W., Wu, X.Y.: An extended discrete gradient formula for oscillatory Hamiltonian systems. J. Phys. A. Math. Theor. **46**(165203), 1–19 (2013)
21. Liu, C., Shi, W., Wu, X.Y.: An efficient high-order explicit scheme for solving Hamiltonian nonlinear wave equations. Appl. Math. Comput. **246**, 696–710 (2014)
22. Liu, K., Wu, X.Y.: Multidimensional ARKN methods for general oscillatory second-order initial value problems. Comput. Phys. Commun. **185**, 1999–2007 (2014)
23. Liu, K., Wu, X.Y.: High-order symplectic and symmetric composition methods for multi-frequency and multi-dimensional oscillatory Hamiltonian systems. J. Comput. Math. **33**, 356–378 (2015)
24. Mei, L.J., Liu, C., Wu, X.Y.: An essential extension of the finite-energy condition for ERKN integrators when applied to nonlinear wave equations. Commun. Comput. Phys. **22**, 742–764 (2017)
25. Schiesser, W.E., Griffiths, G.W.: A Compendium of Partial Differential Equation Models: Method of Lines Analysis with Matlab. Cambridge University Press, Cambridge (2009)
26. Shi, W., Wu, X.Y., Xia, J.: Explicit multi-symplectic extended leap-frog methods for Hamiltonian wave equations. J. Comput. Phys. **231**, 7671–7694 (2012)
27. Trefethen, L.N.: Spectral Methods in MATLAB. SIAM, Philadelphia (2000)
28. Vanden Berghe, G., De Meyer, H., Van Daele, M., Van Hecke, T.: Exponentially fitted Runge–Kutta methods. J. Comput. Appl. Math. **125**, 107–115 (2000)
29. Verwer, J.G., Sanz-Serna, J.M.: Convergence of method of lines approximations to partial differential equations. Computing **33**, 297–313 (1984)
30. Wang, B., Iserles, A., Wu, X.Y.: Arbitrary-order trigonometric Fourier collocation methods for multi-frequency oscillatory systems. Found. Comput. Math. **16**, 151–181 (2016)

31. Wang, B., Wu, X.Y.: Explicit multi-frequency symmetric extended RKN integrators for solving multi-frequency and multidimensional oscillatory reversible systems. CALCOLO **52**, 207–231 (2015)
32. Wang, B., Wu, X.Y., Xia, J.: Error bounds for explicit ERKN integrators for systems of multi-frequency oscillatory second-order differential equations. Appl. Numer. Math. **74**, 17–34 (2013)
33. Wu, X.Y., Liu, K., Shi, W.: Structure-Preserving Algorithms for Oscillatory Differential Equations II. Springer, Berlin (2015)
34. Wu, X.Y., Liu, C.: An integral formula adapted to different boundary conditions for arbitrarily high-dimensional nonlinear Klein-Gordon equations with its applications. J. Math. Phys. **57**, 021504 (2016)
35. Wu, X.Y., Mei, L.J., Liu, C.: An analytical expression of solutions to nonlinear wave equations in higher dimensions with Robin boundary conditions. J. Math. Anal. Appl. **426**, 1164–1173 (2015)
36. Wu, X.Y., Wang, B., Liu, K., Zhao, H.: ERKN methods for long-term integration of multidimensional orbital problems. Appl. Math. Model. **37**, 2327–2336 (2013)
37. Wu, X.Y., Wang, B., Shi, W.: Efficient energy-preserving integrators for oscillatory Hamiltonian systems. J. Comput. Phys. **235**, 587–605 (2013)
38. Wu, X.Y., Wang, B., Xia, J.: Explicit symplectic multidimensional exponential fitting modified Runge–Kutta–Nyström methods. BIT **52**, 773–795 (2012)
39. Wu, X.Y., You, X., Wang, B.: Structure-Preserving Algorithms for Oscillatory Differential Equations. Springer, Berlin (2013)
40. Wu, X.Y., You, X., Shi, W., Wang, B.: ERKN integrators for systems of oscillatory second-order differential equations. Comput. Phys. Commun. **181**, 1873–1887 (2010)
41. Wu, X.Y., You, X., Xia, J.: Order conditions for ARKN methods solving oscillatory systems. Comput. Phys. Commun. **180**, 2250–2257 (2009)
42. Yan, J., Zhang, Z.: New energy-preserving schemes using Hamiltonian Boundary Value and Fourier pseudospectral methods for the numerical solution of the "good" Boussinesq equation. Comput. Phys. Commun. **201**, 33–42 (2016)
43. Yang, H., Wu, X.Y., You, X., Fang, Y.: Extended RKN-type methods for numerical integration of perturbed oscillators. Comput. Phys. Commun. **180**, 1777–1794 (2009)
44. Yang, H., Zeng, X., Wu, X.Y., Ru, Z.: A simplified Nyström-tree theory for extended Runge–Kutta–Nyström integrators solving multi-frequency oscillatory systems. Comput. Phys. Commun. **185**, 2841–2850 (2014)

Index

X. Wu and B. Wang, *Recent Developments in Structure-Preserving Algorithms for Oscillatory Differential Equations*, https://doi.org/10.1007/978-981-10-9004-2

CPI Antony Rowe
Eastbourne, UK
February 15, 2019